Chemical Reaction Engineering

Chemical Reaction Engineering

GT MILLER

ME (Chemical Engineering)

CBS

CBS Publishers & Distributors Pvt Ltd

New Delhi • Bengaluru • Chennai • Kochi • Kolkata • Mumbai • Pune
Hyderabad • Nagpur • Patna • Vijayawada

Chemical
Reaction
Engineering

ISBN: 978-81-239-2831-9

First Edition: 2016

Published by Satish Kumar Jain and produced by Varun Jain for

CBS Publishers & Distributors Pvt Ltd

4819/XI Prahlad Street, 24 Ansari Road, Daryaganj, New Delhi 110 002, India.

Ph: 23289259, 23266861, 23266867   Website: www.cbspd.com

Fax: 011-23243014   e-mail: delhi@cbspd.com; cbspubs@airtelmail.in.

Corporate Office: 204 FIE, Industrial Area, Patparganj, Delhi 110 092

Ph: 4934 4934   Fax: 4934 4935   e-mail: publishing@cbspd.com; publicity@cbspd.com

Branches

- **Bengaluru:** Seema House 2975, 17th Cross, K.R. Road, Banasankari 2nd Stage, Bengaluru 560 070, Karnataka
  Ph: +91-80-26771678/79   Fax: +91-80-26771680   e-mail: bangalore@cbspd.com

- **Chennai:** 7, Subbaraya Street, Shenoy Nagar, Chennai 600 030, Tamil Nadu
  Ph: +91-44-26680620, 26681266   Fax: +91-44-42032115   e-mail: chennai@cbspd.com

- **Kochi:** Ashana House, No. 39/1904, AM Thomas Road, Valanjambalam, Eranakulam 682 018, Kochi Kerala
  Ph: +91-484-4059061-65   Fax: +91-484-4059065   e-mail: kochi@cbspd.com

- **Kolkata:** 6/B, Ground Floor, Rameswar Shaw Road, Kolkata-700 014, West Bengal
  Ph: +91-33-22891126, 22891127, 22891128   e-mail: kolkata@cbspd.com

- **Mumbai:** 83-C, Dr E Moses Road, Worli, Mumbai-400018, Maharashtra
  Ph: +91-22-24902340/41   Fax: +91-22-24902342   e-mail: mumbai@cbspd.com

- **Pune:** Bhuruk Prestige, Sr. No. 52/12/2+1+3/2 Narhe, Haveli (Near Katraj-Dehu Road Bypass), Pune 411 041, Maharashtra
  Ph: +91-20-64704058, 64704059, 32392277   Fax: +91-20-24300160 e-mail: pune@cbspd.com

Representatives

- **Hyderabad** 0-9885175004 • **Nagpur** 0-9021734563
- **Patna** 0-9334159340 • **Vijayawada** 0-9000660880

Printed at India Binding House, Noida, UP

Preface

Chemical reaction engineering (CRE) is concerned with the exploitation of chemical reactions on a commercial scale. Its goal is to successfully design and operation of chemical reactors. The text emphasises qualitative arguments, simple design methods, graphical procedures and frequent comparison of capabilities of the major reactor types. Simple ideas are treated first, and then extended to the more complex. The role of the chemical reactor is crucial for the industrial conversion of raw materials into products and numerous factors must be considered when selecting an appropriate and efficient chemical reactor. Chemical reaction engineering (CRE) and reactor technology defines the qualitative aspects that affect the selection of an industrial chemical reactor and couples various reactor models to case-specific kinetic expression for chemical processes.

This book is an introduction to the quantitative treatment of chemical reaction engineering. The text provides a balanced approach to the understanding of: (i) both homogeneous and heterogeneous reacting systems, and (ii) both chemical reaction engineering and chemical reactor engineering. Completion of the entire text will give the reader a good introduction to the fundamentals of chemical reaction engineering and provide a basis for extensions into other nontraditional uses of these analyses. The emphasis on chemical reaction engineering as opposed to chemical reactor engineering is in the appropriate context for training future chemical engineers who will confront issues in diverse sectors of employment.

In this reference textbook each chapter covers an important aspect with an accurate and up-to-date account of each topic. Chapter 1 is devoted to basic concepts of chemical reactions engineering. Chapter 2 concentrates on stoichiometry and kinetics and educates the readers to calculate the quantities such as the amount of products produced with given reactants and per cent yield obtained. Chapter 3 discusses reactor design and operations. Chapter 4 acquaints the readers with multiphase reactions and various parallel series and independent reactions. The real reactors are almost always and unavoidably operated nonisothermally because the reactions generate or absorb large amount of heat and reaction rate vary strongly with temperature. Considering this chapter 5 focuses on steady-state nonisothermal reactor design. Chapter 6 makes an in-depth study of fundamentals of catalytic processes. Various types of catalysts, their mechanism and significance is discussed in detail. Chapter 7 is devoted to diffusion and chemical kinetics and discusses Fick's laws of diffusion and diffusion controlled reactions. Chapter 8 concentrates on residence time distribution in chemical reactors and discusses its measurements and applications. Chapter 9 focuses on models for nonideal reactors. Chapter 10 deals with chain reactions and combustion reactors. Chapter 11 is devoted to polymerisation reactions and reactors and discusses various kinetic aspects of polymer reactions. A biological process is a process of living organism. Biological processes are made up of any number of chemical reactions and/or other events that results in a transformation. Keeping this in mind chapter 12 focuses on biological reactions engineering and discusses enzyme kinetics and fermentation methods and systems. Chapter 13 concentrates on environmental reactions engineering.

This compact yet comprehensive book covers the materials required for a basic understanding of chemical reaction engineering. The principles of reaction engineering are simply and clearly presented and illustrative problems are used to demonstrate how these principles are practically applied.

This reference textbook of *Chemical Reaction Engineering* is essential reading for all students, teachers, professionals, researchers and industrialists involved with chemical engineering, biochemical engineering, environmental science, microbiology, biotechnology and life sciences. The reference textbook also caters to the requirement of the syllabus prescribed by various Indian universities for undergraduate and postgraduate students pursuing these courses. Constructive suggestions are always welcome from users of this book.

Diagrams, figures, tables and index supplement the text. All the topics have been covered in a cogent and lucid style to help the reader grasp the information quickly and easily.

GT Miller

Contents at a Glance

Contents

Chapter 1

Basic Concepts of Chemical Reactions

INTRODUCTION

A chemical reaction is a process that leads to the transformation of one set of chemical substances to another. Chemical reactions are studied by chemists under a field of science called chemistry. Chemical reactions can be either spontaneous, requiring no input of energy, or non-spontaneous, often coming about only after the input of some type of energy, viz. heat, light or electricity. Classically, chemical reactions encompass changes that strictly involve the motion of electrons in the forming and breaking of chemical bonds, although the general concept of a chemical reaction, in particular the notion of a chemical equation, is applicable to transformations of elementary particles, as well as nuclear reactions.

The substance/substances initially involved in a chemical reaction are called reactants. Chemical reactions are usually characterised by a chemical change, and they yield one or more products, which usually have properties different from the reactants.

Different chemical reactions are used in combination in chemical synthesis in order to get a desired product. In biochemistry, series of chemical reactions catalysed by enzymes form metabolic pathways, by which syntheses and decompositions ordinarily impossible in conditions within a cell are performed.

Chemical reaction engineering (reaction engineering or reactor engineering) is a specialty in chemical engineering or industrial chemistry dealing with chemical reactors. Frequently the term relates specifically to catalytic reaction systems where either a homogeneous or heterogeneous catalyst is present in the reactor. Sometimes a reactor *per se* is not present by itself, but rather is integrated into a process, for example in reactive separations vessels, certain fuel cells and photocatalytic surfaces.

In other words chemical reaction engineering is concerned with the exploitation of chemical reactions on a commercial scale. It's goal is to successful design and operation of chemical reactors.

REACTION MECHANISMS

The mechanism of a reaction is the explanation of how a reaction takes place. For example, we might find out which bonds are broken, what happens in the transition state, and whether the reaction takes place in one, or more than one, stage. Often it is possible to think of several mechanisms for a single reaction. The task of the chemist is to perform experiments that will help to discover which mechanism gives the better explanation. One of the fascinations of studies of mechanisms is that molecules appear to have the remarkable facility to change the way they react depending on cirumstances. For example, a mechanism that explains the reactions of gases at moderate pressures will not fit at low pressures; a mechanism for a reaction is water may not explain the same reaction if it is carried out in another solvent.

Working out a mechanism for a reaction can be very challenging, and fascinating for those people who like puzzles. However, it is not only the intrinsic interest of mechanisms that drives chemists to study them. Advances in biochemistry, medicine and industrial chemistry often rely on an understanding of mechanisms to produce new chemicals or to make known chemicals more efficiently.

Bonds can Break in Two Ways

There are two ways a covalent bond can break. They are called heterolysis and homolysis.

Heterolysis

This when one of the atoms gains both electrons forming the bond. This atom now has one more electron than it started with, and becomes negatively charged. The other atom is left with one less electron than it started with, and becomes positively charged. (We assume that the covalent bond was made by each atom supplying one electron.) We can show heterolysis like this:

$$A \overset{.}{x} B \rightarrow A^+ + B\overset{.}{x}^-$$

or like this:

$$A \overset{.}{x} B \rightarrow A\overset{.}{x}^- + B^+$$

All other things being equal, the atom that has the negative charge will be the most electronegative of the two. An example of heterolysis is where 2-methyl-2-iodopropane undergoes the change:

$$\underset{H_3C}{\overset{H_3C}{\underset{H_3C}{}}}C \overset{.}{x} I \longrightarrow \underset{CH_3}{\overset{H_3C \oplus CH_3}{C}} + I\overset{.}{x}^\ominus$$

Homolysis

Homolysis is a more 'democratic' way of bond breaking. Here both atoms keep one of the two electrons. As a result they end up as neutral atoms:

$$A \overset{.}{x} B \rightarrow A\cdot + Bx$$

The electron left over on each atom is not paired with another electron. Atoms like this are called free radicals. We shall show a free radical by putting a single dot next to the symbol of the atom, like this: $X\cdot$. Free radicals are very reactive. Owing to its unpaired electron, it often seems that a radical's one purpose in life is to react with other atoms or molecules.

The study of free radicals has a long history. One of the earliest methods of detecting them was invented by the German chemist Paneth in 1929. He showed that, if free radicals were passed over a thin layer of lead, the layer would disappear. The apparatus he used is illustrated in Fig. 1.1.

Tetramethyl-lead(iv), $Pb(CH_3)_4$, was carried in to the apparatus by a stream of gas. When the tube was heated, a layer of lead appeared on the glass. He proposed that this was due to the tetramethyl-lead(IV) undergoing homolysis:

$$Pb(CH_3)_4(g) \rightarrow Pb(s) + 4CH_3\cdot(g)$$
Methyl radicals

If the position of heating was moved closer to the entrance to the apparatus, a new layer of lead appeared at the point of heating. However, equally interesting was the fact that the first layer disappeared. This was the result of the methyl radicals recombining with the lead:

$$Pb(s) + 4CH_3\cdot(g) \rightarrow Pb(CH_3)_4(g)$$

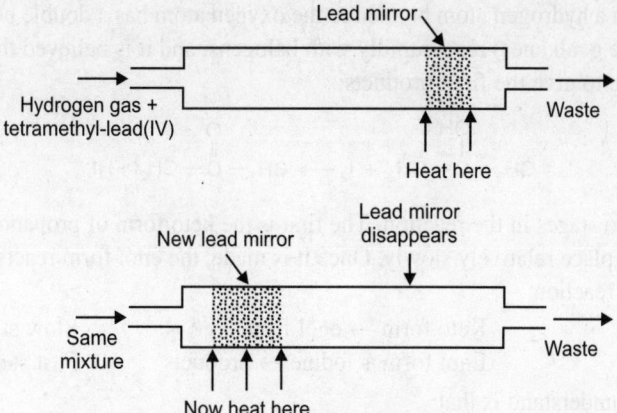

Fig. 1.1. The experiment performed by F. Paneth in 1929, which provided evidence for the existence of free radicals.

There are more sophisticated ways of detecting radicals now. One of the most important is electron spin resonance spectroscopy.

Slowest Step in a Reaction Governs the Rate

A reaction that appears straightforward from its chemical equation often turns out to contain some surprises in its mechanism. A good example is the reaction of iodine with propanone. The chemical equation is

$$CH_3COCH_3(aq) + I_2 (aq) \rightarrow CH_3COCH_2I(aq) + HI(aq)$$

This tells us that one mole of propanone will react with one mole of iodine to give one mole of iodopropanone and one mole of hydrogen iodide. Please be sure to notice that the equation gives us absolutely no information about how fast the reaction takes place. Nor does it tell us anything about the mechanism. It is a golden rule that:

Looking at the equation, it is tempting to think that the reaction of iodine with propanone takes place by molecules of iodine colliding with molecules of propanone. If so, we would expect the rate to change if we changed the concentrations of either of the reactants. We might predict the rate law to be

$$Rate = k[(CH_3COCH_3)(aq)] \; [I_2(aq)]$$

However, the experimental evidence is that

$$Rate = k[(CH_3COCH_3)(aq)]$$

That is, the rate is independent of the concentration of iodine. For example, if the temperature is kept constant, the rate of the reaction is the same if we use 0.01 mol dm^{-3} or 0.1 mol dm^{-3} iodine solution. We shall now consider a mechanism that can explain the observations.

A nuclear magnetic resonance spectrometer detects the presence of hydrogen atoms in a molecule. The spectrometer shows that a sample of propanone contains hydrogen atoms attached to carbon atoms (which is what we expect). However, it also shows that a small proportion of molecules have a hydrogen atom attached to the oxygen atom. The reason for this is thought to be that, of its own accord, propanone exists in two forms (called tautomers), which are in equilibrium with each other:

$$CH_3-\overset{\overset{O}{\|}}{C}-CH_3 \rightleftharpoons CH_3-\overset{\overset{OH}{\|}}{C}-CH_2$$

Keto form   Enol form

The molecule with a hydrogen atom bonded to the oxygen atom has a double bond. It is well known that such molecules (e.g. alkenes) react rapidly with halogens, and it is believed that it is the enol form that reacts with iodine to give the final products:

$$CH_3-\overset{\overset{OH}{|}}{C}=CH_2 + I_2 \longrightarrow CH_3-\overset{\overset{O}{\|}}{C}-CH_2I + HI$$

Thus, there are two stages in the reaction. The first is the keto form of propanone changing into the enol form. This takes place relatively slowly. Once it is made, the enol form reacts rapidly with iodine. This is a much faster reaction:

Stage I	Keto form → enol form	Slow step
Stage II	Enol form + iodine → products	Fast step

The next thing to understand is that:

We can see why this is by comparing this reaction with the way a computer works. The central processor might execute five million instructions each second, but it might take one second to type in a single instruction. So the effective rate would be one instruction each second. If the computer relies on receiving instructions from the keyboard, it would be pointless buying a new computer that executed ten million instructions each second. The rate of performing instructions would still be limited by the rate of typing: the slowest step in the process.

Returning to our reaction, the conversion of the keto to the enol form of propanone is the slowest, rate determining, step. Adding more iodine has no effect on the rate because it cannot increase the rate at which the enol form is made. However, if we increase the concentration of propanone, the greater is the concentration of the enol form, and the more product will be made. This is why the rate depends on the concentration of propanone.

Free Radical Reaction

Reactions between gases often involve free radicals. An example is the reaction between chlorine and a hydrocarbon like ethane. A mixture of chlorine and ethane can be kept for long periods of time at room temperature, provided it is guarded from sunlight. If sunlight, or even better (or worse, depending on your point of view) ultraviolet light, enters the mixture, there is an immediate explosion. When a reaction is sensitive to ultraviolet light it is a sure sign that free radicals are involved. The ultraviolet light has the effect of bringing about the homolytic fission (i.e. breaking) of bonds. The is the initiation step of the reaction. Once free radicals are let loose, the reaction proceeds very rapidly. These radicals can attack other molecules, which give rise to new radicals, which then go on the give further reactions, and so on. This is the propagation stage. From time to time the free radicals combine to give normal molecules. The removal of radicals from the reaction eventually brings the reaction to an end. This is the termination stage. We can show each of these three stages using some sample reactions:

Initiation

$$Cl_2 \rightarrow 2Cl\cdot$$
$$CH_3CH_3 \rightarrow 2CH_3\cdot$$

Propagation

$$CH_3CH_3 + Cl\cdot \rightarrow CH_3CH_2Cl + H\cdot$$
$$CH_3CH_3 + CH_3\cdot \rightarrow CH_4 + CH_3CH_2\cdot$$

$$CH_3CH_2 Cl + Cl \cdot \rightarrow CH_2ClCH_2Cl + H \cdot$$
$$CH_2ClCH_2Cl + Cl \cdot \rightarrow CH_2ClCHCl_2 + H \cdot$$
$$Cl_2 H \cdot \rightarrow HCl + Cl \cdot$$

Termination

$$CH_3CH_2 \cdot + H \cdot \rightarrow CH_3CH_3$$
$$CH_3CH_2 \cdot + Cl \cdot \rightarrow CH_3CH_2Cl$$
$$CH_3 \cdot + Cl \cdot \rightarrow CH_3Cl$$

The contents of the reaction flask (assuming it survives the explosion) will contain a range of chloroalkanes in which the hydrogen atoms of ethane have been substituted by chlorine atoms. A reaction like this is called a chain reaction because one reaction is linked to another like the links in a chain.

Free radicals are involved in many of the chemical reactions that occur in the Earth's atmosphere. Especially they are involved in the way in which chlorofluorocarbons (CFCs) interact with ozone (trioxygen) in the stratosphere. One of the CFCs to be found in aerosol propellants is known commercially as freon-12. Its formula is CF_2Cl_2. Ultraviolet light can break the carbon–chlorine bonds:

$$CF_2Cl_2 \overset{hf}{\rightarrow} \cdot CF_2Cl + Cl \cdot$$

The chlorine radicals then attack ozone molecules:

$$O_3 + Cl \cdot \rightarrow ClO \cdot + O_2$$

However, ozone is also disrupted by ultraviolet light, which provides a supply of free oxygen atoms:

$$O_3 \overset{hf}{\rightarrow} O_2 + O$$

These take part in the reaction:

$$ClO \cdot + O \rightarrow O_2 + Cl \cdot$$

It is this last reaction that makes CFCs so dangerous to the ozone layer. It regenerates a chlorine radical, which can react with another ozone molecule, thus repeating the entire cycle. The production of one chlorine radical from a CFC can be responsible for destroying many ozone molecules.

Mechanisms of the Hydrolysis of Halogenoalkanes

If iodomethane, CH_3I, reacts with hydroxide ions, it is converted into methanol, CH_3OH. The chemical equation is

$$OH^- + CH_3I \rightarrow CH_3OH + I^-$$

The rate law is found to be

$$Rate = k[CH_3I][OH^-]$$

The hydroxide ion is not only negatively charged, it carries three lone pairs of electrons. Ions or molecules with one or more lone pairs very often seek out centres of positive charge. They are called nucleophiles. Hydroxide ions are powerful nucleophiles, and the attack by a hydroxide ion on iodomethane is an example of a nucleophilic attack. One of the lone pairs on the hydroxide ion begins to make a bond to the carbon atom, and at the same time the carbon to iodine bond begins to weaken. Mid-way through the process, a transition state is formed like that shown in Fig. 1.2.

This transition state is made from two molecules (we shall use the word 'molecule' to stand for any reacting particle, be it an atom, ion or true molecule). A single step reaction that has a transition state made from two molecules is said to have a molecularity of 2.

In some cases the transition state breaks apart, returning to reactant molecules. In other cases the iodine leaves the transition state as an iodide ion, leaving the OH group firmly bonded to the carbon atom. This conversion of the transition state into products is assisted by virtue of the iodine ion being a good leaving group. Iodine tends to leave a molecule more easily than, say, a chlorine atom partly because the carbon-iodine bond is relatively weak, and partly because the iodide ion often fits neatly into the surrounding solvent molecules.

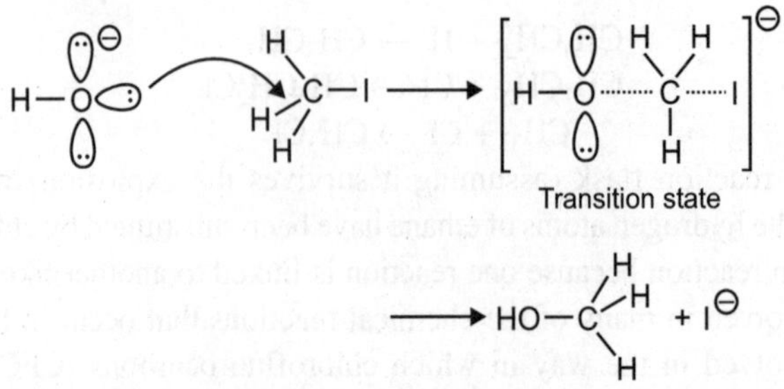

Transition state

Fig. 1.2. The broken lines in the transition state show bonds that are in the process of making or breaking. If the C—OH bond strengthens, then products will be made. If, on the other hand, it weakens and the C—I bond strengthens, then the reactants will be made again.

The result of the reaction is that the iodide atom is substituted by an OH group. We new have three pieces of information about this reaction: it involves a substitution, a nucleophilic attack and has molecularity of 2. This is summarised by calling the reaction an S_N2 reaction.

Now compare this with the reaction of 2-iodo-2-methylpropane, $(CH_3)_3CI$, with sodium hydroxide. The equation for the reaction is not unlike the previous one:

$$OH^- + (CH_3)_3CI \rightarrow (CH_3)_3COH + I^-$$

However, the rate law is

$$Rate = k[(CH_3)_3CI]$$

The fact that the rate does not depend on the concentration of hydroxide ions tells us that hydroxide ions cannot take part in the rate determining step. The 2-iodo-2-methylpropane molecule spontaneously ionises. The positive ion produced is called a carbocation. (Carbonations were once called carbonium ions.) The formation of the carbocation is the slow step in the reaction (Fig. 1.3). Therefore, the ionisation step determines the rate of the reaction. Once the carbocation appears in the solution, it can be attacked quickly by neighbouring negatively charged hydroxide ions.

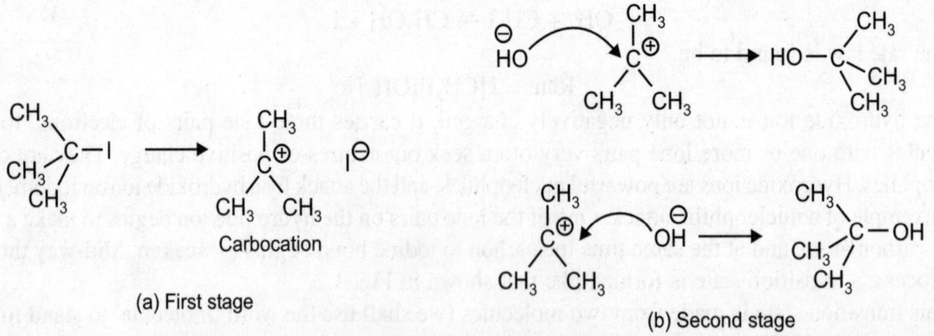

(a) First stage

Carbocation

(b) Second stage

Fig. 1.3. The mechanism of the reaction between hydroxide ions and 2-iodo-2-methylpropane. (a) The first stage is the autoionisation of the 2-iodo-2-methylpropane. (b) The second stage is attack on the carbocation by hydroxide ions. This can take place from either side of the carbocation. In this example the same product is made; in other cases, optical isomers are formed.

This result of the reaction is still a substitution of an iodide ion by an OH group. Similarly it involves a nucleophilic attack by hydroxide ions. However, the transition state consists of just one species, so the molecularity is 1. All this is summed up by calling this an S_N1 reaction.

Influence of Catalysts

It is well known that catalysts provide an alternative pathway for a reaction, and that the new route has a lower activation energy than the original reaction. In this section we shall take a closer look at how some catalysts achieve this feat. As already discussed there are two broad categories of catalyst: heterogeneous and homogeneous. Solids are heterogeneous catalysts. The reactions they catalyse take place on their surface. A good example is the reaction between ethene and hydrogen:

$$C_2H_4(g) + H_2(g) \rightarrow C_2H_6(g)$$

This reaction is catalysed by nickel. At particular sites on the surface of a piece of nickel the atoms are arranged in such a way that the π cloud of electrons can overlap with an empty d orbital (Fig. 1.4). (Nickel is a transition metal, and these metals often have empty d orbitals). The ethene molecule is held to the surface, where it reacts with a hydrogen molecule.

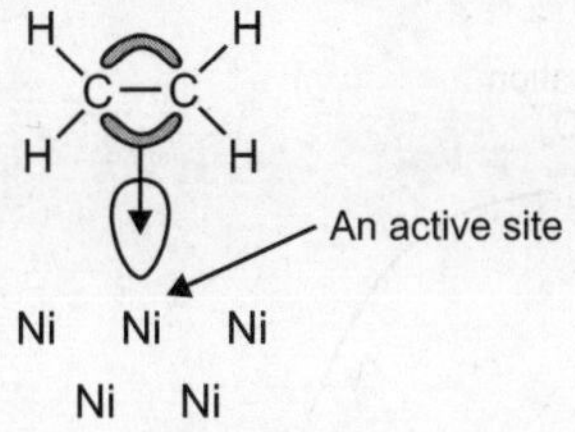

Fig. 1.4. Ethene molecules can bond through their π electrons. Vacant d orbitals on transition elements such as nickel can bond to ethene molecules. The π orbital on an ethene molecule overlaps with the d orbital.

The places on the surface where the geometry is just right for the molecules to sit are called active sites. The effectiveness of a solid catalyst is increased if it is present as a powder. A powder has a much larger surface area than a large lump. By increasing the surface area, the number of active sites is increased.

Enzymes are extremely efficient biologically active catalysts. They are homogeneous catalysts, reacting in solution in body fluids. Enzymes are proteins, which, from a distance, appear as a long tangled chain of atoms consisting mainly of carbon, hydrogen, nitrogen and oxygen. On closer inspection, using X-ray diffraction, the structure of an enzyme shows up. The geometry is always very complicated, but a major feature that they have in common is a region into which only molecules of a very particular shape and size will fit. This region is the active site of the enzyme. Generally, only one type of molecule will fit the active site. This means that enzymes are much more specific than other catalysts. For example, only hydrogen peroxide molecules will fit the active site of catalase. The molecule that fits the active site is called the substrate. Figure 1.5 will give you a visual impression of how enzymes work.

Enzymes have a feature in common with many other catalysts. They can be poisoned. A catalyst is poisoned if its active site(s) become clogged by an unwanted molecule. Hydrogen sulphide is a very efficient poison for metal catalysts, and metal ions will often poison enzymes. A great deal of money has to be spent in industry to ensure that reactants are free of poisons before they are admitted to the reaction chamber containing a catalyst. Also, enzymes are particularly susceptible to damage from too much heat. They are not designed to work at temperatures much above body temperature, 37°C. As the temperature rises, the structure of the protein chain around the active site changes. Very soon the change becomes irreversible. The delicate geometry of the atoms that hold the substrate in position is wrecked, and the enzyme stops working; it is said to be denatured. In living things, changes of this nature will lead to death (Fig. 1.6).

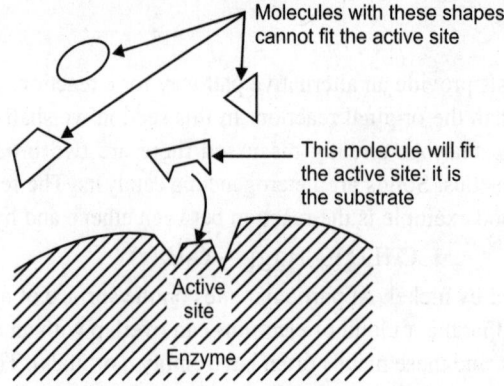

Molecules with these shapes cannot fit the active site

This molecule will fit the active site: it is the substrate

Active site

Enzyme

Fig. 1.5. A visual illustration of how the geometry of an active site will only fit one type of substrate (in most cases).

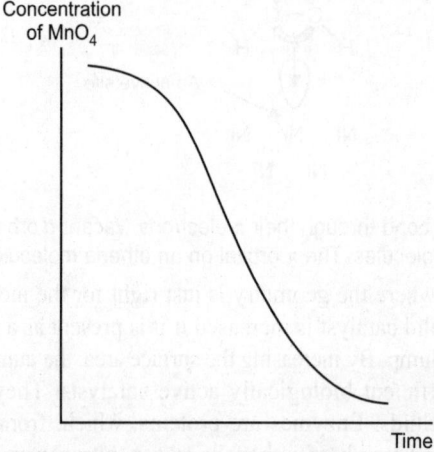

Concentration of MnO₄

Time

Fig. 1.6. The shape of the concentration against time graph for the oxidation of ethanedioic acid by potassium manganate (VII).

Kinetics of Enzyme Reactions

Enzyme reactions are so important that it is sensible to know something of their kinetics. The first thing that you will discover if you use an enzyme in a reaction is that the rate increases with the concentration of the enzyme. Similarly, the rate increases with the substrate concentration, but only up to a certain point. This behaviour is shown in Fig. 1.7. There is a limit to the rate of the reaction. The reason for this behaviour is that the active sites become saturated with substrate molecules. When, on average, all the available active sites have substrate molecules in place, adding more substrate to the solution will have no effect. Enzyme reactions are also sensitive to pH. Usually there is a optimum pH at which the rate is a maximum. If the pH is increased or decreased from the optimum value, the rate decreases. The reason for this is that enzymes, like other proteins, are built from amino acids. These acids have the ability to lose or gain protons. If an enzyme gains a proton, its charge increases by +1; if it loses a proton, its charge increases by –1. Even small changes in charge around the active site can prevent the substrate entering or leaving.

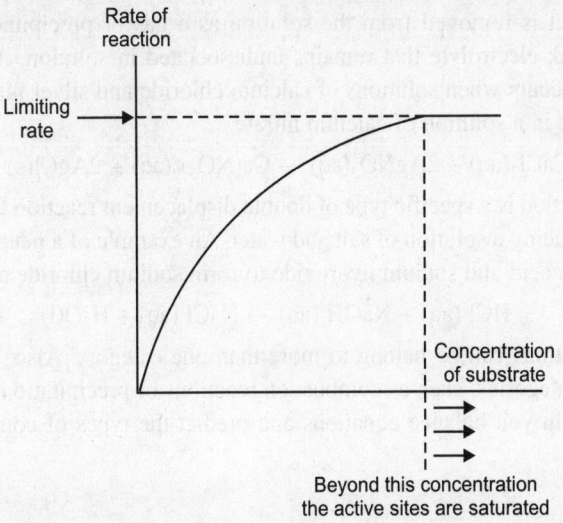

Fig. 1.7. Variation of an enzyme reaction with the concentration of substrate molecules.

Types of Inorganic Chemical Reactions

Inorganic reactions

Elements and compounds react with each other in numerous ways. Memorising every type of reaction would be challenging and also unncecessary, since nearly every inorganic chemical reaction falls into one or more of four broad categories.

Combination reactions

Two or more reactants form one product in a combination reaction. An example of a combination reaction is the formation of sulphur dioxide when sulphur is burned in air:

$$S(s) + O_2 (g) \rightarrow SO_2 (g)$$

Decomposition reactions

In a decomposition reaction, a compound breaks down into two or more substances. Decomposition usually results from electrolysis or heating. An example of a decomposition reaction is the breakdown of mercury (II) oxide into its component elements.

$$2HgO(s) + heat \rightarrow 2Hg (l) + O_2(g)$$

Single displacement reactions

A single displacement reaction is characterised by an atom or ion of a single compound replacing an atom of another element. An example of a single displacement reaction is the displacement of copper ions in a copper sulphate solution by zinc metal, forming zinc sulphate:

$$Zn(s) + CuSO_4(aq) \rightarrow Cu(s) + ZnSO_4(aq)$$

Single displacement reactions are often subdivided into more specific categories (e.g. redox reactions).

Double displacement reactions

Double displacement reactions also may be called metathesis reactions. In this type of reaction, elements from two compounds displace each other to form new compounds. Double displacement reactions may occur when one product is removed from the solution as a gas or precipitate or when two species combine to form a weak electrolyte that remains undissociated in solution. An example of a double displacement reaction occurs when solutions of calcium chloride and silver nitrate are reacted to form insoluble silver chloride in a solution of calcium nitrate.

$$CaCl_2(aq) + 2AgNO_3(aq) \rightarrow Ca(NO_3)_2(aq) + 2AgCl(s)$$

A neutralisation reaction is a specific type of double displacement reaction that occurs when an acid reacts with a base, producing a solution of salt and water. An example of a neutralisation reaction is the reaction of hydrochloric acid and sodium hydroxide to form sodium chloride and water:

$$HCl\ (aq) + NaOH\ (aq) \rightarrow NaCl\ (aq) + H_2O(l)$$

Remember that reactions can be belong to more than one category. Also, it would be possible to present more specific categories, such as combustion reactions or precipitation reactions. Learning the main categories will help you balance equations and predict the types of compounds formed from a chemical reaction.

Organic reactions

Organic reactions are chemical reactions involving organic compounds. The basic organic chemistry reaction types are addition reactions, elimination reactions, substitution reactions, pericyclic reactions, rearrangement reactions and redox reactions. In organic synthesis, organic reactions are used in the construction of new organic molecules. The production of many man-made chemicals such as drugs, plastics, food additives, fabrics depend on organic reactions.

Homogeneous and heterogeneous reactions

Homogeneous reactions are chemical reactions in which the reactants are in the same phase, while heterogeneous reactions have reactants in two or more phases. Reactions that take place on the surface of a catalyst of a different phase are also heterogeneous. A reaction between two gases, two liquids or two solids is homogeneous. A reaction between a gas and a liquid, a gas and a solid or a liquid and a solid is heterogeneous. Practical applications of heterogeneous reactions are in catalytic converters, fuel cells and chemical vapour deposition among others. Recently, manufacturing engineers have used surface reactions for synthesis of micro and nanoscale features in biomedical devices.

Likewise:
1. Homogeneous mixture: Substance in which components are evenly mixed.
2. Heterogeneous mixture: Substance in which components are not evenly mixed.

You can tell a mixture is homogeneous, when everything is settle and equal, the liquid, gas, object is one colour or same form. Such as if you add objects or substances to another substance and it does not change. There are various models which have been proposed over the years to model the concentrations in different phases. The phenomena to be considered are mass rates and reaction rates. Surface area affects the reaction rate of heterogeneous reactions but not homogeneous reactions.

Chemical Kinetics

The rate of a chemical reaction is a measure of how the concentration or pressure of the involved substances changes with time. Analysis of reaction rates is important for several applications, such as in chemical engineering or in chemical equilibrium study. Rates of reaction depends basically on:

1. Reactant concentrations, which usually make the reaction happen at a faster rate if raised through increased collisions per unit time.
2. Surface area available for contact between the reactants, in particular solid ones in heterogeneous systems. Larger surface area leads to higher reaction rates.
3. Pressure, by increasing the pressure, you decrease the volume between molecules. This will increase the frequency of collisions of molecules.
4. Activation energy, which is defined as the amount of energy required to make the reaction start and carry on spontaneously. Higher activation energy implies that the reactants need more energy to start than a reaction with a lower activation energy.
5. Temperature, which hastens reactions if raised, since higher temperature increases the energy of the molecules, creating more collisions per unit time.
6. The presence or absence of a catalyst. Catalysts are substances which change the pathway (mechanism) of a reaction which in turn increases the speed of a reaction by lowering the activation energy needed for the reaction to take place. A catalyst is not destroyed or changed during a reaction, so it can be used again.
7. For some reactions, the presence of electromagnetic radiation, most notably ultraviolet, is needed to promote the breaking of bonds to start the reaction. This is particularly true for reactions involving radicals.

Reaction rates are related to the concentrations of substances involved in reactions, as quantified by the rate law of each reaction. Note that some reactions have rates that are independent of reactant concentrations. These are called zero order reactions.

Reactions and energy

Chemical energy is part of all chemical reactions. Energy is needed to break chemical bonds in the starting substances. As new bonds form in the final substances, energy is released. By comparing the chemical energy of the original substances with the chemical energy of the final substances, you can decide if energy is released or absorbed in the overall reaction.

Exothermic reactions

A chemical reaction in which energy is released is called an exothermic reaction. *Exo* means 'go out' or 'exit'. *Thermic* means 'heat' or 'energy'. Exothermic reactions can give off energy in several forms. If heat is released in an exothermic reaction, the nearby matter will become warmer. The nearby matter absorbs the heat released by the reaction. The reaction between gasoline and oxygen in a car's engine is an exothermic reaction.

FACTORS AFFECTING THE RATE OF REACTIONS

Following factor affect the rate of reactions: (i) temperature, (ii) concentrations of reactants, (iii) catalysts, (iv) surface area of a solid reactant, and (v) pressure of gaseous reactants or products.

Temperature

When two chemicals react, their molecules have to collide with each other with sufficient energy for the reaction to take place. This is collision theory. The two molecules will only react if they have enough energy. By heating the mixture, you will raise the energy levels of the molecules involved in the reaction. Increasing temperature means the molecules move faster. This is kinetic theory. If your reaction is between atoms rather than molecules you just substitute 'atom' for 'molecule' in your explanation.

Catalysts

Catalysts speed up chemical reactions. Only very minute quantities of the catalyst are required to produce a dramatic change in the rate of the reaction. This is really because the reaction proceeds by a different pathway when the catalyst is present. Adding extra catalyst will make absolutely no difference.

Concentration

Increasing the concentration of the reactants will increase the frequency of collisions between the two reactants. So this is collision theory again. You also need to discuss kinetic theory in an experiment where you vary the concentration. Although you keep the temperature constant, kinetic theory is relevant. This is because the molecules in the reaction mixture have a range of energy levels. When collisions occur, they do not always result in a reaction. If the two colliding molecules have sufficient energy they will react.

Surface Area

If one of the reactants is a solid, the surface area of the solid will affect how fast the reaction goes. This is because the two types of molecule can only bump into each other at the liquid solid interface, i.e. on the surface of the solid. So the larger the surface area of the solid, the faster the reaction will be.

Smaller particles have a bigger surface area than larger particle for the same mass of solid. There is a simple way to visualise this. Take a loaf of bread and cut it into slices. Each time you cut a new slice, you get an extra surface onto which you can spread butter and jam. The thinner you cut the slices, the more slices you get and so the more butter and jam you can put on them. This is 'bread and butter theory'. You should have come across the idea in your biology lessons. By chewing your food you increase the surface area so that digestion can go faster.

Pressure

You should already know that the atoms or molecules in a gas are very spread out. For the two chemicals to react, there must be collisions between their molecules. By increasing the pressure, you squeeze the molecules together so you will increase the frequency of collisions between them. This is collision theory again.

In a diesel engine, compressing the gaseous mixture of air and diesel also increases the temperature enough to produce combustion. Increasing pressure also results in raising the temperature. It is not enough in a petrol engine to produce combustion, so petrol engines need a spark plug. When the petrol air mixture has been compressed, a spark from the plug ignites the mixture. In both cases the reaction (combustion) is very fast. This is because once the reaction has started, heat is produced and this will make it go even faster.

REACTION RATES

Concentration Changes

Chemical kinetics is the study of the speed with which a chemical reaction occurs and the factors that affect this speed. This information is especially useful for determining how a reaction occurs.

What is meant by the speed of a reaction? The speed of a reaction is the rate at which the concentrations of reactants and products change.

Consider the following hypothetical example. The letters A, B, and C represent chemical species (in this context, the letters do not represent elements). Suppose the following imaginary reaction occurs:

$$A + 2B \rightarrow 3C$$

The simulation below illustrates how this reaction can be studied. The apparatus at the left is called a stopped-flow apparatus. Each syringe contains a solution filled with a different reactant (A or B). When the two solutions are forced out of the syringes, they are quickly mixed in a mixing block and the reaction starts. The reacting solution passes through the tube at the bottom. An analytical technique such as spectrophotometry is used to measure the concentrations of the species in the reaction mixture (which is in the tube at the bottom) and how those concentrations change with time.

In this example, the syringe at the left contains a solution of species A, which has a black colour. The syringe at the right contains a solution of species B, which has a gray colour. The product C has a black colour. The graph at the right shows how the concentration of each species changes as time progresses. Run the simulation and observe the stopped-flow experiment and the shape of the concentration-time plots (Fig. 1.8).

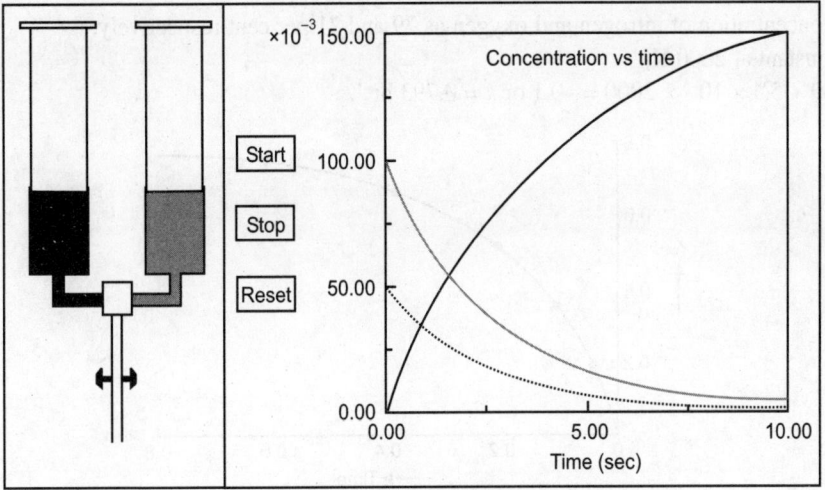

Fig. 1.8. Analytical technique spectrophotometry.

Notice that the colour of the reaction mixture changes as the reaction progresses. The reactants are black and gray, which when mixed produce a gray colour. As the reactants are consumed, the product, which is red, is produced.

This behaviour is reflected in the concentration-time plots. The concentration of A, shown by the doted line, decreases as time progresses, because the reaction consumes A. The same behaviour is observed for B (gray line). Conversely, there is initially no product C (black line) present. The reaction produces C, however, so the concentration of C increases as time goes by. (In this simulation, A and B were initially present in stoichiometric amounts.)

Reaction Rate

The rate of change in the concentrations of the reactants and products can be used to characterise the rate of a chemical reaction. The rate of change in the concentration corresponds with the slope of the concentration-time plot.

The simulation below is the same as that presented above except that the slope of the concentration-time curves are also plotted on the graph. Select the species (A, B or C) whose slope is shown and use the controls to step through the points on the graph.

SOLVED EXAMPLES

Example 1.1. Find the reaction rate constants for synthesis of nitric oxide from air at 2000 K. Experiments give the following relationship between the conversion of nitric oxide from reaction mixture with the time.

Time (hr.)	0.035	0.12	0.25	0.425	0.61
X_A (% by volume)	0.2	0.35	0.5	0.64	0.7

The rate constant k for the reverse reaction is given by the equation:

$$\ln k = -10.9 + 5.4 \times 10^{-3}\, T \qquad \qquad \text{... (1.1)}$$

Solution:

Let reaction be, $$N_2 + O_2 \rightleftharpoons 2NO \qquad\qquad \text{... (1.2)}$$

Consider concentration of nitrogen and oxygen as 79 and 21 per cent respectively.
The rate constant at 2000 K,
$\ln k = -10.9 + 5.4 \times 10^{-3} \times 2000 = -0.1$ or $k = 0.793$ hr^{-1}.

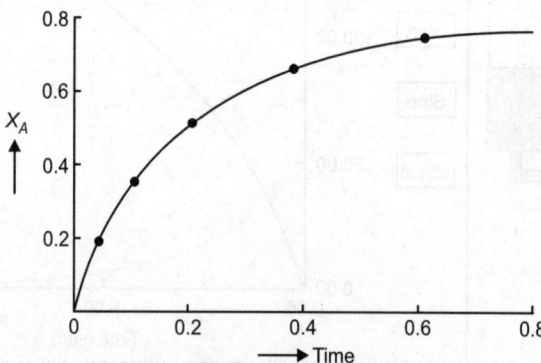

Fig. 1.9. Relationship between the conversion of nitric oxide from reaction mixture with the time.

For choosen values for time the X_A and k values were obtained as follows:

Time (hr.)	X_A (%)	$k\ (hr^{-1})$
0.24	45	7.15×10^{-4}
0.42	60	5.53×10^{-4}
0.11	30	9.10×10^{-4}

The above approximate values of k differ considerably owing to the small yield of nitric oxide. Hence low concentration of reaction mixture. The mean value of k is 7.25×10^{-4} sec^{-1} which may satisfactory fit the data.

Example 1.2. Estimate the activation energy for the decomposition of $C_6H_6N_2Cl$ to form chlorobenzene and nitrogen. Using:

The following information for first order reaction:

k (hr^{-1}) $\times 10^3$	0.43	1.03	1.8	3.55	7.17
$T(K)$	313.0	319.0	323.0	328.0	333.0

Solution:

Using Arrhenius' law:

$$k = k_0\, e^{-E/RT} \qquad \qquad ...(1.3)$$

$$\ln k = \ln k_0 - \frac{E}{RT}$$

The data given can be transformed in following forms:

$k \times 10^3$ (hr)$^{-1}$	0.43	1.03	1.8	3.55	7.17
$1000/T$ (K^{-1})	3.2	3.19	3.23	3.28	3.33

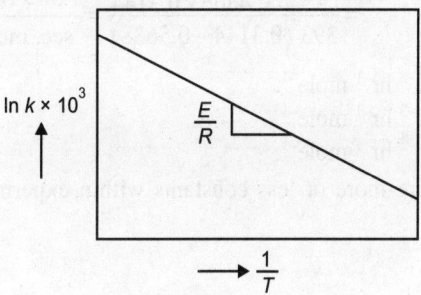

Fig. 1.10. Plot of ln k versus $1/T$.

$$\text{Slope} = 14.433 = \frac{E}{R}$$

$$\therefore \quad E = 14.433R$$

$$= 120\frac{kJ}{mole}$$

$$\text{or } 28.7\frac{kcal}{mole}.$$

A plot of ln k versus $1/T$ is shown in Fig. 1.10 to predict E.

Example 1.3. In a reaction between NaOH and $CH_3COOC_2H_5$, the 0.5638 moles and 0.3114 moles of NaOH and $CH_3COOC_2H_5$ were used respectively. With time the amounts of reactants in reactor were found as follows:

Time (t), hr.	0.0	369	669	1010	1265
$CH_3COOC_2H_3$ (moles)	0.5638	0.4866	0.4467	0.4113	0.3879
NaOH (moles)	0.3144	0.2342	0.1943	0.1589	0.1354

Find the order of reaction and reaction rate constant.

Solution:

$$CH_3COOC_2H_5 + NaOH \rightarrow C_2H_5OH + CH_3COONa \qquad \text{... (1.4)}$$

Assume that reaction is irreversible and the rate of formation of ethanol is proportional to concentrations of reactants, i.e. second order reaction.

It is well known that for second order reaction:

$$\ln\frac{C_B C_{A0}}{C_A C_{B0}} = (C_{B0} - C_{A0})kt \qquad \text{... (1.5)}$$

or

$$k = \frac{1}{t\left(C_{B0} \cdot C_{A0}\right)} \ln\frac{C_B \cdot C_{A0}}{C_A \cdot C_{B0}} \qquad \text{... (1.6)}$$

$$= \frac{1}{393\left(0.3144 - 0.5638\right)} \ln\frac{0.2342 \times 0.5638}{0.4866 \times 0.3144} = \frac{1.38 \times 10^{-3}}{\text{sec. mole}}$$

Similarly, $k = 1.42 \times 10^{-3}$ hr^{-1} mole^{-1}.

$k = 1.40 \times 10^{-3}$ hr^{-1} mole^{-1}.

$k = 1.43 \times 10^{-3}$ hr^{-1} mole^{-1}.

It can be seen that k values are more or less constants within experimental error. This supports the assumptions of second order.

Chapter 2
Stoichiometry and Kinetics

INTRODUCTION

Stoichiometry (sometimes called reaction stoichiometry to distinguish it from composition stoichiometry) is the calculation of quantitative (measurable) relationships of the reactants and products in a balanced chemical reaction. It can be used to calculate quantities such as the amount of products that can be produced with given reactants and per cent yield (the percentage of the given reactant that is made into the product). A simple example of the principle of the term would be of a typical box-shaped room that always has four walls, one floor and one ceiling giving a stoichiometry of 4:1:1 (walls, floor, ceiling).

Stoichiometry rests upon the law of conservation of mass, the law of definite proportions (i.e. the law of constant composition) and the law of multiple proportions. In general, chemical reactions combine in definite ratios of chemicals. Since chemical reactions can neither create nor destroy matter, nor transmute one element into another, the amount of each element must be the same throughout the overall reaction. For example, the amount of element X on the reactant side must equal the amount of element X on the product side.

Stoichiometry is often used to balance chemical equations. For example, the two diatomic gases, hydrogen and oxygen, can combine to form a liquid, water, in an exothermic reaction, as described by the following equation:

$$2H_2 + O_2 \rightarrow 2H_2O$$

The term stoichiometry is also often used for the molar proportions of elements in stoichiometric compounds. For example, the stoichiometry of hydrogen and oxygen in H_2O is 2:1. In stoichiometric compounds, the molar proportions are whole numbers (that is what the law of definite proportions is about).

Stoichiometry is not only used to balance chemical equations but also used in conversion, i.e. converting from grams to moles, or from grams to millilitres. For example, to find the number of moles in 2.00 g of NaCl, one would do the following:

$$\frac{2.00 \text{ g NaCl}}{58.44 \text{ g NaCl mol}^{-1}} = 0.034 \text{ mol}$$

In the above example, when written out in fraction form, the units of grams form a multiplicative identity, which is equivalent to one (g/g=1), with the resulting amount of moles (the unit that was needed), is shown in the following equation,

$$\left(\frac{2.00 \text{ g NaCl}}{1}\right)\left(\frac{1 \text{ mol NaCl}}{58.44 \text{ g NaCl}}\right) = 0.034 \text{ mol}$$

Stoichiometry is also used to find the right amount of reactants to use in a chemical reaction. An example is shown below using the thermite reaction,

$$Fe_2O_3 + 2Al \rightarrow Al_2O_3 + 2Fe$$

This equation shows that 1 mole of aluminium oxide and 2 moles of iron will be produced with 1 mole of iron(III) oxide and 2 moles of aluminium. So, to completely react with 85.0 g of iron(III) oxide (0.532 mol), 28.7 g (1.06 mol) of aluminium are needed.

$$mAl = \left(\frac{85.0 \text{ g Fe}_2O_3}{1}\right)\left(\frac{1 \text{ mol Fe}_2O_3}{159.7 \text{ g Fe}_2O_3}\right)\left(\frac{2 \text{ mol Al}}{1 \text{ mol Fe}_2O_3}\right)\left(\frac{27.0 \text{ g Al}}{1 \text{ mol Al}}\right)$$

DIFFERENT STOICHIOMETRIES IN COMPETING REACTIONS

Often, more than one reaction is possible given the same starting materials. The reactions may differ in their stoichiometry. For example, the methylation of benzene (C_6H_6) may produce singly-methylated ($C_6H_5CH_3$), doubly-methylated ($C_6H_4(CH_3)_2$), or still more highly-methylated ($C_6H_{6-n}(CH_3)_n$) products, as shown in the following example,

$$C_6H_6 + CH_3Cl \rightarrow C_6H_5CH_3 + HCl$$
$$C_6H_6 + 2CH_3Cl \rightarrow C_6H_4(CH_3)_2 + 2HCl$$
$$C_6H_6 + nCH_3Cl \rightarrow C_6H_{6-n}(CH_3)_n + nHCl$$

In this example, which reaction takes place is controlled in part by the relative concentrations of the reactants.

Stoichiometric Coefficient

The stoichiometric coefficient in a chemical reaction system of the i–th component is defined as:

$$v_i = \frac{dN_i}{d\xi}$$

or

$$dN_i = v_i d\xi$$

where, N_i is the number of molecules of i, and ξ is the progress variable or extent of reaction.

The extent of reaction ξ can be regarded as a real (or hypothetical) product, one molecule of which is produced each time the reaction event occurs. It is the extensive quantity describing the progress of a chemical reaction equal to the number of chemical transformations, as indicated by the reaction equation on a molecular scale, divided by the Avogadro constant (it is essentially the amount of chemical transformations). The change in the extent of reaction is given by:

$$d\xi = dn_B/n_B$$

where, n_B is the stoichiometric number of any reaction entity B (reactant or product) and dn_B is the corresponding amount.

The stoichiometric coefficient v_i represents the degree to which a chemical species participates in a reaction. The convention is to assign negative coefficients to reactants (which are consumed) and positive ones to products. However, any reaction may be viewed as 'going' in the reverse direction, and all the coefficients then change sign (as does the free energy). Whether a reaction actually will go in the arbitrarily-selected forward direction or not depends on the amounts of the substances present at any given time, which determines the kinetics and thermodynamics, i.e. whether equilibrium lies to the right or the left.

If one contemplates actual reaction mechanisms, stoichiometric coefficients will always be integers, since elementary reactions always involve whole molecules. If one uses a composite representation of an 'overall' reaction, some may be rational fractions. There are often chemical species present that do not participate in a reaction; their stoichiometric coefficients are therefore zero. Any chemical species that is regenerated, such as a catalyst, also has a stoichiometric coefficient of zero.

The simplest possible case is an isomerism:

$$A \Leftrightarrow B$$

in which $v_B = 1$ since one molecule of B is produced each time the reaction occurs, while $v_A = -1$ since one molecule of A is necessarily consumed. In any chemical reaction, not only is the total mass conserved, but also the numbers of atoms of each kind are conserved, and this imposes a corresponding number of constraints on possible values for the stoichiometric coefficients. Of course, only a small subset of the possible atomic rearrangements will occur.

There are usually multiple reactions proceeding simultaneously in any natural reaction system, including those in biology. Since any chemical component can participate in several reactions simultaneously, the stoichiometric coefficient of the i–th component in the k–th reaction is defined as

$$v_{ik} = \frac{\partial N_i}{\partial \xi_k}$$

so that the total (differential) change in the amount of the i–th component is:

$$dN_i = \sum_k v_{ik} d\xi_k.$$

Extents of reaction provide the clearest and most explicit way of representing compositional change, although they are not yet widely used.

With complex reaction systems, it is often useful to consider both the representation of a reaction system in terms of the amounts of the chemicals present $\{N_i\}$ (state variables), and the representation in terms of the actual compositional degrees of freedom, as expressed by the extents of reaction $\{\xi_k\}$. The transformation from a vector expressing the extents to a vector expressing the amounts uses a rectangular matrix whose elements are the stoichiometric coefficients $[v_{ik}]$.

The maximum and minimum for any ξ_k occur whenever the first of the reactants is depleted for the forward reaction; or the first of the 'products' is depleted if the reaction as viewed as being pushed in the reverse direction. This is a purely kinematic restriction on the reaction simplex, a hyperplane in composition space, or N-space, whose dimensionality equals the number of linearly-independent chemical reactions. This is necessarily less than the number of chemical components, since each reaction manifests a relation between at least two chemicals. The accessible region of the hyperplane depends on the amounts of each chemical species actually present, a contingent fact. Different such amounts can even generate different hyperplanes, all of which share the same algebraic stoichiometry.

In accord with the principles of chemical kinetics and thermodynamic equilibrium, every chemical reaction is reversible, at least to some degree, so that each equilibrium point must be an interior point of the simplex. As a consequence, extrema for the ξ's will not occur unless an experimental system is prepared with zero initial amounts of some products. The number of physically-independent reactions can be even greater than the number of chemical components, and depends on the various reaction mechanisms. For example, there may be two (or more) reaction paths for the isomerism above. The reaction may occur by itself, but faster and with different intermediates, in the presence of a catalyst.

The (dimensionless) 'units' may be taken to be molecules or moles. Moles are most commonly used, but it is more suggestive to picture incremental chemical reactions in terms of molecules. The N's and ξ's are reduced to molar units by dividing by Avogadro's number. While dimensional mass units may be used, the comments about integers are then no longer applicable.

Stoichiometry Matrix

In complex reactions, stoichiometries are often represented in a more compact form called the stoichiometry matrix. The stoichiometry matrix is denoted by the symbol, N.

If a reaction network has n reactions and m participating molecular species then the stoichiometry matrix will have corresponding n columns and m rows.

For example, consider the system of reactions shown below:

$$S_1 \rightarrow S_2$$
$$5S_3 + S_2 \rightarrow 4S_3 + 2S_2$$
$$S_3 \rightarrow S_4$$
$$S_4 \rightarrow S_5.$$

This systems comprises four reactions and five different molecular species. The stoichiometry matrix for this system can be written as:

$$N = \begin{bmatrix} -1 & 0 & 0 & 0 \\ 1 & 1 & 0 & 0 \\ 0 & -1 & -1 & 0 \\ 0 & 0 & 1 & -1 \\ 0 & 0 & 0 & 1 \end{bmatrix}$$

where the rows correspond to S_1, S_2, S_3, S_4 and S_5, respectively. Note that the process of converting a reaction scheme into a stoichiometry matrix can be a lossy transformation, for example, the stoichiometries in the second reaction simplify when included in the matrix. This means that it is not always possible to recover the original reaction scheme from a stoichiometry matrix.

Often the stoichiometry matrix is combined with the rate vector, v to form a compact equation describing the rates of change of the molecular species:

$$\frac{dS}{dt} = N{\bullet}v$$

Gas Stoichiometry

Gas stoichiometry is the quantitative relationship (ratio) between reactants and products in a chemical reaction with reactions that produce gases. Gas stoichiometry applies when the gases produced are assumed to be ideal, and the temperature, pressure, and volume of the gases are all known. The ideal gas law is used for these calculations. Often, but not always, the standard temperature and pressure (STP) are taken as 0°C and 1 bar and used as the conditions for gas stoichiometric calculations.

Gas stoichiometry calculations solve for the unknown volume or mass of a gaseous product or reactant. For example, if we wanted to calculate the volume of gaseous NO_2 produced from the combustion of 100 g of NH_3, by the reaction:

$$4NH_3 \text{ (g)} + 7O_2 \text{ (g)} \rightarrow 4NO_2 \text{ (g)} + 6H_2O$$

we would carry out the following calculations:

$$100 \text{ g } NH_3 \cdot \frac{1 \text{ mol } NH_3}{17.034 \text{ g } NH_3} = 5.871 \text{ mol } NH_3$$

There is a 1:1 molar ratio of NH_3 to NO_2 in the above balanced combustion reaction, so 5.871 mol of NO_2 will be formed. We will employ the ideal gas law to solve for the volume at 0°C (273.15 K) and 1 atmosphere using the gas law constant of R = 0.08206 L $\cdot$ atm $\cdot$ K^{-1} $\cdot$ mol^{-1} :

$$PV = nRT$$

$$V = \frac{nRT}{P}$$

Gas stoichiometry often involves having to know the molar mass of a gas, given the density of that gas. The ideal gas law can be rearranged to obtain a relation between the density and the molar mass of an ideal gas:

$$\rho = \frac{m}{V} \quad \text{and} \quad n = \frac{m}{M}$$

and thus:

$$\rho = \frac{MP}{RT}$$

where,

P = absolute gas pressure
V = gas volume
n = number of moles
R = universal ideal gas law constant
T = absolute gas temperature
ρ = gas density at T and P
m = mass of gas
M = molar mass of gas

Stoichiometric Air-fuel Ratios of Common Fuels

Stoichiometric air-fuel ratios of common fuels are shown in Table 2.1.

Table 2.1. Stoichiometric air-fuel ratios of common fuels.

Fuel	By mass	By volume	Per cent fuel by mass
Gasoline	14.7:1	–	6.8%
Natural gas	7.2 :1	9.7:1	5.8%

(Contd ...)

Fuel	By mass	By volume	Per cent fuel by mass
Propane (LP)	15.5:1	23.9:1	6.45%
Ethanol	9:1	–	11.1%
Methanol	6.4:1	–	15.6%
Hydrogen	34:1	2.39:1	2.9%
Diesel	14.6:1	0.094:1	6.8%

RATE LAWS

Basic Definition

Homogeneous and heterogeneous reactions

Homogeneous reactions are chemical reactions in which the reactants are in the same phase, while heterogeneous reactions have reactants in two or more phases. Reactions that take place on the surface of a catalyst of a different phase are also heterogeneous. A reaction between two gases, two liquids or two solids is homogeneous. A reaction between a gas and a liquid, a gas and a solid or a liquid and a solid is heterogeneous. Practical applications of heterogeneous reactions are in catalytic converters, fuel cells and chemical vapour deposition among others. Recently, manufacturing engineers have used surface reactions for synthesis of micro and nanoscale features in biomedical devices. Likewise:

1. Homogeneous mixture: Substance in which components are evenly mixed.
2. Heterogeneous mixture: Substance in which components are not evenly mixed. (Do not blend together)(example: toothpaste, milk, perfume, steel, etc.).

You can tell a mixture is homogeneous, when everything is settle and equal, the liquid, gas, object is one colour or same form. Such as if you add objects or substances to another substance and it does not change. There are various models which have been proposed over the years to model the concentrations in different phases. The phenomena to be considered are mass rates and reaction rates.

Surface area affects the reaction rate of heterogeneous reactions but not homogeneous reactions.

Molecularity

Molecularity in chemistry is the number of colliding molecular entities that are involved in a single reaction step. While the order of a reaction is derived experimentally, the molecularity is a theoretical concept and can only be applied to elementary reactions. In elementary reactions, the reaction order, the molecularity and the stoichiometric coefficient are the same, although only numerically, because they are different concepts.

1. A reaction involving one molecular entity is called unimolecular.
2. A reaction involving two molecular entities is called bimolecular.
3. A reaction involving three molecular entities is called termolecular. Termolecular reactions in solutions or gas mixtures are very rare, because of the improbability of three molecular entities simultaneously colliding. However, the term termolecular is also used to refer to three body association reactions of the type:

$$A + B \xrightarrow{\ M\ } C$$

where, the M over the arrow denotes that to conserve energy and momentum a second reaction with a third body is required. After the initial bimolecular collision of A and B an energetically excited reaction intermediate is formed, then, it collides with a M body, in a second bimolecular reaction, transferring the excess energy to it.

The reaction can be explained as two consecutive reactions:

$$A + B \rightarrow AB*$$
$$AB* + M \rightarrow C + M$$

These reactions frequently have a pressure and temperature dependence region of transition between second and third order kinetics.

Kinetics

Kinetics is the area of chemistry concerned with reaction rates. The rate can be expressed as:

Rate = Change in substance/time for change to occur (usually in M/s)

There are several factors that determine the rate of a specific reaction and those are expressed in the 'collision theory' that states that for molecules to react, they must:

1. Collide.
2. Have the right energy.
3. Have the right geometry.

To increase the rate, you must make the above more likely to occur. This is possible by changing other factors such as:

1. Increasing the surface area (of solids)—this allows for more collisions and gives more molecules the right geometry.
2. Increasing the temperature—this gives more molecules the right energy (also called the activation energy, E_a) (Fig. 2.1).
3. Increasing the concentration (of gases and solutions)—this allows for more collisions and more correct geometry.
4. Using a catalyst—helps molecules achieve the correct geometry by providing a different way to react (Fig. 2.2).

The reaction rate can also be expressed by using a 'rate law' and is written as follows:

For the general reaction: $aA + bB + ... \rightarrow gG + hH + ...$

the reaction rate can be calculated by:

$$\text{Reaction rate} = k[A]^m[B]^n ....$$

where:

[A], [B], etc. are the concentrations of the reactants.

k is the rate constant or rate coefficient, a value dependent on temperature.

m, n, etc. are exponents that correspond to a, b, etc. The concentration is raised to the power of its coefficient in the balanced equation.

Reaction order is a topic that comes with reaction rates. If you have a reaction in that A, B, and C are possible reactants, then we can describe the order of the reaction following this chart:

Rate = kC_a	1st order
Rate = kC_a^2 or kC_aC_b	2nd order
Rate = kC_a^3 or $kC_aC_bC_c$	3rd order

The order of the reaction is defined as the sum of the exponents of the coefficients. In general, first order reactions are most commonly seen, but reactions of other orders are also important. Zero-order reactions—those for that the change in the reaction is independent of the concentration of the reaction— are also possible.

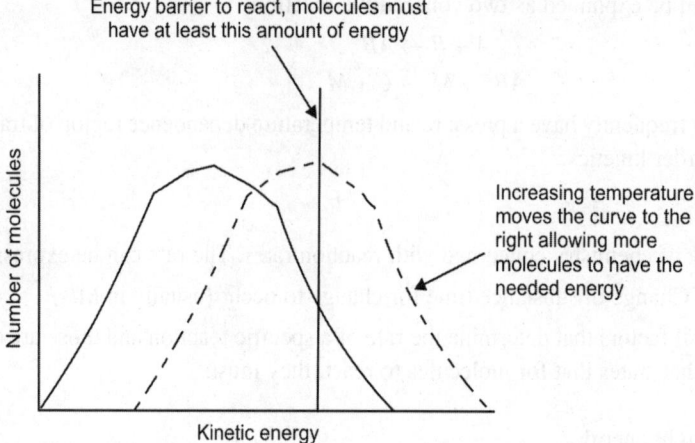

Fig. 2.1. Increasing the temperature gives more molecules the right energy (activation energy, E_a).

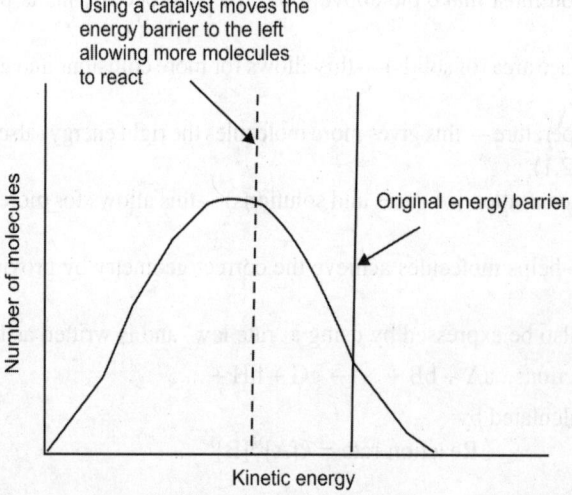

Fig. 2.2. Catalyst helps molecules achieve the correct geometry by providing a different way to react.

It is possible to determine the order of a reactant, and eventually the reaction rate, using initial rate information that includes the concentration of the reactants and the rate at that the product is formed. If you double the concentration of reactant X and the rate increases by 2^a, then the order of reactant X is 'a'. If you triple the concentration of reactant Y and the rate increases by 3^b, then the order of Y is 'b'. For example, if you have a reaction with one reactant, A, and you double [A] and the rate doubles, then the rate = $k[A]^1$. If, instead, you double [A] and the rate quadruples, the rate = $k[A]^2$. If you double [A] and the rate stays the same, then the rate = $k[A]^0$.

To find the rate constant, k, using initial rate information, just plug in from the experiment one of the concentrations and rate into the rate law and solve. The units of k are trickier:

$$\text{Units of } k = \text{units of rate/(units of concentration)}^{\text{reaction order}}$$

Ex: for 2nd order reaction, $k = (M/s)/M^2 = M^{-1}s^{-1}$

Example problem: Find the rate law and rate constant of $A + B \rightarrow C$ using the following data:

Ex#	[A] (M)	[B] (M)	Initial rate of C (M/s)
1	0.100	0.100	4.0×10^{-5}
2	0.100	0.200	4.0×10^{-5}
3	0.300	0.100	3.6×10^{-4}

Answer: Doubling [B] had no effect on the rate so B is zero order.
Tripling [A] caused the rate to multiply by 9 or by 3^2, so A is 2nd order.
Rate $= k[A]^2[B]^0 = k[A]^2$
$4.0 \times 10^{-5} = k (0.100)^2$ $k = 4.0 \times 10^{-3} M^{-3} s^{-1}$

When dealing with reaction rates, it is sometimes important to know how to graph a straight line with the data you have. When graphing concentration versus time, there are two ways to graph a line. If you have a first order reaction, then the graph of ln[A] vs. time is a line. If you have a second order reaction, then the graph of 1/[A] vs. time produces a line.

A quantitative way to examine reaction rates is through Arrhenius Equation that states:

$$k = Ae^{-E_a/RT}$$

where,

A is a constant related to the geometry needed
e is a constant, approximately 2.7281
E_a is the activation energy
R is the gas law constant, 8.314 J/mol-K
T is the temperature in kelvins.

If it is a simple geometry to attain, A will be large. If a large E_a is needed then the exponent becomes more negative and therefore decreases k. If the temperature increases then the exponent becomes less negative and therefore increases k. A pop-up calculator is available to help practice using Arrhernius' Equation to make calculations. The following (Fig. 2.3) are two energy profile graphs that help demonstrate energy changes during a reaction.

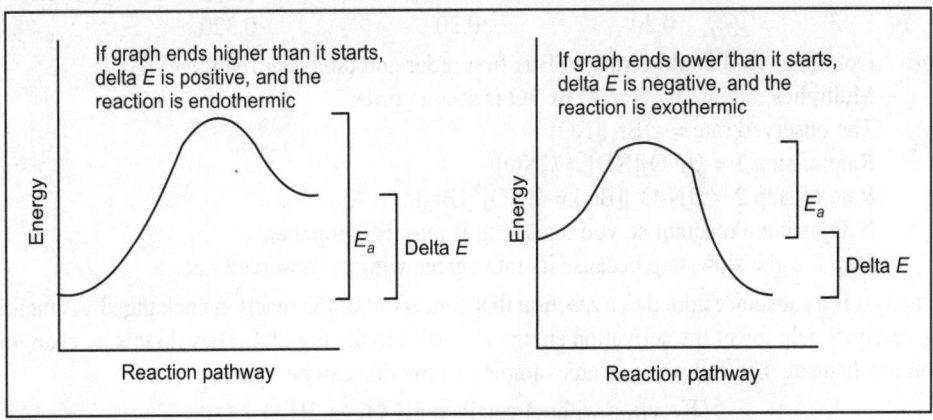

Fig. 2.3. Two energy profile graphs that help demonstrate energy changes during a reaction.

Not all reactions happen exactly as they are written. Most, in fact, go through an intermediate step. Reaction mechanism studies look at how a reaction actually occurs. Defined, a reaction mechanism is a series of elementary reactions that are proposed to account for the rate law (kinetics) of a particular reaction. The diagram below shows the two steps involved in a particular mechanism, and it shows how we get the reaction from the mechanism.

Mechanism:

Step 1: $NO + NO \rightarrow N_2O_2$

Step 2: $N_2O_2 + Br_2 \rightarrow 2NOBr$

Now we add up all the terms on both sides of the reaction and cancel out those that appear on both sides.

$NO + NO + \cancel{N_2O_2} + Br_2 \rightarrow \cancel{N_2O_2} + 2NOBr$

reaction:

$2NO + Br_2 \rightarrow 2NOBr$

It is helpful to remember certain terms and facts when dealing with mechanisms. You cannot derive a mechanism from the equation and when you combine the steps of a mechanism, you end up with the reaction. The molecularity of a step tells how many molecules are involved (most involve two molecules so they are bimolecular). An intermediate product is a molecule formed in one step and then used in another. One of the most important concepts to keep in mind is that the steps are not equally important. To speed up the reaction, you must speed up the slowest step (also called the rate-determining step).

Rate law of slow step = rate law of reaction

When determining the rate of a step, simply make the exponent of the reactant's concentration in the rate law the same as the coefficient of the reactant in the step.

Example problem: Find the slow step of the following reaction mechanism.

Reaction: $2NO + Br_2 \rightarrow 2NOBr$

Mechanism: Step 1: $NO + NO \rightarrow N_2O_2$

                        Step 2: $N_2O_2 + Br_2 \rightarrow 2NOBr$

Ex#	[NO] (M)	[Br$_2$] (M)	Initial rate of NOBr (M/s)
1	0.10	0.10	0.040
2	0.10	0.20	0.080
3	0.20	0.20	0.320

Answer: Doubling [Br$_2$] doubles rate so B is first order and doubling [NO] and [Br$_2$]

Multiplies the rate by 8 or 2^3 so NO is second order

The observed rate = $k[Br_2][NO]^2$

Rate of step 1 = $k[NO][NO] = k[NO]^2$

Rate of step 2 = $k[N_2O_2][Br_2] = k[NO]^2 [Br_2]$

N_2O_2 is not a reactant so you must split it into its components

Step 2 is the slow step because its rate agrees with the observed rate.

A catalyst is a substance added to a reaction that comes out of the reaction unchanged. As mentioned earlier, catalysts help lower the activation energy as shown in the Fig. 2.4. They do this by changing the reaction mechanism. The following is an example of how this can be done.

Reaction without catalyst: $2H_2O_2 \rightarrow 2H_2O + O_2$

this is one step reaction

Reaction with catalyst: $2H_2O_2 \xrightarrow{Br_2} 2H_2O + O_2$

The Br_2 allows this to be a two step reaction:

1. $H_2O_2 + Br_2 \rightarrow 2Br^- + 2H^+ + O_2$
2. $H_2O_2 + 2Br^- + 2H^+ \rightarrow 2Br_2 + 2H_2O$

This two step reaction is faster than the one step reaction.

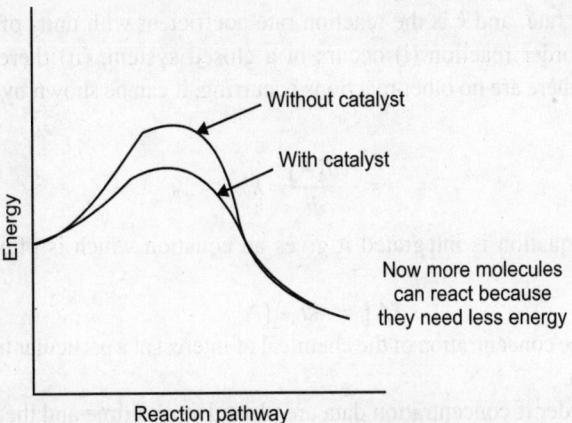

Fig. 2.4. Catalysts help lower the activation energy.

Rate Equation

The rate law or rate equation for a chemical reaction is an equation which links the reaction rate with concentrations or pressures of reactants and constant parameters (normally rate coefficients and partial reaction orders). To determine the rate equation for a particular system one combines the reaction rate with a mass balance for the system. For a generic reaction $mA + nB \rightarrow C$ with no intermediate steps in its reaction mechanism (that is, an elementary reaction), the rate is given by:

$$r = k[A]^m[B]^n$$

where, [A] and [B] express the concentration of the species A and B, respectively (usually in moles per litre (molarity)); m and n are not the respective stoichiometric coefficients of the balanced equation; they must be determined experimentally. k is the rate coefficient or rate constant of the reaction. The value of this coefficient k depends on conditions such as temperature, ionic strength, surface area of the adsorbent or light irradiation. For elementary reactions, the rate equation can be derived from first principles using collision theory. Again, m and n are NOT always derived from the balanced equation.

The rate equation of a reaction with a multi-step mechanism cannot, in general, be deduced from the stoichiometric coefficients of the overall reaction; it must be determined experimentally. The equation may involve fractional exponential coefficients, or it may depend on the concentration of an intermediate species. The rate equation is a differential equation, and it can be integrated to obtain an integrated rate equation that links concentrations of reactants or products with time.

If the concentration of one of the reactants remains constant (because it is a catalyst or it is in great excess with respect to the other reactants) its concentration can be included in the rate constant, obtaining a pseudo constant: if B is the reactant whose concentration is constant then $r = k[A][B] = k'[A]$. The second order rate equation has been reduced to a pseudo first order rate equation. This makes the treatment to obtain an integrated rate equation much easier.

Zero-order reactions

A zero-order reaction has a rate which is independent of the concentration of the reactant(s). Increasing the concentration of the reacting species will not speed up the rate of the reaction. Zero-order reactions are typically found when a material that is required for the reaction to proceed, such as a surface or a catalyst, is saturated by the reactants. The rate law for a zero-order reaction is:

$$r = k$$

where, r is the reaction rate, and k is the reaction rate coefficient with units of concentration/time. If, and only if, this zero-order reaction (i) occurs in a closed system, (ii) there is no net build-up of intermediates, and (iii) there are no other reactions occurring, it can be shown by solving a mass balance for the system that:

$$r = -\frac{d[A]}{dt} = k$$

If this differential equation is integrated it gives an equation which is often called the integrated zero-order rate law.

$$[A]_t = -kt + [A]_0$$

where, $[A]_t$ represents the concentration of the chemical of interest at a particular time, and $[A]_0$ represents the initial concentration.

A reaction is zero order if concentration data are plotted versus time and the result is a straight line. The slope of this resulting line is the negative of the zero-order rate constant k. The half-life of a reaction describes the time needed for half of the reactant to be depleted (same as the half-life involved in nuclear decay, which is a first-order reaction). For a zero-order reaction the half-life is given by:

$$t_{1/2} = \frac{[A]}{2k}$$

Example of a zero-order reaction:

Reversed Haber process: $2NH_3\ (g) \rightarrow 3H_2(g) + N_2(g)$

It should be noted that the order of a reaction cannot be deduced from the chemical equation of the reaction.

First-order reactions

A first-order reaction depends on the concentration of only one reactant (a unimolecular reaction). Other reactants can be present, but each will be zero-order. The rate law for an elementary reaction that is first order with respect to a reactant A is:

$$r = -\frac{d[A]}{dt} = k[A]$$

k is the first order rate constant, which has units of 1/time.

The integrated first-order rate law is

$$\ln [A] = -kt + \ln [A]_0$$

A plot of $\ln[A]$ vs. time t gives a straight line with a slope of $-k$.

The half-life of a first-order reaction is independent of the starting concentration and is given by:

$$t_{1/2} = \frac{\ln(2)}{k}$$

Examples of reactions that are first-order with respect to the reactant:

1. $H_2O_2\ (l) \rightarrow H_2O(l) + \frac{1}{2}O_2(g)$.
2. $SO_2Cl_2(l) \rightarrow SO_2(g) + Cl_2(g)$.
3. $2N_2O_5(g) \rightarrow 4NO_2(g) + O_2(g)$.

An alternative view of first order kinetics

The integrated first-order rate law:

$$\ln[A] = -kt + \ln[A]_0$$

is usually written in the form of the exponential decay equation:

$$A = A_0 e^{-kt}$$

While this is the traditional solution for the integration of the first-order process, the use of Eulers constant in this situation introduces a systematic error in the reported kinetic constants.

In a first-order process, the rate (r) of disappearance of a reactant is proportional to the amount of the reactant present (A_0).

$$r = kA_0$$

Therefore, the kinetic constant must represent the fraction of the population of reactant present that will breakdown in a given time period and the fraction must be less than one. For rates that are very small in comparison to the total population the traditional equation works fairly well.

For example, for simplicity, if the initial population is assumed to be 1 $(A_0 = 1)$ and the time period is restricted to the first interval $(t = 1)$, and a rate of 5 per cent conversion per time period is examined and assuming a remainder of 95 per cent of the original reactant after the first time period the result is:

$$A = A_0 e^{-kt}$$
$$A = e^{-0.05}$$
$$A = 95.12\%$$

To get the expected remainder of 95 per cent this equation requires the kinetic constant to be increased to 5.129 per cent.

$$A = e^{-0.05129}$$
$$A = 0.95$$

The problem becomes more apparent the higher the rate observed, for example if the rate was 95 per cent, then the remaining reactant after one time period would be expected to be 5 per cent of the initial starting population however,

$$A = e^{-0.95}$$
$$A = 0.3867$$

To produce the expected 5 per cent remainder the rate constant must be increased to $k = 2.99573$ a rate of approximately 300 per cent, which may be difficult to observe.

$$A = e^{-2.99573}$$
$$A = 0.05$$

The problems introduced by the use of this equation can be overcome by recognising the artificial splitting of the constant as both eulers constant and the kinetic constant are constant and do not change, so a constant raised to a constant is a constant. This constant represents the fraction of the reactant population remaining (%RP) per time period so can also be rewritten to incorporate the rate of the fraction of the population that will breakdown (%BD) per time period as well.

$$e^{-k} = \%RP = 1 - \%BD$$

Therefore, the kinetic equation for first order kinetics can be rewritten as:

$$A = A_0 (1 - [\%BD)]^t$$

Where the fraction of the population that will breakdown ($\%BD$ = the 1st order kinetic constant) can be expressed as the observed reaction rate (r) divided by the initial reactant concentration A_0.

$$A = A_0 (1 - [r/A_0)]^t$$

This notation relates the kinetic constant directly to the observed rate and recognises the kinetic constant cannot exceed 1 as the rate can never be a value greater than the number in initial starting reactants.

Second-order reactions

A second-order reaction depends on the concentrations of one second-order reactant, or two first-order reactants. For a second order reaction, its reaction rate is given by:

$$r = k[A]^2 \text{ or } r = k[A][B] \text{ or } r = k[B]^2$$

The integrated second-order rate laws are respectively:

$$\frac{1}{[A]} = kt + \frac{1}{[A]_0}$$

(The stoichiometric factor of 2 should not be included as a part of the rate constant for an elementary reaction of the type $2A \rightarrow B$. Unlike it is presented in several popular kinetics books, the proper definition of the rate law for second-order reactions is.

$$-\frac{d[A]}{dt} = 2k[A]^2$$

This proper definition is used in most peer-reviewed literature, tables of rate constants, and simulation software.)

or

$$\frac{[A]}{[B]} = \frac{[A]_0}{[B]_0} e^{([A]_0 - [B]_0)kt}$$

$[A]_0$ and $[B]_0$ must be different to obtain that integrated equation.

The half-life equation for a second-order reaction dependent on one second-order reactant is:

$$t_{1/2} = \frac{1}{k[A]_0}.$$

For a second-order reaction half-lives progressively double.

Another way to present the above rate laws is to take the log of both sides:

$$\ln r = \ln k + 2 \ln[A]$$

Examples of a second-order reaction:

$$2NO_2(g) \rightarrow 2NO(g) + O_2(g)$$

Pseudo first-order

Measuring a second-order reaction rate with reactants A and B can be problematic: the concentrations of the two reactants must be followed simultaneously, which is more difficult; or measure one of them and calculate the other as a difference, which is less precise. A common solution for that problem is the pseudo first-order approximation.

If either [A] or [B] remain constant as the reaction proceeds, then the reaction can be considered pseudo first-order because in fact it only depends on the concentration of one reactant. If for example [B] remains constant then:

$$r = k[A][B] = k'[A]$$

where, $k' = k[B]_0$ (k' or k_{obs} with units s^{-1}) and an expression is obtained identical to the first-order expression above.

One way to obtain a pseudo first order reaction is to use a large excess of one of the reactants ([B]>>[A] would work for the previous example) so that, as the reaction progresses only a small amount of the reactant is consumed and its concentration can be considered to stay constant. By collecting k' for many reactions with different (but excess) concentrations of [B]; a plot of k' versus [B] gives k (the regular second-order rate constant) as the slope.

Equilibrium reactions or opposed reactions

A pair of forward and reverse reactions may define an equilibrium process. For example A and B react into X and Y and vice versa (s, t, u and v are the stoichiometric coefficients):

$$sA + tB \rightleftharpoons uX + vY$$

The reaction rate expression for the above reactions (assuming they each are elementary) can be expressed as:

$$r = k_1[A]^s[B]^t - k_2[X]^u[Y]^v$$

where, k_1 is the rate coefficient for the reaction which consumes A and B; k_2 is the rate coefficient for the backwards reaction, which consumes X and Y and produces A and B (Fig. 2.5).

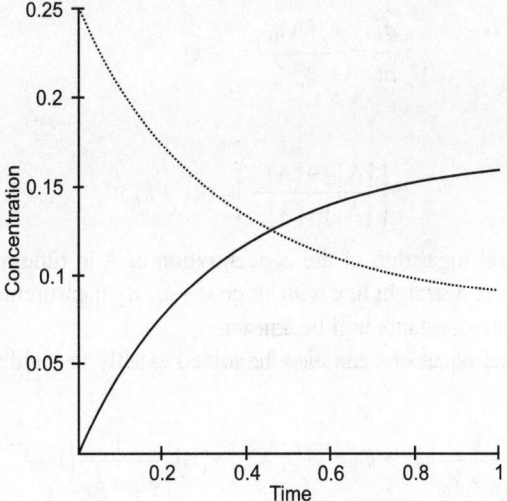

Fig. 2.5. Concentration of A ($A_0 = 0.25$ mole/l) and B versus time reaching equilibrium $k_f = 2$ min^{-1} and $k_r = 1$ min^{-1}.

The constants k_1 and k_2 are related to the equilibrium coefficient for the reaction (K) by the following relationship (set $r = 0$ in balance):

$$k_1[A]^s[B]^t = k_2[X]^u[Y]^v$$

$$K = \frac{[X]^u[Y]^v}{[A]^s[B]^t} = \frac{k_1}{k_2}$$

In a simple equilibrium between two species:

$$A \rightleftharpoons B$$

the constant K at equilibrium is expressed as:

$$K \overset{def}{=} \frac{k_f}{k_b} = \frac{[B]_e}{[A]_e}$$

When the concentration of A at equilibrium is that of the concentration at time 0 minus the conversion in moles:

$$[A]_e = [A]_0 - x$$

with x equal to the concentration of B at equilibrium:

$$[B]_e = x$$

then it follows that

$$[B]_e = x = \frac{k_f}{k_f + k_b}[A]_0$$

and

$$[A]_e = [A]_0 - x = \frac{k_b}{k_f + k_b}[A]_0$$

The reaction rate becomes:

$$\frac{dx}{dt} = \frac{k_f[A]_0}{x_e}(x_e - x)$$

which results in:

$$\ln\left(\frac{[A]_0 - [A]_e}{[A_t] - [A]_e}\right) = (k_f + k_b)t$$

A plot of the negative natural logarithm of the concentration of A in time minus the concentration at equilibrium versus time t gives a straight line with slope $k_f + k_b$. By measurement of A_e and B_e the values of K and the two reaction rate constants will be known.

The system of differential equations can also be solved exactly to yield the following generalised expressions:

$$[A] = [A]_0 \frac{1}{k_f + k_b}\left(k_b + k_f e^{-(k_f + k_b)t}\right) + [B]_0 \frac{1}{k_f + k_b}\left(1 - e^{-(k_f + k_b)t}\right)$$

$$[B] = [A]_0 \frac{k_f}{k_f + k_b}\left(1 - e^{-(k_f + k_b)t}\right) + [B]_0 \frac{1}{k_f + k_b}\left(k_f + k_b e^{-(k_f + k_b)t}\right)$$

When the equilibrium constant is close to unity and the reaction rates very fast for instance in conformational analysis of molecules, other methods are required for the determination of rate constants for instance by complete lineshape analysis in NMR spectroscopy.

Consecutive reactions

If the rate constants for the following reaction are k_1 and k_2; $A \rightarrow B \rightarrow C$, then the rate equation is:

For reactant A: $\dfrac{d[A]}{dt} = -k_1[A]$

For reactant B: $\dfrac{d[B]}{dt} = k_1[A] - k_2[B]$

For product C: $\dfrac{d[C]}{dt} = k_2[B]$

With the individual concentrations scaled by the total population of reactants to become probabilities, linear systems of differential equations such as these can be formulated as a master equation. The differential equations can be solved analytically and the integrated rate equations are:

$$[A] = [A]_0 e^{-k_1 t}$$

$$[B] = \begin{cases} [A]_0 \dfrac{k_1}{k_2 - k_1} (e^{-k_1 t} - e^{-k_2 t}) + [B]_0 e^{-k_2 t} & k_1 \neq k_2 \\[2ex] [A]_0 k_1 t e^{-k_1 t} + [B]_0 e^{-k_1 t} & \text{Otherwise} \end{cases}$$

$$[C] = \begin{cases} [A]_0 \left(1 + \dfrac{k_1 e^{-k_2 t} - k_2 e^{-k_1 t}}{k_2 - k_1} \right) + [B]_0 (1 - e^{-k_2 t}) + [C]_0 & k_1 \neq k_2 \\[2ex] [A]_0 (1 - e^{-k_1 t} - k_1 t e^{-k_1 t}) + [B]_0 (1 - e^{-k_1 t}) + [C]_0 & \text{Otherwise} \end{cases}$$

The steady-state approximation leads to very similar results in an easier way.

Parallel or competitive reactions

When a substance reacts simultaneously to give two different products, a parallel or competitive reaction is said to take place.

1. Two first-order reactions:

   $A \rightarrow B$ and $A \rightarrow C$, with constants k_1 and k_2 and rate equations $-d[A]/dt = (k_1 + k_2)[A]$, $d[B]/dt = k_1[A]$ and $d[C]/dt = k_2[A]$.

   The integrated rate equations are then:

   $$[A] = [A]_0 e^{-(k_1 + k_2)t};$$

   $$[B] = \frac{k_1}{k_1 + k_2} [A]_0 (1 - e^{-(k_1 + k_2)t})$$

   and

$$[C] = \frac{k_2}{k_1 + k_2}[A]_0(1 - e^{-(k_1 + k_2)t}).$$

One important relationship in this case is: $\dfrac{[B]}{[C]} = \dfrac{k_1}{k_2}$

2. One first order and one second-order reaction: This can be the case when studying a bimolecular reaction and a simultaneous hydrolysis (which can be treated as pseudo order one) takes place: the hydrolysis complicates the study of the reaction kinetics, because some reactant is being 'spent' in a parallel reaction. For example A reacts with R to give our product C, but meanwhile the hydrolysis reaction takes away an amount of A to give B, a by-product: $A + H_2O \rightarrow B$ and $A + R \rightarrow C$. The rate equations are: $d[B]/dt = k_1[A][H_2O] = k_1'[A]$ and $d[C]/dt = k_2[A][R]$. Where k_1' is the pseudo first-order constant.

The integrated rate equation for the main product [C] is $[C] = [R]_0 \left[1 - e^{-k_2/k_1'[A]_0(1 - e^{-k_1't})} \right]$,

which is equivalent to $\ln \dfrac{[R]_0}{[R]_0 - [C]} = \dfrac{k_2[A]_0}{k_1'}(1 - e^{-k_1't})$. Concentration of B is related to that

of C through $[B] = -\dfrac{k_1'}{k_2} \ln \left(1 - \dfrac{[C]}{[R]_0} \right)$.

The integrated equations were analytically obtained but during the process it was assumed that $[A]_0 - [C] \approx [A]_0$ therefore, previous equation for [C] can only be used for low concentrations of [C] compared to $[A]_0$.

Power Law

A power law is a special kind of mathematical relationship between two quantities. When the number or frequency of an object or event varies as a power of some attribute of that object (e.g. its size), the number or frequency is said to follow a power law. For instance, the number of cities having a certain population size is found to vary as a power of the size of the population, and hence follows a power law. Power laws govern a wide variety of natural and man-made phenomena, including frequencies of words in most languages, frequencies of family names, sizes of craters on the moon and of solar flares, the sizes of power outages, earthquakes, and wars, the popularity of books and music, and many other quantities (Fig. 2.6).

Technical definition

A power law is any polynomial relationship that exhibits the property of scale invariance. The most common power laws relate two variables and have the form:

$$f(x) = ax^k + o(x^k)$$

where, a and k are constants, and $o(x^k)$ is an asymptotically small function of x^k. Here, k is typically called the scaling exponent, where the word 'scaling' denotes the fact that a power-law function satisfies $f(cx) \propto f(x)$, where c is a constant. Thus, a rescaling of the function's argument changes the constant of proportionality but preserves the shape of the function itself.

This point becomes clearer if we take the logarithm of both sides:

$$\log [f(x)] = k \log x + \log a$$

Notice that this expression has the form of a linear relationship with slope k. Rescaling the argument produces a linear shift of the function up or down but leaves both the basic form and the slope k unchanged.

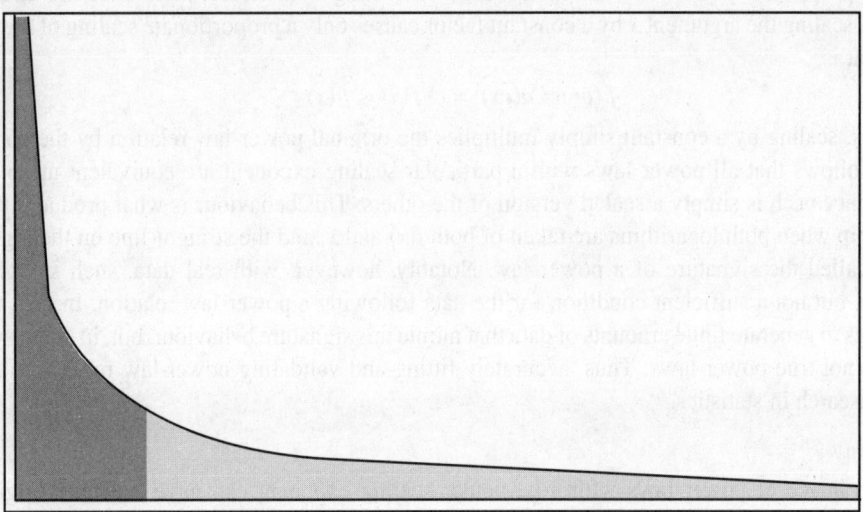

Fig. 2.6. An example power law graph, being used to demonstrate ranking of popularity. To the right is the long tail, to the left are the few that dominate (also known as the 80–20 rule).

Power-law relations characterise a staggering number of naturally occurring phenomena, and this is one of the principal reasons why they have attracted such wide interest. For instance, inverse-square laws, such as gravitation and the Coulomb force, are power laws, as are many common mathematical formulae such as the quadratic law of area of the circle. However, much of the recent interest in power laws comes from the study of probability distributions: it's now known that the distributions of a wide variety of quantities seem to follow the power-law form, at least in their upper tail (large events). The behaviour of these large events connects these quantities to the study of theory of large deviations (also called extreme value theory), which considers the frequency of extremely rare events like stock market crashes and large natural disasters. It is primarily in the study of statistical distributions that the name 'power law' is used; in other areas the power-law functional form is more often referred to simply as a polynomial form or polynomial function.

Scientific interest in power law relations stems partly from the ease with which certain general classes of mechanisms generate them. The demonstration of a power-law relation in some data can point to specific kinds of mechanisms that might underlie the natural phenomenon in question, and can indicate a deep connection with other, seemingly unrelated systems (see the reference by Simon and the subsection on universality below). The ubiquity of power-law relations in physics is partly due to dimensional constraints, while in complex systems, power laws are often thought to be signatures of hierarchy or of specific stochastic processes. A few notable examples of power laws are the Gutenberg-Richter law for earthquake sizes, Pareto's law of income distribution, structural self-similarity of fractals, and scaling laws in biological systems. Research on the origins of power-law relations, and efforts to observe and validate them in the real world, is an active topic of research in many fields of science, including physics, computer science, linguistics, geophysics, sociology, economics and more.

Properties of power laws

Scale invariance

The main property of power laws that makes them interesting is their scale invariance. Given a relation $f(x) = ax^k$, scaling the argument x by a constant factor causes only a proportionate scaling of the function itself. That is,

$$f(cx) = a(cx)^k = c^k f(x) \propto f(x)$$

That is, scaling by a constant simply multiplies the original power-law relation by the constant c^k. Thus, it follows that all power laws with a particular scaling exponent are equivalent up to constant factors, since each is simply a scaled version of the others. This behaviour is what produces the linear relationship when both logarithms are taken of both $f(x)$ and x, and the straight-line on the log-log plot is often called the signature of a power law. Notably, however, with real data, such straightness is necessary, but not a sufficient condition for the data following a power-law relation. In fact, there are many ways to generate finite amounts of data that mimic this signature behaviour, but, in their asymptotic limit, are not true power laws. Thus, accurately fitting and validating power-law models is an active area of research in statistics.

Universality

The equivalence of power laws with a particular scaling exponent can have a deeper origin in the dynamical processes that generate the power-law relation. In physics, for example, phase transitions in thermodynamic systems are associated with the emergence of power-law distributions of certain quantities, whose exponents are referred to as the critical exponents of the system. Diverse systems with the same critical exponents—that is, which display identical scaling behaviour as they approach criticality—can be shown, via renormalisation group theory, to share the same fundamental dynamics. For instance, the behaviour of water and CO_2 at their boiling points fall in the same universality class because they have identical critical exponents. In fact, almost all material phase transitions are described by a small set of universality classes. Similar observations have been made, though not as comprehensively, for various self-organised critical systems, where the critical point of the system is an attractor. Formally, this sharing of dynamics is referred to as universality, and systems with precisely the same critical exponents are said to belong to the same universality class.

Power-law functions

The general power-law function follows the polynomial form given above, and is a ubiquitous form throughout mathematics and science. Notably, however, not all polynomial functions are power laws because not all polynomials exhibit the property of scale invariance. Typically, power-law functions are polynomials in a single variable, and are explicitly used to model the scaling behaviour of natural processes. For instance, allometric scaling laws for the relation of biological variables are some of the best known power-law functions in nature. In this context, the $o(x^k)$ term is most typically replaced by a deviation term ε, which can represent uncertainty in the observed values (perhaps measurement or sampling errors) or provide a simple way for observations to deviate from the no power-law function (perhaps for stochastic reasons):

$$y = ax^k + \varepsilon.$$

Examples of power law functions

The examples of power law functions are given below:
1. The Stevens' power law of psychophysics.
2. The Stefan–Boltzmann law.
3. The Ramberg–Osgood stress-strain relationship.
4. The inverse-square laws of Newtonian gravity and electrostatics.
5. Electrostatic potential and Gravitational potential.
6. Model of van der Waals force.
7. Force and potential in simple harmonic motion.
8. Kepler's third law.
9. The initial mass function.
10. Gamma correction relating light intensity with voltage.
11. Kleiber's law relating animal metabolism to size, and allometric laws in general.
12. Behaviour near second-order phase transitions involving critical exponents.
13. Proposed form of experience curve effects.
14. The differential energy spectrum of cosmic-ray nuclei.
15. Square-cube law (ratio of surface area to volume).
16. Constructal law.
17. Fractals.
18. The Pareto principle also called the '80–20 rule'.
19. Zipf's Law in corpus analysis and population distributions amongst others, where frequency of an item or event is inversely proportional to its frequency rank (i.e. the second most frequent item/event occurring half as often the most frequent item and so on).
20. Weight vs. length models in fish.

Power-law distributions

A power-law distribution is any that, in the most general sense, has the form:

$$p(x) \propto L(x)x^{-\alpha}$$

where, $\alpha > 1$, and $L(x)$ is a slowly varying function, which is any function that satisfies $\lim_{\alpha \to \infty} L(tx)/L(x)$ = 1 with t constant. This property of $L(x)$ follows directly from the requirement that $p(x)$ be asymptotically scale invariant; thus, the form of $L(x)$ only controls the shape and finite extent of the lower tail. For instance, if $L(x)$ is the constant function, then we have a power-law that holds for all values of x. In many cases, it is convenient to assume a lower bound x_{min} from which the law holds. Combining these two cases, and where x is a continuous variable, the power law has the form

$$p(x) = \frac{\alpha - 1}{x_{min}} \left(\frac{x}{x_{min}} \right)^{-\alpha}$$

where the prefactor to $x^{-\alpha}$ is the normalising constant. We can now consider several properties of this distribution. For instance, its moments are given by:

$$\left\langle x^m \right\rangle = \int_{x_{min}}^{\infty} x^m p(x) \mathrm{d}x = \frac{\alpha - 1}{\alpha - 1 - m} x_{min}^m$$

which is only well defined for $m < \alpha - 1$. That is, all moments $m \geq \alpha - 1$ diverge: when $\alpha < 2$, the average and all higher-order moments are infinite; when $2 < \alpha < 3$, the mean exists, but the variance and higher-order moments are infinite, etc. For finite-size samples drawn from such distribution, this behaviour implies that the central moment estimators (like the mean and the variance) for diverging moments will never converge—as more data is accumulated, they continue to grow.

Another kind of power-law distribution, which does not satisfy the general form above, is the power law with an exponential cutoff:

$$p(x) \propto L(x)x^{-\alpha}e^{-\lambda x}.$$

In this distribution, the exponential decay term $e^{-\lambda x}$ eventually overwhelms the power-law behaviour at very large values of x. This distribution does not scale and is thus not asymptotically a power-law; however, it does approximately scale over a finite region before the cutoff. (Note that the pure form above is a subset of this family, with $\lambda = 0$.) This distribution is a common alternative to the asymptotic power-law distribution because it naturally captures finite-size effects. For instance, although the Gutenberg–Richter law is commonly cited as an example of a power-law distribution, the distribution of earthquake magnitudes cannot scale as a power law in the limit $x \to \infty$ because there is a finite amount of energy in the earth's crust and thus there must be some maximum size to an earthquake. As the scaling behaviour approaches this size, it must taper off.

Plotting power-law distributions

In general, power-law distributions are plotted on doubly logarithmic axes, which emphasises the upper tail region. The most convenient way to do this is via the (complementary) cumulative distribution (cdf), $P(x) = \Pr(X > x)$,

$$P(x) = \Pr(X > x) = C\int_x^\infty p(X)\mathrm{d}X = \frac{\alpha - 1}{x_{\min}^{-\alpha+1}}\int_x^\infty X^{-\alpha}\mathrm{d}X = \left(\frac{x}{x_{\min}}\right)^{(-\alpha+1)}$$

Note that the cdf is also a power-law function, but with a smaller scaling exponent. For data, an equivalent form of the cdf is the rank-frequency approach, in which we first sort the n observed values in ascending

order, and plot them against the vector $\left[1, \dfrac{n-1}{n}, \dfrac{n-2}{n}, ..., \dfrac{1}{n}\right]$.

Although it can be convenient to log-bin the data, or otherwise smooth the probability density (mass) function directly, these methods introduce an implicit bias in the representation of the data, and thus should be avoided. The cdf, on the other hand, introduces no bias in the data and preserves the linear signature on doubly logarithmic axes.

Estimating the exponent from empirical data

There are many ways of estimating the value of the scaling exponent for a power-law tail, however not all of them yield unbiased and consistent answers. The most reliable techniques are often based on the method of maximum likelihood.

Alternative methods are often based on making a linear regression on either the log-log probability, the log-log cumulative distribution function, or on log-binned data, but these approaches should be avoided as they can all lead to highly biased estimates of the scaling exponent.

For real-valued data, we fit a power-law distribution of the form:

$$p(x) = \frac{\alpha - 1}{x_{\min}}\left(\frac{x}{x_{\min}}\right)^{-\alpha}$$

to the data $x \geq x_{min}$. Given a choice for x_{min}, a simple derivation by this method yields the estimator equation.

$$\hat{\alpha} = 1 + n \left[\sum_{i=1}^{n} \ln \frac{x_i}{x_{min}} \right]^{-1}$$

where, $\{x_i\}$ are the n data points $x_i \geq x_{min}$. This estimator exhibits a small finite sample-size bias of order $O(n^{-1})$, which is small when $n > 100$. Further, the uncertainty in the estimation can be derived from the maximum likelihood argument, and has the form $\sigma = \dfrac{\alpha - 1}{\sqrt{n}}$. This estimator is equivalent to the popular Hill estimator from quantitative finance and extreme value theory.

For a set of n integer-valued data points $\{x_i\}$, again where each $x_i \geq x_{min}$, the maximum likelihood exponent is the solution to the transcendental equation:

$$\frac{\varsigma'(\hat{\alpha}, x_{min})}{\varsigma(\hat{\alpha}, x_{min})} = -\frac{1}{n} \sum_{i=1}^{n} \ln \frac{x_i}{x_{min}}$$

where, $\zeta(\alpha, x_{min})$ is the incomplete zeta function. The uncertainty in this estimate follows the same formula as for the continuous equation. However, the two equations for $\hat{\alpha}$ are not equivalent, and the continuous version should not be applied to discrete data, nor vice versa.

Further, both of these estimators require the choice of x_{min}. For functions with a non-trivial $L(x)$ function, choosing x_{min} too small produces a significant bias in $\hat{\alpha}$, while choosing it too large increases the uncertainty in $\hat{\alpha}$, and reduces the statistical power of our model. In general, the best choice of x_{min} depends strongly on the particular form of the lower tail, represented by $L(x)$ above.

More about these methods, and the conditions under which they can be used, can be found in the Clauset reference below. Further, this comprehensive review article provides usable code (Matlab and R) for estimation and testing routines for power-law distributions.

SECOND-ORDER REACTIONS

Second-order reactions are characterised by the property that their rate is proportional to the product of two reactant concentrations (or the square of one concentration). Suppose that A $\rightarrow$ products is second order in A, or suppose that A + B $\rightarrow$ products is first order in A and also first order in B. Then the differential rate laws in these two cases are given by Differential Rate Laws:

$$d[A]/dt = -k\,[A]^2 \text{ (for 2A} \rightarrow \text{products)}$$

$$\text{or } d\xi/dt = -k\,[A][B] \text{ (for A + B} \rightarrow \text{products)}$$

In mathematical language, these are first order differential equations because they contain the first derivative and no higher derivatives. A chemist calls them second order rate laws because the rate is proportional to the product of two concentrations. By elementary integration of these differential equations Integrated Rate Laws can be obtained:

$$1/[A] - 1/[A]_0 = kt \text{ (for 2A} \rightarrow \text{products)}$$

$$\text{or } (1/(a-b))\,[\ln((a-\xi)/(b-\xi)) - \ln(a/b)] = kt \text{ (for A + B} \rightarrow \text{ products)}$$

where, a and b are the initial concentrations of A and B (assuming a not equal to b), and ξ is the extent of reaction at time t. Note that the latter can also be written:

$$(a-x)/(b-x) = (a/b)\exp[(a-b)kt].$$

A common way for a chemist to discover that a reaction follows second order kinetics is to plot $1/[A]$ versus the time in the former case, or $\ln(b(a-\xi)/a(b-\xi))$ versus t in the latter case.

Data analysis: $1/[A] = 1/[A]_0 + kt$

A plot of $1/[A]$ versus t is a straight line with slope k.

Software Tools for Second-order Reactions

Computer software tools can be used to solve chemical kinetics problems. In second-order reactions it is often useful to plot and fit a straight line to data. One tool for this is the 'slope(x,y)' command in the product MathCad. Here is a mathcad file that can serve as template for second order kinetics data analysis.

Reversible Reaction

A reversible reaction is a chemical reaction that results in an equilibrium mixture of reactants and products. For a reaction involving two reactants and two products this can be expressed symbolically as

$$aA + bB \rightleftharpoons cC + dD$$

A and B can react to form C and D or, in the reverse reaction, C and D can react to form A and B. This is distinct from reversible process in thermodynamics.

The concentrations of reactants and products in an equilibrium mixture are determined by the analytical concentrations of the reagents (A and B or C and D) and the equilibrium constant, K. The magnitude of the equilibrium constant depends on the Gibbs free energy change for the reaction. So, when the free energy change is large (more than about 30 kJ mol^{-1}), then the equilibrium constant is large (log $K > 3$) and the concentrations of the reactants at equilibrium are very small. Such a reaction is sometimes considered to be an irreversible reaction, although in reality small amounts of the reactants are still expected to be present in the reacting system. A truly irreversible chemical reaction is usually achieved when one of the products exits the reacting system, for example, as does carbon dioxide (volatile) in the reaction:

$$CaCO_3 + 2HCl \rightarrow CaCl_2 + H_2O + CO_2\uparrow$$

It is a common observation that most of the reactions when carried out in closed vessels do not go to completion, under a given set of conditions of temperature and pressure. In fact in all such cases, in the initial state, only the reactants are present but as the reaction proceeds, the concentration of reactants decreases and that of products increases. Finally a stage is reached when no further change in concentration of the reactants and products is observed.

This state at which the concentration of reactants and products do not change with time is called a state of chemical equilibrium.

The amount of reactants unused depend on the experimental conditions such as concentration, temperature, pressure and the nature of the reaction.

If a mixture of gaseous hydrogen and iodine vapours is made to react at 717K in a closed vessel for about 2–3 hours, gaseous hydrogen iodide is produced according to the following equation:

$$H_2(g) + I_2(g) \rightarrow 2HI(g)$$

But along with gaseous hydrogen iodide, there will be some amount of unreacted gaseous hydrogen and gaseous iodine left.

On the other hand if gaseous hydrogen iodide is kept at 717K in a closed vessel for about 2–3 hours it decomposes to give gaseous hydrogen and gaseous iodine.

$$2HI(g) \rightarrow H_2(g) + I_2(g)$$

In this case also some amount of gaseous hydrogen iodide will be left unreacted. This means that the products of certain reactions can be converted back to the reactants. These types of reactions are called reversible reactions.

Thus, in reversible reactions the products can react with one another under suitable conditions to give back the reactants. In other words, in reversible reactions the reaction takes place in both the forward and backward directions. The reversible reaction may be expressed as:

$$H_2(g) + I_2(g) \Leftrightarrow 2HI(g)$$

These reversible reactions never go to completion if performed in a closed container. For a reversible chemical reaction, an equilibrium state is attained when the rate at which a chemical reaction is proceeding in forward direction equals the rate at which the reverse reaction is proceeding.

At equilibrium,

Rate of forward reaction = Rate of reverse reaction

Consider the reversible reaction:

$$N_2(g) + 3H_2(g) \Leftrightarrow 2NH_3(g)$$

When this reaction is performed at high pressure and temperature in a close container, at equilibrium,

Rate of formation of ammonia = Rate of decomposition of ammonia

Now, the question arises whether all the ammonia molecules are remaining intact and not decomposing? Are all the molecules of nitrogen and hydrogen becoming inactive and not combining?

If this is the case, we would say a static equilibrium is attained. To understand the concept of static equilibrium, let us consider two children sitting on a see-saw. At balance point (i.e. the equilibrium position) no movement of children on the see-saw occurs.

Static equilibrium: In the case of reversible reaction, however a static equilibrium is not being established. In the case of ammonia, using deuterium, D (an isotope of hydrogen) it has been proved that even at equilibrium, decomposition of ammonia into hydrogen and nitrogen and combination of hydrogen and nitrogen into ammonia continues. This equilibrium is dynamic in nature and is, therefore, called dynamic equilibrium.

A dynamic steady state can be compared with the equilibrium of water in a reservoir, which is being simultaneously filled and discharged. If the rate of water flowing in is equal to the rate of water flowing out, the quantity of water in the reservoir will remain unchanged like the quantities of substances in a state of chemical equilibrium.

Dynamic equilibrium:

Rate of water entering = Rate of water leaving

Hence the level of water is constant. Similarly, some other reversible reactions are:

$$N_2(g) + O_2(g) \Leftrightarrow 2NO(g)$$

$$2SO_2(g) + O_2(g) \Leftrightarrow 2SO_3(g)$$

On the other hand, the chemical reaction in which the products formed do not combine to give the reactants are known as irreversible reactions.

For e.g. potassium chlorate decomposes on heating to form potassium chloride and oxygen.

$$2K\,Cl\,O_3(s) \xrightarrow{\Delta} 2K\,Cl(s) + 3O_2(g)$$

However, the products cannot combine to form potassium chlorate. In case of irreversible physical and chemical processes, the change occurs only in one direction and the processes go to completion. However, the reversible processes do not go to completion and appear to stop (attain state of chemical equilibrium) even though some starting materials are remaining.

Some examples of irreversible reactions are:

$$AgNO_3(aq) + NaCl(aq) \rightarrow AgCl(s) + NaNO_3(aq)$$

$$2Mg(s) + O_2(g) \rightarrow 2MgO(s)$$

It may be noted that for reversible reactions the symbol $\Leftrightarrow$ is used between the reactants and products. For the irreversible reactions, single headed arrow $\rightarrow$ is used.

REACTION RATE

The reaction rate or rate of reaction for a reactant or product in a particular reaction is intuitively defined as how fast a reaction takes place. For example, the oxidation of iron under the atmosphere is a slow reaction which can take many years, but the combustion of butane in a fire is a reaction that takes place in fractions of a second.

Chemical kinetics is the part of physical chemistry that studies reaction rates. The concepts of chemical kinetics are applied in many disciplines, such as chemical engineering, enzymology and environmental engineering.

Formal definition of reaction rate: Consider a typical chemical reaction:

$$aA + bB \rightarrow pP + qQ$$

The lowercase letters (a, b, p, and q) represent stoichiometric coefficients, while the capital letters represent the reactants (A and B) and the products (P and Q).

According to the IUPAC's Gold Book definition the reaction rate v (also r or R) for a chemical reaction occurring in a closed system under constant-volume conditions, without a build-up of reaction intermediates, is defined as:

$$v = -\frac{1}{a}\frac{d[A]}{dt} = -\frac{1}{b}\frac{d[B]}{dt} = \frac{1}{p}\frac{d[P]}{dt} = \frac{1}{q}\frac{d[Q]}{dt}$$

(Note: Rate of a reaction is always positive. '–' sign is present in the reactant involving terms because the reactant concentration is decreasing.) The IUPAC recommends that the unit of time should always be the second. In such a case the rate of reaction differs from the rate of increase of concentration of a product P by a constant factor (the reciprocal of its stoichiometric number) and for a reactant A by minus the reciprocal of the stoichiometric number. Reaction rate usually has the units of mol dm^{-3} s^{-1}. It is important to bear in mind that the previous definition is only valid for a single reaction, in a closed system of constant volume. This most usually implicit assumption must be stated explicitly, otherwise the definition is incorrect: If water is added to a pot containing salty water, the concentration of salt decreases, although there is no chemical reaction.

For any system in general the full mass balance must be taken into account: IN – OUT + Generation = Accumulation.

$$F_{A0} - F_A + \int_0^v v dV = \frac{dN_A}{dt}$$

When applied to the severe case stated previously this equation reduces to: $v = d[A]/dt$.

For a single reaction in a closed system of varying volume the so-called rate of conversion can be is used, in order to avoid handling concentrations. It is defined as the derivative of the extent of reaction with respect to time.

$$\dot{\xi} = \frac{d\xi}{dt} = \frac{1}{v_i}\frac{dn_i}{dt} = \frac{1}{v_i}\left(V\frac{dC_i}{dt} + C_i\frac{dV}{dt} \right)$$

v_i is the stoichiometric coefficient for substance i, V is the volume of reaction and C_i is the concentration of substance i.

When side products or reaction intermediates are formed, the IUPAC recommends the use of the terms rate of appearance and rate of disappearance for products and reactants, properly.

Reaction rates may also be defined on a basis that is not the volume of the reactor. When a catalyst is used the reaction rate may be stated on a catalyst weight (mol g^{-1} s^{-1}) or surface area (mol m^{-2} s^{-1}) basis. If the basis is a specific catalyst site that may be rigorously counted by a specified method, the rate is given in units of s^{-1} and is called a turnover frequency.

Factors Influencing Rate of Reaction

1. The nature of the reaction: Some reactions are naturally faster than others. The number of reacting species, their physical state (the particles that form solids move much more slowly than those of gases or those in solution), the complexity of the reaction and other factors can influence greatly the rate of a reaction.
2. Concentration: Reaction rate increases with concentration, as described by the rate law and explained by collision theory. As reactant concentration increases, the frequency of collision increases.
3. Pressure: The rate of gaseous reactions increases with pressure, which is, in fact, equivalent to an increase in concentration of the gas. For condensed-phase reactions, the pressure dependence is weak.
4. Order: The order of the reaction controls how the reactant concentration (or pressure) affects reaction rate.
5. Temperature: Usually conducting a reaction at a higher temperature delivers more energy into the system and increases the reaction rate by causing more collisions between particles, as explained by collision theory. However, the main reason that temperature increases the rate of reaction is that more of the colliding particles will have the necessary activation energy resulting in more successful collisions (when bonds are formed between reactants). The influence of temperature is described by the Arrhenius equation. As a rule of thumb, reaction rates for many reactions double for every 10°C increase in temperature, though the effect of temperature may be very much larger or smaller than this.

For example, coal burns in a fireplace in the presence of oxygen but it doesn't when it is stored at room temperature. The reaction is spontaneous at low and high temperatures but at room temperature its rate is so slow that it is negligible. The increase in temperature, as created by a match, allows the reaction to start and then it heats itself, because it is exothermic. That is valid for many other fuels, such as methane, butane, hydrogen... .

Reaction rates can be independent of temperature (non-Arrhenius) or decrease with increasing temperature (anti-Arrhenius). Reactions without an activation barrier (e.g. some radical reactions), tend to have anti Arrhenius temperature dependence: the rate constant decreases with increasing temperature.

1. Solvent: Many reactions take place in solution and the properties of the solvent affect the reaction rate. The ionic strength also has an effect on reaction rate.
2. Electromagnetic radiation and intensity of light: Electromagnetic radiation is a form of energy. As such, it may speed up the rate or even make a reaction spontaneous as it provides the particles of the reactants with more energy. This energy is in one way or another stored in the reacting particles (it may break bonds, promote molecules to electronically or vibrationally excited states) creating intermediate species that react easily. As the intensity of light increases, the particles absorb more energy and hence the rate of reaction increases.

For example when methane reacts with chlorine in the dark, the reaction rate is very slow. It can be sped up when the mixture is put under diffused light. In bright sunlight, the reaction is explosive.

1. A catalyst: The presence of a catalyst increases the reaction rate (in both the forward and reverse reactions) by providing an alternative pathway with a lower activation energy.

For example, platinum catalyses the combustion of hydrogen with oxygen at room temperature.

1. Isotopes: The kinetic isotope effect consists in a different reaction rate for the same molecule if it has different isotopes, usually hydrogen isotopes, because of the mass difference between hydrogen and deuterium.
2. Surface area: In reactions on surfaces, which take place for example during heterogeneous catalysis, the rate of reaction increases as the surface area does. That is due to the fact that more particles of the solid are exposed and can be hit by reactant molecules.
3. Stirring: Stirring can have a strong effect on the rate of reaction for heterogeneous reactions.

All the factors that affect a reaction rate, except for concentration and reaction order, are taken into account in the rate equation of the reaction.

Rate Equation

For a chemical reaction $nA + mB \rightarrow C + D$, the rate equation or rate law is a mathematical expression used in chemical kinetics to link the rate of a reaction to the concentration of each reactant. It is of the kind:

$$r = k(T)[A]^{n'}[B]^{m'}$$

In this equation $k(T)$ is the reaction rate coefficient or rate constant, although it is not really a constant, because it includes all the parameters that affect reaction rate, except for concentration, which is explicitly taken into account. Of all the parameters described before, temperature is normally the most important one. The exponents n' and m' are called reaction orders and depend on the reaction mechanism.

Stoichiometry, molecularity (the actual number of molecules colliding) and reaction order only coincide necessarily in elementary reactions, that is, those reactions that take place in just one step. The reaction equation for elementary reactions coincides with the process taking place at the atomic level, i.e. n molecules of type A are colliding with m molecules of type B (n plus m is the molecularity).

For gases the rate law can also be expressed in pressure units using, e.g. the ideal gas law.

By combining the rate law with a mass balance for the system in which the reaction occurs, an expression for the rate of change in concentration can be derived. For a closed system with constant volume such an expression can look like:

$$\frac{d[C]}{dt} = k(T)[A]^{n'}[B]^{m'}$$

Temperature Dependence

Each reaction rate coefficient k has a temperature dependency, which is usually given by the Arrhenius equation:

$$k = Ae^{-\frac{E_a}{RT}}$$

E_a is the activation energy and R is the gas constant. Since at temperature T the molecules have energies given by a Boltzmann distribution, one can expect the number of collisions with energy greater than E_a to be proportional to $e^{-\frac{E_a}{RT}}$. A is the pre-exponential factor or frequency factor.

The values for A and E_a are dependent on the reaction. There are also more complex equations possible, which describe temperature dependence of other rate constants which do not follow this pattern.

Pressure Dependence

The pressure dependence of the rate constant for condensed-phase reactions (i.e. when reactants and products are solids or liquid) is usually sufficiently weak in the range of pressures normally encountered in industry that it is neglected in practice.

The pressure dependence of the rate constant is associated with the activation volume. For the reaction proceeding through an activation-state complex:

$$A + B \rightleftharpoons |A \cdots B|^{\ddagger} \rightarrow P$$

the activation volume, $\Delta V^{\ddagger}$, is:

$$\Delta V^{\dagger} = \bar{V}_{\dagger} - \bar{V}_A - \bar{V}_B$$

where, $\bar{V}$ denote the partial molar volumes of the reactants and products and indicates the activation-state complex.

For the above reaction, one can expect the change of the reaction rate constant (based either on mole-fraction or molar-concentration) with pressure at constant temperature to be:

$$-RT \left(\frac{\partial \ln k_x}{\partial P} \right)_T = \Delta V^{\dagger}$$

In practice, the matter can be complicated because the partial molar volumes and the activation volume can themselves be a function of pressure.

Reactions can increase or decrease their rates with pressure, depending on the value of $\Delta V^{\ddagger}$. As an example of the possible magnitude of the pressure effect, some organic reactions were shown to double the reaction rate when the pressure was increased from atmospheric (0.1 MPa) to 50 MPa (which gives $\Delta V^{\ddagger} = -0.025$ L/mol).

Examples: For the reaction:

$$2H_2(g) + 2NO(g) \rightarrow N_2(g) + H_2O(g)$$

The rate equation is:

$$r = k|H_2|^1 [NO]^2$$

The rate equation does not simply reflect the reactants stoichiometric coefficients in the overall reaction: it is first order in H_2, although the stoichiometric coefficient is 2 and it is second order in NO.

In chemical kinetics the overall reaction is usually proposed to occur through a number of elementary steps. Not all of these steps affect the rate of reaction; normally it is only the slowest elementary step that affect the reaction rate. For example, in:

1. Fast equilibrium.
2. $N_2O_2 + H_2 \rightarrow N_2O + H_2O$ (slow).
3. $N_2O_2 + H_2 \rightarrow N_2 + H_2O$ (fast).

Reactions 1 and 3 are very rapid compared to the second, so it is the slowest reaction that is reflected in the rate equation. The slow step is considered the rate determining step. The orders of the rate equation are those from the rate determining step.

Reaction Rate Constant

In chemical kinetics a reaction rate constant k or λ quantifies the speed of a chemical reaction.

For a chemical reaction where substance A and B are reacting to produce C, the reaction rate has the form:

Reaction: $$A + B \rightarrow C$$

$$\frac{d[C]}{dt} = k(T)[A]^m[B]^n$$

$k(T)$ is the reaction rate constant that depends on temperature.

[C] is the concentration of substance C in moles per volume of solution assuming the reaction is taking place throughout the volume of the solution (for a reaction taking place at a boundary it would denote something like moles of C per area).

The exponents m and n are called orders and depend on the reaction mechanism. They can be determined experimentally.

A single-step reaction can also be written as:

$$\frac{d[C]}{dt} = Ae^{-E_a/RT}[A]^m[B]^n$$

E_a is the activation energy and R is the Gas constant. Since at temperature T the molecules have energies according to a Boltzmann distribution, one can expect the proportion of collisions with energy greater than E_a to vary with $e^{-Ea/RT}$. A is the pre-exponential factor or frequency factor.

The Arrhenius equation gives the quantitative basis of the relationship between the activation energy and the reaction rate at which a reaction proceeds.

The unit of the rate coefficient depend on the global order of reaction:

1. For order zero, the rate coefficient has unit of $mol \cdot L^{-1} \cdot s^{-1}$.
2. For order one, the rate coefficient has unit of s^{-1}.
3. For order two, the rate coefficient has unit of $L \cdot mol^{-1} \cdot s^{-1}$.
4. For order n, the rate coefficient has unit of $mol^{1-n} \cdot L^{n-1} \cdot s^{-1}$.

Plasma and gases

Calculation of rate constants of the processes of generation and relaxation of electronically and vibrationally excited particles are of great importance. It is used in the computer simulation of processes in plasma chemistry, microelectronics. First-principle based models should be used for such calculation. It can be done with the help of computer simulation software.

Activation Energy

In chemistry, activation energy is a term introduced in 1889 by the Swedish scientist Svante Arrhenius, that is defined as the energy that must be overcome in order for a chemical reaction to occur. Activation energy may also be defined as the minimum energy required to start a chemical reaction. The activation energy of a reaction is usually denoted by E_a, and given in units of kilojoules per mole.

Activation energy can be thought of as the height of the potential barrier (sometimes called the energy barrier) separating two minima of potential energy (of the reactants and products of a reaction). For a chemical reaction to proceed at a reasonable rate, there should exist an appreciable number of molecules with energy equal to or greater than the activation energy.

Negative activation energy

In some cases rates of reaction decrease with increasing temperature. When following an approximately exponential relationship so the rate constant can still be fit to an Arrhenius expression, this results in a negative value of E_a. Reactions exhibiting these negative activation energies are typically barrierless reactions, in which the reaction proceeding relies on the capture of the molecules in a potential well. Increasing the temperature leads to a reduced probability of the colliding molecules capturing one another (with more glancing collisions not leading to reaction as the higher momentum carries the colliding particles out of the potential well), expressed as a reaction cross section that decreases with increasing temperature. Such a situation no longer leads itself to direct interpretations as the height of a potential barrier.

Temperature independence and the relation to the Arrhenius equation

The Arrhenius equation gives the quantitative basis of the relationship between the activation energy and the rate at which a reaction proceeds. From the Arrhenius equation, the activation energy can be expressed as:

$$E_a = -RT \ln\left(\frac{k}{A}\right)$$

where, A is the frequency factor for the reaction, R is the universal gas constant, T is the temperature (in kelvins), and k is the reaction rate coefficient. While this equation suggests that the activation energy is dependent on temperature, in regimes in which the Arrhenius equation is valid this is cancelled by the temperature dependence of k. Thus E_a can be evaluated from the reaction rate coefficient at any temperature (within the validity of the Arrhenius equation).

Catalysis

A substance that modifies the transition state to lower the activation energy is termed a catalyst; a biological catalyst is termed an enzyme. It is important to note that a catalyst increases the rate of reaction without being consumed by it. In addition, while the catalyst lowers the activation energy, it does not change the energies of the original reactants nor products. Rather, the reactant energy and the product energy remain the same and only the activation energy is altered (lowered) (Fig. 2.7).

BATCH REACTORS

The batch reactor (BR) is the most frequently used type of reactor in biotechnological productions. Virtually all food processing, pharmaceutical and agricultural bioprocesses are carried out in batch reactors. Batch reactor applications in environmental processes are however relatively rare and very much limited to small scale applications.

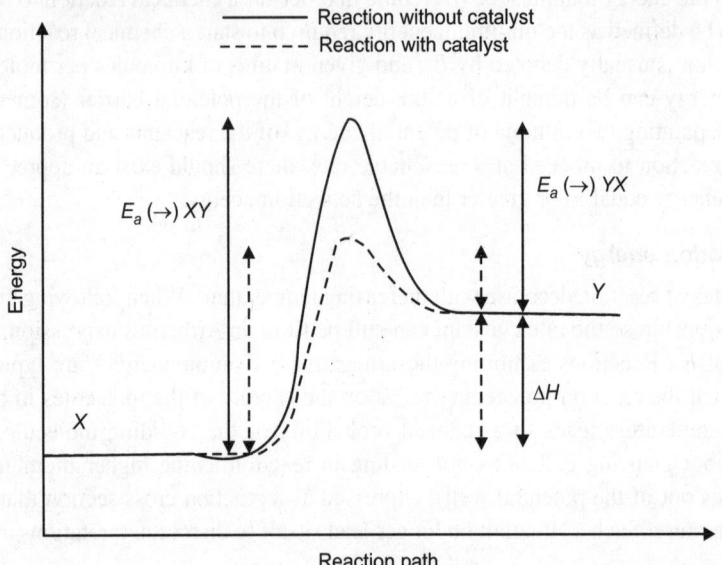

Fig. 2.7. The relationship between activation energy (E_a) and enthalpy of formation (ΔH) with and without a catalyst. The highest energy position (peak position) represents the transition state. With the catalyst, the energy required to enter transition state decreases, thereby decreasing the energy required to initiate the reaction.

One of the reasons for the difference in usage frequency between biotechnological and environmental technological applications of the BR is, no doubt, related to the fact that the BR is very well suited to operate under sterile conditions, something of great importance in biotechnology and often irrelevant in environmental processes. A BR is usually a closed, well mixed vessel (Fig. 2.8). In the laboratory, the Erlenmeyer flask of a few millilitre is often used as a BR. In industrial applications BR's can have a volume of over 100 m^3.

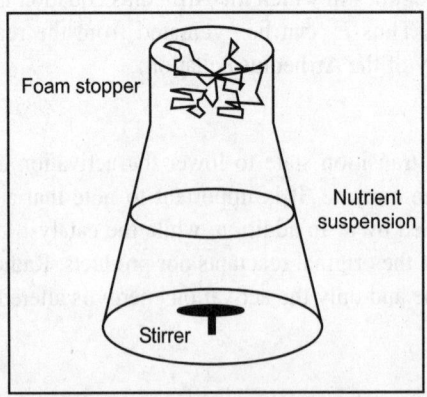

Fig. 2.8. Batch reactor as used in the laboratory.

We will begin this section by giving a brief description of batch reactors and by describing the batch process cycle including a discussion of the batch growth curve. We will then write balance equations for these types of reactors before moving onto a description of fed-batch reactors.

Batch reactors are usually regarded as closed systems: Operational intricacies are best related with the description of a full batch process cycle. Before inoculation, the BR contains a certain volume of nutrient in suspension. After inoculation, the process is left untouched, i.e. no material is added to or removed from the reactor. However, the reactor might be aerated, is virtually always stirred and process control, e.g. pH control, might be applied. Batch reactors are often referred to as closed systems.

The term 'closed' refers to the fact that material can neither enter nor leave the reactor. In that sense, the use of this term is often not entirely accurate because in aerobic processes, oxygen is allowed into the reactor (in laboratory conditions, through an air filter of cotton or foam) and carbon dioxide allow to leave. Obviously if a pH control system is used this will also lead to some additions to the reactor. Thus, in many cases batch reactors should not be regarded strictly as dosed systems.

Batch Growth Curve

The biological conversion process in a BR usually proceeds through a series of phases (Fig. 2.9).

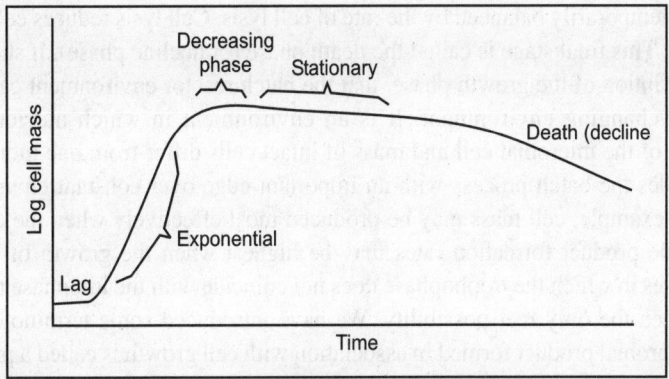

Fig. 2.9. Typical bacterial growth curve.

After inoculation, micro-organisms need time to adjust to the new environment, i.e. to prepare the enzymatic machinery for the specific cellular functions involved in the process of growth. Virtually no growth occurs in this so-called lag phase although it is a period of intense metabolic activity. Once adjusted to their environment, the organisms start to grow and, with all necessary growth factors plentifully available, growth occurs in an exponential (logarithmic) fashion. Hence, this phase is called the exponential or logarithmic growth phase. Growth in the log phase is referred to as balanced growth, defined as growth in which every extensive property of the growing system increases by the same factor.

Extensive and intensive properties

Intensive properties are all related to a system quality. These properties are not additive. Extensive properties relate to the system quantities and therefore are additive. Table 2.2 gives a summary of the features of these types of properties. Thus, the extensive properties you could have used were such features as mass of cells present, mass of DNA or protein and so on. Colour and turbidity are examples of intensive properties.

Table 2.2. Features of intensive and extensive properties.

Intensive properties	Extensive properties
Relates to system quality	Relates to system quantities
Cannot be added up	Can be added up
No balance possible	Balance allowed
Examples	Examples
Temperature, pressure	Volume, mass
Colour, concentration	Energy, electric charge

Returning to our description of the batch growth curve, during logarithmic (balanced) growth, each extensive property increases by the same factor and the composition of cells remains constant.

At some point, however, one or more nutrients start to become depleted and, hence, growth-limiting. As a result, growth first slows down (decreasing phase) and then virtually ceases, the beginning of the stationary phase. After growth has halted for some time the process of cell death and cell lysis sets in. Cell lysis is the breaking open of the cell membrane and the subsequent release of cell contents into the medium. Cell contents may provide nutrients for intact cells. Therefore, cellular growth rate in the stationary phase is temporarily balanced by the rate of cell lysis. Cell lysis reduces cell numbers until no living cells are left. This final stage is called the death phase (or decline phase). It should be clear from the preceding description of the growth phase, that the batch reactor environment can be characterised as a continuously changing environment. It is an environment in which nutrient concentrations, physiological state of the microbial cell and mass of intact cells differ from one moment to another.

This fact provides the batch process with an important edge over constant environment processes (e.g. CFSTR). For example, cell mass may be produced most effectively when the cells are in the log growth phase, while product formation rates may be highest when the growth of cells has virtually stalled. For processes in which the trophophase does not coincide with the idiophase the batch reactor is well suited and often the only real possibility. We have introduced some terminology here so let us explain them. A microbial product formed in association with cell growth is called a primary metabolite, a product formed independent of cell growth is called a secondary metabolite.

Products are referred to by a variety of names, i.e. extracellular polymeric substances (EPS), glycocalix and extracellular products while the word exopolymer is also used. Idiophase refers to the phase of product formation. Trophophase means cell production phase. The idiophase and trophophase may or may not coincide. Figure 2.10 shows the growth cycles of batch cultures of two organisms. Examine these carefully and answer the following questions. Products A and B are extracellular products (Note that these figures are somewhat stylised).

1. What is the duration of the idiophase in each culture?
2. What is the duration of the trophophase in each culture?
3. Which product is probably a primary metabolite and which a secondary metabolite?
4. For the production of which compound would a batch reactor be better than a CFSTR?

Balance Equations for the BR

For a three species system of $5 \to M + P$. Mass balance equations for the batch reactor (BR) can be derived from those for the CFSTR by omitting the transport terms. Because part of the batch processes may occur at low substrate concentrations, a maintenance term should be included.

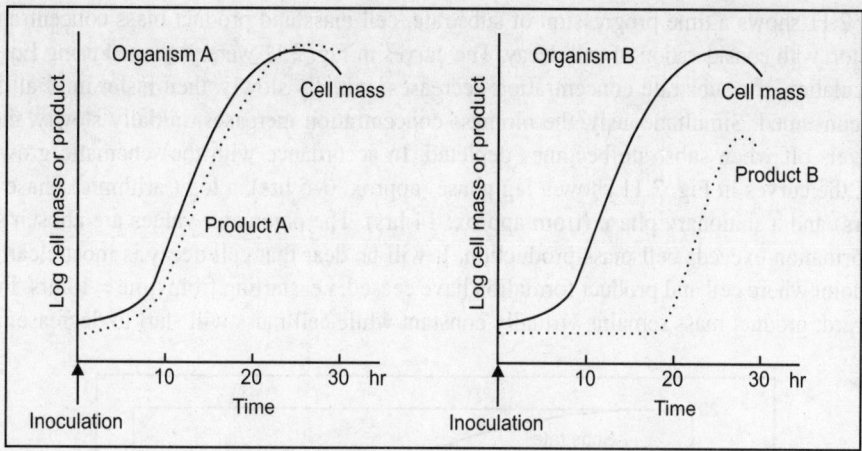

Fig. 2.10. The growth cycles of batch cultures of two organisms.

We remind you of some of the relevant equations for CFSTRs. In a CFSTR, for species S:

$$V\frac{d(C_{S1})}{dt} = F(C_{Si} - C_{S1}) - \mu C_{M1} \cdot V \cdot \frac{1}{Y_{MSo}} - r_P \cdot C_{M1} \cdot V \cdot \frac{1}{Y_{PS}} \qquad \text{... (2.1)}$$

For species M:

$$V \cdot \frac{d(C_{M1})}{dt} = F(C_{Mi} - C_{M1}) - \mu \cdot C_{M1} \cdot V \qquad \text{... (2.2)}$$

For species P:

$$V\frac{d(C_{P1})}{dt} = F(C_{Pi} - C_{P1}) + r_P \cdot C_{M1} \cdot V \qquad \text{... (2.3)}$$

When we considered a situation in which part of the substrate in the CFSTR was used for providing maintenance energy, we wrote:

$$V\frac{d(C_{S1})}{dt} = F(C_{Si} - C_{S1}) - \mu\frac{C_{M1}}{Y_{MSg}}V - r_P\frac{C_{Mi}}{Y_{PS}}V - r_m \cdot C_{M1} \cdot V \qquad \text{... (2.4)}$$

These equations may be applied to BRs. From Eqs 2.1–2.3, the BR mass balance equations for a three species system can be derived for $F = 0$ and including the I maintenance term of Eq. 2.4:

For species S:

$$V \cdot \frac{d(C_{S1})}{dt} = -\mu \cdot C_{M1} \cdot V \cdot \frac{1}{Y_{MSg}} - r_P \cdot C_{M1} \cdot V \cdot \frac{1}{Y_{PS}} - r_m \cdot C_{M1} \cdot V \qquad \text{... (2.5)}$$

For species M:

$$\frac{V d(C_{M1})}{dt} = \mu \cdot C_{M1} \cdot V \qquad \text{... (2.6)}$$

For species P:

$$\frac{V d(C_{P1})}{dt} = r_P \cdot C_{Mi} \cdot V \qquad \text{... (2.7)}$$

Figure 2.11 shows a time progression of substrate, cell mass and product mass concentrations in a batch reactor, with consideration of cell decay. The curves in Fig. 2.11 were generated using Eqs 2.5–2.7. After inoculation, the substrate concentration decreases, initially slowly, then faster until all substrate has been consumed. Simultaneously, the biomass concentration increases, initially slowly, then faster until it levels off when substrate becomes depleted. In accordance with the schematic growth curve (Fig. 2.9), the curves in Fig. 2.11 show a lag phase (approx. 0–5 hrs), a log (arithmic) phase (approx. 6–12 hours) and a stationary phase (from approx. 14 hrs). The parameter values are chosen such that product formation exceeds cell mass production. It will be dear that cell decay is most clearly visible from the point where cell and product formation have ceased, i.e. starting from time = 14 hrs. From This point onward, product mass remains virtually constant while cell mass will start to decrease.

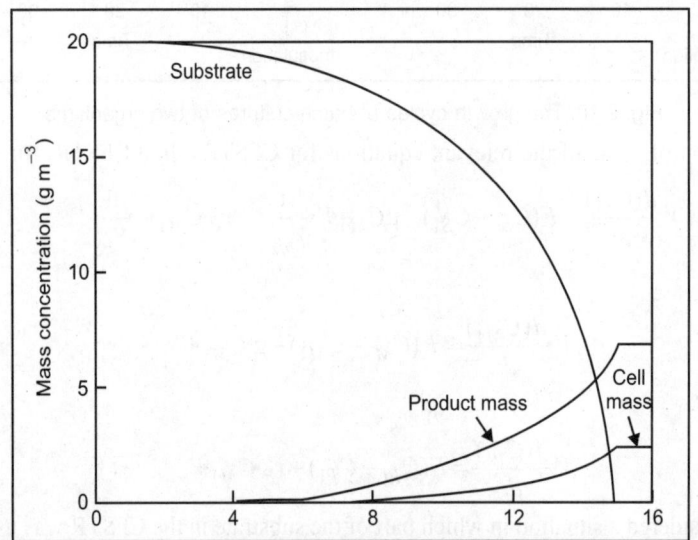

Fig. 2.11. Time progression of cell mass, product mass and substrate in a batch process with consideration of cell decay (see text).

Fed-Batch Reactor

Operating a batch reactor involves a large series of preliminary activities including cleaning, filling with nutrient suspension and, generally, sterilisation of the reactor. After inoculation the bioprocess proceeds through the lag, exponential and stationary phase. The death phase marks the end of the process. The cell mass and/or product are harvested, the reactor is dismantled.

To obtain a second batch of cell and/or product mass requires passing through the entire cycle of events again. Thus, the batch reactor goes through a considerable amount of downtime (time in which the reactor is not operational).

Let us do another little calculation. Let us assume that it takes 24 hrs to down-load a batch reactor, clean it, fill it with fresh medium, sterilise it, cool it and inoculate it. Let us also assume that the culture has no lag phase, but immediately enters the log phase on inoculation and has a mean generation time of 1 hr. Let us also assume that a normal cycle involves 24 generations. We can represent this sequence of events like this.

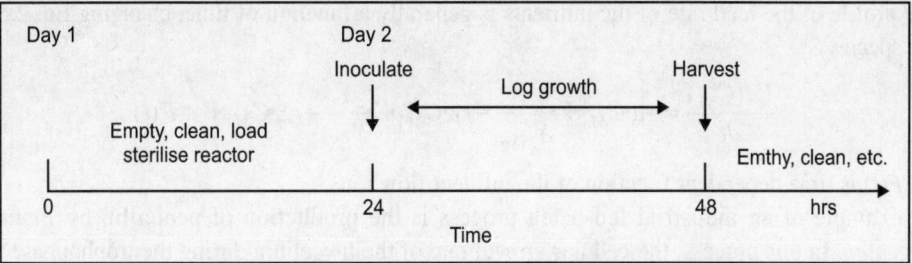

If the product we are interested in is biomass (or a product directly linked to biomass), half of the total quantity which will be produced in the 2-day cycle will be produced in the period 47–48 h[1] (remember the exponential nature of growth $2 \rightarrow 4 \rightarrow 8$, etc.). Another way of thinking about this is that the catalyst for producing biomass or products is biomass itself. Thus the reactor is only effectively making biomass (or products) at a high rate when it contains significant quantities of biomass. A feature of a batch system is that for most of the time the reactor is either empty or containing only small quantities of biomass.

In contrast, the CFSTR does not incorporate the need of restarting a new process after completion of the old one (once started CFSTR can operate, theoretically, for indefinite lengths of time and can be run with fairly high biomass concentrations). However, the CFSTR is characteristic for a constant environment and therefore unsuited for bioprocesses in which the idiophase is, time wise, independent of the trophase. Moreover, the CFSTR has the disadvantage of operational complexity (e.g. sterility). The fed-batch process is developed to meet both the requirement of a more or less continuous operation and of a continuously changing environment (Fig. 2.12).

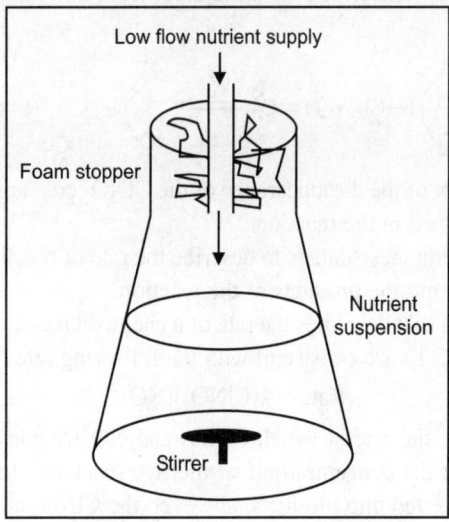

Fig. 2.12. Fed-batch reactor.

The fed-batch reactor (FBR) is a batch reactor to which, when the nutrients approach depletion, fresh nutrients are added. In other words, the reactor is fed. It is assumed that the concentration of the nutrients added is so high that volume changes are negligible (justifying the batch part of the name).

The profile of the feed rate of the nutrients is generally a function of time, changing Eq. 2.5 to:
For species S:

$$\frac{Vd(C_{S1})}{dt} = -\mu \cdot C_{M1} \cdot V \cdot \frac{1}{Y_{MSg}} - r_P \cdot C_{M1} \cdot V \cdot \frac{1}{Y_{PS}} - r_m \cdot C_{M1} \cdot V + F(t) \qquad \ldots (2.8)$$

where, $F(t)$ is time dependent function of the influent flow rate.

An example of an industrial fed-batch process is the production of penicillin by *Penicillium chrysogenum*. In this process, the cellular growth rate of the mycelium during the trophophase is high allowing this phase to be short. On the other hand, the cellular growth rate during the idiophase is very low and the emphasis in this phase is on the production of penicillin.

A sanitary engineering example of a batch process is a landfill to which new garbage (nutrients) is added constantly while no material is removed. Or the 'fill and draw' sedimentation tank, a tank that is filled with water, allowed to stand for some time so that suspended material can settle after which the clarified supernatant (top water layer) is 'drawn' off.

GAS-PHASE REACTION

The simplest chemical reactions are those that occur in the gas-phase in a single step, such as the transfer of a chlorine atom from $ClNO_2$ to NO to form NO_2 and ClNO.

$$ClNO_2(g) + NO(g) \rightleftharpoons NO_2(g) + ClNO(g)$$

This reaction can be understood by writing the Lewis structures for all four components of the reaction. Both NO and NO_2 contain an odd number of electrons. Both NO and NO_2 can, therefore, combine with a neutral chlorine atom to form a molecule in which all of the electrons are paired. This reaction therefore involves the transfer of a chlorine atom from one molecule to another, as shown in the figure below.

Figure 2.13 combines a plot of the disappearance of the $ClNO_2$ consumed in this reaction with a plot of the appearance of NO_2 formed in the reaction.

One of the goals of collecting these data is to describe the rate of reaction, which is the rate at which the reactants are transformed into the products of the reaction.

The mathematical equation that describes the rate of a chemical reaction is called the rate law for the reaction. The data in the Fig. 2.13 are consistent with the following rate law for this reaction.

$$\text{Rate} = k(ClNO_2)(NO)$$

According to this rate law, the rate at which $ClNO_2$ and NO are converted into NO_2 and ClNO is proportional to the product of the concentrations of the two reactants. Initially, the rate of reaction is fast. As the reactants are converted into products, however, the $ClNO_2$ and NO concentrations become smaller, and the reaction slows down. We might expect the reaction to stop when it runs out of either $ClNO_2$ or NO. In practice, the reaction stops before this happens.

This is a very fast reaction—the concentration of $ClNO_2$ drops by a factor of two in less than a second. And yet, no matter how long we wait, some residual $ClNO_2$ and NO remains in the reaction flask.

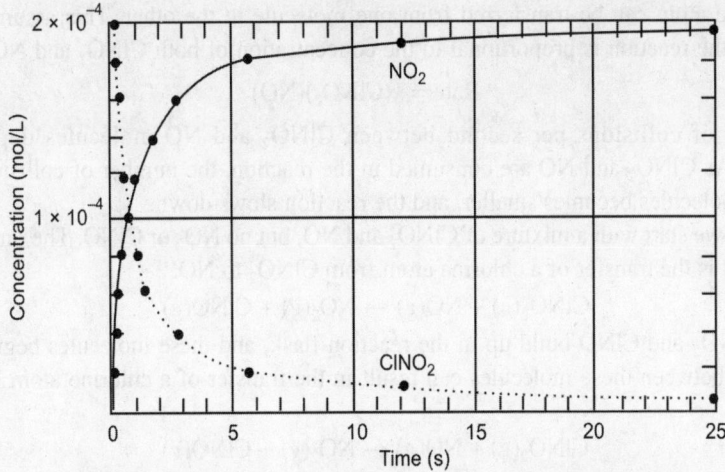

Fig. 2.13. Combines a plot of the disappearance of the $CINO_2$.

The Fig. 2.14 divides the plot of the change in the concentrations of NO_2 and $CINO$ into a kinetic region and an equilibrium region. By definition, the kinetic region is the period during which the concentrations of the components of the reaction are constantly changing. The equilibrium region is the period after which the reaction seems to stop, when there is no further change in the concentrations of the components of the reaction.

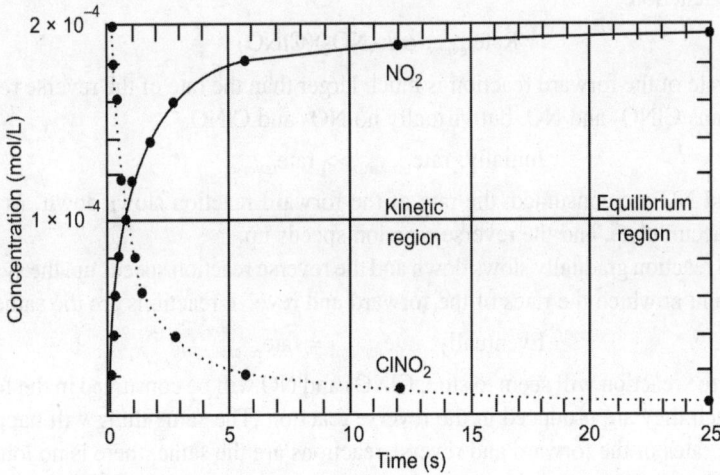

Fig. 2.14. Divides the plot of the change in the concentrations of NO_2 and $CINO$ into a kinetic region and an equilibrium region.

Collision Theory Model for Gas-Phase Reactions

The fact that the following reaction:

$$CINO_2(g) + NO(g) \rightleftharpoons NO_2(g) + CINO(g)$$

seems to stop before all of the reactants are consumed can be explained with a model of chemical reactions known as the collision theory. This model assumes that $CINO_2$ and NO molecules must collide before a chlorine atom can be transferred from one molecule to the other. This assumption explains why the rate of the reaction is proportional to the concentration of both $ClNO_2$ and NO.

$$Rate = k(ClNO_2)(NO)$$

The number of collisions per second between $ClNO_2$ and NO molecules depends on their concentrations. As $ClNO_2$ and NO are consumed in the reaction, the number of collisions per second between these molecules becomes smaller, and the reaction slows down.

Suppose that we start with a mixture of $ClNO_2$ and NO, but no NO_2 or ClNO. The only reaction that can occur at first is the transfer of a chlorine atom from $ClNO_2$ to NO.

$$ClNO_2(g) + NO(g) \rightarrow NO_2(g) + ClNO(g)$$

Eventually, NO_2 and ClNO build up in the reaction flask, and these molecules begin to collide as well. Collisions between these molecules can result in the transfer of a chlorine atom in the opposite direction.

$$ClNO_2(g) + NO(g) \leftarrow NO_2(g) + ClNO(g)$$

The collision theory model of chemical reactions assumes that the rate of a simple, one-step reaction is proportional to the product of the concentrations of the ions or molecules consumed in that reaction. The rate of the forward reaction is, therefore, proportional to the product of the concentrations of the two 'reactants'.

$$Rate_{forward} = k_f(ClNO_2)(NO)$$

The rate of the reverse reaction, on the other hand, is proportional to the concentrations of the 'products' of the reaction.

$$Rate_{reverse} = k_r(NO_2)(ClNO)$$

Initially, the rate of the forward reaction is much larger than the rate of the reverse reaction, because the system contains $ClNO_2$ and NO, but virtually no NO_2 and ClNO.

$$\text{Initially: } rate_{forward} \gg rate_{reverse}$$

As $ClNO_2$ and NO are consumed, the rate of the forward reaction slows down. At the same time, NO_2 and ClNO accumulate, and the reverse reaction speeds up.

If the forward reaction gradually slows down and the reverse reaction speeds up, the system eventually has to reach a point at which the rates of the forward and reverse reactions are the same.

$$\text{Eventually: } rate_{forward} = rate_{reverse}$$

At this point, the reaction will seem to stop. $ClNO_2$ and NO will be consumed in the forward reaction at the rate at which they are produced in the reverse reaction. The same thing will happen to NO_2 and ClNO. When the rates of the forward and reverse reactions are the same, there is no longer any change in the concentrations of the reactants or products of the reaction. In other words, the reaction is at equilibrium.

We can now see that there are two definitions of equilibrium:
1. A system in which there is no apparent change in the concentrations of the reactants and products of a reaction.
2. A system in which the rates of the forward and reverse reactions are equal.

The first definition is based on the results of experiments that tell us that some reactions seem to stop prematurely—they reach a point at which no more reactants are converted into products before the limiting reagent is consumed.

The other definition is based on a theoretical model of chemical reactions that explains why reactions reach equilibrium.

We can now distinguish between reactions that go to completion and those that reach equilibrium. Reactions that are not reversible, or that strongly favour the products, are assumed to go to completion and are represented by equations that contain a single arrow.

$$2\ \text{Mg}(s) + \text{O}_2(g) \rightarrow 2\ \text{MgO}(s)$$

Reversible reactions that reach equilibrium are indicated by a pair of arrows between the two sides of the equation.

$$\text{ClNO}_2(g) + \text{NO}(g) \rightleftharpoons \text{NO}_2(g) + \text{ClNO}(g)$$

Molecular Dynamics

Molecular dynamics (MD) is a form of computer simulation in which atoms and molecules are allowed to interact for a period of time by approximations of known physics, giving a view of the motion of the particles. This kind of simulation is frequently used in the study of proteins and biomolecules, as well as in materials science. It is tempting, though not entirely accurate, to describe the technique as a 'virtual microscope' with high temporal and spatial resolution. Whereas it is possible to take 'still snapshots' of crystal structures and probe features of the motion of molecules through NMR, no conventional experiment allows access to all the time scales of motion with atomic resolution, recent developments in atto-second lasers might give an opportunities in the near future. Richard Feynman once said that 'If we were to name the most powerful assumption of all, which leads one on and on in an attempt to understand life, it is that all things are made of atoms, and that everything that living things do can be understood in terms of the jigglings and wigglings of atoms.' Molecular dynamics lets scientists peer into the motion of individual atoms in a way which is not possible in laboratory experiments.

In all kinds of molecular dynamics simulations, the simulation box size must be large enough to avoid boundary condition artifacts. Boundary conditions are often treated by choosing fixed values at the edges (which may cause artifacts), or by employing periodic boundary conditions in which one side of the simulation loops back to the opposite side, mimicking a bulk phase.

Microcanonical ensemble (NVE)

In the microcanonical, or NVE ensemble, the system is isolated from changes in moles (N), volume (V) and energy (E). It corresponds to an adiabatic process with no heat exchange. A microcanonical molecular dynamics trajectory may be seen as an exchange of potential and kinetic energy, with total energy being conserved. For a system of N particles with coordinates X and velocities V, the following pair of first order differential equations may be written in Newton's notation as:

$$E(X) = -\nabla U(X) = M\dot{V}(t)$$

$$V(t) = \dot{X}(t)$$

The potential energy function $U(X)$ of the system is a function of the particle coordinates X. It is referred to simply as the 'potential' in Physics, or the 'force field' in Chemistry. The first equation comes from Newton's laws; the force F acting on each particle in the system can be calculated as the negative gradient of $U(X)$.

For every timestep, each particle's position X and velocity V may be integrated with a symplectic method such as Verlet. The time evolution of X and V is called a trajectory. Given the initial positions (e.g. from theoretical knowledge) and velocities (e.g. randomised Gaussian), we can calculate all future (or past) positions and velocities.

Canonical ensemble (NVT)

In the canonical ensemble, moles (N), volume (V) and temperature (T) are conserved. It is also sometimes called constant temperature molecular dynamics (CTMD). In NVT, the energy of endothermic and exothermic processes is exchanged with a thermostat. A variety of thermostat methods is available to add and remove energy from the boundaries of an MD system in a more or less realistic way, approximating the canonical ensemble. Popular techniques to control temperature include velocity rescaling, the Nosé-Hoover thermostat, Nosé-Hoover chains, the Berendsen thermostat and Langevin dynamics. Note that the Berendsen thermostat might introduce the flying ice cube effect, which leads to unphysical translations and rotations of the simulated system. It is not trivial to obtain a canonical distribution of conformations and velocities using these algorithms. How this depends on system size, thermostat choice, thermostat parameters, time step and integrator is the subject of many articles in the field.

Transition State Theory

Transition state theory (TST) explains the reaction rates of elementary chemical reactions. The theory assumes a special type of chemical equilibrium (quasi-equilibrium) between reactants and activated transition state complexes. TST is used primarily to understand qualitatively how chemical reactions take place. TST has been less successful in its original goal of calculating absolute reaction rate constants because the calculation of absolute reaction rates requires precise knowledge of potential energy surfaces, but it has been successful in calculating the standard enthalpy of activation ($\Delta^{\ddagger}H^{\ominus}$), the standard entropy of activation ($\Delta^{\ddagger}S^{\ominus}$), and the standard Gibbs energy of activation ($\Delta^{\ddagger}G^{\ominus}$) for a particular reaction if its rate constant has been experimentally determined. (The $\ddagger$ notation refers to the value of interest at the transition state.) (Fig. 2.15).

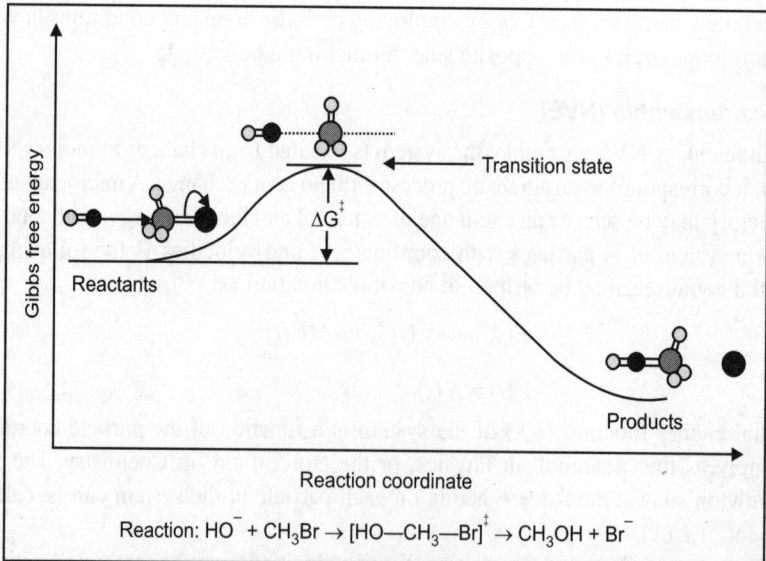

Fig. 2.15. Reaction coordinate diagram for the bimolecular nucleophilic substitution (S_N2) reaction between bromomethane and the hydroxide anion.

Theory

Basic ideas behind the transition state theory are as follows:

1. Rates of the reactions are studied by studying activated complexes which lie at the col (saddle point) of a potential energy surface. The details of how the complexes are formed are not important.
2. The activated complexes are in a special equilibrium (quasi-equilibrium) with the reactant molecules.
3. The activated complexes can convert into products which allows kinetic theory to calculate the rate of this conversion.

In the development of TST, three approaches were taken as summarised below.

Thermodynamic treatment

In 1884, Jacobus van't Hoff proposed the Van't Hoff equation describing the temperature dependence of the equilibrium constant for a reversible reaction:

$$A \rightleftharpoons B$$

$$\frac{d \ln K}{dT} = \frac{\Delta U}{RT^2}$$

where, ΔU is the change in internal energy, K is the equilibrium constant of the reaction, R is the universal gas constant, and T is thermodynamic temperature. Based on experimental work, in 1889, Svante Arrhenius proposed a similar expression for the rate constant of a reaction, given as follows:

$$\frac{d \ln k}{dT} = \frac{\Delta E}{RT^2}$$

Integration of this expression leads to the Arrhenius equation

$$k = A e^{\frac{-E}{RT}}$$

A was referred to as the frequency factor (now called the pre-exponential coefficient), and E is regarded as the activation energy. By the early 20th century many had accepted the Arrhenius equation, but the physical interpretation of A and E remained vague. This led many researchers in chemical kinetics to offer different theories of how chemical reactions occurred in an attempt to relate A and E to the molecular dynamics directly responsible for chemical reactions.

In 1910, Rene Marcelin introduced the concept of standard Gibbs energy of activation. His equation can be written as:

$$k \propto \exp\left(\frac{-\Delta^{\ddagger} G^{\ominus}}{RT}\right)$$

At about the same time as Marcelin was working on his formulation, Dutch chemists Philip Abraham Kohnstamm, Frans Eppo Cornelis Scheffer, and Wiedold Frans Brandsma introduced for the first time standard entropy of activation and the standard enthalpy of activation. They proposed the following rate constant equation

$$k \propto \exp\left(\frac{-\Delta^{\ddagger} S^{\ominus}}{R}\right) \exp\left(\frac{-\Delta^{\ddagger} H^{\ominus}}{RT}\right)$$

Kinetic-theory treatment

In early 1900, Max Trautz and William Lewis studied the rate of the reaction using collision theory, based on the kinetic theory of gases. Collision theory treats reacting molecules as hard spheres colliding with one another; this theory neglects entropy changes.

Lewis applied his treatment to the following reaction and obtained good agreement with experimental result.

$$2HI \rightarrow H_2 + I_2$$

However, later when the same treatment was applied to other reactions, there were large discrepancies between theoretical and experimental results.

Statistical-mechanical treatment

Statistical mechanics played a significant role in the development of TST. However, the application of statistical mechanics to TST was developed very slowly given the fact that in mid 1800s, James Clerk Maxwell, Ludwig Boltzmann, and Leopold Pfaundler published several papers discussing reaction equilibrium and rates in terms of molecular motions and the statistical distribution of molecular speeds.

It was not until 1912 when the French chemist A. Berthoud used Maxwell-Boltzmann distribution law to obtain an expression for the rate constant.

$$\frac{d \ln k}{dT} = \frac{a - bT}{RT^2}$$

where, a and b are constants related to energy terms.

Two years later, Marcelin made an essential contribution by treating the progress of a chemical reaction as a motion of a point in phase space. He then applied Gibbs' statistical-mechanical procedures and obtained an expression similar to the one which he had obtained earlier from thermodynamic consideration. In 1915, another important contribution came from British physicist James Rice. Based on his statistical analysis, he concluded that the rate constant is proportional to the 'critical increment'. His ideas were further developed by Tolman. In 1919, Austrian physicist Karl Ferdinand Herzfeld applied statistical mechanics to the equilibrium constant and kinetic theory to the rate constant of the reverse reaction, k_{-1}, for the reversible dissociation of a diatomic molecule.

$$AB \underset{k_{-1}}{\overset{k_1}{\rightleftharpoons}} A + B$$

He obtained the following equation for the rate constant of the forward reaction:

$$k_1 = \frac{k_B T}{h} \left(1 - \exp\left(\frac{-h\nu}{k_B T} \right) \right) \exp\left(\frac{-E^{\ominus}}{RT} \right)$$

where, $E^{\ominus}$ is the dissociation energy at absolute zero, k_B is the Boltzmann constant, h is the Planck constant, T is thermodynamic temperature, Θ is vibrational frequency of the bond. This expression is very important since it is the first time that the factor $k_B T/h$, which is a critical component of TST, has appeared in a rate equation.

Derivation of the Eyring equation

The only important feature added by Eyring, Polanyi and Evans was the notion that activated complexes are in quasi-equilibrium with the reactants. The rate is then directly proportional to the concentration of these complexes multiplied by the frequency ($k_B T/h$) with which they are converted into products.

Quasi-equilibrium assumption

It should be noted that quasi-equilibrium is different from classical chemical equilibrium, but can be described using the same thermodynamic treatment. Consider the reaction below:

$$A + B \rightleftharpoons [AB]^{\ddagger} \rightarrow P$$

where complete equilibrium is achieved between all the species in the system including activated complexes, $[AB]^{\ddagger}$. Using statistical mechanics, concentration of $[AB]^{\ddagger}$ can be calculated in terms of the concentration of A and B.

TST assumes that even when the reactants and products are not in equilibrium with each other, the activated complexes are in quasi-equilibrium with the reactants. As illustrated in Fig. 2.16, at any instant of time, there will be a few activated complexes, some were reactant molecules in the immediate past, which are designated $[AB_l]^{\ddagger}$ (since they are moving from left to right). The remainder of them were product molecules in the immediate past $[AB_r]^{\ddagger}$. Since the system is in complete equilibrium, the concentrations of $[AB_l]^{\ddagger}$ and $[AB_r]^{\ddagger}$ are equal, so that each concentration is equal to one-half of the total concentration of activated complexes:

$$[AB_r]^{\ddagger} = 1/2[AB]^{\ddagger} \text{ and } [AB_l]^{\ddagger} = 1/2[AB]^{\ddagger}$$

If the product molecules are suddenly removed from the reaction system, the flow of those activated complexes that began as products ($[AB_r]^{\ddagger}$) will stop; however, there will still be a flow from left to right.

Therefore, the assumption is that the rate of flow from left to right is unaffected by the removal of the products; in other words, the flux in the two directions are assumed to be independent of each other.

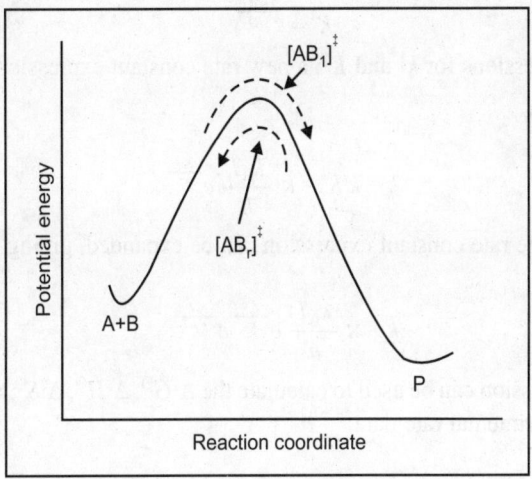

Fig. 2.16. Potential energy diagram.

In TST, it is important to realise that when it is said that the activated complexes are in equilibrium with the reactants, it is referred only to those activated complexes($[AB_l]^{\ddagger}$) that were reactant molecules in the immediate past.

The equilibrium constant $K^{\ddagger\ominus}$ for the quasi-equilibrium can be written as:

$$K^{\ddagger\ominus} = \frac{[AB]^{\ddagger}}{[A][B]}$$

So, the concentration of the transition state $AB^{\ddagger}$ is:

$$[AB]^{\ddagger} = K^{\ddagger\ominus}[A][B]$$

Therefore, the rate equation for the production of product is

$$\frac{d[P]}{dt} = k^{\ddagger\ominus}[AB]^{\ddagger} = k^{\ddagger}K^{\ddagger}[A][B] = k[A][B]$$

where the rate constant k is given by:

$$k = k^{\ddagger}K^{\ddagger}$$

$k^{\ddagger}$ is directly proportional to the frequency of the vibrational mode responsible for converting the activated complex to the product; the frequency of this vibrational mode is ν. Every vibration does not necessarily lead to the formation of product, so a proportionality constant κ, referred to as the transmission coefficient, is introduced to account for this effect. So $k^{\ddagger}$ can be rewritten as

$$k^{\ddagger} = \kappa\nu$$

For the equilibrium constant $K^{\ddagger}$, statistical mechanics leads to a temperature dependent expression given as

$$K^{\ddagger} = \frac{k_{B}T}{h\nu} K^{\ddagger'}$$

where,

$$K^{\ddagger'} = e^{\frac{-\Delta G^{\ddagger}}{RT}}$$

Combining the new expressions for $k^{\ddagger}$ and $K^{\ddagger}$, a new rate constant expression can be written, which is given as:

$$k = k^{\ddagger}K^{\ddagger} = \kappa\, \frac{k_{B}T}{h}\, e^{\frac{-\Delta G^{\ddagger}}{RT}}$$

Since $\Delta G = \Delta H - T\Delta S$, the rate constant expression can be expanded, giving the Eyring equation:

$$k = \kappa\, \frac{k_{B}T}{h}\, e^{\frac{-\Delta S^{\ddagger}}{R}}\, e^{\frac{-\Delta H^{\ddagger}}{RT}}$$

TST's rate constant expression can be used to calculate the $\Delta^{\ddagger}G^{\ominus}$, $\Delta^{\ddagger}H^{\ominus}$, $\Delta^{\ddagger}S^{\ominus}$, and even $\Delta^{\ddagger}V$ (the volume of activation) using experimental rate data.

Chapter 3

Reactor Design and Operations

INTRODUCTION

The general balance equation is a fundamental concept of chemical engineering which is based on the principles of conservation of mass and conservation of energy. The general balance equation states that the total or component mass or energy of any system can be modelled by:

$$\text{In} - \text{Out} + \text{Generation} - \text{Consumption} = \text{Accumulation}$$

The equation is used to describe and design almost every process unit within chemical engineering.

When the general balance equation is applied to the mass of a component or of the entire system, it is called a mass balance. Mass balances are the simplest forms of the general balance equation and one of the first forms a chemical engineer learns. One of the ways chemical engineers ensure their calculations on linked process units have been performed correctly is to perform a mass balance over the entire system.

USING MASS BALANCES

Below is a general description of how to use mass balances for different situations.

Steady-State

If the system is at steady state then there are no terms which vary over time. This means that the in, out, generation, and consumption terms are constants, and the accumulation term is 0. Steady-state can only be applied to flow systems, or batch systems in which no changes are taking place. The mass balance for a steady-state system is:

$$\text{In} - \text{Out} + \text{Generation} - \text{Consumption} = 0$$

Batch Systems

A batch system is one in which all of the materials of the system are placed into a fixed vessel and left to react for an amount of time. Since there is no flow in or out, the in and out terms are both 0. Thus the general balance equation applied to a batch system is:

$$-\text{Consumption} + \text{Generation} = \text{Accumulation}$$

If the system is at steady-state, then the accumulation term is 0. The balance equation then becomes

$$\text{Consumption} = \text{Generation}$$

This type of situation is also known as chemical equilibrium.

Overall Mass Balance

An overall mass balance is a mass balance which models the entire mass of the system, as opposed to the mass of one component of the system. Since mass cannot be created or destroyed, the consumption and generation terms for an overall mass balance are 0. Thus the overall balance is shown by:

In – Out = Accumulation

If the system is also at steady state, then the equation reduces to:

In = Out

Note that an overall mass balance cannot be performed on a batch system, as all terms would equal 0.

MOLE BALANCE

Another way the general balance equation is used is the mole balance, that is, a balance on the number of moles of one component of the system. Since each mole of any given chemical must weigh the same, a mole balance is simply a direct product of the mass balance. However, mole balances still have advantages over mass balances in some situations (for example, if the flow rates are given in moles per time). The same principles used to form mass balances can be used to form mole balances.

CONTINUOUS STIRRED-TANK REACTOR

The continuous stirred-tank reactor (CSTR), also known as vat- or backmix reactor, is a common ideal reactor type in chemical engineering. A CSTR often refers to a model used to estimate the key unit operation variables when using a continuous agitated-tank reactor to reach a specified output. The mathematical model works for all fluids: liquids, gases, and slurries.

The behaviour of a CSTR is often approximated or modelled by that of a continuous ideally stirred-tank reactor (CISTR). All calculations performed with CISTRs assume perfect mixing. If the residence time is 5–10 times the mixing time, this approximation is valid for engineering purposes. The CISTR model is often used to simplify engineering calculations and can be used to describe research reactors. In practice it can only be approached, in particular in industrial size reactors.

Integral mass balance on number of moles N_i of species i in a reactor of volume V.

[accumulation] = [in] – [out] + [generation]

$$\frac{dN_i}{dt} = F_{io} - F_i + Vv_i r_i \qquad \qquad ...(3.1)$$

where, F_{io} is the molar flow rate inlet of species i, F_i the molar flow rate outlet, and v_i stoichiometric coefficient. The reaction rate, r, is generally dependent on the reactant concentration and the rate constant (k). The rate constant can be figured by using the Arrhenius temperature dependence. Generally, as the temperature increases so does the rate at which the reaction occurs. Residence time, τ, is the average amount of time a discrete quantity of reagent spends inside the tank.

Assume:
1. Constant density (valid for most liquids; valid for gases only if there is no net change in the number of moles or drastic temperature change).
2. Isothermal conditions, or constant temperature (k is constant).
3. Steady-state.

4. Single, irreversible reaction ($v_A = -1$).
5. First-order reaction ($r = kC_A$).

$$A \rightarrow products$$

$N_A = C_A V$ (where, C_A is the concentration of species A, V is the volume of the reactor, N_A is the number of moles of species A.)

$$C_A = \frac{C_{Ao}}{1 + kT} \qquad \text{... (3.2)}$$

The values of the variables, outlet concentration and residence time, in Eq. 3.2 are major design criteria. To model systems that do not obey the assumptions of constant temperature and a single reaction, additional dependent variables must be considered. If the system is considered to be in unsteady-state, a differential equation or a system of coupled differential equations must be solved.

CSTR's are known to be one of the systems which exhibit complex behaviour such as steady-state multiplicity, limit cycles and chaos. In environmental engineering, a continuous or continuously stirred tank reactor (CSTR) is a system that has the following properties:

1. There is inflow and outflow of matter.
2. Chemical reactions occur within the system's boundary.
3. The accumulation rate of any substance.
4. The system is in a steady-state, i.e. the concentration of any substance remains constant or equivalently, the accumulation rate is zero.
5. Any substance on the system is assumed to be homogeneously distributed.

With the above assumptions the law of conservation of mass can be written in the generic form:

Accumulation rate = 0 = input rate − output rate + reaction rate

The Damköhler numbers (Da) are dimensionless numbers used in chemical engineering to relate chemical reaction timescale to other phenomena occurring in a system. It is named after German chemist Gerhard Damköhler. There are several Damköhler numbers, and their definition varies according to the system under consideration.

For a general chemical reaction $A \rightarrow B$ of nth order, the Damköhler number is defined as

$$Da = kC_0^{n-1}t$$

where,

k = kinetics reaction rate constant.
C_0 = initial concentration.
n = reaction order.
t = time.

and it represents a dimensionless reaction time. It provides a quick estimate of the degree of conversion (X) that can be achieved in continuous flow reactors.

Generally, if $Da < 0.1$, then $X < 0.1$. Similarly, if $Da > 10$, then $X > 0.9$.

In continuous or semibatch chemical processes, the general definition of the Damköhler number is:

$$Da = \frac{\text{Reaction rate}}{\text{Convective mass transport rate}}$$

or as

$$Da = \frac{\text{Characteristic fluid time}}{\text{Characteristic chemical reaction time}}$$

For example, in a continuous reactor, the Damköhler number is:

$$Da = \frac{k_c C_0^n}{C_0 / \tau} = k_c C_0^{(n-1)} \tau$$

where, τ is the mean residence time or space time.

In reacting systems that include also interphase mass transport, the second Damköhler number (Da_{II}) is defined as the ratio of the chemical reaction rate to the mass transfer rate

$$Da_{II} = \frac{k C_0^{n-1}}{k_g a}$$

where,

k_g = is the global mass transport coefficient.

a = is the interfacial area.

CONTINUOUS FLOW TUBULAR REACTORS

Tubular reactors are always used in a continuous flow mode with reagents flowing in and products being removed. They can be the simplest of all reactor designs. Tubular reactors are often referred to by other names:

1. Pipe reactors.
2. Packed-bed reactors.
3. Trickle-bed reactors.
4. Bubble-column reactors.
5. Ebulating-bed reactors.

Single-phase flow in a tubular reactor can be upward or downward. Two-phase flow can be co-current up-flow, counter-current (liquid down, gas up) or, most commonly, cocurrent down-flow.

Tubular reactors can have a single wall and be heated with an external furnace or they can be jacketed for heating or cooling with a circulating heat transfer fluid. External furnaces can be rigid, split-tube heaters or be flexible mantle heaters. Tubular reactors are used in a variety of industries:

1. Petroleum.
2. Petrochemical.
3. Polymer.
4. Pharmaceutical.
5. Waste treatment.
6. Speciality chemical.

Tubular reactors are used in a variety of applications:

1. Carbonylation.
2. Dehydrogenation.
3. Hydrogenation.
4. Hydrocracking.
5. Hydroformulation.

6. Oxidation destruction.
7. Partial oxidation.
8. Polymerisation.
9. Reforming.

Tubular reactors may be empty for homogenous reactions or packed with catalyst particles for heterogeneous reactions. Packed reactors require upper and lower supports to hold particles in place. Uppermost packing is often of inert material to serve as a preheat section. Preheating can also be done with an internal spiral channel to keep incoming reagents close to the heated wall during entry (Fig 3.1).

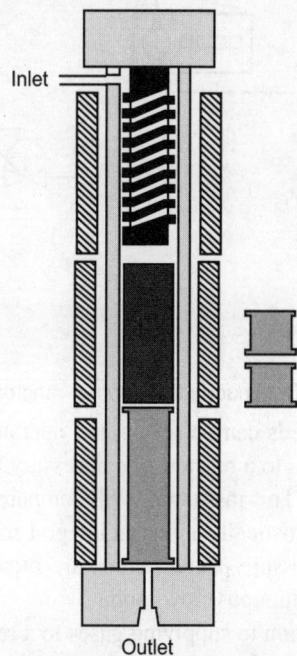

Inlet

Outlet

Fig. 3.1. Tubular reactor.

It is often desirable to size a tubular reactor to be large enough to fit 8 to 10 catalyst particles across the diameter and be at least 40–50 particle diameters long. The length to diameter ratio can be varied to study the effect of catalyst loading by equipping the reactor with 'spools' to change this ratio.

Temperature is typically controlled by thermocouples located on the outer wall of an externally heated tubular reactor. A moveable internal thermocouple is often employed to observe the temperature changes occurring as the reaction proceeds through the reactor. Tubular reactor systems are highly customisable and can be made to various lengths and diameters and engineered for various pressures and temperatures not specified above. We provide a split-tube furnace for heating these vessels. Insulation is provided at each end so that the end caps are not heated to the same temperature as the core of the reactor. The heater length is normally divided into two or three separate heating zones, although it can be split into as many zones as required. We can furnish either a fixed internal thermocouple in each zone or a single moveable thermocouple that can be used to measure the temperature at points along the catalyst bed. External thermocouples are typically provided for control of each zone of the heater. Figure 3.2 shows the continuous flow tubular reactor system.

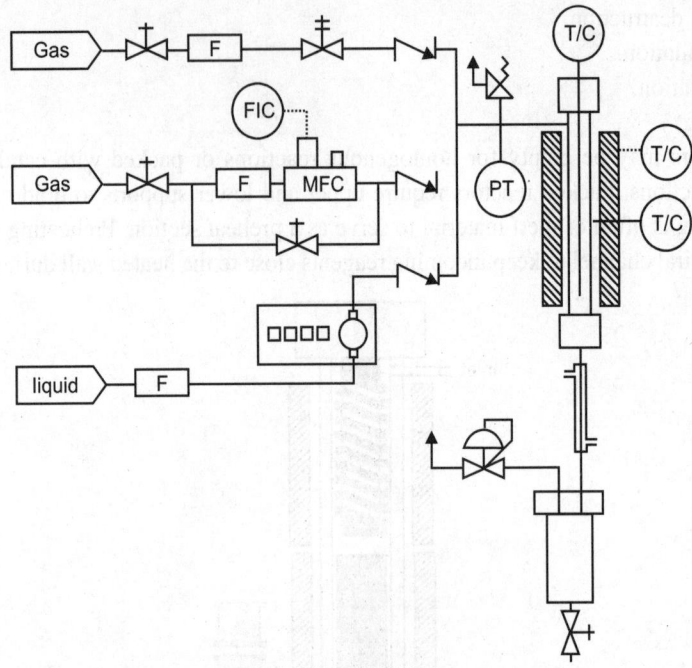

Fig. 3.2. Continuous flow tubular reactor system.

Gas feed systems: Various gas feeds can be set up and operated from a gas distribution panel. In order to deliver a constant flow of gas to a reactor, it is necessary to provide gas at a constant pressure to an electronic mass flow controller. This instrument will compare the actual flow rate delivered to the set point chosen by the user, and automatically adjust an integral control valve to assure a constant flow.

Liquid metering pumps: High pressure piston pumps are most often used to inject liquids into a pressurised reactor operating in a continuous-flow mode.

Back pressure regulators: In addition to supplying gases to a reaction through electronic mass flow controllers, the reactor is kept at a constant pressure by installing a back pressure regulator (BPR) downstream of the reactor. This style of regulator will release products only when the reactor pressure exceeds a preset value.

Cooling condensers: It is often desired to cool the products of the reaction prior to handling them. For this purpose, tube-and-shell heat exchangers are available to act as the cooling condensers. An adaptation of our standard condensers provides an excellent design.

Gas/liquid separators: Tubular reactors operating in continuous-flow mode with both gas and liquid products will also require a gas/liquid separator for smooth operation. The separator is placed downstream of the reactor, often separated from the reactor by a cooling condenser. In the separator vessel, liquids are condensed and collected in the bottom of the vessel. Gases and non-condensed vapours are allowed to leave the top of the vessel and pass on to the back pressure regulator. It is important to operate the BPR with a single fluid phase to prevent oscillation of the reactor pressure.

Control and data acquisition systems: A variety of solutions exist to meet the needs of system operators. System accessories such as heaters, mass flow controllers, and pumps can be obtained with individual control packages to create a manual, distributed control system. Each of the various controllers that are equipped with an output signal can be incorporated into a PC-based data acquisition system.

Tubular reactor system with three-zone heater, 300 ml separator vessel and 4871 controller are shown in Fig. 3.3.

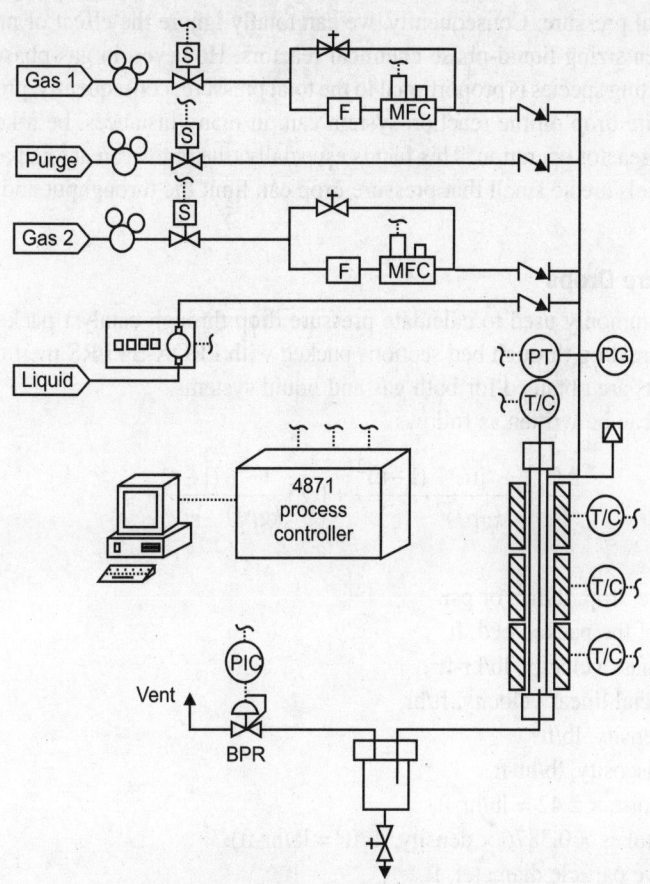

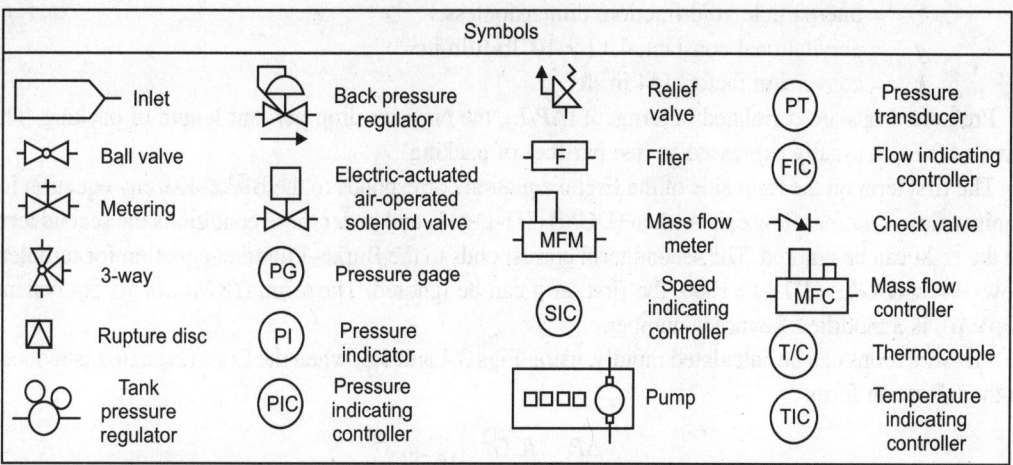

Fig. 3.3. Tubular reactor system with three-zone heater, 300 ml separator vessel and 4871 controller.

PRESSURE DROP IN REACTORS

In liquid-phase reactions, the concentration of reactants in insignificantly affected by even relatively large changes in the total pressure. Consequently, we can totally ignore the effect of pressure drop on the rate of reaction when sizing liquid-phase chemical reactors. However, in gas-phase reactions, the concentration of the reacting species is proportional to the total pressure; consequently, proper accounting for the effects of pressure drop on the reaction system can, in many instances, be a key factor in the success or failure of the reactor operation. This fact is especially true in microreactors packed with solid catalyst. Here the channels are so small that pressure drop can limit the throughput and conversion for gas-phase reactions.

Calculation of Pressure Drops

The Ergun equation, commonly used to calculate pressure drop through catalyst packed beds, can be used to calculate pressure drop through bed sections packed with PROX-SVERS inert catalyst support balls. Satisfactory results are obtained for both gas and liquid systems.

The ergun equation can be written as follows:

$$\frac{\Delta P}{L} 150 \frac{\mu G}{kg\rho D^2} \frac{(1-\varepsilon)^2}{\varepsilon^3} + 1.75 \frac{G^2}{kg\rho D} \frac{(1-\varepsilon)}{\varepsilon^3}$$

where,

ΔP = pressure drop, lb/in^2, or psi.

L = depth of the packed bed, ft.

G = ρV = mass velocity, lb/hr-ft^2.

V = superficial linear velocity, ft/hr.

ρ = fluid density, lb/ft^3

μ = fluid viscosity, lb/hr-ft.

(centipoise $\times$ 2.42 = lb/hr-ft).

(centistokes $\times$ 0.3876 $\times$ density, lb/ft^3 = lb/hr-ft).

D = effective particle diameter, ft.

ε = interparticle void fraction, dimensionless.

g = gravitational constant, 4.17 $\times$ 10^8 lb-ft/lb-hr^2.

k = conversion factor, 144 in^2/ft^2.

Pressure drops are correlated in terms of $(\Delta P/L)$, the pressure drop per unit length of packing. The term, $(\Delta P/L)$, is usually expressed as 'psi per foot of packing'.

The first term on the right side of the Ergun equation corresponds to the Blake–Kozeny equation for laminar flow. Laminar flow exists when (DG/μ) $(1/1-\varepsilon)<10$, and under these conditions the second term on the right can be ignored. The second term corresponds to the Burke–Plummer equation for turbulent flow. When (DG/μ) $(1/1-\varepsilon) >1000$, the first term can be ignored. The term, (DG/μ), or its equivalent, $(D\rho V/\mu)$, is a modified Reynolds number.

Pressure drops can be calculated rapidly, using Figs 3.4 and 3.5, when the Ergun equation is reduced to the following form:

$$\frac{\Delta P}{L} = \frac{fCG^2}{\rho D} \times 10^{-10}$$

where,

$$C = \frac{1.75 \times 10^{10}}{144g} \left(\frac{1-\varepsilon}{\varepsilon^3} \right)$$

$$f = 1 + \frac{150}{1.75} \left(\frac{DG}{\mu} \right)^{-1} (1-\varepsilon)$$

Values of C are given in Fig. 3.4 for ε in the range 0.30 to 0.50. Values of f as a function of modified Reynolds number, (DG/μ), for selected values of ε, are given in Fig. 3.5.

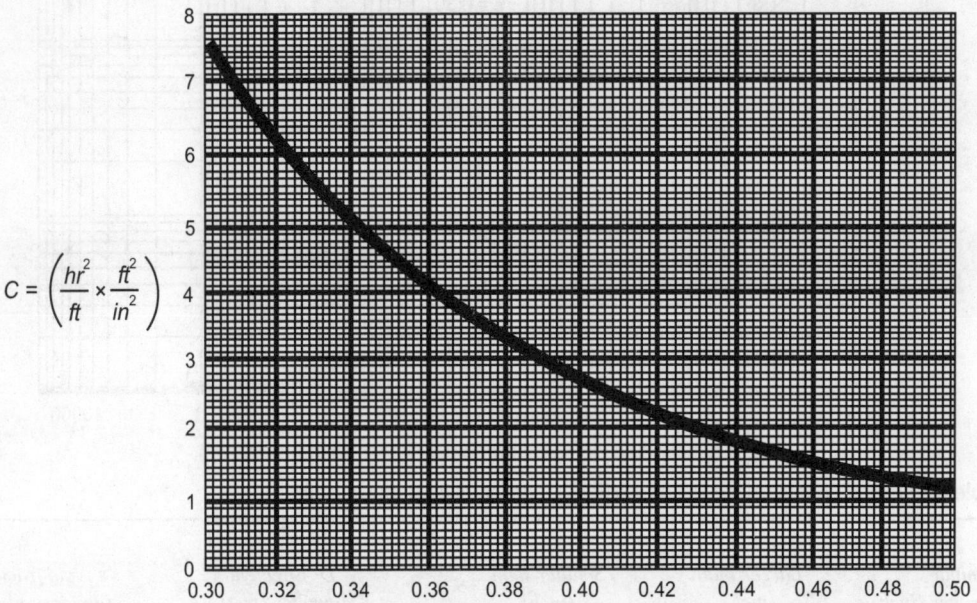

$$C = \left(\frac{hr^2}{ft} \times \frac{ft^2}{in^2} \right)$$

Fig. 3.4. Interparticle void fraction, ε, (dimensionless).

Typical values of void fraction, ε, and effective particle diameter, D, for PROX-SVERS inert catalyst support balls of various sizes are presented in the accompanying Table 3.1.

Flow through Packed Beds and Fluidised Beds

Chemical engineering operations commonly involve the use of packed and fluidised beds. These are devices in which a large surface area for contact between a liquid and a gas (absorption, distillation) or a solid and a gas or liquid (adsorption, catalysis) is obtained for achieving rapid mass and heat transfer, and particularly in the case of fluidised beds, catalytic chemical reactions.

Packed beds

A typical packed bed is a cylindrical column that is filled with a suitable packing material. You can learn about different types of packing materials from Perry's Handbook. The liquid is distributed as uniformly as possible at the top of the column and flows downward, wetting the packing material. A gas is admitted at the bottom, and flows upward, contacting the liquid in a countercurrent fashion. An example of a packed bed is an absorber. Here, the gas contains some carrier species that is insoluble in the liquid (such as air) and a soluble species such as carbon dioxide or ammonia. The soluble species is absorbed in the liquid, and the lean gas leaves the column at the top. The liquid rich in the soluble species is taken out at the bottom.

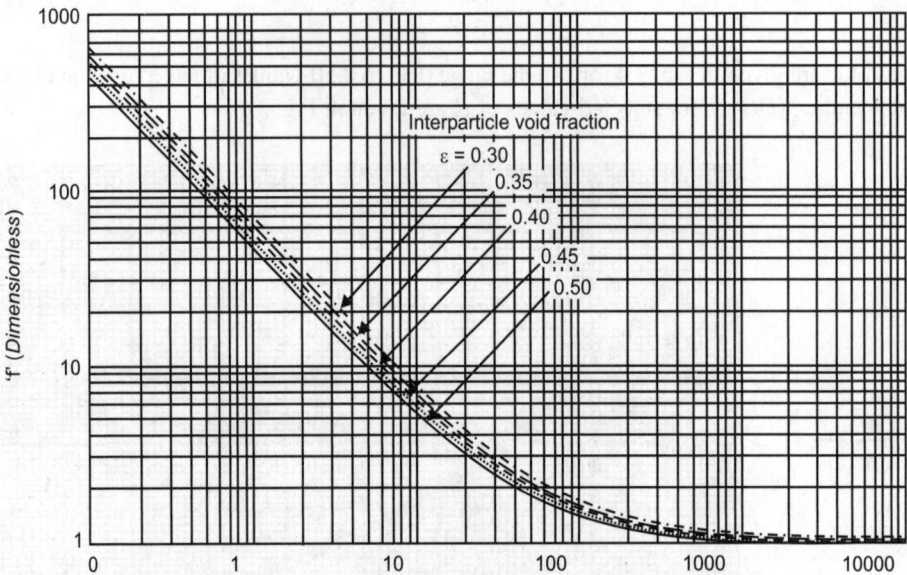

Fig. 3.5. Modified reynolds number, (DG/μ), (dimensionless).

Table 3.1. Ergun constants for PROX-SVERS® inert catalyst support balls.

| Nominal diameter (inch) | Tolerance | | 'D' Effective diameter (feet) | 'ε' void fraction (dimensionless) |
	Larger than (inch)	Smaller than (inch)		
1/8	1/8	3/16	0.013	0.40
1/4	3/16	3/8	0.023	0.42
3/8	1/4	1/2	0.031	0.43
1/2	3/8	5/8	0.041	0.43
5/8	1/2	3/4	0.051	0.44
3/4	5/8	7/8	0.062	0.45
1	7/8	1–1/8	0.082	0.45
1–1/4	1–1/8	1–3/8	0.104	0.46
1–1/2	1–3/8	1–5/8	0.125	0.46
2	1–7/8	2–1/8	0.167	0.46
3	2–7/8	3–1/8	0.250	0.46

From a fluid mechanical perspective, the most important issue is that of the pressure drop required for the liquid or the gas to flow through the column at a specified flow rate. To calculate this quantity we rely on a friction factor correlation attributed to Ergun. Other fluid mechanical issues involve the proper distribution of the liquid across the cross-section, and developing models of the velocity profile in the liquid film around a piece of packing material so that heat/mass transfer calculations can be made. Design of packing materials to achieve uniform distribution of the fluid across the cross-section throughout the column is an important subject as well. Here, we only focus on the pressure drop issue.

The Ergun equation that is commonly employed is given below.

$$f_p = \frac{150}{Re_p} + 1.75$$

Here, the friction factor p_p for the packed bed, and the Reynolds number Re_p, are defined as follows.

$$f_p = \frac{\Delta p}{L} \frac{Dp}{\rho V_s^2} \left(\frac{\varepsilon^3}{1-\varepsilon} \right) \qquad\qquad Re_p = \frac{D_p V_s \rho}{(1-\varepsilon)\mu}$$

The various symbols appearing in the above equations are defined as follows.
Δp : Pressure drop.
L : Length of the bed.

D_p : Equivalent spherical diameter of the particle defined by $D_p = 6 \dfrac{\text{Volume of the particle}}{\text{Surface area of the particle}}$.

ρ : Density of the fluid.
μ : Dynamic viscosity of the fluid.
V_s : Superficial velocity ($V_s = \dfrac{Q}{A}$ where Q is the volumetric flow rate of the fluid and is A the cross-sectional area of the bed).
ε : Void fraction of the bed (ε is the ratio of the void volume to the total volume of the bed).

Sometimes, we may use the concept of the interstitial velocity V_i, which is related to the superficial velocity by $V_i = \dfrac{V_s}{\varepsilon}$. The interstitial velocity is the average velocity that prevails in the pores of the column.

Two simpler results, each obtained by ignoring one or the other term in the Ergun equation also are in use. One is the Kozeny-Carman equation, used for flow under very viscous conditions.

$$f_p = \frac{150}{Re_p}, \qquad Re_p \leq 1000$$

The other is the Burke-Plummer equation, used when viscous effects are not as important as inertia.

$$f_p = 1.75, \qquad Re_p \geq 1000$$

It is suggested that the student simply use the Ergun equation. There is no need to use other two approximate results, even though they continue to be reported in textbooks.

Fluidised beds

A fluidised bed is a packed bed through which fluid flows at such a high velocity that the bed is loosened and the particle-fluid mixture behaves as though it is a fluid. Thus, when a bed of particles is fluidised, the entire bed can be transported like a fluid, if desired. Both gas and liquid flows can be used to fluidise a bed of particles. The most common reason for fluidising a bed is to obtain vigorous agitation of the solids in contact with the fluid, leading to excellent contact of the solid and the fluid and the solid and the wall. This means that nearly uniform temperatures can be maintained even in highly exothermic reaction situations where the particles are used to catalyse a reaction in the species contained in the fluid. In fact, fluidised beds were used in catalytic cracking in the petroleum industry in the past. The catalyst is suspended in the fluid by fluidising a bed of catalytic particles so that intimate contact can be achieved between the particles and the fluid. Nowadays, you will find fluidised beds used in catalyst regeneration, solid-gas reactors, combustion of coal, roasting of ores, drying, and gas adsorption operations.

First, we consider the behaviour of a bed of particles when the upward superficial fluid velocity is gradually increased from zero past the point of fluidisation, and back down to zero (Fig. 3.6).

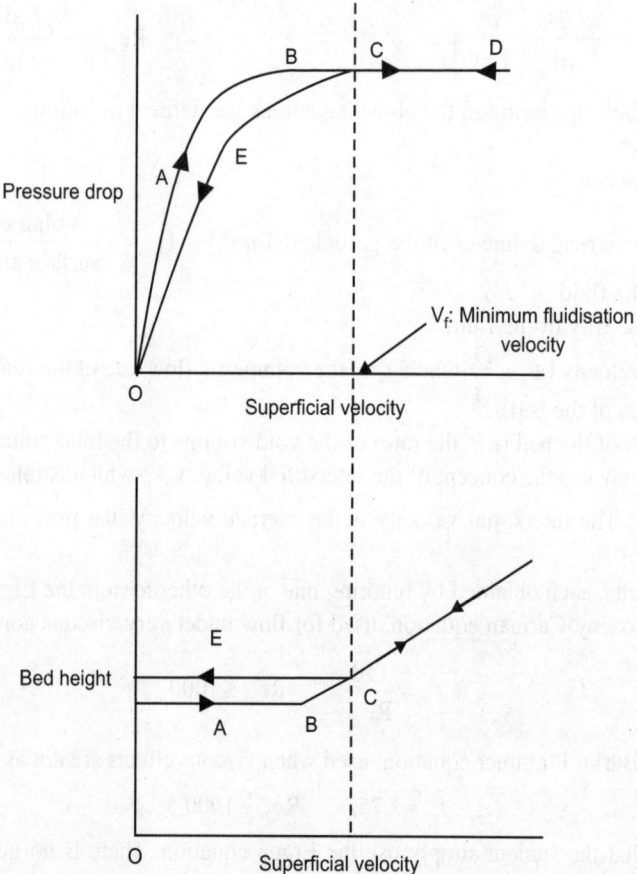

Fig. 3.6. Fluidised beds showing supervelocity, pressure drop and bed height.

At first, when there is no flow, the pressure drop zero, and the bed has a certain height. As we proceed along the right arrow in the direction of increasing superficial velocity, tracing the path ABCD, at first, the pressure drop gradually increases while the bed height remains fixed. This is a region where the Ergun equation for a packed bed can be used to relate the pressure drop to the velocity. When the point B is reached, the bed starts expanding in height while the pressure drop levels off and no longer increases as the superficial velocity is increased.

This is when the upward force exerted by the fluid on the particles is sufficient to balance the net weight of the bed and the particles begin to separate from each other and float in the fluid. As the velocity is increased further, the bed continues to expand in height, but the pressure drop stays constant. It is possible to reach large superficial velocities without having the particles carried out with the fluid at the exit. This is because the settling velocities of the particles are typically much larger than the largest superficial velocities used.

Now, if we trace our path backward, gradually decreasing the superficial velocity, in the direction of the reverse arrows in the figure, we find that the behaviour of the bed follows the curves DCE. At first, the pressure drop stays fixed while the bed settles back down, and then begins to decrease when the point C is reached. The bed height no longer decreases while the pressure drop follows the curve CEO. A bed of particles, left alone for a sufficient length of time, becomes consolidated, but it is loosened when it is fluidised. After fluidisation, it settles back into a more loosely packed state; this is why the constant bed height on the return loop is larger than the bed height in the initial state. If we now repeat the experiment by increasing the superficial velocity from zero, we will follow the set of curves ECD in both directions. Because of this reason, we define the velocity at the point C in the figure as the minimum fluidisation velocity V_f. We can calculate it by balancing the net weight of the bed against the upward force exerted on the bed, namely the pressure drop across the bed Δp multiplied by the cross-sectional area of the bed A. In doing this balance, we ignore the small frictional force exerted on the wall of the column by the flowing fluid.

$$\text{Upward force on the bed} = \Delta p\, A$$

If the height of the bed at this point is L and the void fraction is ε, we can write:

$$\text{Volume of particles} = (1-\varepsilon)\, AL$$

If the acceleration due to gravity is g, the net gravitational force on the particles (net weight) is net weight of the particles = $(1-\varepsilon)(\rho_p-\rho_f)\, ALg$.

Balancing the two yields:

$$\Delta p = (1-\varepsilon)(\rho_p-\rho_f)Lg$$

By using an expression relating Δp to the superficial velocity, which is the fluidisation velocity at this point, we can obtain a result for the latter.

Typically, for a bed of small particles ($D_p \leq 0.1$ mm), the flow conditions at this stage are such that the Reynolds number is relatively small (Re ≤ 10) so that we can use the Kozeny–Carman equation, applicable to the viscous flow regime, for establishing the point of onset of fluidisation. This yields:

$$V_f = \frac{(\rho_p-\rho_f)gD_p^2}{150\mu}\frac{\varepsilon^3}{1-\varepsilon}$$

When the superficial velocity V_s is equal to V_f, we refer to the state of the bed as one of incipient fluidisation. The void fraction ε at this state depends upon the material, shape, and size of the particles. For nearly spherical particles, McCabe, Smith, and Harriott suggest that ε lies in the range 0.40–0.45, increasing a bit with particle size.

For large particles ($D_p \geq 1$ mm), inertial effects are important, and the full Ergun equation must be used to determine V_f. When in doubt, use the Ergun equation instead of a simplified version of it.

Now, we consider the condition we must impose on the superficial velocity so that particles are not carried out with the fluid at the exit. This would occur if the superficial velocity is equal to the settling velocity of the particles. Restricting attention to small particles so that Stokes law can be used to calculate their settling velocity, we can write:

$$V_{settling} = \frac{(\rho_p - \rho_f)gD_p^2}{18\mu}$$

If we now use the result for the minimum fluidisation velocity for the case of small particles, given above, we see that the ratio:

$$\frac{V_{settling}}{V_f} = \frac{25}{3}\frac{1-\varepsilon}{\varepsilon^3}$$

For ε lying in the range 0.40–0.45, this yields a ratio ranging from 78–50. McCabe suggests that it is common to operate fluidised beds at velocities as high as 30 V_f, and values as large as 100 V_f are used on occasion. Recognising that not all particles are of the same size and that D_p is only an average size, we see that fine particles are likely to be carried out with the exiting fluid in such a situation. They can be recovered by filters or cyclone separators and returned, in order to obtain the benefits of operating a bed at such large superficial velocities.

Fluidisation can be broadly classified into particulate fluidisation or bubbling fluidisation. Particulate fluidisation occurs in liquids. As the velocity of the liquid is increased past the minimum fluidisation velocity, the bed expands uniformly, and uniform conditions prevail in the liquid solid mixture. In contrast, bubbling fluidisation occurs in gas-fluidised beds. Here, when the bed is fluidised, large pockets of gas, free of particles, are seen to rise through the bed. Where there are particles, the bed void fraction is approximately at the value that prevails at the point of incipient fluidisation. The bubbles grow until they fill the cross-section, and then successive bubbles move up the column, a condition known as slugging.

The above classification should not be interpreted rigidly. Sometimes, very dense particles in a liquid can show 'bubbling' and gases at high pressure when flowing through beds of fine particles, can give rise to particulate fluidisation. Usually, this occurs at lower velocities, and at higher velocities, the bed shows bubbling. Pressure drop in pipes is caused by:

1. Friction.
2. Vertical pipe difference or elevation.
3. Changes of kinetic energy.
4. Calculation of pressure drop caused by friction in circular pipes.

To determine the fluid (liquid or gas) pressure drop along a pipe or pipe component, the following calculations, in the following order.

Fluid Pressure Drop Along Pipe Length of Uniform Diameter

Determine Reynolds number:

$$Re = \frac{\bar{w} \times D}{v}$$

where,

Re = Reynolds number.
$\bar{w}$ = Velocity of flow.
D = Diameter of pipe.
v = Kinematics viscosity.

If the Reynolds number <2320, than you have laminar flow. Laminar flow is characterised by the gliding of concentric cylindrical layers past one another in orderly fashion. The velocity of the fluid is at its maximum at the pipe axis and decreases sharply to zero at the wall. The pressure drop caused axis by friction of laminar flow does not depend of the roughness of pipe.

If the Reynolds number >2320, we have turbulent flow. There is an irregular motion of fluid particles in directions transverse to the direction of the main flow. The velocity distribution of turbulent flow is more uniform across the pipe diameter than in laminar flow. The pressure drop caused by friction of turbulent flow depends on the roughness of pipe. Select pipe friction coefficient: The pipe friction coefficient is a dimensionless number. The friction factor for laminar flow condition is a function of Reynolds number only, for turbulent flow it is also a function of the characteristics of the pipe wall.

Determine pipe friction coefficient at laminar flow:

$$\lambda = \frac{64}{Re}$$

where,

λ = Pipe friction coefficient.

Re = Reynolds number.

Note: Perfectly smooth pipes will have a roughness of zero.

Determine pipe friction coefficient at turbulent flow (in the most cases):

$$\frac{1}{\sqrt{\lambda}} = 2 \ \lg\left[\frac{2.51}{Re \times \sqrt{\lambda}} + \frac{k}{D} \times 0.269\right]$$

where,

λ = Pipe friction coefficient.

g = Acceleration of gravity.

Re = Reynolds number.

k = Absolute roughness.

D = Diameter of pipe.

lg = Log.

The solutions to this calculation is plotted vs. the Reynolds number to create a moody chart. Determine pressure drop in circular pipes:

$$\Delta p = \lambda \times \frac{L}{D} \times \frac{\rho}{2} \times \bar{w}^2$$

where,

Δp = Pressure drop.

λ = Pipe friction coefficient.

L = Length of pipe.

D = Pipe diameter.

ρ = Density.

$\bar{w}$ = Flow velocity.

If you have valves, elbows and other elements along your pipe then you calculate the pressure drop with resistance coefficients specifically for the element. The resistance coefficients are in most cases found through practice tests and through vendor specification documents. If the resistance coefficiet is known, then we can calculate the pressure drop for the element:

$$\Delta p = \zeta \times \frac{\rho}{2} \times \bar{w}^2$$

where,

Δp = Pressure drop.

ζ = Resistance coefficient (determined by test or vendor specification).

ρ = Density.

$\overline{w}$ = Flow velocity.

Pressure drop by gravity or vertical elevation:

$$\Delta p = \rho \times g \times \Delta H$$

where,

Δp = Pressure drop.

ρ = Density.

g = Acceleration of gravity.

ΔH = Vertical elevation or drop.

Pressure drop of gases and vapour: Compressible fluids expands caused by pressure drops (friction) and the velocity will increase. Therefore, is the pressure drop along the pipe not constant,

$$\frac{p_1^2 - p_2^2}{2 \times p_1} = \lambda \times \frac{L}{D} \times \rho_1 \times \frac{\overline{w}_1^2}{2} \times \frac{\overline{T}}{T_1}$$

where,

p_1 = Pressure incoming.

T_1 = Temperature incoming.

p_2 = Pressure leaving.

T_2 = Temperature leaving.

$$\overline{T} = \frac{T_1 + T_2}{2}$$

We set the pipe friction number as a constant and calculate it with the input-data. The temperature, which is used in the equation, is the average of entrance and exit of pipe. Ergun equation, derived by the Turkish chemical engineer Sabri Ergun, expresses the friction factor in a packed column as a function of the Reynolds number:

$$f_p = \frac{150}{\mathrm{Re}_p} + 1.75$$

where, f_p and Re_p are defined as

$$f_p = \frac{\Delta p}{L} \frac{Dp}{\rho V_s^2} \left(\frac{\varepsilon^3}{1 - \varepsilon} \right) \quad \text{and} \quad \mathrm{Re}_p = \frac{D_p V_s \rho}{(1 - \varepsilon)\mu}$$

where,

Δp = is the pressure drop across the bed.

L = is the length of the bed (not the column).

D_p = is the equivalent spherical diameter of the packing.

ρ = is the density of fluid.

μ = is the dynamic viscosity of the fluid.

V_s = is the superficial velocity (i.e. the velocity that the fluid would have through the empty tube at the same volumetric flow rate).

ε = is the void fraction of the bed (bed porosity at any time).

Spherical Packed-Bed Reactor

In a spherical reactor the cross-section varies as we move through the reactor and is greater than in a normal packed-bed reactor.

Another advantage of spherical reactors is that they are the most economical shape for high pressures. As a first approximation we will assume that the fluid moves down through the reactor in plug flow. Consequently, because of the increase in cross-sectional area, A_c, as the fluid enters the sphere, the superficial velocity, $G = \dot{m}/A_c$, will decrease. From the Ergun equation (Eq. 3.3),

$$\frac{dP}{dz} = -\frac{G(1-\phi)}{\rho g_c D_P \phi^3}\left[\frac{150(1-\phi)\mu}{D_P} + 1.75G\right] \qquad \text{... (3.3)}$$

we know that by decreasing G, the pressure drop will be reduced significantly, resulting in higher conversions.

Because the cross-sectional area of the reactor is small near the inlet and outlet, the presence of catalyst there would cause substantial pressure drop, thereby reducing the efficiency of the spherical reactor. To solve this problem, screens to hold the catalyst are placed near the reactor entrance and exit (Figs 3.7 and 3.8).

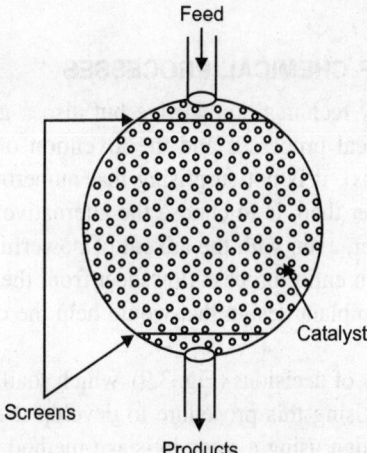

Fig. 3.7. Schematic drawing of the inside of a spherical reactor.

Here L is the location of the screen from the center of the reactor. We can use elementary geometry and integral calculus to derive the following expressions for cross-sectional area and catalyst weight as a function of the variables defined in Fig. 3.8:

$$A_c = \pi[R^2 - (z - L)^2] \qquad \text{... (3.4)}$$

$$W = \rho_c(1-\phi)V = \rho_c(1-\phi)\pi\left[R^2 z \frac{1}{3}(z - L)^3 - \frac{1}{3}L^3\right] \qquad \text{... (3.5)}$$

By using these formulas and the standard pressure drop algorithm, one can solve a variety of spherical reactor problems.

Note that (Eqs 3.4 and 3.5) make use of L and not L'. Thus, one does not need to adjust these formulas to treat spherical reactors that have different amounts of empty space at the entrance and exit (i.e. $L \neq L'$). Only the upper limit of integration needs to be changed, $z_f = L + L'$.

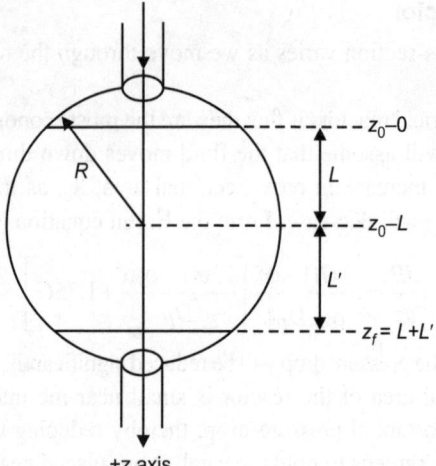

Fig. 3.8. Coordinate system and variables used with a spherical reactor. The initial and final integration values are shown as z_0 and z_f.

SYNTHESIS AND DESIGN OF CHEMICAL PROCESSES

Process design involves not only technical knowledge but also a good dose of creativity. With the increasing complexity of chemical processes and the invention of novel operations (e.g. reactive distillation, membrane separations), it is not surprising that numerous alternatives can be generated. The goal of the chemical engineer then, is to choose the alternative that brings the largest economic benefits to the company. However, even with the advent of powerful process simulation tools, it still remains a challenge to develop an entire process flowsheet from the drawing block. Hence arises the need for a systematic approach to plant design that would help the chemical engineer to arrive at the best design in a logical fashion.

Douglas suggested a hierarchy of decisions (Fig. 3.9), which shall be referred to in this paper as the conventional design procedure. Using this procedure to develop the process flowsheet will help the designer to arrive at the final design using a stage-by-stage method. At each stage, he generates new alternatives and evaluates those using selected economic criteria. He then chooses the best alternative and proceeds to the next stage where he will have to make another decision among the new alternatives that are generated. Finally, he will arrive at the last stage where the final choice will lead to the 'best' alternative.

Stanislav suggested a new hierarchy, also shown in Fig. 3.9, which shall be referred to in this paper as the modified design procedure. There are two key modifications from the conventional procedure. Firstly, the reactor design is considered as a separate stage of the design procedure rather than as part of the recycle structure. Secondly, the decision to recycle is left to the last stage of the hierarchy, just before the heat integration.

The objective of this study is, therefore, to present a comparison of the conventional and modified design procedures, the aim being to see which design hierarchy would help the engineer to arrive at the 'best' process flowsheet in the shortest time.

The HDA process is used as the example in this study, which involves the extensive use of a process simulator as well as sizing, costing and profitability analysis.

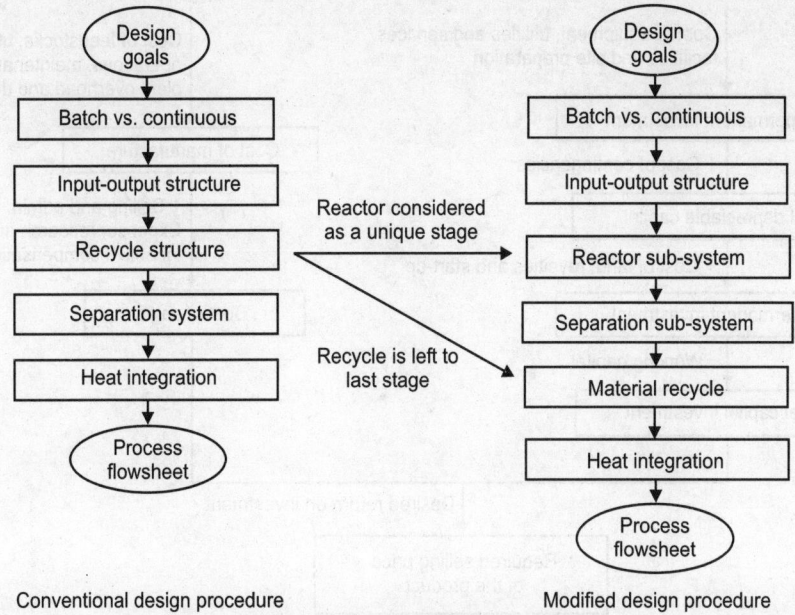

Conventional design procedure   Modified design procedure

Fig. 3.9. Conventional and modified design procedures.

Flowsheet Evaluation

In order to apply either of the two procedures listed in Fig. 3.9, it is necessary to evaluate the profitability of the different flowsheets generated at each stage of the procedure so that we can select the most profitable alternative and proceed to the next stage. To determine the profitability, the total capital investment and the annual cost of sales have to be computed. These two components must in turn be calculated from various subcomponents including equipment cost, cost of land, site preparation, feedstock cost, utility cost, labour cost, maintenance, operating overhead, depreciation, selling expense, etc. A summary of these computations is shown in Fig. 3.10 and further details can be obtained from Seider as well as Turton.

The problem however, is that these computations are lengthy and even with the advent of commercial simulation software, it still takes a substantial amount of time to arrive at the profitability of a particular flowsheet, e.g. the product selling price required to give a return on investment of 20 per cent. This was the reason why Douglas, in demonstrating the application of his proposed design hierarchy, used shortcut calculations and estimates to determine the profitability of the different flowsheets generated at each stage of the hierarchy.

This method, though time-saving, is undesirable because it might lead to the engineer eliminating a potentially more economical flowsheet.

To circumvent this problem, a special program was developed in this study to automate the cumbersome process of determining the required selling price for a particular flowsheet. This program is essentially an interface, which combines the process simulation power of Hysys with the spreadsheet capabilities of Excel by linking the object libraries of these two applications through Visual Basic as shown in Fig 3.11 with this interface program, the time taken to evaluate a process flowsheet is thus greatly reduced and as a result, the use of the design procedures becomes simpler and more accurate.

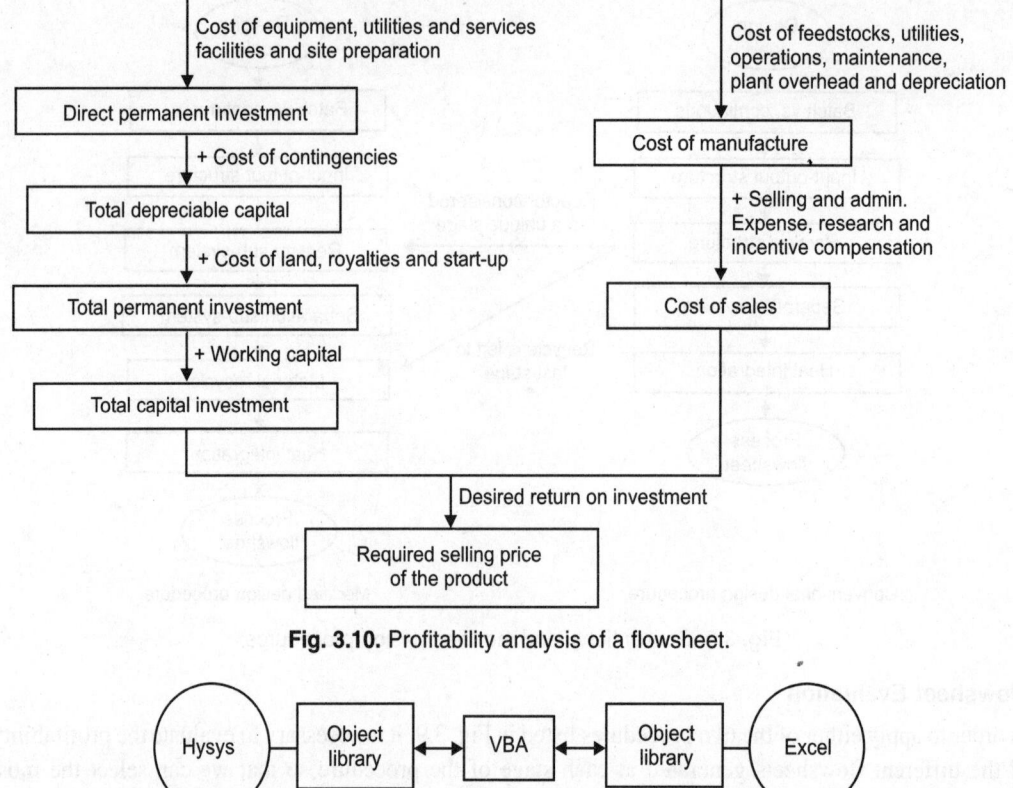

Fig. 3.10. Profitability analysis of a flowsheet.

Fig. 3.11. Linking object libraries of Hysys and Excel.

The Hysys-Excel interface is based in the Excel environment. The interface captures the key process simulation results from Hysys and displays it in a user-friendly interface, allowing the user to see at one glance whether all the key process constraints are satisfied. This interface also allows the user to change the simulation parameters in Hysys directly from Excel. For example, the user can change the desired reaction conversion rate on the Excel interface, which will automatically transmit the new input to Hysys, which will in turn run the simulation based on the new input and then send the new simulation results back to the Excel interface.

The main feature of the interface that allows it to speed up the evaluation of a process flowsheet is its ability to size and cost equipment using macros in the Excel interface. For example, sizing and costing a distillation column manually requires a lot of work because iterative computations have to be done to obtain the column diameter which can give a vapour velocity that will not bring about flooding or entrainment. In addition, calculations will also have to factor in the tray efficiencies. However, all these calculations can be done automatically using macros in the Hysys-Excel interface, which will capture the necessary data from the Hysys simulation, perform the iterative calculations and compute the cost based on pre-entered cost equations. A similar algorithm is followed for the other process equipment. Similarly, macros can also be written for all the cost components in Fig. 3.10 and hence, the profitability of a flowsheet can be obtained at the click of a button.

REACTION EQUILIBRIUM IN THE GAS PHASE

In beginning our study of the reactions of gases, we will assume a knowledge of the physical properties of gases as described by the Ideal gas law and an understanding of these properties as given by the postulates and conclusions of the Kinetic molecular theory. We assume that we have developed a dynamic model of phase equilibrium in terms of competing rates. We will also assume an understanding of the bonding, structure, and properties of individual molecules.

In performing stoichiometric calculations, we assume that we can calculate the amount of product of a reaction from the amount of the reactants we start with. For example, if we burn methane gas, $CH_4(g)$, in excess oxygen, the reaction

$$CH_4 (g) + 2O_2 (g) \rightarrow CO_2 (g) + 2H_2O (g)$$

occurs, and the number of moles of $CO_2(g)$ produced is assumed to equal the number of moles of $CH_4(g)$ we start with.

From our study of phase transitions we have learned the concept of equilibrium. We observed that, in the transition from one phase to another for a substance, under certain conditions both phases are found to coexist, and we refer to this as phase equilibrium. It should not surprise us that these same concepts of equilibrium apply to chemical reactions as well. In the reaction, therefore, we should examine whether the reaction actually produces exactly one mole of CO_2 for every mole of CH_4 we start with or whether we wind up with an equilibrium mixture containing both CO_2 and CH_4. We will find that different reactions provide us with varying answers. In many cases, virtually all reactants are consumed, producing the stoichiometric amount of product. However, in many other cases, substantial amounts of reactant are still present when the reaction achieves equilibrium, and in other cases, almost no product is produced at equilibrium. Our goal will be to understand, describe and predict the reaction equilibrium.

An important corollary to this goal is to attempt to control the equilibrium. We will find that varying the conditions under which the reaction occurs can vary the amounts of reactants and products present at equilibrium. We will develop a general principle for predicting how the reaction conditions affect the amount of product produced at equilibrium.

Observation 1: Reaction Equilibrium

We begin by analysing a significant industrial chemical process, the synthesis of ammonia gas, NH_3, from nitrogen and hydrogen:

$$N_2(g) + 3H_2(g) \rightarrow 2NH_3(g) \qquad \qquad \text{... (3.6)}$$

If we start with 1 mole of N_2 and 3 moles of H_2, the balanced equation predicts that we will produce 2 moles of NH_3. In fact, if we carry out this reaction starting with these quantities of nitrogen and hydrogen at 298 K in a 100.0 L reaction vessel, we observe that the number of moles of NH_3 produced is 1.91 mol. This 'yield' is less than predicted by the balanced equation, but the difference is not due to a limiting reagent factor. Recall that, in stoichiometry, the limiting reagent is the one that is present in less than the ratio of moles given by the balanced equation. In this case, neither N_2 nor H_2 is limiting because they are present initially in a 1:3 ratio, exactly matching the stoichiometry. Note also that this seeming deficit in the yield is not due to any experimental error or imperfection, nor is it due to poor measurements or preparation. Rather, the observation that, at 298 K, 1.91 moles rather than 2 moles are produced is completely reproducible: every measurement of this reaction at this temperature in this volume starting with 1 mole of N_2 and 3 moles of H_2 gives this result. We conclude that the reaction achieves reaction equilibrium in which all three gases are present in the gas mixture. We can determine the amounts of each gas at equilibrium from the stoichiometry of the reaction. When $n_{NH_3} = 1.91$ mol are created, the number of moles of N_2 remaining at equilibrium is $n_{N_2} = 0.045$ mol and $n_{H_2} = 0.135$ mol.

It is important to note that we can vary the relative amount of NH_3 produced by varying the temperature of the reaction, the volume of the vessel in which the reaction occurs, or the relative starting amounts of N_2 and H_2. We shall study and analyse this observation in detail in later sections. For now, though, we demonstrate that the concept of reaction equilibrium is general to all reactions.

Consider the reaction

$$H_2 (g) + I_2 (g) \rightarrow 2HI (g) \qquad \qquad ... (3.7)$$

If we begin with 1.00 mole of H_2 and 1.00 mole of I_2 at 500 K in a reaction vessel of fixed volume, we observe that, at equilibrium, $n_{HI} = 1.72$ mol, leaving in the equilibrium mixture $n_{H_2} = 0.14$ mol and $n_{I_2} = 0.14$ mol.

Similarly, consider the decomposition reaction

$$N_2O_4 (g) \rightarrow 2NO_2 (g) \qquad \qquad ... (3.8)$$

At 298 K in a 100.0 L reaction flask, 1.00 mol of N_2O_4 partially decomposes to produce, at equilibrium, $n_{NO_2} = 0.64$ mol and $n_{N_2O_4} = 0.68$ mol.

Some chemical reactions achieve an equilibrium that appears to be very nearly complete reaction. For example,

$$H_2 (g) + Cl_2 (g) \rightarrow 2HCl (g) \qquad \qquad ... (3.9)$$

If we begin with 1.00 mole of H_2 and 1.00 mole of Cl_2 at 298 K in a reaction vessel of fixed volume, we observe that, at equilibrium, n_{HCl} is almost exactly 2.00 mol, leaving virtually no H_2 or Cl_2. This does not mean that the reaction has not come to equilibrium. It means instead that, at equilibrium, there are essentially no reactants remaining.

In each of these cases, the amounts of reactants and products present at equilibrium vary as the conditions are varied but are completely reproducible for fixed conditions. Before making further observations that will lead to a quantitative description of the reaction equilibrium, we consider a qualitative description of equilibrium.

We begin with a dynamic equilibrium description. We know from our studies of phase transitions that equilibrium occurs when the rate of the forward process (e.g. evapouration) is matched by the rate of reverse process (e.g. condensation). Since we have now observed that gas reactions also come to equilibrium, we postulate that at equilibrium the forward reaction rate is equal to the reverse reaction rate. For example, in the reaction here, the rate of decomposition of N_2O_4 molecules at equilibrium must be exactly matched by the rate of recombination (or dimerisation) of NO_2 molecules.

To show that the forward and reverse reactions continue to happen at equilibrium, we start with the NO_2 and N_2O_4 mixture at equilibrium and we vary the volume of the flask containing the mixture. We observe that, if we increase the volume and the reaction is allowed to come to equilibrium, the amount of NO_2 at equilibrium is larger at the expense of a smaller amount of N_2O_4. We can certainly conclude that the amounts of the gases at equilibrium depend on the reaction conditions. However, if the forward and reverse reactions stop once the equilibrium amounts of material are achieved, the molecules would not 'know' that the volume of the container had increased. Since the reaction equilibrium can and does respond to a change in volume, it must be that the change in volume affects the rates of both the forward and reverse processes. This means that both reactions must be occurring at equilibrium, and that their rates must exactly match at equilibrium.

This reasoning reveals that the amounts of reactant and product present at equilibrium are determined by the rates of the forward and reverse reactions. If the rate of the forward reaction (e.g. decomposition of N_2O_4) is faster than the rate of the reverse reaction, then at equilibrium we have more product than reactant. If that difference in rates is very large, at equilibrium there will be much more product than reactant. Of course, the converse of these conclusions is also true. It must also be the case that the rates of these processes depends on, amongst other factors, the volume of the reaction flask, since the amounts of each gas present at equilibrium change when the volume is changed.

Observation 2: Equilibrium Constants

It was noted above that the equilibrium partial pressures of the gases in a reaction vary depending upon a variety of conditions. These include changes in the initial numbers of moles of reactants and products, changes in the volume of the reaction flask, and changes in the temperature. We now study these variations quantitatively.

Consider first the reaction here. Following on our previous study of this reaction, we inject an initial amount of $N_2O_4(g)$ into a 100 L reaction flask at 298 K. Now, however, we vary the initial number of moles of N_2O_4 (g) in the flask and measure the equilibrium pressures of both the reactant and product gases. The results of a number of such studies are given here. We might have expected that the amount of NO_2 produced at equilibrium would increase in direct proportion to increases in the amount of N_2O_4 we begin with. Table 3.2 shows that this is not the case. Note that when we increase the initial amount of N_2O_4 by a factor of 10 from 0.5 moles to 5.0 moles, the pressure of NO_2 at equilibrium increases by a factor of less than 4.

Table 3.2. Equilibrium partial pressures in decomposition reaction.

Initial $n_{N_2O_4}$	$P_{N_2O_4}$ (atm)	P_{NO_2} (atm)
0.1	0.00764	0.033627
0.5	0.071011	0.102517
1	0.166136	0.156806
1.5	0.26735	0.198917
2	0.371791	0.234574
2.5	0.478315	0.266065
3	0.586327	0.294578
3.5	0.695472	0.320827
4	0.805517	0.345277
4.5	0.916297	0.368255
5	1.027695	0.389998

The relationship between the pressures at equilibrium and the initial amount of N_2O_4 is perhaps more easily seen in a graph of the data in Table 3.2, as shown in Fig. 3.12. There are some interesting features here. Note that, when the initial amount of N_2O_4 is less than 1 mol, the equilibrium pressure of NO_2 is greater than that of N_2O_4. These relative pressures reverse as the initial amount increases, as the

N_2O_4 equilibrium pressure keeps track with the initial amount but the NO_2 pressure falls short. Clearly, the equilibrium pressure of NO_2 does not increase proportionally with the initial amount of N_2O_4. In fact, the increase is slower than proportionality, suggesting perhaps a square root relationship between the pressure of NO_2 and the initial amount of N_2O_4.

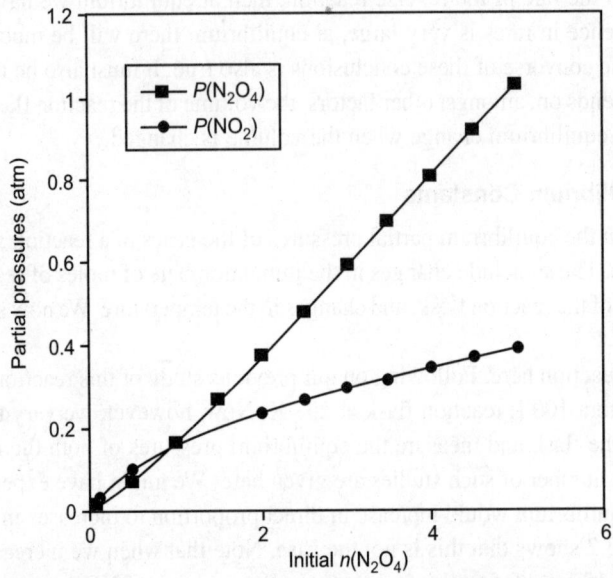

Fig. 3.12. Equilibrium partial pressures in decomposition reaction.

We test this in Fig. 3.13 by plotting P_{NO_2} at equilibrium versus the square root of the initial number of moles of N_2O_4.

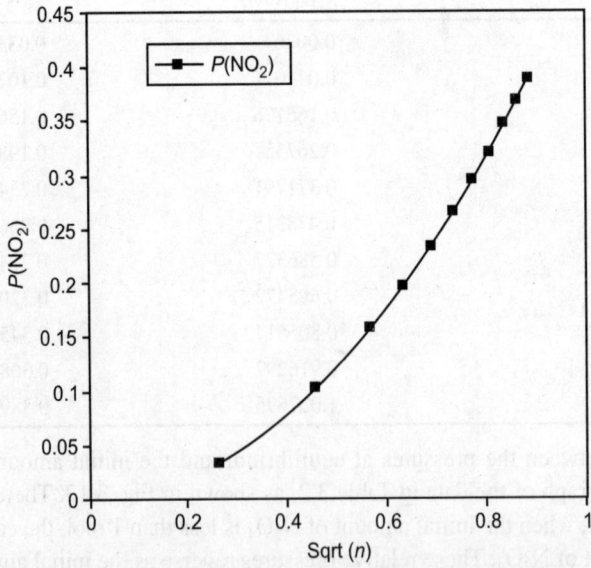

Fig. 3.13. Relationship of pressure of product to initial amount of reactant.

Figure 3.13 makes it clear that this is not a simple proportional relationship, but it is closer. Note in Fig. 3.12 that the equilibrium pressure $P_{N_2O_4}$ increases close to proportionally with the initial amount of N_2O_4. This suggests plotting P_{NO_2} versus the square root of $P_{N_2O_4}$. This is done in Fig. 3.14, where we discover that there is a very simple proportional relationship between the variables plotted in this way. We have thus observed that:

$$P_{NO_2} = c\sqrt{P_{N_2O_4}} \qquad \ldots (3.10)$$

where, c is the slope of the graph. Eq. 3.10 can be rewritten in a standard form

$$K_p = \frac{P_{NO_2}^{\,2}}{P_{N_2O_4}} \qquad \ldots (3.11)$$

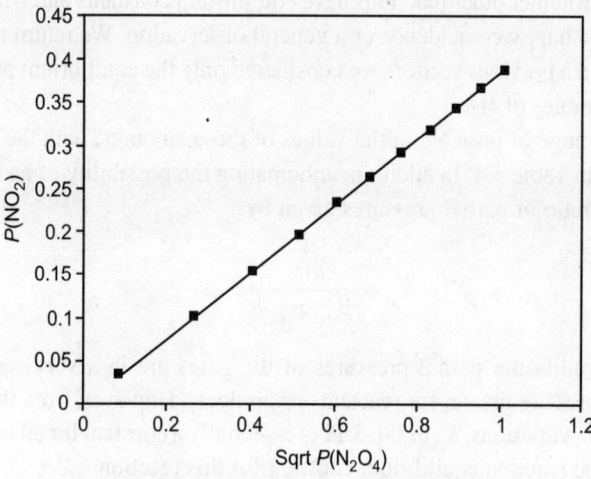

Fig. 3.14. Equilibrium partial pressures.

To test the accuracy of this equation and to find the value of K_p, we return to Table 3.2 and add another column in which we calculate the value of K_p for each of the data points. Table 3.3 makes it clear that the 'constant' in Eq. 3.11 truly is independent of both the initial conditions and the equilibrium partial pressure of either one of the reactant or product. We, thus refer to the constant K_p in Eq. 3.11 as the reaction equilibrium constant.

Table 3.3. Equilibrium partial pressures in decomposition reaction.

Initial $n_{N_2O_4}$	$P_{N_2O_4}$ (atm)	P_{NO_2} (atm)	K_p
0.1	0.00764	0.0336	0.148
0.5	0.0710	0.102	0.148
1	0.166	0.156	0.148
1.5	0.267	0.198	0.148
2	0.371	0.234	0.148
2.5	0.478	0.266	0.148

(Contd ...)

Initial $n_{N_2O_4}$	$P_{N_2O_4}$ (atm)	P_{NO_2} (atm)	K_p
3	0.586	0.294	0.148
3.5	0.695	0.320	0.148
4	0.805	0.345	0.148
4.5	0.916	0.368	0.148
5	1.027	0.389	0.148

It is very interesting to note the functional form of the equilibrium constant. The product NO_2 pressure appears in the numerator, and the exponent 2 on the pressure is the stoichiometric coefficient on NO_2 in the balanced chemical equation. The reactant N_2O_4 pressure appears in the denominator, and the exponent 1 on the pressure is the stoichiometric coefficient on N_2O_4 in the chemical equation.

We now investigate whether other reactions have equilibrium constants and whether the form of this equilibrium constant is a happy coincidence or a general observation. We return to the reaction for the synthesis of ammonia. In a previous section, we considered only the equilibrium produced when 1 mole of N_2 is reacted with 3 moles of H_2.

We now consider a range of possible initial values of these amounts, with the resultant equilibrium partial pressures given in Table 3.4. In addition, anticipating the possibility of an equilibrium constant, we have calculated the ratio of partial pressures given by:

$$K_p = \frac{P_{NH_3}^{\,2}}{P_{N_2}\, P_{H_2}^{\,3}} \qquad \qquad \text{... (3.12)}$$

In Table 3.4, the equilibrium partial pressures of the gases are in a very wide variety, including whether the final pressures are greater for reactants or products. However, from the data in Table 3.4, it is clear that, despite these variations, K_p in Eq. 3.12 is essentially a constant for all of the initial conditions examined and is thus the reaction equilibrium constant for this reaction.

Table 3.4. Equilibrium partial pressures of the synthesis of ammonia.

V (L)	n_{N_2}	n_{H_2}	P_{N_2}	P_{H_2}	P_{NH_3}	K_p
10	1	3	0.0342	0.1027	4.82	6.2×10^5
10	0.1	0.3	0.0107	0.0322	0.467	6.0×10^5
100	0.1	0.3	0.00323	0.00968	0.0425	6.1×10^5
100	3	3	0.492	0.00880	0.483	6.1×10^5
100	1	3	0.0107	0.0322	0.467	6.0×10^5
1000	1.5	1.5	0.0255	0.00315	0.0223	6.2×10^5

Studies of many chemical reactions of gases result in the same observations. Each reaction equilibrium can be described by an equilibrium constant in which the partial pressures of the products, each raised to their corresponding stoichiometric coefficient, are multiplied together in the numerator, and the partial pressures of the reactants, each raised to their corresponding stoichiometric coefficient, are multiplied together in the denominator. For historical reasons, this general observation is sometimes referred to as the law of mass action.

Observation 3: Temperature Dependence of the Reaction Equilibrium

We have previously observed that phase equilibrium, and in particular vapour pressure, depend on the temperature, but we have not yet studied the variation of reaction equilibrium with temperature. We focus our initial study on this reaction and we measure the equilibrium partial pressures at a variety of temperatures. From these measurements, we can compile the data showing the temperature dependence of the equilibrium constant K_p for this reaction in Table 3.5.

Table 3.5. Equilibrium constant for the synthesis of HI.

$T(K)$	K_p
500	6.25×10^{-3}
550	8.81×10^{-3}
650	1.49×10^{-2}
700	1.84×10^{-2}
720	1.98×10^{-2}

Note that the equilibrium constant increases dramatically with temperature. As a result, at equilibrium, the pressure of HI must also increase dramatically as the temperature is increased.

These data do not seem to have a simple relationship between K_p and temperature. We must appeal to arguments based on Thermodynamics, from which it is possible to show that the equilibrium constant should vary with temperature according to the following equation:

$$\ln K_p = -\left(\frac{\Delta H^\circ}{RT}\right) + \frac{\Delta S^\circ}{R} \qquad \dots (3.13)$$

If ΔH° and ΔS° do not depend strongly on the temperature, then this equation would predict a simple straight line relationship between $\ln K_p$ and $1/T$. In addition, the slope of this line should be $-(\Delta H^\circ/R)$. We test this possibility with the graph in Fig. 3.15.

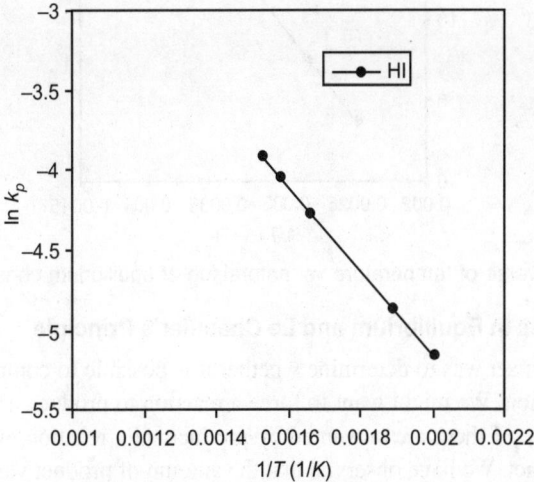

Fig. 3.15. Inverse of temperature vs natural log of equilibrium constant for HI

In fact, we do observe a straight line through the data. In this case, the line has a negative slope. Note carefully that this means that K_p is increasing with temperature. The negative slope via Eq. 3.13 means that $-(\Delta H^\circ/R)$ must be negative, and indeed for this reaction in this temperature range, $\Delta H^\circ = 15.6$ kJ/mol. This value matches well with the slope of the line in Fig. 3.15.

Given the validity of Eq. 3.13 in describing the temperature dependence of the equilibrium constant, we can also predict that an exothermic reaction with ΔH°; should have a positive slope in the graph of ln K_p versus, and thus the equilibrium constant should decrease with increasing temperature. A good example of an exothermic reaction is the synthesis of ammonia for which $\Delta H^\circ = -99.2$ kJ/mol. Equilibrium constant data are given in Table 3.6. Note that, as predicted, the equilibrium constant for this exothermic reaction decreases rapidly with increasing temperature. The data from Table 3.6 is shown in Fig. 3.16, clearly showing the contrast between the endothermic reaction and the exothermic reaction. The slope of the graph is positive for the exothermic reaction and negative for the endothermic reaction. From Eq. 3.13, this is a general result for all reactions.

Table 3.6. Equilibrium constant for the synthesis of ammonia.

T (K)	K_p
250	7×10^8
298	6×10^5
350	2×10^3
400	36

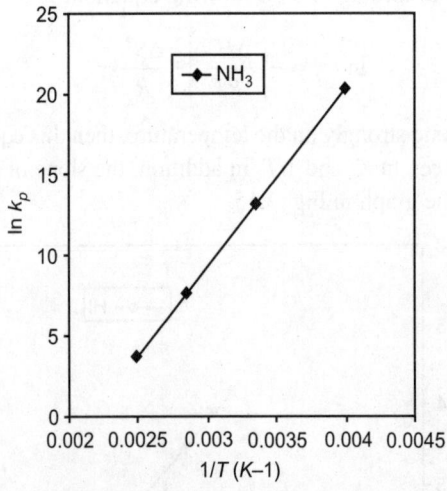

Fig. 3.16. Inverse of temperature vs. natural log of equilibrium constant for NH_3

Observation 4: Changes in Equilibrium and Le Châtelier's Principle

One of our goals at the outset was to determine whether it is possible to control the equilibrium which occurs during a gas reaction. We might want to force a reaction to produce as much of the products as possible. In the alternative, if there are unwanted by-products of a reaction, we might want conditions which minimise the product. We have observed that the amount of product varies with the quantities of initial materials and with changes in the temperature. Our goal is a systematic understanding of these variations. A look back at Table 3.2 and Table 3.3 shows that the equilibrium pressure of the product of the reaction increases with increasing the initial quantity of reaction. This seems quite intuitive. Less intuitive is the variation of the equilibrium pressure of the product of this reaction with variation in the volume of the container, as shown in Table 3.4. Note that the pressure of NH_3 decreases by more than a factor of ten when the volume is increased by a factor of ten. This means that, at equilibrium, there are fewer moles of NH_3 produced when the reaction occurs in a larger volume.

To understand this effect, we rewrite the equilibrium constant in Eq. 3.12 to explicit show the volume of the container. This is done by applying Dalton's Law of Partial Pressures, so that each partial pressure is given by the Ideal Gas Law:

$$K_p = \frac{n_{NH_3}^2 \left(\dfrac{RT}{V}\right)^2}{n_{N_2}\dfrac{RT}{V} n_{H_2}^3 \left(\dfrac{RT}{V}\right)^3}$$

$$= \frac{n_{NH_3}^2}{n_{N_2} n_{H_2}^3 \left(\dfrac{RT}{V}\right)^2} \qquad \ldots (3.14)$$

Therefore,

$$K_p \left(\frac{RT}{V}\right)^2 = \frac{n_{NH_3}^2}{n_{N_2} n_{H_2}^3} \qquad \ldots (3.15)$$

This form of the equation makes it clear that, when the volume increases, the left side of the equation decreases. This means that the right side of the equation must decrease also, and in turn, n_{N_2} must decrease while n_{N_2} and n_{H_2} must increase. The equilibrium is thus shifted from products to reactants when the volume increases for this reaction.

The effect of changing the volume must be considered for each specific reaction, because the effect depends on the stoichiometry of the reaction. One way to determine the consequence of a change in volume is to rewrite the equilibrium constant as we have done in Eq. 3.15.

Finally, we consider changes in temperature. We note that K_p increases with T for endothermic reactions and decreases with T for exothermic reactions. As such, the products are increasingly favoured with increasing temperature when the reaction is endothermic, and the reactants are increasingly favoured with increasing temperature when the reaction is exothermic. On reflection, we note that when the reaction is exothermic, the reverse reaction is endothermic. Putting these statements together, we can say that the reaction equilibrium always shifts in the direction of the endothermic reaction when the temperature is increased. All of these observations can be collected into a single unifying concept known as Le Châtelier's Principle.

Principle 1: Le Châtelier's principle

When a reaction at equilibrium is stressed by a change in conditions, the equilibrium will be reestablished in such a way as to counter the stress.

This statement is best understood by reflection on the types of 'stresses' we have considered in this section. When a reactant is added to a system at equilibrium, the reaction responds by consuming some of that added reactant as it establishes a new equilibrium. This offsets some of the stress of the increase in reactant. When the temperature is raised for a reaction at equilibrium, this adds thermal energy. The system shifts the equilibrium in the endothermic direction, thus absorbing some of the added thermal energy, countering the stress.

The most challenging of the three types of stress considered in this section is the change in volume. By increasing the volume containing a gas phase reaction at equilibrium, we reduce the partial pressures of all gases present and thus reduce the total pressure. Recall that the response of this reaction to the volume increase was to create more of the reactants at the expense of the products. One consequence of this shift is that more gas molecules are created, and this increases the total pressure in the reaction flask. Thus, the reaction responds to the stress of the volume increase by partially offsetting the pressure decrease with an increase in the number of moles of gas at equilibrium.

Le Châtelier's principle is a useful mnemonic for predicting how we might increase or decrease the amount of product at equilibrium by changing the conditions of the reaction. From this principle, we can predict whether the reaction should occur at high temperature or low temperature, and whether it should occur at high pressure or low pressure.

MICROREACTOR

A microreactor or microstructured reactor or microchannel reactor is a device in which chemical reactions take place in a confinement with typical lateral dimensions below 1 mm; the most typical form of such confinement are microchannels. Microreactors are studied in the field of micro process engineering, together with other devices (such as micro heat exchangers) in which physical processes occur. The microreactor is usually a continuous flow reactor (contrast with/to a batch reactor). Microreactors offer many advantages over conventional scale reactors, including vast improvements in energy efficiency, reaction speed and yield, safety, reliability, scalability, on-site/on-demand production, and a much finer degree of process control. These are reactors with around 1 mm size micro channels, micro heat exchangers, mixers, etc. and the entire system can be accommodated in a briefcase. As against in batch reactors, reactions are carried out continuously in microreactors.

Types of Reactors

Conventionally, following types of reactors are used in the chemical industry:
1. Tank reactors including continuous stirred tank reactors (CSTR).
2. Pipe reactors including plug flow reactors.
3. Continuous oscillatory baffled reactors.
4. Fluidised bed reactors.
5. Falling film reactors.
6. Falling film microreactors.
7. Micro-reactors, including T-type reactors.

Advantages of Microreactors

Microreactors are energy efficient; give high yield/mass transfer due to large surface to volume ratio; and can process unstable intermediates. They can have better selectivity; can be operated at temperatures beyond the boiling point through pressure build-up; and facilitate more efficient synthesis. One can have online purification, fast and exothermic reactions.

There are engineering advantages: with small holdup of hazardous reactants, one can avoids batch delays, and consequently decay.

Disadvantages of Microreactors

Microreactors do have some disadvantages. They include:
1. Particulate material is not tolerated as it results in clogging.
2. Lack of separation.
3. Easy to carry and hence terrorists can use it to manufacture chemical weapons (e.g. methyl isocyanate and nerve gases like Sarin, Soman, etc.).
4. Their capital cost is high and therefore use is unlikely for commodity chemical manufacture.

Microreactor Based All Purpose R&D Plant

The basic application of microreactors is in R & D and complete plants are available with pumps, thermostats, sensors, measurement and control units, sample and dosing units, analytical devices, etc. One can also set up a microreactor system on a small base plate with attachments as per requirement. The attachments, among others, include mixers, heat exchangers, reactors, filters, sensors, actuators, neutral bows, inlets and outlets, peripherals, lab automation, thermostats, laboratory pumps, analytics, comb-shaped micro-mixers, slit-plate micromixers, split and recombine micromixers for liquid-liquid mixing, liquid-gas mixing, emulsification, etc.

A number of different type of reactors are also available, including sandwich reactor, cartridge reactor, cascade reactor (fluidic tempered), etc.

Microreactors for Research

Microreactors have proved a boon to academic research. Due to small size and controlled mixing in the reactors, reaction kinetics can be effectively and accurately studied in a laboratory with real time monitoring of reactors using sensors and analytical probes.

Microreactors are also widely used for synthesis of oligomers and production of hydrogen for micro fuel cells, in biochemistry of enzymes, polymerase chain reactions, etc.

Applications

Synthesis

Microreactors can be used to synthesise material more effectively than current batch techniques allow. The benefits here are primarily enabled by the mass transfer, thermodynamics, and high surface area to volume ratio environment as well as engineering advantages in handling unstable intermediates. Microreactors are applied in combination with photochemistry, electrosynthesis, multicomponent reactions and polymerisation (for example that of butyl acrylate). It can involve liquid-liquid systems but also solid-liquid systems with for example the channel walls coated with a heterogeneous catalyst. Synthesis is also combined with online purification of the product.

Following green chemistry principles, microreactors can be used to synthesise and purify extremely reactive organometallic compounds for ALD and CVD applications, with improved safety in operations and higher purity products.

In one microreactor study a Knoevenagel condensation was performed with the channel coated with a zeolite catalyst layer which also serves to remove water generated in the reaction:

$$34 \times (0.3 \times 0.6 \times 25) \text{ mm}$$
$$0.2\text{--}12 \text{ ml/hr}$$

60% conv
3.5 hrs

A Suzuki reaction was examined in another study with a palladium catalyst confined in a polymer network of polyacrylamide and a triarylphosphine formed by interfacial polymerisation:

$$1 \times (0.1 \times 0.04 \times 140) \text{ mm}$$

The combustion of propane was demonstrated to occur at temperatures as low as 300°C in a micro-channel setup filled up with an aluminium oxide lattice coated with a platinum/molybdenum catalyst:

$$C_3H_8 + 5O_2 \xrightarrow[\substack{300°C \\ 14 \times (0.5 \times 0.25 \times 0.025) \text{ mm} \\ 300 \text{ dm}^3/\text{hr. g}}]{\text{Pt/Mo cat. on } Al_2O_3} 3CO_2 + 4H_2O$$

100% conv.

Analysis

Microreactors can also enable experiments to be performed at a far lower scale and far higher experimental rates than currently possible in batch production, while not collecting the physical experimental output. The benefits here are primarily derived from the low operating scale, and the integration of the required sensor technologies to allow high quality understanding of an experiment. The integration of the required synthesis, purification and analytical capabilities is impractical when operating outside of a microfluidic context.

NMR

Researchers at the Radboud University Nijmegen and Twente University, the Netherlands, have developed a microfluidic high-resolution NMR flow probe. They have shown a model reaction being followed in real-time. The combination of the uncompromised (sub-Hz) resolution and a low sample volume can prove to be a valuable tool for flow chemistry.

Membrane Bioreactor

Membrane bioreactor (MBR) is the combination of a membrane process like microfiltration or ultrafiltration with a suspended growth bioreactor, and is now widely used for municipal and industrial waste-water treatment with plant sizes up to 80,000 population equivalent (i.e. 48 MLD). A simple membrane bioreactor is shown in Fig. 3.17.

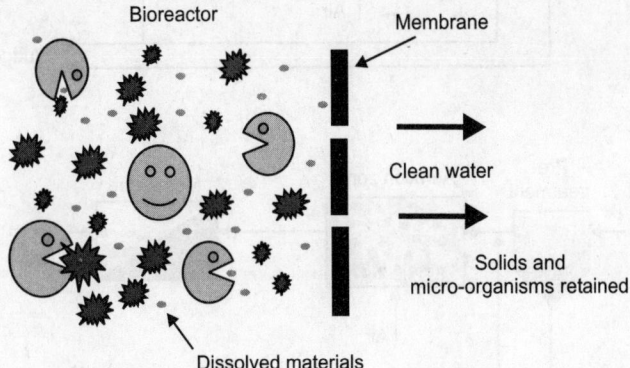

Fig. 3.17. Simple schematic diagram describing the MBR process.

When used with domestic waste-water, MBR processes could produce effluent of high quality enough to be discharged to coastal, surface or brackish waterways or to be reclaimed for urban irrigation. Other advantages of MBRs over conventional processes include small footprint, easy retrofit and upgrade of old waste-water treatment plants. Two MBR configurations exist: internal, where the membranes are immersed in and integral to the biological reactor; and external/sidestream, where membranes are a separate unit process requiring an intermediate pumping step. A conventional activated sludge process and membrane bioreactor is shown in Fig. 3.18.

Recent technical innovation and significant membrane cost reduction have pushed MBRs to become an established process option to treat waste-waters. As a result, the MBR process has now become an attractive option for the treatment and reuse of industrial and municipal waste-waters, as evidenced by their constantly rising numbers and capacity.

Major considerations in MBR

Fouling and fouling control

The MBR filtration performance inevitably decreases with filtration time. This is due to the deposition of soluble and particulate materials onto and into the membrane, attributed to the interactions between activated sludge components and the membrane. This major drawback and process limitation has been under investigation since the early MBRs, and remains one of the most challenging issues facing further MBR development.

In recent reviews covering membrane applications to bioreactors, it has been shown that, as with other membrane separation processes, membrane fouling is the most serious problem affecting system performance. Fouling leads to a significant increase in hydraulic resistance, manifested as permeate flux decline or transmembrane pressure (TMP) increase when the process is operated under constant-TMP or constant-flux conditions respectively. Frequent membrane cleaning and replacement is therefore required, increasing significantly the operating costs.

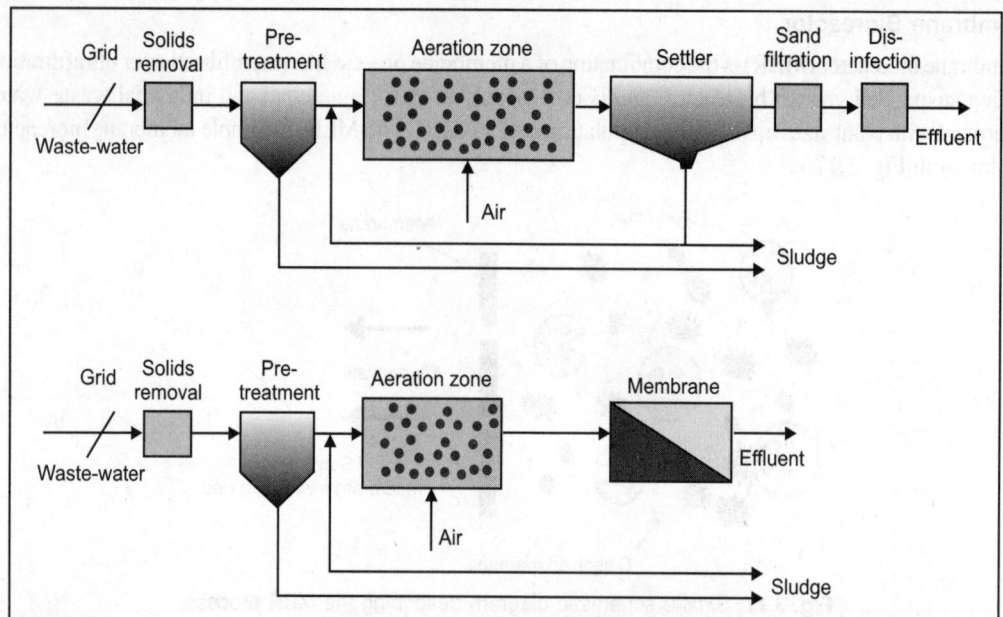

Fig. 3.18. Schematic of conventional activated sludge process (top) and membrane bioreactor (bottom).

Membrane fouling results from interaction between the membrane material and the components of the activated sludge liquor, which include biological flocs formed by a large range of living or dead micro-organisms along with soluble and colloidal compounds. The suspended biomass has no fixed composition and varies both with feed water composition and MBR operating conditions employed. Thus though many investigations of membrane fouling have been published, the diverse range of operating conditions and feedwater matrices employed, the different analytical methods used and the limited information reported in most studies on the suspended biomass composition, has made it difficult to establish any generic behaviour pertaining to membrane fouling in MBRs specifically. Factors influencing and fltration in MBR given in Fig. 3.19.

The air-induced cross flow obtained in submerged MBR can efficiently remove or at least reduce the fouling layer on the membrane surface. A recent review reports the latest findings on applications of aeration in submerged membrane configuration and describes the enhancement of performances offered by gas bubbling. As an optimal air flow-rate has been identified behind which further increases in aeration have no effect on fouling removal, the choice of aeration rate is a key parameter in MBR design.

Many other anti-fouling strategies can be applied to MBR applications. They comprise, for example:

1. Intermittent permeation, where the filtration is stopped at regular time interval for a couple of minutes before being resumed. Particles deposited on the membrane surface tend to diffuse back to the reactor; this phenomena being increased by the continuous aeration applied during this resting period.

2. Membrane backwashing, where permeate water is pumped back to the membrane, and flow through the pores to the feed channel, dislodging internal and external foulants.

3. Air backwashing, where pressurised air in the permeate side of the membrane build up and release a significant pressure within a very short period of time. Membrane modules therefore

need to be in a pressurised vessel coupled to a vent system. Air usually does not go through the membrane. If it did, the air would dry the membrane and a rewet step would be necessary, by pressurising the feed side of the membrane.

4. Proprietary anti-fouling products, such as Nalco's membrane performance enhancer technology.

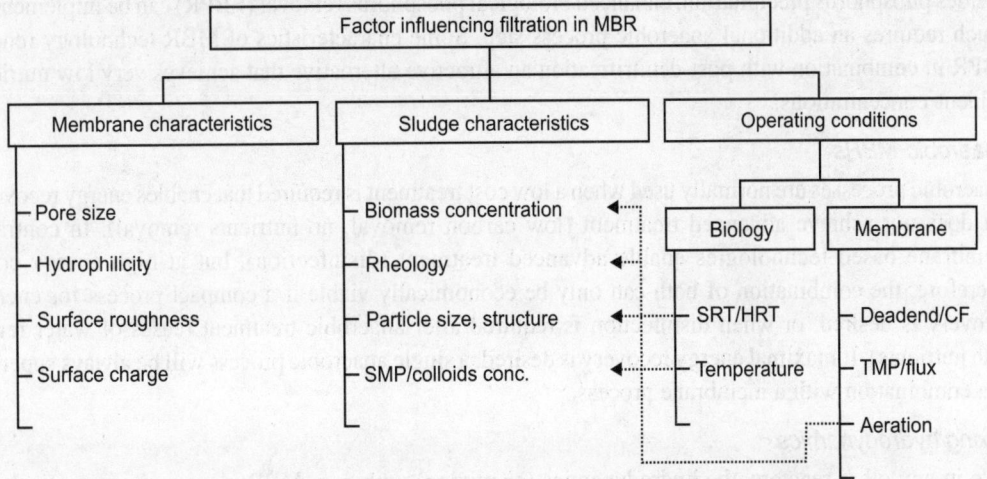

Fig. 3.19. Factors influencing fouling (interactions in dotted line).

In addition, different types/intensities of chemical cleaning may also be recommended:

1. Chemically enhanced backwash (daily).
2. Maintenance cleaning with higher chemical concentration (weekly).
3. Intensive chemical cleaning (once or twice a year).

Intensive cleaning is also carried out when further filtration cannot be sustained because of an elevated transmembrane pressure (TMP).

Biological performances/kinetics

COD removal and sludge yield

Simply due to the high number of micro-organism in MBRs, the pollutants uptake rate can be increased. This leads to better degradation in a given time span or to smaller required reactor volumes. In comparison to the conventional activated sludge process (ASP) which typically achieves 95 per cent, COD removal can be increased to 96–99 per cent in MBRs. COD and BOD5 removal are found to increase with MLSS concentration. Above 15g/L COD removal becomes almost independent of biomass concentration at >96 per cent. Arbitrary high MLSS concentrations are not employed, however, as oxygen transfer is impeded due to higher and Non-Newtonian fluid viscosity. Kinetics may also differ due to easier substrate access. In ASP, flocs may reach several 100 μm in size. This means that the substrate can reach the active sites only by diffusion which causes an additional resistance and limits the overall reaction rate (diffusion controlled). Hydrodynamic stress in MBRs reduces floc size (to 3.5 μm in sidestream MBRs) and thereby increases the apparent reaction rate. Like in the conventional ASP, sludge yield is decreased at higher SRT or biomass concentration. Little or no sludge is produced at sludge loading rates of 0.01 kgCOD/(kg MLSS d). Due to the biomass concentration limit imposed, such low loading rates would result in enormous tank sizes or long HRTs in conventional ASP.

Nutrient removal

Nutrient removal is one of the main concerns in modern waste-water treatment especially in areas that are sensitive to eutrophication. Like in the conventional ASP, currently, the most widely applied technology for N-removal from municipal waste-water is nitrification combined with denitrification. Besides phosphorus precipitation, enhanced biological phosphorus removal (EBPR) can be implemented which requires an additional anaerobic process step. Some characteristics of MBR technology render EBPR in combination with post-denitrification an attractive alternative that achieves very low nutrient effluent concentrations.

Anaerobic MBRs

Anaerobic processes are normally used when a low cost treatment is required that enables energy recovery but does not achieve advanced treatment (low carbon removal, no nutrients removal). In contrast, membrane-based technologies enable advanced treatment (disinfection), but at high energy cost. Therefore, the combination of both can only be economically viable if a compact process for energy recovery is desired, or when disinfection is required after anaerobic treatment (cases of water reuse with nutrients). If maximal energy recovery is desired, a single anaerobic process will be always superior to a combination with a membrane process.

Mixing/hydrodynamics

Like in any other reactors, the hydrodynamics (or mixing) within an MBR plays an important role in determining the pollutant removal and fouling control within an MBR. It has a substantial effect on the energy usage and size requirements of an MBR, therefore the whole life cost of an MBR. The removal of pollutants is greatly influenced by the length of time fluid elements spend in the MBR (i.e. the residence time distribution or RTD). The residence time distribution is a description of the hydrodynamics/mixing in the system and is determined by the design of the MBR (e.g. MBR size, inlet/recycle flowrates, wall/baffle/mixer/aerator positioning, mixing energy input). An example of the effect of mixing is that a continuous stirred-tank reactor will not have as high pollutant conversion per unit volume of reactor as a plug flow reactor. The control of fouling, as previously mentioned, is primarily undertaken using coarse bubble aeration. The distribution of bubbles around the membranes, the shear at the membrane surface for cake removal and the size of the bubble are greatly influenced by the mixing/hydrodynamics of the system. The mixing within the system can also influence the production of possible foulants. For example, vessels not completely mixed (i.e. plug flow reactors) are more susceptible to the effects of shock loads which may cause cell lysis and release of soluble microbial products.

Many factors affect the hydrodynamics of waste-water processes and hence MBRs. These range from physical properties (e.g. mixture rheology and gas/liquid/solid density, etc.) to the fluid boundary conditions (e.g. inlet/outlet/recycle flowrates, baffle/mixer position, etc.). However, many factors are peculiar to MBRs, these cover the filtration tank design (e.g. membrane type, multiple outlets attributed to membranes, membrane packing density, membrane orientation, etc.) and it is operation (e.g. membrane relaxation, membrane back flush, etc.). The mixing modelling and design techniques applied to MBRs are very similar to those used for conventional activated sludge systems. They include the relatively quick and easy compartmental modelling technique which will only derive the RTD of a process (e.g. the MBR) or the process unit (e.g. membrane filtration vessel) and relies on broad assumptions of the mixing properties of each sub unit. Computational fluid dynamics modelling (CFD) on the other hand does not rely on broad assumptions of the mixing characteristics and attempts to predict the hydrodynamics from a fundamental level. It is applicable to all scales of fluid flow and can reveal much information about the mixing in a process, ranging from the RTD to the shear profile on a membrane surface.

ANALYSIS OF THERMAL EFFECTS IN A CSTR REACTOR

Balance Equations

Consider a CSTR reactor with a single inlet feed stream and a outlet feed stream. Let the volume of the reactor be V a constant. The macroscopic species balance equation for a control volume V is:

$$\frac{d}{dt}\int_V C_i\ dV + \int_A C_i\ v\cdot n\ dA = \int_v R_i\ dV \qquad \dots (3.16)$$

We will assume that in volumetric flowrate in and out of the reactor are equal. Thus evaluating the integrals gives:

$$V\frac{dC_i}{dt} = q\left(C_i^0 - C_i\right) + VR_i \qquad \dots (3.17)$$

Here C_i^0 is the concentration of species i in the inlet stream. For a well mixed reactor the outlet concentration is C_i. Note a more precise version of the above equation would involve average quantities for the volume integrals:

$$\langle R_i \rangle = \frac{1}{V}\int_v R_i\ dV, \langle C_i \rangle = \frac{1}{V}\int_v R_i\ dV \qquad \dots (3.18)$$

and an appropriate area average quantities at the inlet and exit streams. For example, at the pth inlet/exit stream we can define the following area average quantity by:

$$\left[C_i\ v_n\right]_p = \frac{1}{A_p}\int_{A_p} C_i\ v\cdot n\ dA, \qquad \dots (3.19)$$

where, v_n is the noramla component of the velocity at the pth inlet/exit. If the flow rate q at the pth inlet/exit is constant, we can approximate the above integral as:

$$q\left[C_i\right]_p = \int_{A_p} C_i\ v\cdot n\ dA \qquad \dots (3.20)$$

This level of detail is usually not needed if we assume the reactor is well mixed. Thus for the rest of these notes we will dispense with writing down average quantities.

Suppose we have the following single reaction with an intrinsic rate of r:

$$v_A A + v_B B \rightarrow v_c C + v_D D \qquad \dots (3.21)$$

Then the species balance becomes:

$$V\frac{dC_i}{dt} = q\left(C_i^0 - C_i\right) + V\ v_i\ r \qquad \dots (3.22)$$

If we have p multiple reactions the species balance is:

$$V\frac{dC_i}{dt} = q\left(C_i^0 - C_i\right) + V\sum_{j=1}^{p} v_{ij}\ r_j \qquad \dots (3.23)$$

where, r_j is the intrinsic rate of the jth reaction. Recall that the use of the intrinsic rate r is simply a bookkeeping device to account for the stoichiometry of the reaction as:

$$R_A = v_A \, r, \, R_B = v_B \, r \qquad \qquad \text{... (3.24)}$$

such that:

$$\frac{R_A}{R_B} = \frac{v_A}{v_B} = \frac{R_A}{R_C} = \frac{v_A}{v_C} = \frac{R_A}{R_D} = \frac{v_A}{v_D}, \text{ etc.} \qquad \qquad \text{... (3.25)}$$

If there are multiple reactions then r_j is the instrinsic reaction of the jth stoichiometric schema.

CSTR Reactor: Energy Balance

As our axiom for the thermal energy balance we will assume that:

$$\begin{pmatrix} \text{Local rate of change} \\ \text{of enthalpy/volume} \end{pmatrix} + \begin{pmatrix} \text{Net flux of enthalpy} \\ \text{per unit volume} \end{pmatrix} = -\begin{pmatrix} \text{Net heat flux } q \\ \text{per unit volume} \end{pmatrix} \qquad \text{... (3.26)}$$

This is a very simplified version of the energy equation. It ignores viscous dissipation and work of compression. This is a reasonable assumption for a constant volume CSTR where the work of mixing is negligible. The word form for the thermal energy balance can be stated as the following point equation:

$$\frac{\partial}{\partial t}\left(\sum_{i=1}^{N} C_i \, \bar{H}_i \right) + \nabla \cdot \left\{ v \sum_{i=1}^{N} C_i \, \bar{H}_i \right\} = \nabla \cdot q \qquad \qquad \text{... (3.27)}$$

Here $\bar{H}_i$ is the partial molar enthalpy for the ith species evaluated at T, the temperature of the reactor taken to be constant and equal to the temperature of the exit stream, Next we integrate this expression over the control volume V (taken to be fixed) of our reactor:

$$\frac{d}{dt} \int_v \left(\sum_{i=1}^{N} C_i \, \bar{H}_i \right) dV + \int_v \left\{ \nabla \cdot v \left(\sum_{i=1}^{N} C_i \, \bar{H}_i \right) \right\} dV = -\int_v \nabla \cdot q \, dV \qquad \text{... (3.28)}$$

Then we apply the divergence theorem to the last two integrals to get:

$$\frac{d}{dt} \int_v \left(\sum_{i=1}^{N} C_i \, \bar{H}_i \right) dV + \int_{A_p} v \left(\sum_{i=1}^{N} C_i \, \bar{H}_i \right) \cdot n_p dA = -\int_A q \cdot n \, dA \qquad \text{... (3.29)}$$

Next we evaluate the surface integrals to get:

$$\frac{d}{dt} \int_v \left(\sum_{i=1}^{N} C_i \, \bar{H}_i \right) dv = q_{in} \left(\sum_{i=1}^{N} C_i^0 \, \bar{H}_i^0 \right) - q_{out} \left(\sum_{i=1}^{N} C_i \, \bar{H}_i \right) + \int_A k \nabla T \cdot n \, dA \qquad \text{... (3.30)}$$

Note we have assumed Fourier's law so that $q = -k \, \nabla T$. Here $\bar{H}_i^0$ is the partial molar enthalpy of the ith species entering the reactor at temperature T_0. Note that the exit stream has the same temperature T as the contents of the reactor which is different from the inlet stream temperature T_0. Again we take the flow rates to be equal so that $q_{in} = q_{out} = q$. Our thermal energy balance becomes.

$$v \frac{d}{dt} \left(\sum_{i=1}^{N} C_i \, \bar{H}_i \right) = q \left\{ \left(\sum_{i=1}^{N} C_i^0 \, \bar{H}_i^0 \right) - \left(\sum_{i=1}^{N} C_i \, \bar{H}_i \right) \right\} + \int_A k \nabla T \cdot n \, dA \qquad \text{... (3.31)}$$

where we have made use of volume averaging to simplify the integrals. The last term on the RHS of Eq. 3.31 is evaluated on the surface of our reactor. So:

$$\dot{Q} = \int_A q \cdot n dA = -\int_A k\nabla T \cdot n \, dA \qquad \ldots (3.32)$$

is the net rate of heat loss from the reactor. If U is the overall heat transfer coefficient for the reactor we can express this result as:

$$\dot{Q} = AU \, (T - T_\infty) \qquad \ldots (3.33)$$

where, T_∞ is a suitable reference temperature. Equation 3.32 must be solved in conjunction with the species balance equations:

$$V \frac{dC_i}{dt} = q\left(C_i^0 - C_i\right) + V \, v_i \, r \qquad \ldots (3.34)$$

Steady-state Analysis

Let us now consider steady-state operation of our continuous flow reactor. At steady-state the species concentration C_i and partial molar enthalpies H_i do not change with time. Thus:

$$\frac{dC_i}{dt} = 0, \quad \frac{d}{dt}\left(\sum_{i=1}^N C_i \, \bar{H}_i\right) = 0 \qquad \ldots (3.35)$$

Thus the design equation, for the reactor are:

$$q\left(C_i^0 - C_i\right) + V \, v_i \, r = 0$$

$$q\left\{\left(\sum_{i=1}^N C_i^0 \, \bar{H}_i^0\right) - \left(\sum_{i=1}^N C_i \, \bar{H}_i\right)\right\} = -\dot{Q} \qquad \ldots (3.36)$$

At this point we can use the species balance equation to eliminate C_i from the thermal energy balance. Solving for C_i gives:

$$C_i = C_i^0 + \frac{V}{q} v_i \, r \qquad \ldots (3.37)$$

Then substituting this result into the thermal energy balance gives:

$$q \sum_{i=1}^N C_i^0 \left(\bar{H}_i^0 - \bar{H}_i\right) - V\left(\sum_{i=1}^N v_i \, \bar{H}_i\right) r = -\dot{Q} \qquad \ldots (3.38)$$

We can simplify this expression by invoking the definition of the heat of reaction:

$$\Delta H_{rxn} = \sum_{i=1}^N v_i \, \bar{H}_i \qquad \ldots (3.39)$$

The final result is:

$$q \sum_{i=1}^N C_i^0 \left(\bar{H}_i - \bar{H}_i^0\right) + V \, \Delta H_{rxn} \, r = \dot{Q} \qquad \ldots (3.40)$$

Note that the heat of reaction is evaluated at the temperature of the exit stream T, not the inlet stream temperature T_0. If we have p multiple independent reactions taking place then we need to account for the heats of reaction for the independent stoichiometric schema. The result is:

$$q \sum_{i=1}^{N} C_i^0 \left(\bar{H}_i - \bar{H}_i^0 \right) + V \sum_{j=1}^{P} \left(\Delta H_{rxn, j} \right) r_j = \dot{Q} \qquad \text{... (3.41)}$$

SOLVED EXAMPLES

Example 3.1. Calculate the time required to convert 80 per cent of urea to NH_3 and CO_2 in a 0.5 litre batch reactor. The initial concentration of urea being 0.1 mole/litre and that of urease is 0.001 gm/litre. The reaction is to be carried out isothermally. Take $K_m = 0.0266$ g·mole/litre.

Solution

$$t = \frac{K_M}{V_{max}} \ln \frac{1}{1 - X_A} + \frac{C_{urea,0} X_A}{V_{max}} \qquad \text{... (3.42)}$$

$$\therefore \qquad V_{max} = 1.33 \frac{g.\,mole}{litre}$$

As $V_{max} = E_t . k_3$, V_{max} for the second reaction concentration is:

$$V_{max\,2} = \frac{E_{t2}}{E_{t1}} V_{max\,1} = \frac{0.001}{5} \times 1.33$$

$$= 2.66 \times 10^{-4} \text{ mole/litre sec.}$$

$$t = \frac{0.0266}{0.000266} \ln \frac{1}{0.2} + \frac{(0.8)(0.1)}{0.000266}$$

$$= 160.9 + 300.8 = 461.7 \text{ seconds}$$

$$= 7 \text{ minutes } 41.7 \text{ seconds.}$$

Example 3.2. The first-order isomerisation $A \rightarrow R$ is being carried out isothermally in a batch reactor on a catalyst which is decaying due to ageing. Derive an equation for conversion as a function of time.

Solution

Design equation

$$N_{A0} \frac{dX_A}{dt} = -r_A{}'W \qquad \text{... (3.43)}$$

Reaction rate equation

$$-r_A' = k' C_A \alpha(t) \qquad \text{... (3.44)}$$

Activity equation

$$\alpha(t) = \frac{1}{1 + k_d t} \qquad \text{... (3.45)}$$

Stoichiometric equation

$$C_A = C_{A0}(1 - X_A) = \frac{N_{A0}}{V}(1 - X_A) \qquad \text{... (3.46)}$$

Combining all the above equations give:

$$\frac{dX_A}{dt} = \frac{W}{V}k'\alpha(t)(1-X_A)$$ ... (3.47)

Let

$$K = k'\frac{W}{V}$$

∴

$$\frac{dX_A}{1-X_A} = K\alpha(t).dt$$ ... (3.48)

or

$$\int_0^{X_A} \frac{dX_A}{1-X_A} = k\int_0^t \frac{dt}{1-k_dt}$$

$$\ln\frac{1}{1-X_A} = \frac{k}{k_d}\ln(1+k_dt)$$

or

$$X_A = \left[\frac{1}{(1+k_dt)^{k/k_d}}\right].$$

Example 3.3. Nitrogen and oxygen are reacting in a small batch reactor to produce nitrous oxide. At 2700 K and pressure 20 atm. If 80 per cent of equilibrium is reached in 150 micro second, determine the forward reaction rate constant, consider the initial concentration of $N_2 = 77$ per cent, $O_2 = 15$ per cent and remaining inert gases.

Solution

Consider the reaction be

$$N_2 + O_2 \rightleftharpoons 2NO$$ ... (3.49)

$$A + B \rightleftharpoons 2R$$ ... (3.50)

For 80 per cent of equilibrium conversion

$$X_A = 0.8\,(X_e)$$
$$= 0.016$$

For the batch reactor the design equation gives

$$k = \frac{1}{tC_{A0}}\int_0^{X_A} \frac{dX_A}{(1-X_A)(M-X_A) - \frac{4X_A^2}{K_e}}$$ ... (3.51)

Substituting the values of the different terms and using Simpson's rules with $h = 0.008$ one get

$$k = 1.11 \times 10^4 \text{ m}^3/\text{mole sec.}$$

Example 3.4. A reaction was carried out in a batch reactor and the result are reported below. Calculate the order of reaction.

t (min)	0	10	30.0
% conversion	19.8	46.7	74.0

Solution

Assume: First-order reaction.

The rate constant for first-order reaction is given by

$$k = \frac{1}{t}\ln\left(-\frac{C_A}{C_{A0}}\right) \qquad \text{... (3.52)}$$

$$= \frac{1}{t}(-)\ln(1 - X_A) \qquad \text{... (3.53)}$$

For time = 10 minutes

$$k = \frac{1}{10}(-)\ln(1 - X_A)$$

$$= \frac{1}{10}(-)\ln(1 - 0.467) = 0.088 \text{ min}^{-1}$$

For time = 30 minutes

$$k = \frac{1}{30}(-)\ln(1 - 0.74)$$

$$= 0.086 \text{ min}^{-1}.$$

Since k values in both the cases are almost same, hence assumption on first-order reaction is valid

$$\therefore \qquad k = 0.087 \text{ min}^{-1}.$$

Example 3.5. In a batch recycle reactor the dissolution of oxygen in acrilamide solution was studied. The kinetics of disappearance of oxygen with time was noted as below:

Time, sec	100	300	500	700	1000
$C_0 \times 10^4$ k·mole/m³	2.31	2.2	2.09	1.9	1.75

The following additional data are also available:

$$Q = 46.7 \times 10^{-6} \text{ m}^3/\text{sec.}$$
$$V_0 = 6.6 \times 10^{-3} \text{ m}^3.$$
$$V_R = 0.232 \times 10^{-3} \text{ m}^3.$$
$$C_0 = 2.4 \times 10^{-4} \text{ k·mol/m}^3 \text{ at } t = 0.$$

From the available informations calculate:

1. Rate of reaction as a function of oxygen concentration.
2. Concentration difference through reactor.

Solution

Let us plot the concentration of oxygen with time to get slope to predict the order of reaction, as shown in Fig. 3.20.

Since slope is constant it means zero order reaction dependency on oxygen concentration over the range of oxygen concentration

$$(-r_A) = \frac{(0.232 + 6.60) \times 10^{-3}}{(0.232 \times 10^{-3})}(-6.79 \times 10^{-8})$$

$$\therefore$$

$$= -2.0 \times 10^{-6} \frac{k \cdot \text{mole}}{m^3 \sec}.$$

$$C_0 - C = -\frac{V_R}{Q} \cdot \frac{V_0}{V_R + V_0} (-r_A)$$

$$= -\frac{0.232 \times 10^{-3}}{46.7 \times 10^{-6}} \cdot \frac{6.6 \times 10^{-3}}{(0.232 + 6.6)10^{-3}} (-2.0 \times 10^{-6})$$

$$= 9.6 \times 10^{-6} \, k \cdot \text{mole/m}^3.$$

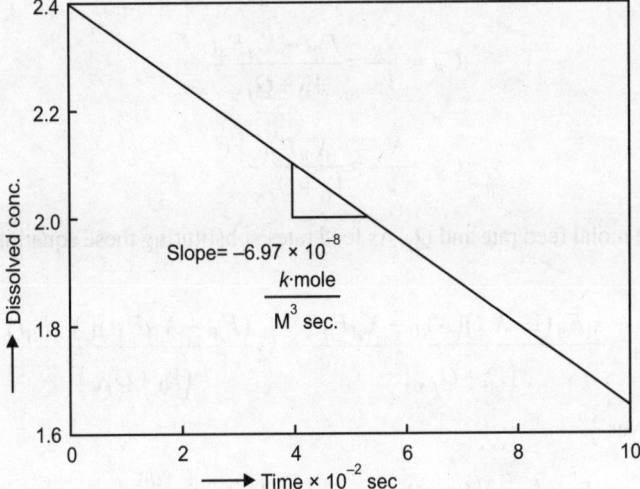

Fig. 3.20. Concentration of oxygen with time.

Note: This is very small compared to C_0. Therefore, differential reactor operation will be preferred. The highest conversion/pass may be achieved if C_0 is lowest at this jucture.

$$(\Delta X_A)_{\text{pass}} = -\frac{C_0 - C}{C_0} = \frac{9.6 \times 10^{-6}}{1.73 \times 10^{-4}} = 0.056 \text{ or } 5.6\%.$$

Very low conversion.

Example 3.6. The esterification of acetic acid and ethyl alcohol is to be carried out in a semi batch CSTR at constant temperature of 100°C. Initially alcohol is added as 400 gm pure C_2H_5OH. The aqueous solution of acetic acid is followed at a rate of 3.92 g/min. for 120 minutes. The solution contains 42.6 per cent by weight acid. The following reversible reaction takes place with specific reaction rate as below:

$$CH_3COOH + C_2H_5OH \rightleftharpoons CH_3COOC_2H_5 + H_2O$$

$$A + B \rightleftharpoons R + S$$

$k_1 = 4.76 \times 10^{-4}$ litre/gm mol. min.
$k_2 = 1.63 \times 10^{-4}$ litre/gm mol. min.
Compute the conversion of acetic acid as a function of time for complete reaction.

Solution

$$N_A = F_A t (1 - X_A) \qquad \ldots (3.54)$$

For the above reaction

$$-r_A = k_1 C_A C_B - k_2 C_R \cdot C_S \qquad \ldots (3.55)$$

Thus the concentration in the reactor

$$C_A = \frac{N_A}{V} = \frac{F_A t (1 - X_A)}{V_0 + Q_{f,t}} \qquad \ldots (3.56)$$

$$C_B = \frac{N_B}{V} = \frac{F_{B0} - X_A F_A t}{V_0 + Q_{f\tau}} \qquad \ldots (3.57)$$

$$C_R = \frac{N_R}{V} = \frac{F_R t - X_A F_A t}{V_0 + Q_{f\tau}} \qquad \ldots (3.58)$$

$$C_S = \frac{N_S}{V} = \frac{X_A F_A t}{V_0 + Q_{f\tau}} \qquad \ldots (3.59)$$

Here F represents the molal feed rate and Q_f. As feed rates substituting these equations for concentration in Eq. 3.55 one gets:

$$-r_A = \frac{k_1 F_A (1 - X_A)(N_{B0} - X_A F_A t)}{(V_0 + Q_{f\tau})^2} - k_2 \frac{(F_A t + X_A F_A t)(X_A F_A t)}{(V_0 + Q_{f\tau})^2} \qquad \ldots (3.60)$$

The mass balance gives

$$F_A + (-r_A)(V_0 + Q_{f\tau}) = F_A (1 - X_A) - F_A t \frac{dX_A}{dt} \qquad \ldots (3.61)$$

or

$$\frac{dX_A}{dt} = -r_A = \left\{ \frac{-r_A (V_0 + Q_{ft})}{F_A} - X_A \right\} \qquad \ldots (3.62)$$

which can be solved by Runge-Kutta method

or
$$V = 6.69 + 0.0656t.$$

The variation of V and $(-r_A)$ is tabulated in Table 3.7.

Table 3.7. Calculated results for semi-batch C.S.T.R. design.

t, min	V, cm³	Conversion, X_A	$(-r_A) \dfrac{\text{gm mole}}{\text{min cm}^3} \times 10^4$
0	6.69	0.000	0.000
5	7.02	0.024	1.820
10	7.34	0.045	3.200
20	8.00	0.083	5.200

(Contd ...)

t, min	V, cm³	Conversion, X_A	$(-r_A)\dfrac{\text{gm mole}}{\text{min cm}^3}\times 10^4$
40	9.31	0.141	7.000
60	10.60	0.184	7.300
80	11.90	0.215	6.900
100	13.20	0.240	6.300
120	14.50	0.259	5.700

Note: $(-r_A)$ increases with time from 0 to 60 minutes and then decreases continuously. This gives a feeling that acetic acid is not reacting as fast as added. This may be due to the relatively high feed rate of acid.

Example 3.7. In a batch reactor liquid A decomposes by first-order kinetics. The conversion is 50 per cent of A in a 5-minute run. What will be the time taken for 80 per cent conversion of A?

Solution:

Data: Batch reactor, liquid phase (constant density system), 1st order:

$$t = C_{Ao} \int_0^{X_A} \frac{dX_A}{(-r_A)}; -r_A = K C_A = K C_{Ao}(1 - X_A)$$

$$t = \frac{C_{Ao}}{K C_{Ao}} \int_0^{X_A} \frac{dX_A}{(1 - X_A)}$$

$$t = \frac{1}{K} \ln \frac{1}{(1 - X_A)}$$

Case 1: $5 = \dfrac{1}{K} \ln \dfrac{1}{(1-0.5)}$ $\therefore K = \dfrac{1}{7.21}\text{min}^{-1}$

$$= 0.1386 \text{ min}^{-1}$$

Case 2: $t = \dfrac{1}{0.1386} \ln \dfrac{1}{(1-0.8)}$

∴ $t = 11.8$ min.

Example 3.8. A zero order homogeneous gas-reaction $A \rightarrow rR$ proceeds in a constant volume bomb, 20 per cent inerts, and the pressure rises from 1.0 to 1.3 atmosphere in 2 minutes. If the same reaction takes place in a constant pressure batch reactor, what is the fractional volume change in 4 minutes, if the feed is at 3-atmosphere and consists of 40 per cent inerts?

Solution: Data; zero order, gas phase

Case 1: Constant volume

$$-r_A = \frac{-dC_A}{dt} = K$$

$$-\int_{C_{Ao}}^{C_A} d\,C_A = K\int_0^t dt$$

$$C_{Ao} - C_A = kt$$

$$P_{Ao} - P_A = (KRT)t \qquad \qquad \text{... (3.63)}$$

Also,
$$P_A = P_{Ao} - \frac{\alpha}{\Delta n}(\pi - \pi_o) \qquad \qquad \text{... (3.64)}$$

From Eq. 3.63 and 3.64

$$\frac{\alpha}{\Delta n}(\pi - \pi_o) = (KRT)t \qquad \qquad \text{... (3.65)}$$

As per the stoichiometry

$$A + inert \rightarrow rR + inert$$

$$(0.8 + 0.2) \quad (0.8\ r + 0.2)$$

∴
$$\Delta n = (0.8\ r + 0.2) - (0.8 + 0.2) = 0.8\ (r - 1)$$

also,
$$\alpha = 1.0$$

Now Eq. 3.65.
$$\Rightarrow \frac{1.0}{0.8(r-1)} = \frac{2KRT}{(1.3-1)}$$

∴
$$KRT\ (r - 1) = 0.187 \qquad \qquad \text{... (3.66)}$$

Case 2: For zero order constant pressure and variable volume system:

$$-r_A = \frac{-dC_A}{dt} = K; \qquad C_A = \frac{C_{Ao}(1 - X_A)}{(1 + \varepsilon_A X_A)}$$

$$\Rightarrow C_{Ao} \int_0^{X_A} \frac{dX_A}{(1 + \varepsilon_A X_A)} = K\int_0^t dt$$

$$\Rightarrow \frac{C_{Ao}}{\varepsilon_A} \ln\ (1 + \varepsilon_A X_A) = Kt$$

$$\Rightarrow \frac{C_{Ao}}{\varepsilon_A} \ln\left(\frac{V}{U_o}\right) = Kt \qquad \qquad \text{... (3.67)}$$

In this case
$$A + inert \rightarrow rR + inert$$

$$(0.6 + 0.4) \quad (0.6r + 0.4)$$

$$\varepsilon_A = \frac{(0.6r + 0.4) - 1.0}{1.0} = 0.6(r - 1)$$

Now Eq. 3.67

$$\Rightarrow \frac{C_{Ao}}{\varepsilon_A} \ln\left(\frac{V}{v_o}\right) = Kt$$

$$\Rightarrow \frac{P_{Ao}}{RT \times 0.6(r-1)} \ln\frac{V}{v_o} = Kt$$

$$\Rightarrow \ln\frac{V}{v_o} = \frac{0.187 \times 0.6}{1.8} = 0.25$$

$$\Rightarrow V = 1.284 \, v_o$$

∴Fractional change in volume

$$= \frac{1.284 \, v_o - v_o}{v_o} = 0.284$$

$$= 28.4 \text{ per cent.}$$

Example 3.9. The gas discharged from the reactor for NH_3 oxidation is rapidly cooled for condensation of steam in it. The gas contains on mole per cent basis: 82 per cent N_2, 9 per cent NO_2, 1 per cent NO and 8 per cent O_2. The gas is oxidised to get $NO_2 : NO : : 5 : 1$ and fed to the absorption column to produce HNO_3. The temperature is maintained at 20°C by providing sufficient cooling. The consumption of gas is 10,000 m^3/hr. at pressure of 10^5 Pa. Estimate the volume of P.F.R. The reaction being carried out is: $2NO + O_2 \rightarrow 2NO_2$. Take reaction rate constant as 1.4×10^4 litre/mole^2. sec.

Solution: Design equation for P.F.R.

$$\tau = \frac{\upsilon}{V} = C_{A0} \int_{x_0}^{x_A} \frac{dX_A}{(-r_A)} \qquad \text{... (3.68)}$$

or

$$\upsilon = G_1 \int_{x_0}^{x_A} \frac{dX_A}{(-r_A)} \qquad \text{... (3.69)}$$

$$(-r_A) = kC_A^2 \, C_B$$

Component	Initial moles	Final mole
NO	0.09	$(0.09 - X_A)$
NO_2	0.01	$(0.01 + X_A)$
O_2	0.08	$(0.08 - 0.5X_A)$
N_2	0.82	(0.82)
		Total $(1 - 0.5X_A)$

The volume of $(1 - 0.5X_A)$ moles of reaction mixture at temperature 'T' and pressure 'P' is:

$$V = (1 - 0.5X_A) \times 22.4T \times \frac{1}{P} \times \frac{1}{273} \qquad \text{... (3.70)}$$

The concentration of reactants (in moles per litre):

$$C_A = \frac{0.09 - X_A}{(1 - 0.5X_A)\dfrac{22.4T}{273P}} \qquad \text{... (3.71)}$$

$$C_R = \frac{0.08 - 0.5X_A}{(1 - 0.5X_A)\frac{22.4T}{273P}} \qquad \text{... (3.72)}$$

Substituting the values of concentration in Eq. 3.68:

$$\upsilon = C_1 \int_0^{X_A} \frac{dX_A}{\left[\frac{0.05 - X_A}{(1 - 0.5X_A)\frac{22.4T}{273P}}\right]^2 \left[\frac{0.08 - 0.5X_A}{(1 - 0.5X_A)\frac{22.4T}{273P}}\right]} \qquad \text{... (3.73)}$$

or

$$\upsilon = \frac{C_1}{k} \int_0^{X_A} \frac{(0.082)^3 T^3 (1 - 0.5X_A)^3 dX_A}{P^3 (0.09 - X_A)^2 (0.08 - 0.5X_A)} \qquad \text{... (3.74)}$$

At time t the reaction mixture contains $(0.01 + X_A)$ moles of NO_2 and $(0.09 - X_A)$ moles of NO.

$$\therefore \qquad \left(\frac{0.01 + X_A}{0.09 - X_A}\right) = \frac{5}{1}. \text{ Giving } X_A = 0.0733 = 7.33\%$$

$$\text{Hence } \upsilon = \frac{G_1}{k} \int_0^{0.0733} \frac{(0.082)^3 T^3 (1 - 0.5X_A)^3 dX_A}{P^3 (0.09 X_A)^2 (0.08 - 0.5X_A)} \qquad \text{... (3.75)}$$

Substituting the values of G_1, k, T and P Eq. 3.75 reduces to:

$$\upsilon = \frac{10^4 \times (0.083)^3 \times 293^3}{3.6 \times 22.4 \times 1.4 \times 10^4} \int_0^{0.0733} \frac{(1 - 0.5X_A)^3 dX_A}{(0.09 - X_A)^2 (0.08 - 0.5X_A)} \qquad \text{... (3.76)}$$

Equation 3.76 should be integrated graphically as shown in Fig. 3.21 for selected values of X_A.

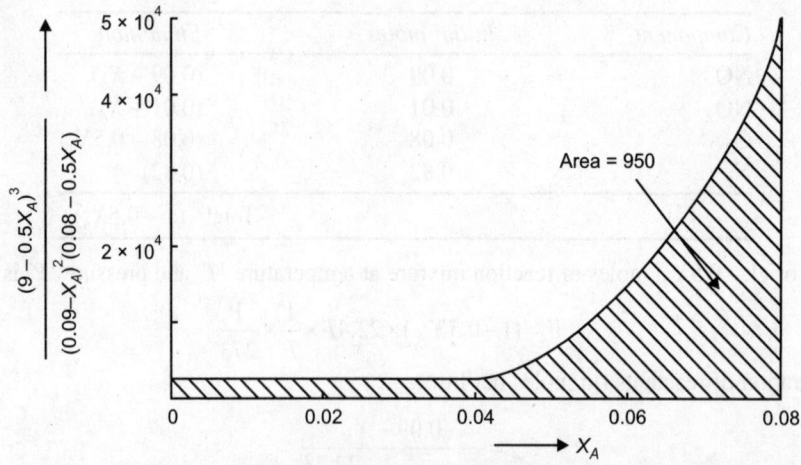

Fig. 3.21. For selected values of X_A.

$$t = \frac{u}{V}$$

$$V = \frac{v}{t} = \frac{10^4}{950} = 1.1 \times 10^5 \text{ litres}$$

$$= 110 \text{ m}^3.$$

Example 3.10. Phosphine decomposes in homogeneous gas phase as follows:

$$4PH_3 (g) \rightarrow P_4(g) + 6H_2(g)$$

$$4A \rightarrow R + 6S$$

The reaction proceeds at 650°C with rate $-r_A = (10/hr) C_A$. Calculate the size of P.F.R. needed to give 80 per cent conversion of feed at 650°C and 4.6 atm. pressure if the feed consists of 4 kg mole of pure phosphine/hour.

Solution

$$V = F_{A0} \int_0^{X_A} \frac{dX_A}{(-r_A)} \qquad \qquad \text{... (3.77)}$$

$$= F_{A0} \int_0^{X_A} \frac{dX_A}{(kC_A)} \qquad \qquad \text{... (3.78)}$$

At constant pressure

$$C_A = C_{A0} \left(\frac{1-X_A}{1+\varepsilon_A X_A} \right) \qquad \qquad \text{... (3.79)}$$

Hence

$$V = \frac{F_{A0}}{kC_{A0}} \int_0^{X_A} \left(\frac{1+\varepsilon_A X_A}{1-X_A} \right) dX_A \qquad \qquad \text{... (3.80)}$$

or

$$V = \frac{F_{A0}}{kC_{A0}} \left\{ (1+\varepsilon_A) \ln \frac{1}{1-X_A} - \varepsilon_A X_A \right\} \qquad \qquad \text{... (3.81)}$$

$$C_{A0} = \frac{p_{A0}}{R_T} = \frac{4.6 \text{ atm.}}{82.06 \dfrac{\text{cm}^3 \cdot \text{atm.}}{\text{g mole } °K} \cdot 923 \ °K}$$

$$= 6.07 \times 10^{-5} \frac{\text{g mole}}{\text{cm}^3} = 6.07 \times 10^{-2} \frac{\text{g mole}}{\text{litre}}$$

$$\varepsilon_A = \frac{7-4}{4} = 0.75$$

Substituting these values in Eq. 3.81 gives:

$$V = \frac{4 \text{ kg mole/hr}}{(10/hr) \times 6.07 \times 10^{-5} \dfrac{\text{kg mole}}{\text{litre}}} \left\{ (1+0.75) \ln \frac{1}{0.2} - 0.75 \times 0.8 \right\}$$

$$= 6589 \{1.75 \times 1.609 - 0.6\} = 14,600 \text{ litres.}$$

Example 3.11. An elementary liquid-phase reaction; A + B $\rightleftharpoons$ R + S takes place in a 150 litre capacity steady state back mixed reactor. The reaction rate constants are 7 and 3 litre/mole min for forward and backward reactions, respectively. The feed rate of A and B are 2.8 and 1.6 moles/litre, respectively. The volume of both the reactants are same. If 75 per cent conversion of limiting reactant is desired estimate the flow rates of each stream. Consider constant density reaction. The initial concentration of reactants and products are:

$$C_{A0} = 1.4 \text{ moles/litre, } C_{B0} = 0.8 \text{ moles/litre, } C_{R0} = C_{S0} = 0.$$

Solution: For 75 per cent conversion of B (limiting reactant) and $\varepsilon_A = 0$ the composition of each streams in the reactor are:

$$C_A = 1.4 - 0.6 = 0.8 \text{ moles/litre}$$

$$C_B = 0.8 - 0.6 = 0.2 \text{ moles/litre}$$

$$C_R = C_S = 0.6 \text{ moles/litre}$$

$$-r_A = -r_B = k_1 C_A C_B - k_2 C_R C_S \qquad \qquad \text{... (3.82)}$$

$$= 7 \times 0.8 \times 0.2 - 3 \times (0.6)^2$$

$$= 0.04 \text{ mole/litre min.}$$

Design equation for back mixed reactor gives:

$$\tau = \frac{V}{\upsilon} = \frac{C_{A0} - C_A}{-r_A} = \frac{C_{B0} - C_B}{-r_B} \qquad \qquad \text{... (3.83)}$$

or

$$\upsilon = \frac{V(-r_A)}{C_{A0} - C_A} = \frac{V(-r_B)}{C_{B0} - C_B} \qquad \qquad \text{... (3.84)}$$

$$= \frac{150 \times 0.04}{0.6}$$

$$= 8 \text{ litre/min.}$$

Note: 4 litre/min. of each of two feed stream.

Example 3.12. Propylene glycol is produced by the hydrolysis of propylene oxide as follows in a new glass lined 10 gallon batch reactor.

$$CH_2 - CH - CH_3 + H_2O \xrightarrow{H_2SO_4} CH_2 - CH - CH_3$$

with O bridging the first two carbons on the left, and OH groups on the first two carbons on the right.

One gallon of methanol and 5 gal of water containing 0.1 weight % H_2SO_4 was charged at 15°C. Calculate the time needed to give 51.5 per cent conversion. Estimate the required temperature.

Solution: The design equation,

$$t = C_{A0} \frac{dX_A}{-r_A} \qquad \qquad \text{... (3.85)}$$

$$-r_A = k C_A \qquad \qquad \text{... (3.86)}$$

$$C_A = C_{A0} (1 - X_A) \qquad \qquad \text{... (3.87)}$$

Combining the three equations of above:

$$\frac{dX_A}{dt} = k(1 - X_A) \qquad \qquad \text{... (3.88)}$$

$$k = \left(4.71 \times 10^9\right) \exp.\left\{\frac{-32,400}{1.987\,T}\right\} \sec^{-1} \qquad \qquad \text{... (3.89)}$$

or

$$k = \left(4.71 \times 10^9\right) \exp.\left\{\frac{32,800}{1.987}\left(\frac{1}{535} - \frac{1}{T}\right)\right\} \qquad \qquad \text{... (3.90)}$$

Integrating Eq. 3.88 with Simpson's rules and the results are tabulated in Table 3.8.

Table 3.8. Results of numerical integration.

X_A	$T°K$	$k \times 10^4\ sec^{-1}$	$\dfrac{1}{k(1 - X_A)}\,sec$
0	348	2.73	3663
0.1288	360	5.33	2154
0.2575	371	9.59	1404
0.3863	383	17.70	921
0.5150	395	32.00	644

Hence

$t = 833$ sec or 13.9 min.
$T = 67°C$

$$\tau = \frac{V}{v_0} = \frac{X_A}{k(1 - X_A)} = \frac{0.515}{\left(3.2 \times 10^{-3}\right)(0.485)} = 382 \text{ sec.}$$

Note: The residence time in the C.S.T.R. for this conversion is far less than it would be in a batch or tubular reactor.

Chapter 4

Multiphase Reactions

INTRODUCTION

Seldom is the reaction of interest the only one that occurs in a chemical reactor. Typically, multiple reactions will occur, some desired and some undesired. One of the key factors in the economic success of a chemical plant is the minimisation of undesired side reactions that occur along with the desired reaction.

In this chapter, we discuss reactor selection and general mole balances for multiple reactions. First, we describe the four basic types of multiple reactions: series, parallel, independent, and complex. Next we define the selectivity parameter and discuss how it can be used to minimise unwanted side reactions by proper choice of operating conditions and reactor selection. We then develop the algorithm that can be used to solve reaction engineering problems when multiple reactions are involved. Finally, a number of examples are given that show how the algorithm is applied to a number of real reactions.

DEFINITIONS

Types of Reactions

There are four basic types of multiple reactions: parallel, series, complex, and independent. These types of multiple reactions can occur by themselves, in pairs, or all together. When there is a combination and series reactions, they are often referred to as complex reactions.

Parallel reactions

Parallel reactions (also called competing reactions) are reactions where the reactant is consumed by two different reaction pathways to form different products:

Two reactions compete for the same reactant A. Usually, one of these products is much more valuable than the other.

Example: Oxidation reaction of ethylene to ethylene oxide:

Multiple reactions in series

Series reaction (also called consecutive reactions) are reactions where the reactant forms an intermediate product, which reacts further to form another product:

$$A \xrightarrow{k_1} B \xrightarrow{k_2} C$$

Typically, product B is much more valuable than product C:

$$A \xrightarrow{k_1} B \xrightarrow{k_2} C$$

| Ethylbenzene | Styrene | Phenylacetylene |

Complex reactions

Complex reactions are multiple reactions that involve a combination of both series and parallel reactions, such as:

$$A + B \xrightarrow{k_1} C + D$$

$$A + C \xrightarrow{k_2} E$$

Some species behave in series, e.g. $A \rightarrow C \rightarrow E$, and some species react in parallel, e.g. $A \rightarrow C, A \rightarrow E$.

Example 1: Series-parallel hydrogenation of phenylacetylene

Example 2: 'Dry' reforming of methane.

$$CH_4 + O_2 \xrightarrow{k_1} CO_2 + H_2O$$

$$CH_4 + H_2O \xrightarrow{k_2} CO_2 + H_2$$

Example 3: Cyclic Wei-Prater reaction (great for theoretical problems)

$$A \underset{k_3}{\overset{k_1}{\rightleftharpoons}} B \overset{k_2}{\to} C$$

Independent reaction

Independent reaction are reactions that occur at the same time but neither the products nor reactants react with themselves or one another.

$$A + B \overset{k_1}{\longrightarrow} C$$

$$D \overset{k_2}{\longrightarrow} E + F$$

Neither products or reactants react themselves or each other. Kinda simple, but in reality means lots of reactions to model in one volume.

Example: Crude oil cracking reactions:

$$C_{15}H_{32} + O_2 \overset{k_1}{\longrightarrow} C_{12}H_{26} + C_3H_6$$

$$C_8H_{18} + H_2O \overset{k_2}{\longrightarrow} C_6H_{14} + C_2H_4$$

[This system in reality has O(10) independent reactions taking place].

Desired vs. Undesired reactions

Of particular interest are reactants that are consumed in the formation of a desired product, D, and the formation of an undesired product, U, in a competing or side reaction. In the parallel reaction sequence:

$$A \overset{k_D}{\underset{k_U}{\longrightarrow}} \begin{matrix} D \\ U \end{matrix}$$

The product that goes to market should be pure D. So, we will need a reactor and a separator.

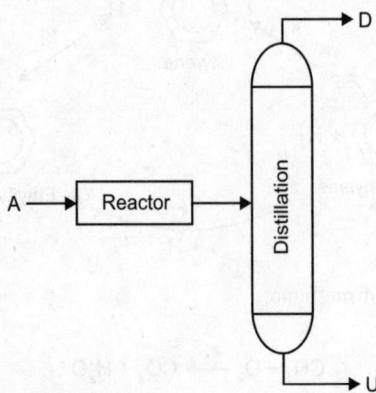

1. If the reactor is cheap and sloppy, (i.e. makes lot of U), then you wind up spending a lot of money on separating D from U.

2. If the reactor is fancy, expensive, even exotic (i.e. makes lots of D), then the separator is pretty cheap.

3. As in life , there is a balance (Fig. 4.1).

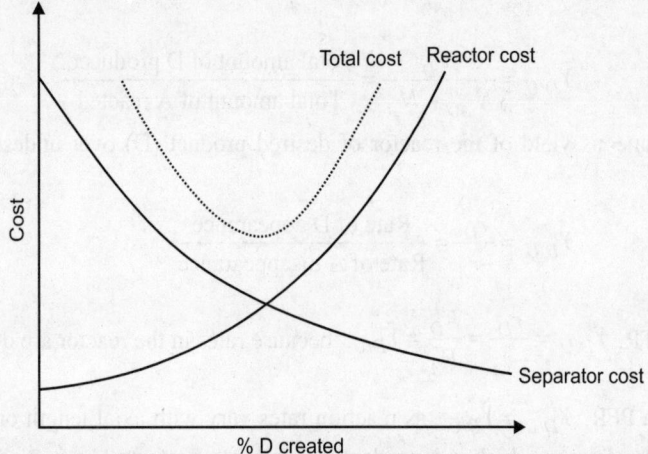

Fig. 4.1. The balance between an expensive reactor that maximises selectivity, and a separator that purifies the reactor product.

Terminology: selectivity and yield

Selectivity: A measure of how selective the reactor is towards producing one species over another. There are two ways to define selectivity:

1. The overall selectivity of the reactor towards desired product (D) over undesired product (U) is defined as:

$$\tilde{S}_{D/U} = \frac{F_D}{F_U} = \frac{\text{Exit molar flow of D}}{\text{Exit molar flow of U}}$$

or

$$\tilde{S}_{D/U} = \frac{N_D}{N_U} = \frac{\text{Final no. of moles of D}}{\text{Final no. of moles of U}}$$

2. The instantaneous selectivity of the reactor is defined as:

$$S_{D/U} = \frac{r_D}{r_U} = \frac{\text{Rate of formation of D}}{\text{Rate of formation of U}}$$

For a CSTR, $S_{D/U} = \frac{r_D}{r_U} = \frac{F_D}{F_U} = \tilde{S}_{D/U}$ because rates in the reactor are defined by the outlet concentration.

For a PFR, $S_{D/U} \neq \tilde{S}_{D/U}$, as reaction rates vary with axial length or reactor volume or catalyst weight (depending on which independent variable you prefer).

Yield: A measure of how much of one product is obtained for a given amount of reagent consumed. Again, can be defined in two ways:

1. The overall yield of the reactor of desired product (D) over undesired product (U) is defined as:

$$\tilde{Y}_{D/U} = \frac{F_D}{F_{AO} - F_A} = \frac{\text{Total amount of D produced}}{\text{Total amount of A reacted}}$$

or

$$\tilde{Y}_{D/U} = \frac{N_D}{N_{AO} - N_A} = \frac{\text{Total amount of D produced}}{\text{Total amount of A reacted}}$$

2. The instantaneous yield of the reactor of desired product (D) over undesired product (U) is defined as:

$$Y_{D/U} = \frac{r_D}{-r_A} = \frac{\text{Rate of D appearance}}{\text{Rate of A disappearance}}$$

Again, for a CSTR, $Y_{D/U} = \dfrac{r_D}{-r_A} = \dfrac{F_D}{F_A} = \tilde{Y}_{D/U}$ because rates in the reactor are defined by the outlet

concentration. For a PFR, $Y_{D/U} \neq \tilde{Y}_{D/U}$, as reaction rates vary with axial length or reactor volume or catalyst weight (depending on which independent variable you prefer).

Note: The maximum theoretical selectivity is infinity ($r_U = 0$), and the maximum theoretical yield is d/u (ratio of reaction stoichiometry).

MULTIPLE REACTIONS

As the next step up in complexity, we consider the case of multiple reactions. Some analytical solutions are available for simple cases with multiple reactions, and Aris provides a comprehensive list, but the scope of these is limited. We focus on numerical computation as a general method for these problems. Indeed, we find that even numerical solution of some of these problems is challenging for two reasons. First, steep concentration profiles often occur for realistic parameter values, and we wish to compute these profiles accurately. It is not unusual for species concentrations to change by 10 orders of magnitude within the pellet for realistic reaction and diffusion rates. Second, we are solving boundary-value problems because the boundary conditions are provided at the center and exterior surface of the pellet. Boundary-value problems (BVPs) are generally much more difficult to solve than initial-value problems (IVPs).

A detailed description of numerical methods for this problem is out of place here. We use the collocation method. The next example involves five species, two reactions with Hougen-Watson kinetics, and both diffusion and external mass-transfer limitations.

Example 4.1: Catalytic converter.

Consider the oxidation of CO and a representative volatile organic such as propylene in a automobile catalytic converter containing spherical catalyst pellets with particle radius 0.175 cm. The particle is surrounded by a fluid at 1.0 atm pressure and 550 K containing 2 per cent CO, 3 per cent O_2 and 0.05 per cent (500 ppm) C_3H_6. The reactions of interest are:

$$CO + \frac{1}{2}O_2 \longrightarrow CO_2 \qquad \qquad \text{... (4.1)}$$

$$C_3H_6 + \frac{9}{2}O_2 \longrightarrow 3CO_2 + 3H_2O \qquad \qquad \text{... (4.2)}$$

with rate expressions given by Oh.

$$r_1 = \frac{k_1 c_{CO} c_{O_2}}{\left(1 + K_{CO} c_{CO} + K_{C_3H_6} c_{C_3H_6}\right)^2} \qquad \text{... (4.3)}$$

$$r_2 = \frac{k_2 c_{C_3H_6} c_{O_2}}{\left(1 + K_{CO} c_{CO} + K_{C_3H_6} c_{C_3H_6}\right)^2} \qquad \text{... (4.4)}$$

The rate constants and the adsorption constants are assumed to have Arrhenius form. The parameter values are given in Table 4.1. The mass-transfer coefficients are taken from DeAcetis and Thodos. The pellet may be assumed to be isothermal. Calculate the steady-state pellet concentration profiles of all reactants and products.

Solution

We solve the steady-state mass balances for the three reactant species,

$$D_j \frac{1}{r^2} \frac{d}{dr} \left(r^2 \frac{dc_j}{dr} \right) = -R_j \qquad \text{... (4.5)}$$

with the boundary conditions:

$$\frac{dc_j}{dr} = 0 \quad r = 0 \qquad \text{... (4.6)}$$

$$D_j \frac{dc_j}{dr} = k_{mj}(c_{jf} - c_j) \qquad r = R \qquad \text{... (4.7)}$$

$j = \{CO, O_2, C_3H_6\}$. The model is solved using the collocation method. The reactant concentration profiles are shown in Figs 4.2 and 4.3.

Table 4.1. Kinetic and mass-transfer parameters for the catalytic converter example.

Parameter	Value	Units	Parameter	Value	Units
P	1.013×10^5	N/m^2	k_{10}	7.07×10^{19}	mol/cm^3·s
T	550	K	k_{20}	1.47×10^{21}	mol/cm^3·s
R	0.175	cm	K_{CO0}	8.099×10^6	cm^3/mol
E_1	13,108	K	$K_{C_3H_6 0}$	2.579×10^8	cm^3/mol
E_2	15,109	K	D_{CO}	0.0487	cm^2/s
E_{CO}	-409	K	D_{O_2}	0.0469	cm^2/s
$E_{C_3H_6}$	191	K	$D_{C_3H_6}$	0.0487	cm^2/s
c_{COf}	2.0%	–	k_{mCO}	3.90	cm/s
c_{O_2f}	3.0%	–	k_{mO_2}	4.07	cm/s
$c_{C_3H_6f}$	0.05%	–	$k_{mC_3H_6}$	3.90	cm/s

Notice that O_2 is in excess and both CO and C_3H_6 reach very low values within the pellet. The log scale in Fig. 4.3 shows that the concentrations of these reactants change by seven orders of magnitude. Obviously the consumption rate is large compared to the diffusion rate for these species. The external mass-transfer effect is noticeable, but not dramatic.

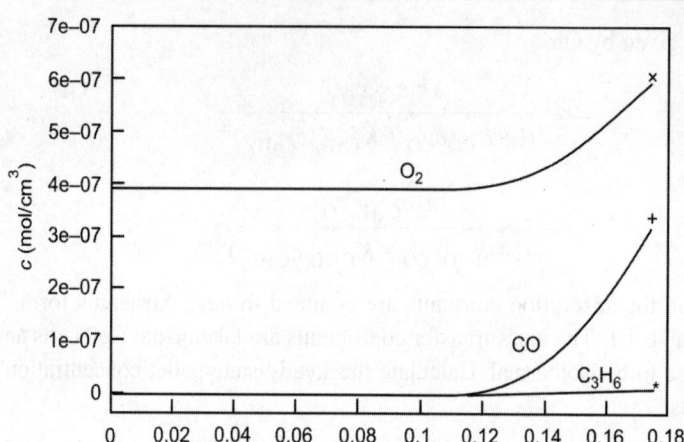

Fig. 4.2. Concentration profiles of reactants; fluid concentration of O_2 (×), CO (+), C_3H_6 (*).

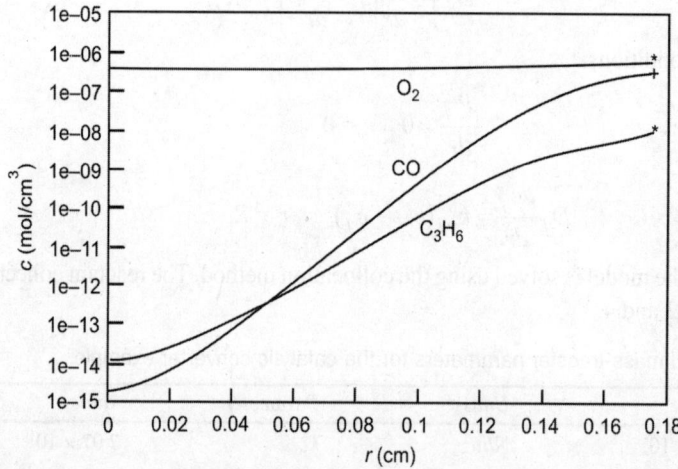

Fig. 4.3. Concentration profiles of reactants (log scale); fluid concentration of O_2 (×), CO (+), C_3H_6 (*).

The product concentrations could simply be calculated by solving their mass balances along with those of the reactants. Because we have only two reactions, however, the concentrations of the products are computable from the stoichiometry and the mass balances. If we take the following mass balances

$$D_{CO}\nabla^2 c_{CO} = -R_{CO} = r_1$$

$$D_{C_3H_6}\nabla^2 c_{C_3H_6} = -R_{C_3H_6} = r_2$$

$$D_{CO_2}\nabla^2 c_{CO_2} = -R_{CO_2} = -r_1 - 3r_2$$

$$D_{H_2O}\nabla^2 c_{H_2O} = -R_{H_2O} = -3r_2$$

and form linear combinations to eliminate the reaction-rate terms for the two products, we obtain:

$$D_{CO_2}\nabla^2 c_{CO_2} = -D_{CO}\nabla^2 c_{CO} - 3D_{C_3H_6}\nabla^2 c_{C_3H_6}$$

$$D_{H_2O}\nabla^2 c_{H_2O} = -3D_{C_3H_6}\nabla^2 c_{C_3H_6}$$

Because the diffusivities are assumed constant, we can integrate these once on $(0, r)$ to obtain for the products:

$$D_{CO_2}\frac{dc_{CO_2}}{dr} = D_{CO}\frac{dc_{CO}}{dr} - 3D_{C_3H_6}\frac{dc_{C_3H_6}}{dr}$$

$$D_{H_2O}\frac{dc_{H_2O}}{dr} = -3D_{C_3H_6}\frac{dc_{C_3H_6}}{dr}$$

The exterior boundary condition can be rearranged to give:

$$c_j - c_{jf} = \frac{D_j}{k_{mj}}\frac{dc_j}{dr}$$

Substituting in the relationships for the products gives:

$$c_{CO_2} = c_{CO_2f} - \frac{1}{k_{mCO_2}}\left[D_{CO}\frac{dc_{CO}}{dr} + 3D_{C_3H_6}\frac{dc_{C_3H_6}}{dr}\right]$$

$$c_{H_2O} = c_{H_2Of} - \frac{1}{k_{mH_2O}}\left[3D_{C_3H_6}\frac{dc_{C_3H_6}}{dr}\right]$$

The right-hand sides are available from the solution of the material balances of the reactants. Plotting these results for the products gives Fig. 4.4. We see that CO_2 is the main product. Note the products flow out of the pellet, unlike the reactants shown in Figs 4.2 and 4.3, which are flowing into the pellet.

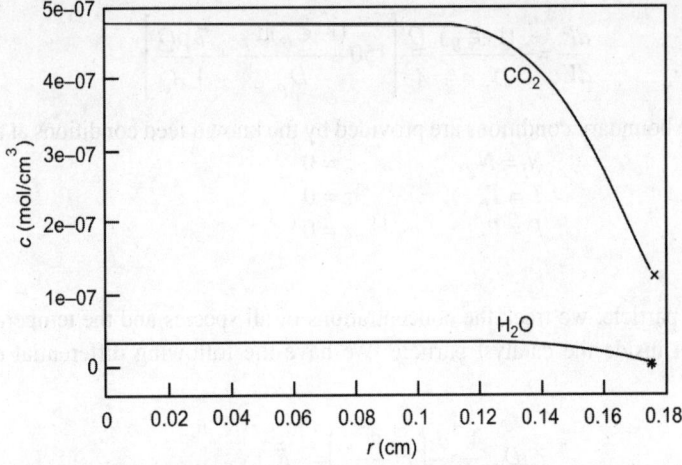

Fig. 4.4. Concentration profiles of the products; fluid concentration of CO_2 (x), H_2O (*).

Fixed-Bed Reactor Design

Given our detailed understanding of the behaviour of a single catalyst particle, we now are prepared to pack a tube with a bed of these particles and solve the fixed-bed reactor design problem. In the fixed-bed reactor, we keep track of two phases. The fluid-phase streams through the bed and transports the reactants and products through the reactor. The reaction-diffusion processes take place in the solid-phase catalyst particles. The two phases communicate to each other by exchanging mass and energy at the catalyst particle exterior surfaces. We have constructed a detailed understanding of all these events, and now we assemble them together.

Coupling the catalyst and fluid

We make the following assumptions:
1. Uniform catalyst pellet exterior. Particles are small compared to the length of the reactor.
2. Plug flow in the bed, no radial profiles.
3. Neglect axial diffusion in the bed.
4. Steady-state.

Fluid

In the fluid phase, we track the molar flows of all species, the temperature and the pressure. We can no longer neglect the pressure drop in the tube because of the catalyst bed. We use an empirical correlation to describe the pressure drop in a packed tube, the well-known Ergun equation. Therefore, we have the following differential equations for the fluid phase:

$$\frac{dN_j}{dV} = R_j \qquad\qquad \text{... (4.8)}$$

$$Q\rho\hat{C}_p \frac{dT}{dV} = -\sum_i \Delta H_{Ri} r_i + \frac{2}{R} U^\circ (T_a - T) \qquad\qquad \text{... (4.9)}$$

$$\frac{dP}{dV} = -\frac{(1-\epsilon_B)}{D_p \epsilon_B^2} \frac{Q}{A_c^2} \left[150 \frac{(1-\epsilon_B)\mu_f}{D_p} + \frac{7}{4} \frac{\rho Q}{A_c} \right] \qquad\qquad \text{... (4.10)}$$

The fluid-phase boundary conditions are provided by the known feed conditions at the tube entrance:

$$N_j = N_{jf}, \qquad z = 0 \qquad\qquad \text{... (4.11)}$$
$$T = T_f, \qquad z = 0 \qquad\qquad \text{... (4.12)}$$
$$P = P_f, \qquad z = 0 \qquad\qquad \text{... (4.13)}$$

Catalyst particle

Inside the catalyst particle, we track the concentrations of all species and the temperature. We neglect any pressure effect inside the catalyst particle. We have the following differential equations for the catalyst particle:

$$D_j \frac{1}{r^2} \frac{d}{dr}\left(r^2 \frac{d\tilde{c}_j}{dr} \right) = -\tilde{R}_j \qquad\qquad \text{... (4.14)}$$

$$\hat{k}\frac{1}{r^2}\frac{d}{dr}\left(r^2\frac{d\tilde{T}}{dr}\right) = \sum_i \Delta H_{Ri}\tilde{r}_i \qquad \ldots (4.15)$$

The boundary conditions are provided by the mass-transfer and heat-transfer rates at the pellet exterior surface, and the zero slope conditions at the pellet center:

$$\frac{d\tilde{c}_j}{dr} = 0 \qquad\qquad r = 0 \qquad\qquad \ldots (4.16)$$

$$D_j\frac{d\tilde{c}_j}{dr} = k_{jm}(c_j - \tilde{c}_j) \quad r = R \qquad\qquad \ldots (4.17)$$

$$\frac{d\tilde{T}}{dr} = 0 \qquad\qquad r = 0 \qquad\qquad \ldots (4.18)$$

$$\tilde{k}\frac{d\tilde{T}}{dr} = k_T(T - \tilde{T}) \qquad r = R \qquad\qquad \ldots (4.19)$$

Coupling equations

Finally, we equate the production rate R_j experienced by the fluid phase to the production rate inside the particles, which is where the reaction takes place. Analogously, we equate the enthalpy change on reaction experienced by the fluid phase to the enthalpy change on reaction taking place inside the particles. These expressions are given below:

$$\underbrace{R_j}_{\text{rate } j/\text{vol}} = -\underbrace{(1-\epsilon_B)}_{\text{vol cat/vol}}\underbrace{\frac{S_p}{V_p}D_j\frac{d\tilde{c}j}{dr}\bigg|_{r=R}}_{\text{rate } j/\text{vol cat}} \qquad \ldots (4.20)$$

$$\underbrace{\sum_i \Delta H_{Ri}r_i}_{\text{rate heat/vol}} = \underbrace{(1-\epsilon_B)}_{\text{vol cat/vol}}\underbrace{\frac{S_p}{V_p}\hat{k}\frac{d\tilde{T}}{dr}\bigg|_{r=R}}_{\text{rate heat/vol cat}} \qquad \ldots (4.21)$$

Notice we require the bed porosity to convert from the rate per volume of particle to the rate per volume of reactor. The bed porosity or void fraction, ϵ_B, is defined as the volume of voids per volume of reactor. The volume of catalyst per volume of reactor is therefore $1 - \epsilon_B$. This information can be presented in a number of equivalent ways. We can easily measure the density of the pellet, ρ_p, and the density of the bed, ρ_B. From the definition of bed porosity, we have the relation:

$$\rho_B = (1 - \epsilon_B)\rho_p$$

or if we solve for the volume fraction of catalyst:

$$1 - \epsilon_B = \rho_B/\rho_p$$

Figure 4.5 shows the particles and fluid, and summarises the coupling relations between the two phases.

Equations 4.8–4.21 provide the full packed-bed reactor model given our assumptions. We next examine several packed-bed reactor problems that can be solved without solving this full set of equations. Finally, we present Example 4.6, which requires numerical solution of the full set of equations.

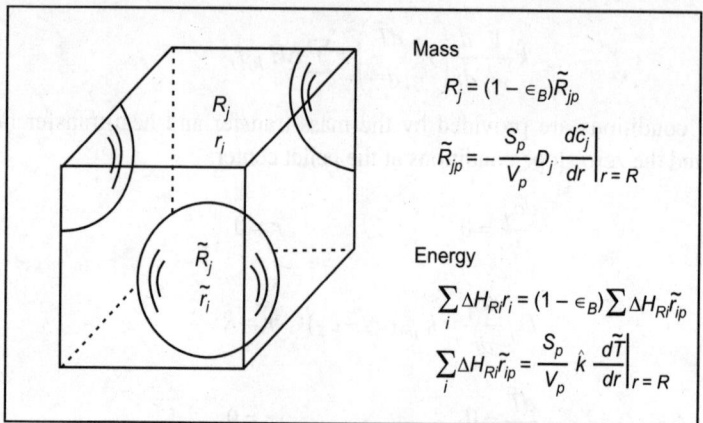

Fig. 4.5. Fixed-bed reactor volume element containing fluid and catalyst particles; the equations show the coupling between the catalyst particle balances and the overall reactor balances.

Example 4.2: First-order, isothermal fixed-bed reactor.

Use the rate data to find the fixed-bed reactor volume and the catalyst mass needed to convert 97 per cent of A. The feed to the reactor is pure A at 1.5 atm at a rate of 12 mol/s. The 0.3 cm pellets are to be used, which leads to a bed density $\rho_B = 0.6$ g/cm^3. Assume the reactor operates isothermally at 450 K and that external mass-transfer limitations are negligible.

Solution: We solve the fixed-bed design equation:

$$\frac{dN_A}{dV} = R_A = -(1-\epsilon_B)\eta k c_A$$

between the limits N_{Af} and 0.03 N_{Af}, in which c_A is the A concentration in the fluid. For the first-order, isothermal reaction, the Thiele modulus is independent of A concentration, and is therefore independent of axial position in the bed:

$$\Phi = \frac{R}{3}\sqrt{\frac{k}{D_A}} = \frac{0.3\ \text{cm}}{3}\sqrt{\frac{2.6\ \text{s}^{-1}}{0.007\,\text{cm}^2/\text{s}}} = 1.93$$

The effectiveness factor is also therefore a constant:

$$\eta = \frac{1}{\Phi}\left[\frac{1}{\tanh 3\Phi} - \frac{1}{3\Phi}\right] = \frac{1}{1.93}\left[1 - \frac{1}{5.78}\right] = 0.429$$

We express the concentration of A in terms of molar flows for an ideal-gas mixture:

$$c_A = \frac{P}{RT}\left(\frac{N_A}{N_A + N_B}\right)$$

The total molar flow is constant due to the reaction stoichiometry so $N_A + N_B = N_{Af}$ and we have:

$$c_A = \frac{P}{RT}\frac{N_A}{N_{Af}}$$

Substituting these values into the material balance, rearranging and integrating over the volume gives:

$$V_R = -(1-\epsilon_B)\left(\frac{RTN_{Af}}{\eta kP}\right)\int_{N_{Af}}^{0.03N_{Af}}\frac{dN_A}{N_A}$$

$$V_R = -\left(\frac{0.6}{0.85}\right)\frac{(82.06)(450)(12)}{(0.429)(2.6)(1.5)}\ln(0.03) = 1.32\times10^5\,\text{cm}^3$$

and

$$W_c = \rho_B V_R = \frac{0.6}{1000}\left(1.32\times10^6\right) = 789\,\text{kg}$$

We see from this example that if the Thiele modulus and effectiveness factors are constant, finding the size of a fixed-bed reactor is no more difficult than finding the size of a plug-flow reactor.

Example 4.3. Mass-transfer limitations in a fixed-bed reactor.

Reconsider Example 4.2 given the following two values of the mass-transfer coefficient:

$$k_{m1} = 0.07\,\text{cm/s}$$
$$k_{m2} = 1.4\,\text{cm/s}$$

Solution: First we calculate the Biot numbers and obtain:

$$B_1 = \frac{(0.07)(0.1)}{(0.007)} = 1$$

$$B_2 = \frac{(1.4)(0.1)}{(0.007)} = 20$$

We expect a significant reduction in the effectiveness factor due to mass-transfer resistance in the first case, and little effect in the second case. Evaluating the effectiveness factors indeed shows:

$$\eta_1 = 0.165$$
$$\eta_2 = 0.397$$

which we can compare to $\eta = 0.429$ from the previous example with no mass-transfer resistance. We can then easily calculate the required catalyst mass from the solution of the previous example without mass-transfer limitations, and the new values of the effectiveness factors:

$$V_{R1} = \left(\frac{0.429}{0.165}\right)(789) = 2051\,\text{kg}$$

$$V_{R2} = \left(\frac{0.429}{0.397}\right)(789) = 852\,\text{kg}$$

As we can see, the first mass-transfer coefficient is so small that more than twice as much catalyst is required to achieve the desired conversion compared to the case without mass-transfer limitations. The second mass-transfer coefficient is large enough that only 8 per cent more catalyst is required.

Example 4.4: Second-order, isothermal fixed-bed reactor.

Estimate the mass of catalyst required in an isothermal fixed-bed reactor for the second-order, heterogeneous reaction.

$$A \xrightarrow{k} B$$

$$r = kc_A^2 \quad k = 2.25 \times 10^5 \text{ cm}^3/\text{mols}$$

The gas feed consists of A and an inert, each with molar flowrate of 10 mol/s, the total pressure is 4.0 atm and the temperature is 550 K. The desired conversion of A is 75 per cent. The catalyst is a spherical pellet with a radius of 0.45 cm. The pellet density is $\rho p = 0.68$ g/cm^3 and the bed density is $\rho_B = 0.60$ g/cm^3. The effective diffusivity of A is 0.008 cm^2/s and may be assumed constant. You may assume the fluid and pellet surface concentrations are equal.

Solution: We solve the fixed-bed design equation.

$$\frac{dN_A}{dV} = R_A = -(1-\epsilon_B)\eta kc_A^2$$

$$N_A(0) = N_{Af} \qquad \text{... (4.22)}$$

between the limits N_{Af} and 0.25 N_{Af}. We again express the concentration of A in terms of the molar flows:

$$c_A = \frac{P}{RT}\left(\frac{N_A}{N_A + N_B + N_I}\right)$$

As in the previous example, the total molar flow is constant and we know its value at the entrance to the reactor:

$$N_T = N_{Af} + N_{Bf} + N_{If} = 2N_{Af}$$

Therefore,

$$c_A = \frac{P}{RT}\frac{N_A}{N_{Af}} \qquad \text{... (4.23)}$$

Next we use the definition of Φ for nth-order reactions:

$$\Phi = \frac{R}{3}\left[\frac{(n+1)kc_A^{n-1}}{2D_e}\right]^{1/2} = \frac{R}{3}\left[\frac{(n+1)k}{2D_e}\left(\frac{P}{RT}\frac{N_A}{2N_{Af}}\right)^{n-1}\right]^{1/2} \qquad \text{... (4.24)}$$

Substituting in the parameter values gives:

$$\Phi = 9.17\left(\frac{N_A}{2N_{Af}}\right)^{1/2} \qquad \text{... (4.25)}$$

For the second-order reaction, Eq. 4.25 shows that Φ varies with the molar flow, which means Φ and η vary along the length of the reactor as N_A decreases. We are asked to estimate the catalyst mass needed to achieve a conversion of A equal to 75 per cent. So for this particular example, Φ decreases from 6.49 to 3.24. The effectiveness factor for the second-order reaction using the analytical result for the first-order reaction:

$$\eta = \frac{1}{\Phi}\left[\frac{1}{\tanh 3\Phi} - \frac{1}{3\Phi}\right] \qquad \text{... (4.26)}$$

Summarising so far, to compute N_A versus V_R, we solve one differential equation, Eq. 4.22, in which we use Eq. 4.23 for c_A, and Eqs 4.25 and 4.26 for Φ and η. We march in V_R until $N_A = 0.25N_{Af}$. The solution to the differential equation is shown in Fig. 4.6. The required reactor volume and mass of catalyst are:

$$V_R = 361 \text{ L}, \qquad W_c = \rho_B V_R = 216 \text{ kg}$$

As a final exercise, given that Φ ranges from 6.49 to 3.24, we can make the large Φ approximation

$$\eta = \frac{1}{\Phi} \qquad\qquad \text{... (4.27)}$$

to obtain a closed-form solution. If we substitute this approximation for η, and Eq. 4.24 into Eq. 4.22 and rearrange we obtain:

$$\frac{dN_A}{dV} = \frac{-(1-\epsilon_B)\sqrt{k}(P/RT)^{3/2}}{(R/3)\sqrt{3/D_A}(2N_{Af})^{3/2}} N_A^{3/2}$$

Separating and integrating this differential equation gives:

$$V_R = \frac{4[(1-x_A)^{-1/2}-1]N_{Af}(R/3)\sqrt{3/D_A}}{(1-\epsilon_B)\sqrt{k}(P/RT)^{3/2}} \qquad \text{... (4.28)}$$

Large Φ approximation

The results for the large Φ approximation also are shown in Fig. 4.6. Notice that we are slightly overestimating the value of η using Eq. 4.27, so we underestimate the required reactor volume. The reactor size and the per cent change in reactor size are:

$$V_R = 333 \text{ L}, \qquad\qquad \Delta = -7.7\%$$

Given that we have a result valid for all Φ that requires solving only a single differential equation, one might question the value of this closed-form solution. One advantage is purely practical. We may not have a computer available. Instructors are usually thinking about in-class examination problems at this juncture. The other important advantage is insight. It is not readily apparent from the differential equation what would happen to the reactor size if we double the pellet size or halve the rate constant, for example. Eq. 4.28, on the other hand, provides the solution's dependence on all parameters. As shown in Fig. 4.6 the approximation error is small. Remember to check that the Thiele modulus is large for the entire tube length, however, before using Eq. 4.28.

Example 4.5. Hougen-Watson kinetics in a fixed-bed reactor.

The following reaction converting CO to CO_2 takes place in a catalytic, fixed-bed reactor operating isothermally at 838 K and 1.0 atm

$$CO + \tfrac{1}{2}O_2 \longrightarrow CO_2 \qquad\qquad \text{... (4.29)}$$

The following rate expression and parameters are adapted from a different model given by Oh. The rate expression is assumed to be of the Hougen-Watson form:

$$r = \frac{kc_{CO}c_{O_2}}{1+Kc_{CO}} \text{ mol/s cm}^3 \text{ pellet}$$

The constants are provided below:

$$k = 8.73 \times 10^{12} \exp(-13,500/T) \text{ cm}^3/\text{mol s}$$
$$K = 8.099 \times 10^6 \exp(409 = T) \text{ cm}^3/\text{mol}$$
$$D_{CO} = 0.0487 \text{ cm}^2/\text{s}$$

in which T is in Kelvin. The catalyst pellet radius is 0.1 cm. The feed to the reactor consists of 2 mol% CO, 10 mol% O_2, zero CO_2 and the remainder inerts. Find the reactor volume required to achieve 95 per cent conversion of the CO.

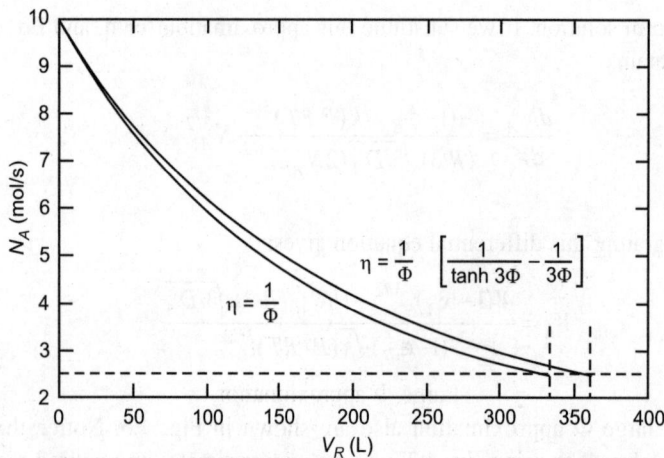

Fig. 4.6. Molar flow of A versus reactor volume for second-order, isothermal reaction in a fixed-bed reactor.

Solution

Given the reaction stoichiometry and the excess of O_2, we can neglect the change in c_{O_2} and approximate the reaction as pseudo-first-order in CO:

$$r = \frac{k'c_{CO}}{1+Kc_{CO}} \text{ mol/s cm}^3 \text{ pellet}$$

$$k' = kc_{O_2f}$$

which is of the form analysed already. We can write the mass balance for the molar flow of CO,

$$\frac{dN_{CO}}{dV} = -(1-\epsilon_B)\eta r(c_{CO})$$

in which c_{CO} is the fluid CO concentration. From the reaction stoichiometry, we can express the remaining molar flows in terms of N_{CO}

$$N_{O_2} = N_{O2f} + 1/2(N_{CO} - N_{COf})$$
$$N_{CO_2} = N_{COf} - N_{CO}$$
$$N = N_{O2f} + 1/2(N_{CO} + N_{COf}).$$

The concentrations follow from the molar flows assuming an ideal-gas mixture:

$$c_j = \frac{P}{RT}\frac{N_j}{N}$$

To decide how to approximate the effectiveness factor, we evaluate $\phi = K_{CO}c_{CO}$, at the entrance and exit of the fixed-bed reactor. With ϕ evaluated, we compute the Thiele modulus and obtain:

$$\phi = 32.0 \qquad \Phi = 79.8, \qquad \text{entrance}$$
$$\phi = 1.74 \qquad \Phi = 326, \qquad \text{exit}$$

It is clear from these values that $\eta = 1/\Phi$ is an excellent approximation for this reactor. Substituting this equation for η into the mass balance and solving the differential equation produces the results shown in Fig. 4.7. The concentration of O_2 is nearly constant, which justifies the pseudo-first-order rate expression. Reactor volume:

$$V_R = 233 \text{ L}$$

is required to achieve 95 per cent conversion of the CO. Recall that the volumetric flowrate varies in this reactor so conversion is based on molar flow, not molar concentration. Figure 4.8 shows how Φ and ϕ vary with position in the reactor.

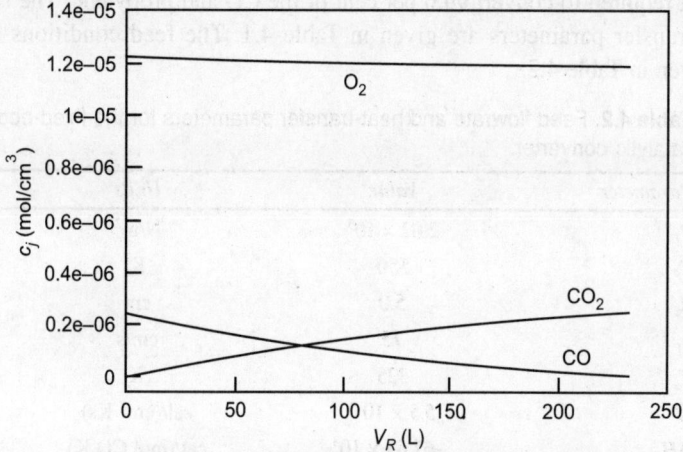

Fig. 4.7. Molar concentrations versus reactor volume.

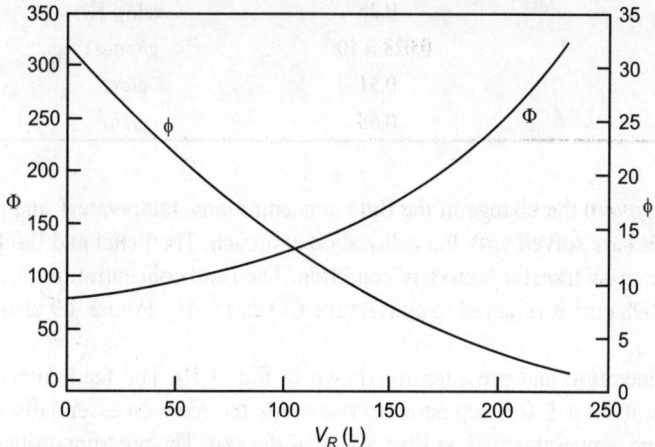

Fig. 4.8. Dimensionless equilibrium constant and Thiele modulus versus reactor volume. Values indicate $\eta = 1/\Phi$ is a good approximation for entire reactor.

In the previous examples, we have exploited the idea of an effectiveness factor to reduce fixed-bed reactor models to the same form as plug-flow reactor models. This approach is useful and solves several important cases, but this approach is also limited and can take us only so far. In the general case, we must contend with multiple reactions that are not first-order, nonconstant thermochemical properties, and nonisothermal behaviour in the pellet and the fluid. For these cases, we have no alternative but to solve numerically for the temperature and species concentrations profiles in both the pellet and the bed. As a final example, we compute the numerical solution to a problem of this type.

We use the collocation method to solve the next example, which involves five species, two reactions with Hougen-Watson kinetics, both diffusion and external mass-transfer limitations, and nonconstant fluid temperature, pressure and volumetric flowrate.

Example 4.6. Multiple-reaction, nonisothermal fixed-bed reactor.

Evaluate the performance of the catalytic converter in converting CO and propylene. Determine the amount of catalyst required to convert 99.6 per cent of the CO and propylene. The reaction chemistry and pellet mass-transfer parameters are given in Table 4.1. The feed conditions and heat-transfer parameters are given in Table 4.2.

Table 4.2. Feed flowrate and heat-transfer parameters for the fixed-bed catalytic converter.

Parameter	Value	Units
P_f	2.02×10^5	N/m^2
T_f	550	K
R_t	5.0	cm
u_f	75	cm/s
T_a	325	K
U^o	5.5×10^{-3}	cal/(cm^2 Ks)
ΔH_{R1}	-67.63×10^3	cal/(mol CO K)
ΔH_{R2}	-460.4×10^3	cal/(mol C$_3$H$_6$ K)
$\hat{C}_p$	0.25	cal/(g K)
μ_f	0.028×10^{-2}	g/(cm s)
ρ_b	0.51	g/cm^3
ρ_p	0.68	g/cm^3

Solution

The fluid balances govern the change in the fluid concentrations, temperature and pressure. The pellet concentration profiles are solved with the collocation approach. The pellet and fluid concentrations are coupled through the mass-transfer boundary condition. The fluid concentrations are shown in Fig. 4.9. A bed volume of 1098 cm^3 is required to convert the CO and C$_3$H$_6$. Figure 4.9 also shows that oxygen is in slight excess.

The reactor temperature and pressure are shown in Fig. 4.10. The feed enters at 550 K, and the reactor experiences about a 130 K temperature rise while the reaction essentially completes; the heat losses then reduce the temperature to less than 500 K by the exit. The pressure drops from the feed value of 2.0 atm to 1.55 atm at the exit. Notice the catalytic converter exit pressure of 1.55 atm must be large enough to account for the remaining pressure drops in the tail pipe and muffler.

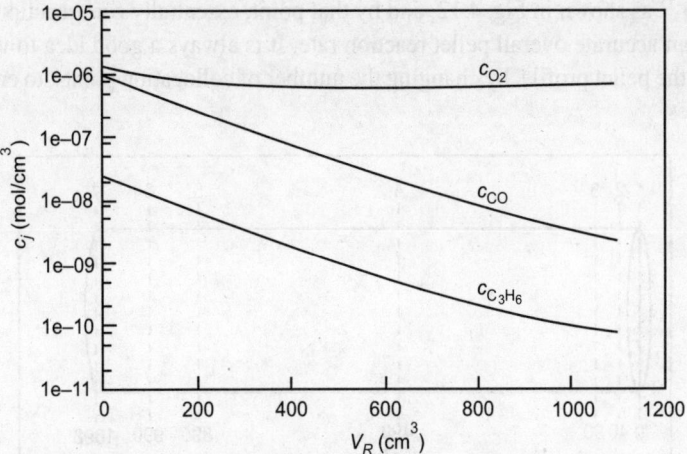

Fig. 4.9. Fluid molar concentrations versus reactor volume.

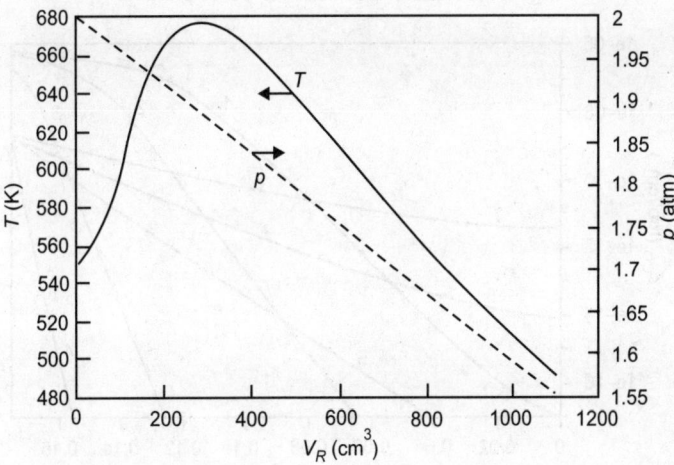

Fig. 4.10. Fluid temperature and pressure versus reactor volume.

In Figs 4.11 and 4.12, the pellet CO concentration profile at several reactor positions is displayed. The feed profile, marked by À in Fig. 4.12, is similar to the one shown in Fig. 4.3 of Example 4.1 (the differences are caused by the different feed pressures). We see that as the reactor heats up, the reaction rates become large and the CO is rapidly converted inside the pellet. By 490 cm³ in the reactor, the pellet exterior CO concentration has dropped by two orders of magnitude, and the profile inside the pellet has become very steep. As the reactions go to completion and the heat losses cool the reactor, the reaction rates drop. At 890 cm³, the CO begins to diffuse back into the pellet. Finally, the profiles become much flatter near the exit of the reactor.

It can be numerically challenging to calculate rapid changes and steep profiles inside the pellet. The good news, however, is that accurate pellet profiles are generally not required for an accurate calculation of the overall pellet reaction rate. The reason is that when steep profiles are present, essentially all of the reaction occurs in a thin shell near the pellet exterior. We can calculate accurately down to concentrations on the order of 10^{-15} as shown in Fig. 4.12, and by that point, essentially zero reaction is occurring, and we can calculate an accurate overall pellet reaction rate. It is always a good idea to vary the numerical approximation in the pellet profile, by changing the number of collocation points, to ensure convergence in the fluid profiles.

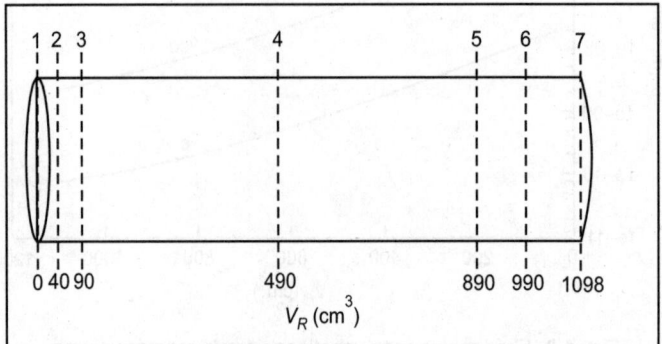

Fig. 4.11. Reactor positions for pellet profiles.

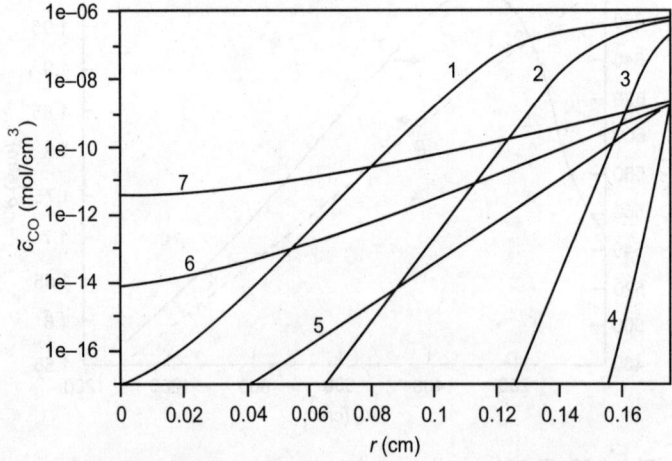

Fig. 4.12. Pellet CO profiles at several reactor positions.

DARCY'S LAW

In fluid dynamics and hydrology, Darcy's law is a phenomenologically derived constitutive equation that describes the flow of a fluid through a porous medium. The law was formulated by Henry Darcy based on the results of experiments on the flow of water through beds of sand. It also forms the scientific basis of fluid permeability used in the earth sciences.

Although Darcy's law (an expression of conservation of momentum) was originally determined experimentally by Henry Darcy, it has since been derived from the Navier-Stokes equations via homogenisation. It is analogous to Fourier's law in the field of heat conduction, Ohm's law in the field of electrical networks, or Fick's law in diffusion theory.

One application of Darcy's law is to water flow through an aquifer. Darcy's law along with the equation of conservation of mass are equivalent to the groundwater flow equation, one of the basic relationships of hydrogeology. Darcy's law is also used to describe oil, water, and gas flows through petroleum reservoirs.

Darcy's law is a simple proportional relationship between the instantaneous discharge rate through a porous medium, the viscosity of the fluid and the pressure drop over a given distance.

$$Q = \frac{-\kappa A}{\mu} \frac{(P_b - P_a)}{L}$$

The total discharge, Q (units of volume per time, e.g. ft³/s or m³/s) is equal to the product of the permeability (κ units of area, e.g. m²) of the medium, the cross-sectional area (A) to flow, and the pressure drop ($P_b - P_a$), all divided by the dynamic viscosity μ [in SI units e.g. kg/(m·s) or Pa·s], and the length L the pressure drop is taking place over (Fig. 4.13).

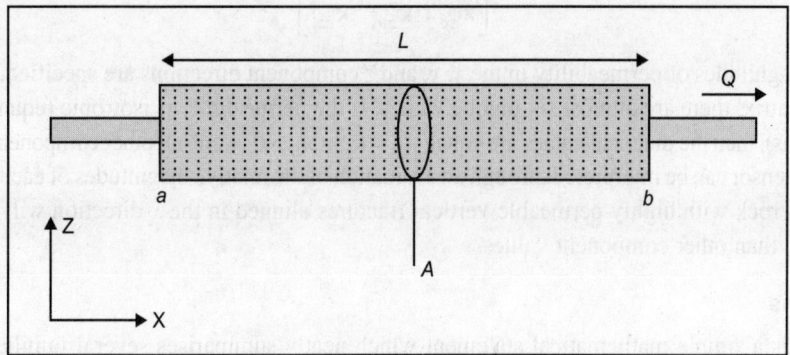

Fig. 4.13. Diagram showing definitions and directions for Darcy's law.

The negative sign is needed because fluids flow from high pressure to low pressure. So if the change in pressure is negative (in the x-direction) then the flow will be positive (in the x-direction). Dividing both sides of the equation by the area and using more general notation leads to:

$$q = \frac{-k}{\mu} \nabla P$$

where, q is the filtration velocity or Darcy flux (discharge per unit area, with units of length per time, m/s) and ∇P is the pressure gradient vector. This value of the filtration velocity (Darcy flux), is not the velocity which the water travelling through the pores is experiencing.

The pore (interstitial) velocity (v) is related to the Darcy flux (q) by the porosity (ϕ). The flux is divided by porosity to account for the fact that only a fraction of the total formation volume is available for flow. The pore velocity would be the velocity a conservative tracer would experience if carried by the fluid through the formation.

$$v = \frac{q}{\phi}$$

In 3D

In three dimensions, gravity must be accounted for, as the flow is not affected by the vertical pressure drop caused by gravity when assuming hydrostatic conditions. The solution is to subtract the gravitational pressure drop from the existing pressure drop in order to express the resulting flow,

$$q = \frac{-\kappa}{\mu}(\nabla P - \rho g \hat{e}_z)$$

where, the flux q is now a vector quantity, κ is a tensor of permeability, ∇ is the gradient operator in 3D, g is the acceleration due to gravity, $\hat{e}_z$ is the unit vector in the vertical direction, pointing downwards and ρ is the density.

Effects of anisotropy in three dimensions are addressed using a symmetric second-order tensor of permeability:

$$k = \begin{bmatrix} \kappa_{xx} & \kappa_{xy} & \kappa_{xz} \\ \kappa_{yx} & \kappa_{yy} & \kappa_{yz} \\ \kappa_{zx} & \kappa_{zy} & \kappa_{zz} \end{bmatrix}$$

where, the magnitudes of permeability in the x, y, and z component directions are specified. Since this a symmetric matrix, there are at most six unique values. If the permeability is isotropic (equal magnitude in all directions), then the diagonal values are equal, $\kappa_{xx} = \kappa_{yy} = \kappa_{zz} > 0$, while all other components are 0. The permeability tensor can be interpreted through an evaluation of the relative magnitudes of each component. For example, rock with highly permeable vertical fractures aligned in the x-direction will have higher values for κ_{xx} than other component values.

Assumptions

Darcy's law is a simple mathematical statement which neatly summarises several familiar properties that groundwater flowing in aquifers exhibits, including:

1. If there is no pressure gradient over a distance, no flow occurs (this of course, is the hydrostatic condition).
2. If there is a pressure gradient, flow will occur from high pressure towards low pressure (opposite the direction of increasing gradient—hence the negative sign in Darcy's law).
3. The greater the pressure gradient (through the same formation material), the greater the discharge rate.
4. The discharge rate of fluid will often be different—through different formation materials (or even through the same material, in a different direction)—even if the same pressure gradient exists in both cases.

A graphical illustration of the use of the steady-state groundwater flow equation (based on Darcy's law and the conservation of mass) is in the construction of flownets, to quantify the amount of groundwater flowing under a dam. Darcy's law is only valid for slow, viscous flow; fortunately, most groundwater flow cases fall in this category.

Typically any flow with a Reynolds number (based on a pore size length scale) less than one is clearly laminar, and it would be valid to apply Darcy's law. Experimental tests have shown that flow regimes with values of Reynolds number up to 10 may still be Darcian. Reynolds number (a dimensionless parameter) for porous media flow is typically expressed as

$$Re = \frac{\rho v d_{30}}{\mu}$$

where, ρ is the density of the fluid (units of mass per volume), v is the specific discharge (not the pore velocity—with units of length per time), d_{30} is a representative grain diameter for the porous medium (often taken as the 30 per cent passing size from a grain size analysis using sieves), and μ is the dynamic viscosity of the fluid.

Derivation

Assuming stationary, creeping, incompressible flow, the Navier-Stokes equation simplify to the Stokes equation:

$$\mu\nabla^2 u_i + \rho g_i - \partial_i p = 0$$

where, μ is the viscosity, u_i is the velocity in the i direction, g_i is the gravity component in the i direction and p is the pressure. Assuming the viscous resisting force is proportional to the velocity, and opposite in direction, we may write

$$-\frac{\mu\phi}{k_i}u_i + \rho g_i - \partial_i p = 0$$

where, ϕ is the porosity. This gives the velocity

$$u_i = -\frac{k_i}{\phi\mu}(\partial_i p - \rho g_i)$$

which gives Darcy's law

$$q_i = -\frac{k_i}{\mu}(\partial_i p - \rho g_i)$$

Additional Forms of Darcy's Law

Time derivative of flux

For very short time scales or high frequency oscillations, a time derivative of flux may be added to Darcy's law, which results in valid solutions at very small times (in heat transfer, this is called the modified form of Fourier's law),

$$\tau\frac{\partial q}{\partial t} + q = -K\nabla h$$

where, τ is a very small time constant which causes this equation to reduce to the normal form of Darcy's law at 'normal' times ($>$ nanoseconds). The main reason for doing this is that the regular groundwater flow equation (diffusion equation) leads to singularities at constant head boundaries at very small times. This form is more mathematically rigorous, but leads to a hyperbolic groundwater flow equation, which is more difficult to solve and is only useful at very small times, typically out of the realm of practical use.

Brinkman term

Another extension to the traditional form of Darcy's law is the Brinkman term, which is used to account for transitional flow between boundaries,

$$\beta\nabla^2 q + q = -K\nabla h$$

where, β is an effective viscosity term. This correction term accounts for flow through medium where the grains of the media are porous themselves, but is difficult to use, and is typically neglected.

Multiphase Flow

For multiphase flow, an approximation is to use Darcy's law for each phase, with permeability replaced by phase permeability, which is the permeability of the rock multiplied with relative permeability. This approximation is valid if the interfaces between the fluids remain static, which is not true in general, but it is still a reasonable model under steady-state conditions.

Assuming that the flow of a phase in the presence of another phase can be viewed as single phase flow through a reduced pore network, we can add the subscript i for each phase to Darcy's law above written for Darcy flux, and obtain for each phase in multiphase flow

$$q_i = -\frac{k_i}{\mu_i}\nabla P_i$$

where, κ_i is the phase permeability for phase i. From this we also define relative permeability κ_{ri} for phase i as:

$$\kappa_{ri} = \kappa_i/\kappa$$

where, κ is the permeability for the porous medium, as in Darcy's law.

Dupuit-Forchheimer equation for non-Darcy flow

For a sufficiently high flow velocity, the flow is nonlinear, and Dupuit and Forchheimer have proposed to generalise the flow equation to

$$-\nabla P = \frac{\mu}{\kappa}V + \beta\rho V^2$$

where, V is the flow velocity and β is a factor to be experimentally deduced.

In membrane operations

In pressure-driven membrane operations, Darcy's law is often used in the form,

$$J = \frac{\Delta P - \Delta\Pi}{\mu(R_f + R_m)}$$

where,

J is the volumetric flux ($m.s^{-1}$).
ΔP is the hydraulic pressure difference between the feed and permeate sides of the membrane (Pa).
$\Delta\Pi$ is the osmotic pressure difference between the feed and permeate sides of the membrane (Pa).
μ is the dynamic viscosity ($Pa.s$).
R_f is the fouling resistance (m^{-1}).
R_m is the membrane resistance (m^{-1}).

Example 4.7. Substance A in a liquid reactor to producer R and S as follows:

A feed ($C_{Ao} = 1$, $C_{Ro} = C_{So} = 0$) enters two mixed reactors in series ($\tau_1 = 2.5$, $\tau_2 = 5$ min). Knowing the composition in the first reactor ($C_{A1} = 0.4$, $C_{R1} = 0.4$, $C_{S1} = 0.2$). Find the composition leaving the 2nd reactor.

Solution

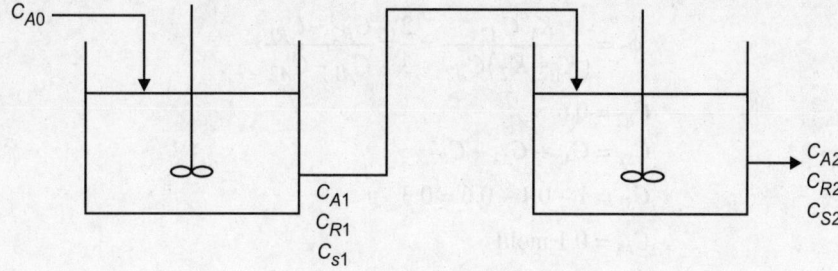

$$\tau_1 = 2.5 \text{ min}, \qquad\qquad \tau_2 = 5 \text{ min}.$$
$$C_{A1} = 0.4, \qquad C_{R1} = 0.4, \qquad C_{S1} = 0.2$$
$$\frac{dC_R}{dt} = K_1 C_A, \quad \frac{dC_S}{dt} = K_2 C_A$$
$$-dC_A/dt = (K_1 + K_2)\,C_A$$

Fractional yield of $R(\phi)$ = moles of R produced/moles of A reacted:

$$= \frac{dC_R/dt}{-dC_A/dt} = \frac{k_1 C_A}{(K_1 + K_2) C_A} = \frac{1}{1 + (K_2/K_1)} \qquad \dots (4.30)$$

Now by mass balance: $\qquad\qquad C_{Ao} = C_{Af} + C_{Rf} + C_{Sf}$

Also overall fractional yield of R:

$$\phi_{CA} = \phi = C_{Af} = \frac{\text{All } R\text{-formed}}{\text{All } A\text{-reacted}} = \frac{C_{Rf}}{C_{Ao} - C_{Af}} = \frac{C_{Rf}}{C_{Rf} + C_{Sf}}$$

For 1st reactor:

$$\phi = \frac{C_{R1}}{C_{R1} + C_{S1}} = \frac{0.4}{0.4 + 0.2} = \frac{2}{3} \qquad \dots (4.31)$$

In case of mixed reactor the composition remains constant everywhere, i.e. ϕ is constant throughout the reactor.

$$\frac{1}{1 + (K_2/K_1)} = \frac{2}{3}; \qquad K_1 = 2K_2 \qquad \dots (4.32)$$

In a liquid phase reaction:

$$\tau_1 = \frac{C_{Ao} - C_{A1}}{(-r_A)} = \frac{C_{Ao} - C_{A1}}{(K_1 + K_2) C_{A1}} = \frac{1 - 0.4}{(K_1 + K_2) 0.4} = 2.5$$

$\therefore$ $\qquad\qquad\qquad\qquad K_1 + K_2 = 0.6 \qquad\qquad \dots (4.33)$

From Eqs 4.32 and 4.33; $\qquad K_1 = 0.4 \text{ min}^{-1}$

$$K_2 = 0.2 \text{ min}^{-1}$$

For the 2nd reactor;

$$\tau_2 = \frac{C_{A1} - C_{A2}}{(K_1 + K_2)C_{A2}} = \frac{0.4 - C_{A2}}{0.6\,C_{A2}}; \qquad C_{A2} = 0.1$$

$$\phi_2 = \frac{K_1\,C_{A2}}{(K_1 + K_2)C_{A2}} = \frac{2}{3} = \frac{C_{R2} - C_{R1}}{C_{A1} - C_{A2}}$$

∴ $C_{R2} = 0.6$ ... (4.34)

Hence $C_{S2} = C_{Ao} - C_{A2} - C_{R2}$

$C_{S2} = 1 - 0.1 - 0.6 = 0.3$

∴ $C_{A2} = 0.1$ mol/l

$C_{R2} = 0.6$ mol/l

$C_{S2} = 0.3$ mol/l.

Chapter 5

Steady-state Nonisothermal Reactor Design

INTRODUCTION

Isothermal is a chemical reaction going to completion at one temperature, not needing a change in temperature to continue reaction to completion of nonisothermal reaction diffusion (RD) systems to control the behaviour of many transport and rate processes in physical, chemical and biological systems. Because most reactions are not carried out isothermally, the attention in this chapter is focused on heat effects in chemical reactors. The basic design equations, rate laws, and stoichiometric relationships derived for isothermal reactor design are valid for the design of nonisothermal reactors. However the major differences lies in the method length of a PFR when heat is removed from CSTR. This chapter shows why we need the energy balance and how it will be used to solve reactor design problems. The chapter also discusses energy balance to a point where it can be applied to different reactors, and also energy balance is applied to design adiabatic reactors. In the end chapter describes multiple reactions with heat effects.

ENERGY BALANCE

Energy balance is a systematic presentation of energy flows and transformations in a system. Theoretical basis for an energy balance is the first law of thermodynamics according to which energy cannot be created or destroyed, only modified in form. Energy sources or wave of energy are, therefore, inputs and outputs of the system under observation.

First Law of Thermodynamics

The first law of thermodynamics, an expression of the principle of conservation of energy, states that energy can be transformed (changed from one form to another), but cannot be created or destroyed.

The increase in the internal energy of a system is equal to the amount of energy added by heating the system minus the amount lost as a result of the work done by the system on its surroundings. The first law of thermodynamics says that energy is conserved in any process involving a thermodynamic system and its surroundings. Frequently it is convenient to focus on changes in the assumed internal energy (U) and to regard them as due to a combination of heat (Q) added to the system and work done by the system (W). Taking dU as an incremental (differential) change in internal energy, one writes:

$$dU = \delta Q - \delta W$$

where, δQ and δW are incremental changes in heat and work, respectively. Note that the minus sign in front of δW indicates that a positive amount of work done by the system leads to energy being lost from the system.

Note, also, that some books formulate the first law as:

$$dU = \delta Q + \delta \tilde{W}$$

where, $\delta \tilde{W} = -\delta W$ is the work done on the system by the surroundings.

When a system expands in a quasistatic process, the work done on the system is $-PdV$ whereas the work done by the system while expanding is PdV. In any case, both give the same result when written explicitly as:

$$dU = \delta Q - PdV.$$

Work and heat are due to processes which add or subtract energy, while U is a particular *form* of energy associated with the system. Thus the term 'heat energy' for δQ means 'that amount of energy added as the result of heating' rather than referring to a particular form of energy. Likewise, 'work energy' for δw means 'that amount of energy lost as the result of work'. Internal energy is a property of the system whereas work done and heat supplied are not. A significant result of this distinction is that a given internal energy change (dU) can be achieved by, in principle, many combinations of heat and work. Informally, the law was first formulated by Germain Hess via Hess's Law, and later by Julius Robert von Mayer.

The first explicit statement of the first law of thermodynamics was given by Rudolf Clausius in 1850: 'There is a state function E, called "energy", whose differential equals the work exchanged with the surroundings during an adiabatic process.'

Mathematical formulation

The infinitesimal heat and work in the equations above are denoted by δ rather than d because, in mathematical terms, they are not exact differentials. In other words, they do not describe the state of any system. The integral of an inexact differential depends upon the particular 'path' taken through the space of thermodynamic parameters while the integral of an exact differential depends only upon the initial and final states. If the initial and final states are the same, then the integral of an inexact differential may or may not be zero, but the integral of an exact differential will always be zero. The path taken by a thermodynamic system through a chemical or physical change is known as a thermodynamic process.

An expression of the first law can be written in terms of exact differentials by realising that the work that a system does is, in case of a reversible process, equal to its pressure times the infinitesimal change in its volume. In other words $\delta w = PdV$ where P is pressure and V is volume. Also, for a reversible process, the total amount of heat added to a system can be expressed as $\delta Q = TdS$ where T is temperature and S is entropy. Therefore, for a reversible process:

$$dU = TdS - PdV.$$

Since U, S and V are thermodynamic functions of state, the above relation holds also for non-reversible changes. The above equation is known as the fundamental thermodynamic relation.

In the case where the number of particles in the system is not necessarily constant and may be of different types, the first law is written:

$$dU = \delta Q - \delta W + \sum_i \mu_i dN_i$$

where dN_i is the (small) number of type i particles added to the system, and μ_i is the amount of energy added to the system when one type i particle is added, where the energy of that particle is such that the volume and entropy of the system remains unchanged. μ_i is known as the chemical potential of the type-i particles in the system. The statement of the first law, using exact differentials is now:

$$dU = TdS - PdV + \sum_i \mu_i dN_i$$

If the system has more external variables than just the volume that can change, the fundamental thermodynamic relation generalises to:

$$dU = TdS - \sum_i X_i dx_i + \sum_i \mu_j dN_j.$$

here, the X_i are the generalised forces corresponding to the external variables x_i.

A useful idea from mechanics is that the energy gained by a particle is equal to the force applied to the particle multiplied by the displacement of the particle while that force is applied. Now consider the first law without the heating term: $dU = -PdV$. The pressure P can be viewed as a force (and in fact has units of force per unit area) while dV is the displacement (with units of distance times area). We may say, with respect to this work term, that a pressure difference forces a transfer of volume, and that the product of the two (work) is the amount of energy transferred as a result of the process.

It is useful to view the TdS term in the same light: With respect to this heat term, a temperature difference forces a transfer of entropy, and the product of the two (heat) is the amount of energy transferred as a result of the process. Here, the temperature is known as a 'generalised' force (rather than an actual mechanical force) and the entropy is a generalised displacement.

Similarly, a difference in chemical potential between groups of particles in the system forces a transfer of particles, and the corresponding product is the amount of energy transferred as a result of the process. For example, consider a system consisting of two phases: liquid water and water vapour. There is a generalised 'force' of evaporation which drives water molecules out of the liquid. There is a generalised 'force' of condensation which drives vapor molecules out of the vapour. Only when these two 'forces' (or chemical potentials) are equal will there be equilibrium, and the net transfer will be zero.

The two thermodynamic parameters which form a generalised force-displacement pair are termed 'conjugate variables'. The two most familiar pairs are, of course, pressure-volume, and temperature-entropy.

Types of thermodynamic processes

Paths through the space of thermodynamic variables are often specified by holding certain thermodynamic variables constant. It is useful to group these processes into pairs, in which each variable held constant is one member of a conjugate pair.

The pressure-volume conjugate pair is concerned with the transfer of mechanical or dynamic energy as the result of work.

1. An isobaric (or quasi equilibrium process) occurs at constant pressure. An example would be to have a movable piston in a cylinder, so that the pressure inside the cylinder is always at atmospheric pressure, although it is isolated from the atmosphere. In other words, the system is dynamically connected, by a movable boundary, to a constant-pressure reservoir. Like a balloon contracting when the gas inside cools. As pressure is constant the work for a isobaric process is $W = PdV$.

2. An isochoric (or isovolumetric) process is one in which the volume is held constant, meaning that the work done by the system will be zero. It follows that, for the simple system of two dimensions, any heat energy transferred to the system externally will be absorbed as internal

energy. An isochoric process is also known as an isometric process. An example would be to place a closed tin can containing only air into a fire. To a first approximation, the can will not expand, and the only change will be that the gas gains internal energy, as evidenced by its increase in temperature and pressure. Mathematically, $\delta Q = dU$. We may say that the system is dynamically insulated, by a rigid boundary, from the environment.

The temperature-entropy conjugate pair is concerned with the transfer of thermal energy as the result of heating.

1. An isothermal process occurs at a constant temperature. An example would be to have a system immersed in a large constant-temperature bath. Any work energy performed by the system will be lost to the bath, but its temperature will remain constant. In other words, the system is thermally connected, by a thermally conductive boundary to a constant-temperature reservoir.

2. An isentropic process occurs at a constant entropy. For a reversible process this is identical to an adiabatic process (see below). If a system has an entropy which has not yet reached its maximum equilibrium value, a process of cooling may be required to maintain that value of entropy.

3. An adiabatic process is a process in which there is no energy added or subtracted from the system by heating or cooling. For a reversible process, this is identical to an isentropic process. We may say that the system is thermally insulated from its environment and that its boundary is a thermal insulator. If a system has an entropy which has not yet reached its maximum equilibrium value, the entropy will increase even though the system is thermally insulated.

The above have all implicitly assumed that the boundaries are also impermeable to particles. We may assume boundaries that are both rigid and thermally insulating, but are permeable to one or more types of particle. Similar considerations then hold for the (chemical potential)-(particle number) conjugate pairs.

Open Systems

In open systems, matter may flow in and out of the system boundaries. The first law of thermodynamics for open systems states: the increase in the internal energy of a system is equal to the amount of energy added to the system by matter flowing in and by heating, minus the amount lost by matter flowing out and in the form of work done by the system. The first law for open systems is given by:

$$dU = dU_{in} + \delta Q - dU_{out} - \delta W$$

where, U_{in} is the average internal energy entering the system and U_{out} is the average internal energy leaving the system (Fig. 5.1).

The region of space enclosed by open system boundaries is usually called a control volume, and it may or may not correspond to physical walls. If we choose the shape of the control volume such that all flow in or out occurs perpendicular to its surface, then the flow of matter into the system performs work as if it were a piston of fluid pushing mass into the system, and the system performs work on the flow of matter out as if it were driving a piston of fluid. There are then two types of work performed: flow work described above which is performed on the fluid (this is also often called PV work) and shaft work which may be performed on some mechanical device. These two types of work are expressed in the equation:

$$\delta W = d(P_{out} V_{out}) - d(P_{in} V_{in}) + \delta W_{shaft}$$

Substitution into the equation above for the control volume cv yields:

$$dU_{cv} = dU_{in} + d(P_{in} V_{in}) - dU_{out} - d(P_{out} V_{out}) + \delta Q - \delta W_{shaft}$$

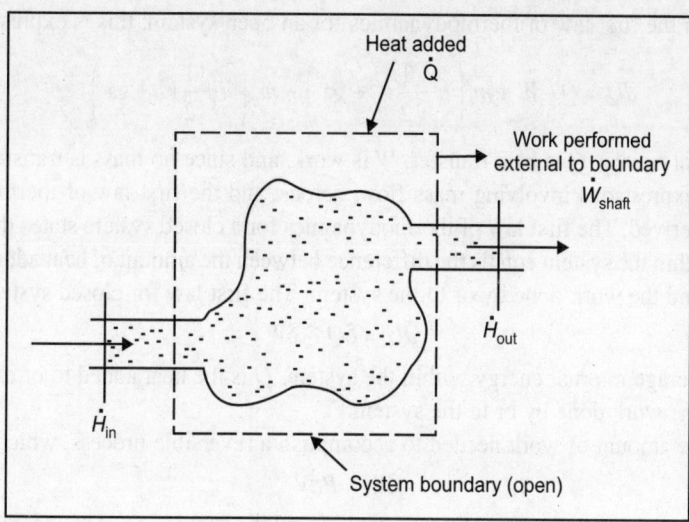

Fig. 5.1. During steady, continuous operation, an energy balance applied to an open system equates shaft work performed by the system to heat added plus net enthalpy added.

The definition of enthalpy, H, permits us to use this thermodynamic potential to account for both internal energy and PV work in fluids for open systems:

$$dU_{cv} + dH_{in} - dH_{out} + \delta Q - \delta W_{shaft}$$

During steady-state operation of a device (see turbine, pump, and engine), any system property within the control volume is independent of time. Therefore, the internal energy of the system enclosed by the control volume remains constant, which implies that dU_{cv} in the expression above may be set equal to zero. This yields a useful expression for the power generation or requirement for these devices in the absence of chemical reactions:

$$\frac{\delta W_{shaft}}{dt} = \frac{dH_{in}}{dt} - \frac{dH_{out}}{dt} + \frac{\delta Q}{dt}$$

This expression is described by the diagram above.

Closed Systems

In a closed system, no mass may be transferred in or out of the system boundaries. The system will always contain the same amount of matter, but heat and work can be exchanged across the boundary of the system. Whether a system can exchange heat, work, or both is dependent on the property of it boundary.

1. Adiabatic boundary—not allowing any heat exchange.
2. Rigid boundary—not allowing exchange of work.

One example is fluid being compressed by a piston in a cylinder. Another example of a closed system is a bomb calorimeter, a type of constant-volume calorimeter used in measuring the heat of combustion of a particular reaction. Electrical energy travels across the boundary to produce a spark between the electrodes and initiates combustion. Heat transfer occurs across the boundary after combustion but no mass transfer takes place either way.

Beginning with the first law of thermodynamics for an open system, this is expressed as:

$$dU = Q - W + m_i \left(h + \frac{1}{2}v^2 + gz \right)_i - m_e \left(h + \frac{1}{2}v^2 + gz \right)_e$$

where, U is internal energy, Q is heat transfer, W is work, and since no mass is transferred in or out of the system, both expressions involving mass flow, zeroes, and the first law of thermodynamics for a closed system is derived. The first law of thermodynamics for a closed system states that the amount of internal energy within the system equals the difference between the amount of heat added to or extracted from the system and the work done by or to the system. The first law for closed systems is stated by:

$$dU = \delta Q - \delta W$$

where, U is the average internal energy within the system, Q is the heat added to or extracted from the system and W is the work done by or to the system.

Substituting the amount of work needed to accomplish a reversible process, which is stated by:

$$\delta W = PdV$$

where, P is the measured pressure and V is the volume, and the heat required to accomplish a reversible process stated by the second law of thermodynamics, the universal principle of entropy, stated by:

$$dQ = TdS$$

where, T is the absolute temperature and S is the entropy of the system, derives the fundamental thermodynamic relationship used to compute changes in internal energy, which is expressed as:

$$\delta U = TdS + PdV$$

The second law of thermodynamics is only true for closed systems. It states that the entropy of an isolated system not in equilibrium will tend to increase over time, approaching maximum value at equilibrium. Overall, in a closed system, the available energy can never increase, and it complement, entropy, can never decrease.

An isolated system is a type of closed system that does not interact with its surroundings in any way. Mass and energy remains constant within the system, and no energy or mass transfer takes place across the boundary.

MATERIAL AND ENERGY BALANCE

Material quantities, as they pass through processing operations, can be described by material balances. Such balances are statements on the conservation of mass. Similarly, energy quantities can be described by energy balances, which are statements on the conservation of energy. If there is no accumulation, what goes into a process must come out. This is true for batch operation. It is equally true for continuous operation over any chosen time interval.

Material and energy balances are very important in an industry. Material balances are fundamental to the control of processing, particularly in the control of yields of the products. The first material balances are determined in the exploratory stages of a new process, improved during pilot plant experiments when the process is being planned and tested, checked out when the plant is commissioned and then refined and maintained as a control instrument as production continues. When any changes occur in the process, the material balances need to be determined again.

The increasing cost of energy has caused the industries to examine means of reducing energy consumption in processing. Energy balances are used in the examination of the various stages of a process, over the whole process and even extending over the total production system from the raw material to the finished product.

Material and energy balances can be simple, at times they can be very complicated, but the basic approach is general. Experience in working with the simpler systems such as individual unit operations will develop the facility to extend the methods to the more complicated situations, which do arise. The increasing availability of computers has meant that very complex mass and energy balances can be set up and manipulated quite readily and therefore used in everyday process management to maximise product yields and minimise costs.

Basic Principles

If the unit operation, whatever its nature is seen as a whole it may be represented diagrammatically as a box, as shown in Fig. 5.2. The mass and energy going into the box must balance with the mass and energy coming out.

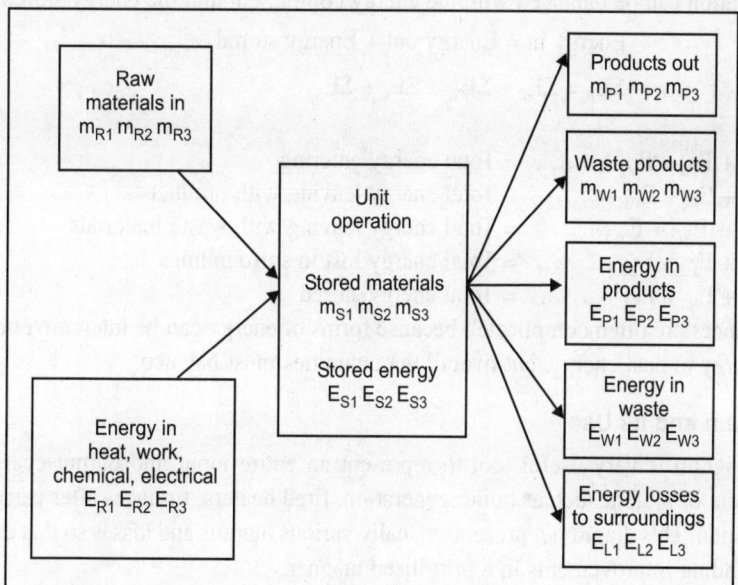

Fig. 5.2. Mass and energy balance.

The law of conservation of mass leads to what is called a mass or a material balance.

Mass in = Mass out + Mass stored

Raw materials = Products + Wastes + Stored materials.

$$\Sigma m_R = \Sigma m_P + \Sigma m_W + \Sigma m_S$$

[where, Σ (sigma) denotes the sum of all terms].

$$\Sigma m_R = \Sigma m_{R1} + \Sigma m_{R2} + \Sigma m_{R3} = \text{Total raw materials}$$

$$\Sigma m_P = \Sigma m_{P1} + \Sigma m_{P2} + \Sigma m_{P3} = \text{Total products.}$$

$$\Sigma m_W = \Sigma m_{W1} + \Sigma m_{W2} + \Sigma m_{W3} = \text{Total waste products}$$

$$\Sigma m_S = \Sigma m_{S1} + \Sigma m_{S2} + \Sigma m_{S3} = \text{Total stored products.}$$

If there are no chemical changes occurring in the plant, the law of conservation of mass will apply also to each component, so that for component A:

$$m_A \text{ in entering materials} = m_A \text{ in the exit materials} + m_A \text{ stored in plant.}$$

For example, in a plant that is producing sugar, if the total quantity of sugar going into the plant is not equalled by the total of the purified sugar and the sugar in the waste liquors, then there is something wrong. Sugar is either being burned (chemically changed) or accumulating in the plant or else it is going unnoticed down the drain somewhere. In this case:

$$M_A = (m_{AP} + m_{AW} + m_{AU})$$

where, m_{AU} is the unknown loss and needs to be identified. So the material balance is now:

$$\text{Raw materials} = \text{Products} + \text{Waste products} + \text{Stored products} + \text{Losses}$$

where Losses are the unidentified materials.

Just as mass is conserved, so is energy conserved in food-processing operations. The energy coming into a unit operation can be balanced with the energy coming out and the energy stored:

$$\text{Energy in} = \text{Energy out} + \text{Energy stored}$$

$$\Sigma E_R = \Sigma E_P + \Sigma E_W + \Sigma E_L + \Sigma E_S$$

where,

$$\Sigma E_R = E_{R1} + E_{R2} + E_{R3} + \dots\dots = \text{Total energy entering}$$
$$\Sigma E_P = E_{P1} + E_{P2} + E_{P3} + \dots\dots = \text{Total energy leaving with products}$$
$$\Sigma E_W = E_{W1} + E_{W2} + E_{W3} + \dots = \text{Total energy leaving with waste materials}$$
$$\Sigma E_L = E_{L1} + E_{L2} + E_{L3} + \dots\dots = \text{Total energy lost to surroundings}$$
$$\Sigma E_S = E_{S1} + E_{S2} + E_{S3} + \dots\dots = \text{Total energy stored}$$

Energy balances are often complicated because forms of energy can be interconverted, for example mechanical energy to heat energy, but overall the quantities must balance.

Sankey Diagram and Its Use

The Sankey diagram is very useful tool to represent an entire input and output energy flow in any energy equipment or system such as boiler generation, fired heaters, furnaces after carrying out energy balance calculation. This diagram represents visually various outputs and losses so that energy managers can focus on finding improvements in a prioritised manner.

Example: The Fig. 5.3 shows a Sankey diagram for a reheating furnace. From the Fig. 5.3, it is clear that exhaust flue gas losses are a key area for priority attention.

Since the furnaces operate at high temperatures, the exhaust gases leave at high temperatures resulting in poor efficiency. Hence a heat recovery device such as air preheater has to be necessarily part of the system. The lower the exhaust temperature, higher is the furnace efficiency.

Material Balances

The first step is to look at the three basic categories: materials in, materials out and materials stored. Then the materials in each category have to be considered whether they are to be treated as a whole, a gross mass balance, or whether various constituents should be treated separately and if so what constituents. To take a simple example, it might be to take dry solids as opposed to total material; this really means separating the two groups of constituents, non-water and water. More complete dissection can separate out chemical types such as minerals, or chemical elements such as carbon. The choice and the detail depend on the reasons for making the balance and on the information that is required. A major factor in industry is, of course, the value of the materials and so expensive raw materials are more likely to be considered than cheaper ones, and products than waste materials.

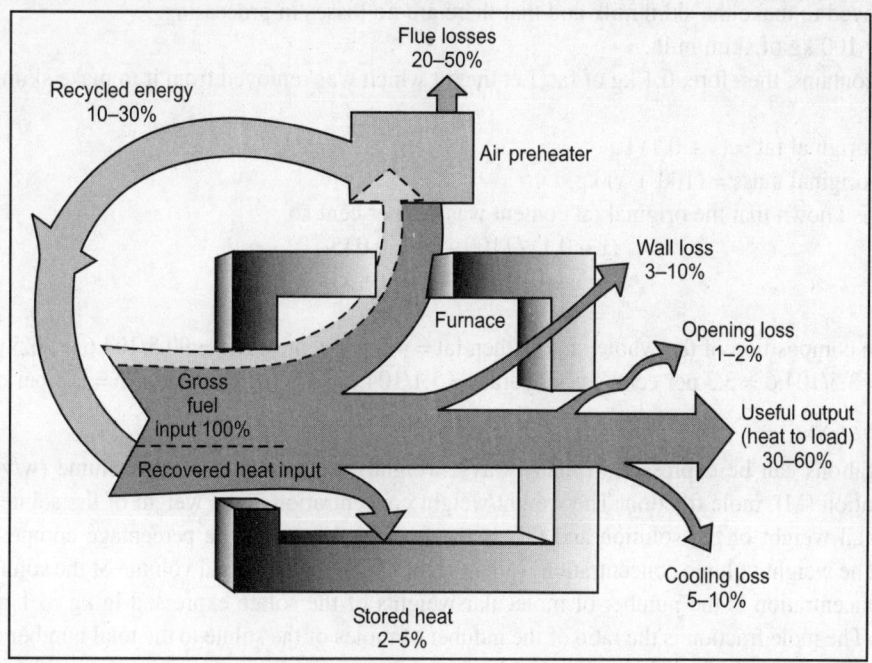

Fig. 5.3. Energy balance for a reheating furnace.

Basis and Units

Having decided which constituents need consideration, the basis for the calculations has to be decided. This might be some mass of raw material entering the process in a batch system, or some mass per hour in a continuous process. It could be: some mass of a particular predominant constituent, for example mass balances in a bakery might be all related to 100 kg of flour entering; or some unchanging constituent, such as in combustion calculations with air where it is helpful to relate everything to the inert nitrogen component; or carbon added in the nutrients in a fermentation system because the essential energy relationships of the growing micro-organisms are related to the combined carbon in the feed; or the essentially inert non-oil constituents of the oilseeds in an oil-extraction process. Sometimes it is unimportant what basis is chosen and in such cases a convenient quantity such as the total raw materials into one batch or passed in per hour to a continuous process are often selected. Having selected the basis, then the units may be chosen such as mass, or concentrations which can be by weight or can be molar if reactions are important.

Total mass and composition

Material balances can be based on total mass, mass of dry solids, or mass of particular components, for example protein.

Example: Constituent balance.

Skim milk is prepared by the removal of some of the fat from whole milk. This skim milk is found to contain 90.5 per cent water, 3.5 per cent protein, 5.1 per cent carbohydrate, 0.1 per cent fat and 0.8 per cent ash. If the original milk contained 4.5 per cent fat, calculate its composition assuming that fat only was removed to make the skim milk and that there are no losses in processing.

Basis: 100 kg of skim milk.

This contains, therefore, 0.1 kg of fat. Let the fat which was removed from it to make skim milk be x kg.

Total original fat = $(x + 0.1)$ kg
Total original mass = $(100 + x)$ kg

and as it is known that the original fat content was 4.5 per cent so

$$(x + 0.1) / (100 + x) = 0.045$$

where,
$$= x + 0.1 = 0.045(100 + x)$$
$$x = 4.6 \text{ kg}$$

So the composition of the whole milk is then fat = 4.5 per cent, water = 90.5/104.6 = 86.5 per cent, protein = 3.5/104.6 = 3.3 per cent, carbohydrate = 5.1/104.6 = 4.9 per cent and ash = 0.8 per cent.

Concentrations

Concentrations can be expressed in many ways: weight/weight (w/w), weight/volume (w/v), molar concentration (M), mole fraction. The weight/weight concentration is the weight of the solute divided by the total weight of the solution and this is the fractional form of the percentage composition by weight. The weight volume concentration is the weight of solute in the total volume of the solution. The molar concentration is the number of molecular weights of the solute expressed in kg in 1 m^3 of the solution. The mole fraction is the ratio of the number of moles of the solute to the total number of moles of all species present in the solution. Notice that in process engineering, it is usual to consider kg moles and in this chapter the term mole means a mass of the material equal to its molecular weight in kilograms. In this chapter percentage signifies percentage by weight (w/w) unless otherwise specified.

Example: Concentrations.

A solution of common salt in water is prepared by adding 20 kg of salt to 100 kg of water, to make a liquid of density 1323 kg/m^3. Calculate the concentration of salt in this solution as a: (a) weight fraction, (b) weight/volume fraction, (c) mole fraction, and (d) molal concentration.

1. Weight fraction:

$$20/(100 + 20) = 0.167: \qquad \% \text{ weight/weight} = 16.7\%$$

2. Weight/volume: A density of 1323kg/m^3 means that 1m^3 of solution weighs 1323 kg, but 1323 kg of salt solution contains

$$(20 \times 1323 \text{ kg of salt})/(100 + 20) = 220.5 \text{ kg salt/m}^3$$

   1 m^3 solution contains 220.5 kg salt.
   Weight/volume fraction = 220.5/1000 = 0.2205
   And so weight/volume = 22.1 per cent

3. Moles of water = 100/18 = 5.56
   Moles of salt $\qquad$ = 20/58.5 = 0.34
   Mole fraction of salt $\quad$ = 0.34/(5.56 + 0.34) = 0.058

4. The molar concentration (M) is 220.5/58.5 = 3.77 moles in m^3.

Note that the mole fraction can be approximated by the (moles of salt/moles of water) as the number of moles of water are dominant, that is the mole fraction is close to $0.34/5.56 = 0.061$. As the solution becomes more dilute, this approximation improves and generally for dilute solutions the mole fraction of solute is a close approximation to the moles of solute/moles of solvent.

In solid/liquid mixtures of all these methods can be used but in solid mixtures the concentrations are normally expressed as simple weight fractions.

With gases, concentrations are primarily measured in weight concentrations per unit volume, or as partial pressures. These can be related through the gas laws. Using the gas law in the form:

$$pV = nRT$$

where, p is the pressure, V the volume, n the number of moles, T the absolute temperature, and R the gas constant which is equal to 0.08206 m^3 atm/mole K, the molar concentration of a gas is then

$$n/V = p/RT$$

and the weight concentration is then nM/V *where M* is the molecular weight of the gas.

The SI unit of pressure is the N/m^2 called the Pascal (Pa). As this is of inconvenient size for many purposes, standard atmospheres (atm) are often used as pressure units, the conversion being 1 atm = 1.013×10^5 Pa, or very nearly 1 atm = 100 kPa.

Example: Air composition.

If air consists of 77 per cent by weight of nitrogen and 23 per cent by weight of oxygen calculate: (a) the mean molecular weight of air, (ii) the mole fraction of oxygen, and (iii) the concentration of oxygen in mole/m^3 and kg/m^3 if the total pressure is 1.5 atmospheres and the temperature is 25°C.

1. Taking the basis of 100 kg of air: it contains 77/28 moles of N$_2$ and 23/32 moles of O$_2$
    Total number of moles   = 2.75 + 0.72 = 3.47 moles.
    So mean molecular weight of air   = 100/3.47 = 28.8
    Mean molecular weight of air   = 28.8
2. The mole fraction of oxygen = 0.72 / (2.75 + 0.72) = 0.72/3.47 = 0.21
    Mole fraction of oxygen   = 0.21
3. In the gas equation, where, n is the number of moles present: the value of R is 0.08206 m^3 atm/mole K and at a temperature of 25°C = 25 + 273 = 298 K, and where $V = 1$ m^3

$$pV = nRT$$

and so, 1.5×1 $= n \times 0.08206 \times 298$
            $n = 0.061$ mole/m^3
weight of air $= n \times$ mean molecular weight
            $= 0.061 \times 28.8 = 1.76$ kg/m^3
and of this 23 per cent is oxygen, so weight of oxygen = $0.23 \times 1.76 = 0.4$ kg in 1 m^3
Concentration of oxygen   = 0.4 kg/m^3
or   0.4/32   = 0.013 mole/m^3

When a gas is dissolved in a liquid, the mole fraction of the gas in the liquid can be determined by first calculating the number of moles of gas using the gas laws, treating the volume as the volume of the liquid, and then calculating the number of moles of liquid directly.

Example: Gas composition.

In the carbonation of a soft drink, the total quantity of carbon dioxide required is the equivalent of 3 volumes of gas to one volume of water at 0°C and atmospheric pressure. Calculate (a) the mass fraction, and (b) the mole fraction of the CO_2 in the drink, ignoring all components other than CO_2 and water.

Basis 1 m^3 of water = 1000 kg

Volume of carbon dioxide added = 3 m^3
From the gas equation, $pV = nRT$
$1 \times 3 = n \times 0.08206 \times 273$
$n = 0.134$ mole
Molecular weight of carbon dioxide = 44
And so weight of carbon dioxide added = $0.134 \times 44 = 5.9$ kg
1. Mass fraction of carbon dioxide in drink = $5.9/(1000 + 5.9) = 5.9 \times 10^{-3}$.
2. Mole fraction of carbon dioxide in drink = $0.134/(1000/18 + 0.134) = 2.41 \times 10^{-3}$.

Types of process situations

Continuous processes

In continuous processes, time also enters into consideration and the balances are related to unit time. Thus in considering a continuous centrifuge separating whole milk into skim milk and cream, if the material holdup in the centrifuge is constant both in mass and in composition, then the quantities of the components entering and leaving in the different streams in unit time are constant and a mass balance can be written on this basis. Such an analysis assumes that the process is in a steady state, that is flows and quantities held up in vessels do not change with time.

Example: Balance across equipment in continuous centrifuging of milk.

If 35,000 kg of whole milk containing 4 per cent fat is to be separated in a 6 hrs period into skim milk with 0.45 per cent fat and cream with 45 per cent fat, what are the flow rates of the two output streams from a continuous centrifuge which accomplishes this separation? Basis 1 hrs flow of whole milk

Mass in: Total mass = 35000/6 = 5833 kg.
Fat = $5833 \times 0.04 = 233$ kg.
And so water plus solids-not-fat = 5600 kg.
Mass out: Let the mass of cream be x kg then its total fat content is $0.45x$. The mass of skim milk is $(5833 - x)$ and its total fat content is $0.0045 (5833 - x)$
Material balance on fat: Fat in = Fat out
$5833 \times 0.04 = 0.0045(5833 - x) + 0.45x$ and so $x = 465$ kg.
So that the flow of cream is 465 kg/hr and skim milk $(5833 - 465) = 5368$ kg/hr.

The time unit has to be considered carefully in continuous processes as normally such processes operate continuously for only part of the total factory time. Usually there are three periods, start up, continuous processing (so-called steady-state) and close down, and it is important to decide what material balance is being studied. Also the time interval over which any measurements are taken must be long enough to allow for any slight periodic or chance variation.

In some instances a reaction takes place and the material balances have to be adjusted accordingly. Chemical changes can take place during a process, for example bacteria may be destroyed during heat processing, sugars may combine with amino acids, fats may be hydrolysed and these affect details of the material balance. The total mass of the system will remain the same but the constituent parts may change, for example in browning the sugars may reduce but browning compounds will increase.

Blending

Another class of situations which arise are blending problems in which various ingredients are combined in such proportions as to give a product of some desired composition. Complicated examples, in which an optimum or best achievable composition must be sought, need quite elaborate calculation methods, such as linear programming, but simple examples can be solved by straightforward mass balances.

Drying

In setting up a material balance for a process a series of equations can be written for the various individual components and for the process as a whole. In some cases where groups of materials maintain constant ratios, then the equations can include such groups rather than their individual constituents. For example in drying vegetables the carbohydrates, minerals, proteins, etc. can be grouped together as 'dry solids', and then only dry solids and water need be taken, through the material balance.

Example: Drying yield.

Potatoes are dried from 14 per cent total solids to 93 per cent total solids. What is the product yield from each 1000 kg of raw potatoes assuming that 8 per cent by weight of the original potatoes is lost in peeling.

Basis 1000 kg potato entering.

As 8 per cent of potatoes are lost in peeling, potatoes to drying are 920 kg, solids 129 kg:

Mass in (kg)		Mass out (kg)	
Potato solids	140 kg	Dried product	92
Water	860 kg	Potato solids	$140 \times (92/100) = 129$ kg
		Associated water	10 kg
		Total product	139 kg
		Losses	
		Peelings-potato	
		Solids	11 kg
		Water	69 kg
		Water evaporated	781 kg
		Total losses	861 kg
		Total	1000 kg

Product yield = 139/1000 = 14 per cent.

Often it is important to be able to follow particular constituents of the raw material through a process. This is just a matter of calculating each constituent.

Energy Balances

Energy takes many forms, such as heat, kinetic energy, chemical energy, potential energy but because of interconversions it is not always easy to isolate separate constituents of energy balances. However, under some circumstances certain aspects predominate. In many heat balances in which other forms of energy are insignificant; in some chemical situations mechanical energy is insignificant and in some mechanical energy situations, as in the flow of fluids in pipes, the frictional losses appear as heat but the details of the heating need not be considered. We are seldom concerned with internal energies.

Therefore, practical applications of energy balances tend to focus on particular dominant aspects and so a heat balance, for example, can be a useful description of important cost and quality aspects of process situation. When unfamiliar with the relative magnitudes of the various forms of energy entering into a particular processing situation, it is wise to put them all down. Then after some preliminary calculations, the important ones emerge and other minor ones can be lumped together or even ignored without introducing substantial errors. With experience, the obviously minor ones can perhaps be left out completely though this always raises the possibility of error.

Energy balances can be calculated on the basis of external energy used per kilogram of product, or raw material processed, or on dry solids or some key component. The energy consumed in food production includes direct energy which is fuel and electricity used on the farm, and in transport and in factories, and in storage, selling, etc. and indirect energy which is used to actually build the machines, to make the packaging, to produce the electricity and the oil and so on. Food itself is a major energy source, and energy balances can be determined for animal or human feeding; food energy input can be balanced against outputs in heat and mechanical energy and chemical synthesis.

In the SI system there is only one energy unit, the joule. However, kilocalories are still used by some nutritionists and British thermal units (Btu) in some heat-balance work.

The two applications used in this chapter are heat balances, which are the basis for heat transfer, and the energy balances used in analysing fluid flow.

Heat balances

The most common important energy form is heat energy and the conservation of this can be illustrated by considering operations such as heating and drying. In these, enthalpy (total heat) is conserved and as with the mass balances so enthalpy balances can be written round the various items of equipment. or process stages, or round the whole plant, and it is assumed that no appreciable heat is converted to other forms of energy such as work.

Enthalpy (H) is always referred to some reference level or datum, so that the quantities are relative to this datum. Working out energy balances is then just a matter of considering the various quantities of materials involved, their specific heats, and their changes in temperature or state (as quite frequently latent heats arising from phase changes are encountered). Figure 5.4 illustrates the heat balance.

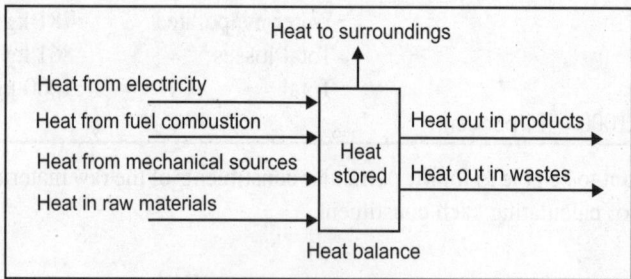

Fig. 5.4. Heat balance.

Heat is absorbed or evolved by some reactions in processing but usually the quantities are small when compared with the other forms of energy entering into food processing such as sensible heat and latent heat. Latent heat is the heat required to change, at constant temperature, the physical state of materials from solid to liquid, liquid to gas or solid to gas. Sensible heat is that heat which when added or subtracted from materials changes their temperature and thus can be sensed. The units of specific heat are J/kg K and sensible heat change is calculated by multiplying the mass by the specific heat by the change in temperature, $(m \times c \times \Delta T)$. The units of latent heat are J/kg and total latent heat change is calculated by multiplying the mass of the material, which changes its phase by the latent heat. Having determined those factors that are significant in the overall energy balance, the simplified heat balance can then be used with confidence in industrial energy studies. Such calculations can be quite simple and straightforward but they give a quantitative feeling for the situation and can be of great use in design of equipment and process.

Example: Dryer heat balance.

A textile dryer is found to consume 4 m³/hr of natural gas with a calorific value of 800 kJ/mole. If the throughput of the dryer is 60 kg of wet cloth per hour, drying it from 55 per cent moisture to 10 per cent moisture, estimate the overall thermal efficiency of the dryer taking into account the latent heat of evaporation only.

   60 kg of wet cloth contains

         60 × 0.55 kg water = 33 kg moisture

         and 60 × (1–0.55) = 27 kg bone dry cloth.

As the final product contains 10 per cent moisture, the moisture in the product is 27/9 = 3 kg.

And so moisture removed/hr = 33 – 3 = 30 kg/hr.

Latent heat of evaporation = 2257 kJ/K.

Heat necessary to supply = 30 × 2257 = 6.8×10^4 kJ/hr.

Assuming the natural gas to be at standard temperature and pressure at which 1 mole occupies 22.4 litres.

Rate of flow of natural gas = 4 m³/hr = (4 × 1000)/22.4 = 179 moles/hr.

Heat available from combustion = 179 × 800 = 14.3×10^4 kJ/hr.

Approximate thermal efficiency of dryer = heat needed/heat used = $6.8 \times 10^4/14.3 \times 10^4$ = 48 per cent.

To evaluate this efficiency more completely it would be necessary to take into account the sensible heat of the dry cloth and the moisture, and the changes in temperature and humidity of the combustion air, which would be combined with the natural gas. However, as the latent heat of evaporation is the dominant term the above calculation gives a quick estimate and shows how a simple energy balance can give useful information.

   Similarly energy balances can be carried out over thermal processing operations, and indeed any processing operations in which heat or other forms of energy are used.

Example: Autoclave heat balance in canning.

An autoclave contains 1000 cans of pea soup. It is heated to an overall temperature of 100°C. If the cans are to be cooled to 40°C before leaving the autoclave, how much cooling water is required if it enters at 15°C and leaves at 35°C?

   The specific heats of the pea soup and the can metal are respectively 4.1 kJ/ kg°C and 0.50 kJ/kg°C. The weight of each can is 60 g and it contains 0.45 kg of pea soup. Assume that the heat content of the autoclave walls above 40°C is 1.6×10^4 kJ and that there is no heat loss through the walls.

   Let w = the weight of cooling water required; and the datum temperature be 40°C, the temperature of the cans leaving the autoclave.

Heat entering

Heat in cans = weight of cans × specific heat × temperature above datum = 1000 × 0.06 × 0.50 × (100–40) kJ = 1.8×10^3 kJ.

Heat in can contents = weight pea soup × specific heat × temperature above datum = 1000 × 0.45 × 4.1 × (100 − 40) = 1.1×10^5 kJ.

Heat in water = weight of water × specific heat × temperature above datum
$$= w \times 4.186 \times (15 - 40)$$
$$= -104.6 \, w \text{ kJ}.$$

Heat leaving

Heat in cans = 1000 × 0.06 × 0.50 × (40 − 40) (cans leave at datum temperature) = 0
Heat in can contents = 1000 × 0.45 × 4.1 × (40 − 40) = 0
Heat in water = $w \times 4.186 \times (35 - 40) = -20.9 \, w$

Table 5.1. Heat-energy balance of cooling process; 40°C as datum line.

Heat entering (kJ)		Heat leaving (kJ)	
Heat in cans	1800	Heat in cans	0
Heat in can contents	110000	Heat in can contents	0
Heat in autoclave wall	16000	Heat in autoclave wall	0
Heat in water	−104.6 w	Heat in water	−20.9 W
Total heat entering	127.800–104.6 w	Total heat leaving	−20.9 W
	Total heat entering =	Total heat leaving	
	127800 − 104.6 w =	−20.9 w	
	w =	1527 kg	

Amount of cooling water required = 1527 kg.

Other forms of energy

Motor power is usually derived, in factories, from electrical energy but it can be produced from steam engines or waterpower. The electrical energy input can be measured by a suitable wattmeter, and the power used in the drive estimated. There are always losses from the motors due to heating, friction and windage; the motor efficiency, which can normally be obtained from the motor manufacturer, expresses the proportion (usually as a percentage) of the electrical input energy, which emerges usefully at the motor shaft and so is available.

When considering movement, whether of fluids in pumping of solids in solids handling, or of foodstuffs in mixers. the energy input is largely mechanical. The flow situations can be analysed by recognising the conservation of total energy whether as energy of motion or potential energy such as pressure energy, or energy lost in friction. Similarly, chemical energy released in combustion can be calculated from the heats of combustion of the fuels and their rates of consumption. Eventually energy emerges in the form of heat and its quantity can be estimated by summing the various sources.

Example: Refrigeration load.

It is desired to freeze 10,000 loaves of bread each weighing 0.75 kg from an initial room temperature of 18°C to a final temperature of −18°C. The bread-freezing operation is to be carried out in an air-blast freezing tunnel. It is found that the fan motors are rated at a total of 80 horsepower and measurements suggest that they are operating at around 90 per cent of their rating, under which conditions their manufacturer's data claims a motor efficiency of 86 per cent. If 1 ton of refrigeration is 3.52 kW, estimate the maximum refrigeration load imposed by this freezing installation assuming: (a) that fans and motors are all within the freezing tunnel insulation, and (b) the fans but not their motors are in the tunnel. The heat-loss rate from the tunnel to the ambient air has been found to be 6.3 kW.

Extraction rate from freezing bread (maximum) $\quad$ = 104 kW

Fan rated horsepower $\quad$ = 80

Now 0.746 kW = 1 horsepower and the motor is operating at 90 per cent of rating, and so (fan + motor) power = (80 × 0.9) × 0.746 = 53.7 kW

1. With motors + fans in tunnel

   Heat load from fans + motors $\quad$ = 53.7 kW

   Heat load from ambient $\quad$ = 6.3 kW

   Total heat load $\quad$ = (104 + 53.7 + 6.3) kW = 164 kW

   $\quad$ = 46 tons of refrigeration

2. With motors outside, the motor inefficiency = (1 − 0.86) does not impose a load on the refrigeration

   Total heat load $\quad$ = (104 + [0.86 × 53.7] + 6.3)

   $\quad$ = 156 kW

   $\quad$ = 44.5 tons of refrigeration

In practice, material and energy balances are often combined as the same stoichiometric information is needed for both.

Method for Preparing Process Flow Chart

The identification and drawing up a unit operation/process is prerequisite for energy and material balance. The procedure for drawing up the process flow diagrams is explained below.

Flow charts are schematic representation of the production process, involving various input resources, conversion steps and output and recycle streams. The process flow may be constructed stepwise, i.e. by identifying the inputs/output/wastes at each stage of the process, as shown in the Fig. 5.5.

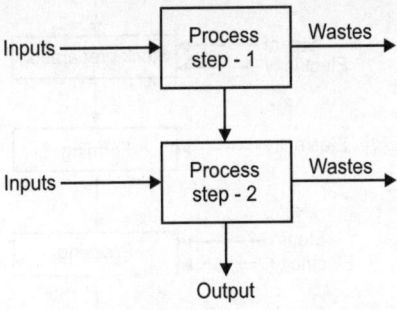

Fig. 5.5. Process flow chart.

Inputs: Inputs of the process could include raw materials, water, steam, energy (electricity, etc.)

Process steps: Should be sequentially drawn from raw material to finished product. Intermediates and any other byproduct should also be represented. The operating process parameters such as temperature, pressure, per cent concentration, etc. should be represented.

The flow rate of various streams should also be represented in appropriate units like m³/hr or kg/hr. In case of batch process the total cycle time should be included.

Wastes: By products could include solids, water, chemicals, energy, etc. For each process steps (unit operation) as well as for an entire plant, energy and mass balance diagram should be drawn.

Output: Of the process is the final product produced in the plant.

Example: Process flow diagram—raw material to finished product: Papermaking is a high energy consuming process. A typical process flow with electrical and thermal energy flow for an integrated waste paper based mill is given in Fig. 5.6.

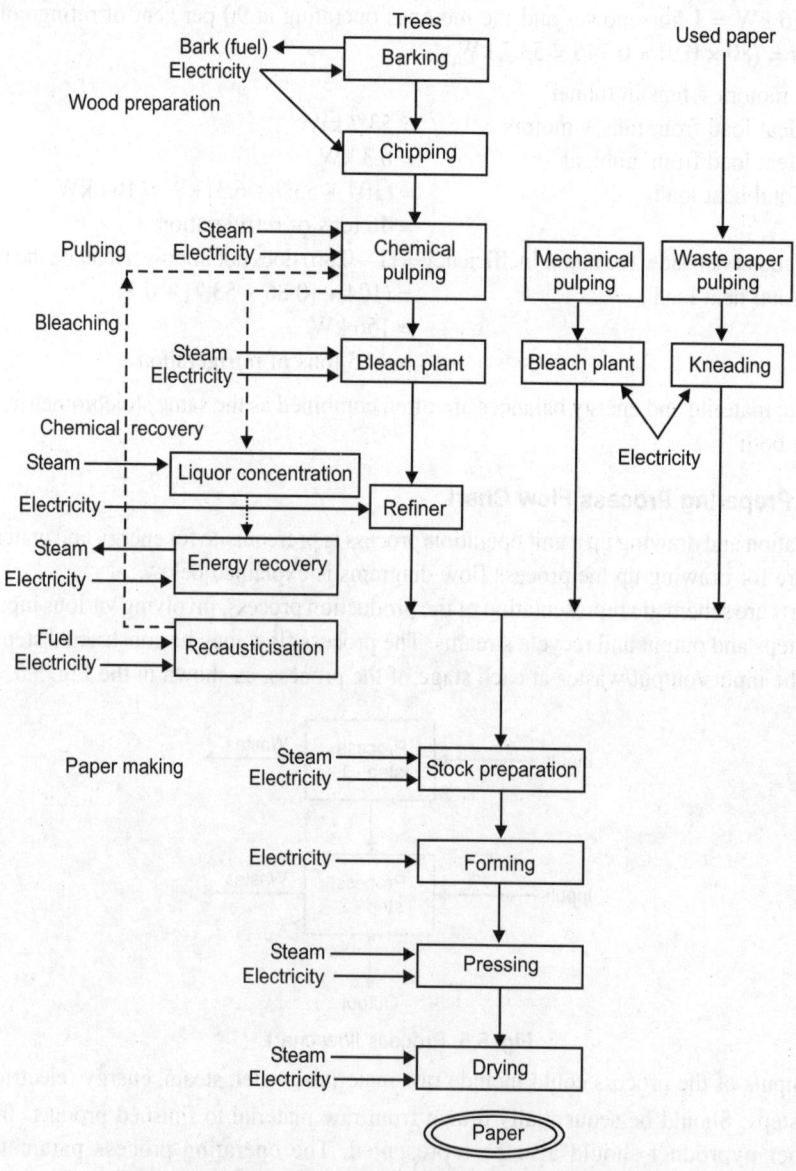

Fig. 5.6. Process flow diagram of pulp and paper industry.

Facility as an Energy System

There are various energy systems/utility services provides the required type of secondary energy such as steam, compressed air, chilled water, etc. to the production facility in the manufacturing plant. A typical plant energy system is shown in Fig. 5.7. Although various forms of energy such as coal, oil, electricity, etc. enters the facility and does its work or heating, the outgoing energy is usually in the form of low temperature heat.

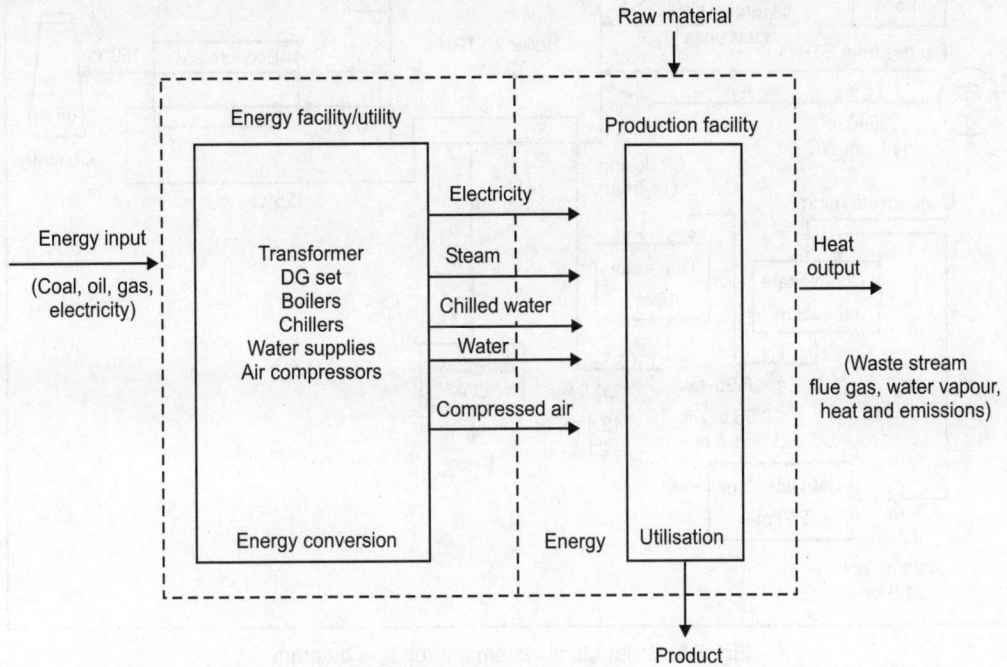

Fig. 5.7. Plant energy system.

The energy usage in the overall plant can be split up into various forms such as:
1. Electrical energy, which is usually purchased as HT and converted into LT supply for end use.
2. Some plants generate their own electricity using DG sets or captive power plants.
3. Fuels such as furnace oil, coal are purchased and then converted into steam or electricity.
4. Boiler generates steam for heating and drying demand.
5. Cooling tower and cooling water supply system for cooling demand.
6. Air compressors and compressed air supply system for compressed air needs.

All energy/utility system can be classified into three areas like generation, distribution and utilisation for the system approach and energy analysis.

A few examples for energy generation, distribution and utilisation are shown below for boiler, cooling tower and compressed air energy system.

Boiler system: Boiler and its auxiliaries should be considered as a system for energy analyses. Energy manager can draw up a diagram as given in Fig. 5.8 for energy and material balance and analysis. This diagram includes many subsystems such as fuel supply system, combustion air system, boiler feed water supply system, steam supply and flue gas exhaust system.

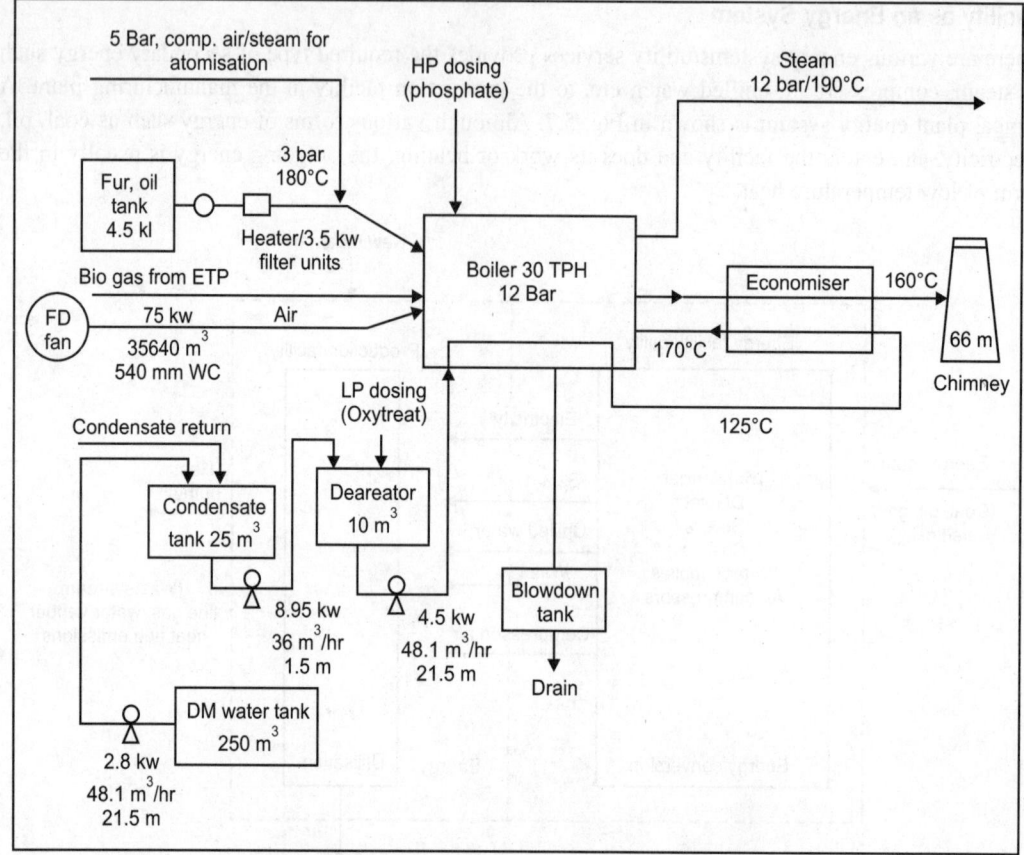

Fig. 5.8. Boiler plant system energy flow diagram.

Cooling tower and cooling water supply system: Cooling water is one of the common utility demands in industry. A complete diagram can be drawn showing cooling tower, pumps, fans, process heat exchangers and return line as given in Fig. 5.9 for energy audit and analysis. All the end use of cooling water with flow quantities should be indicated in the diagram.

Compressed air system

Compressed air is a versatile and safe media for energy use in the plants. A typical compressed air generation, distribution and utilisation diagram is given in Fig. 5.10. Energy analysis and best practices measures should be listed in all the three areas.

How to Carryout Material and Energy (M&E) Balance?

Material and Energy balances are important, since they make it possible to identify and quantify previously unknown losses and emissions. These balances are also useful for monitoring the improvements made in an ongoing project, while evaluating cost benefits. Raw materials and energy in any manufacturing activity are not only major cost components but also major sources of environmental pollution. Inefficient use of raw materials and energy in production processes are reflected as wastes.

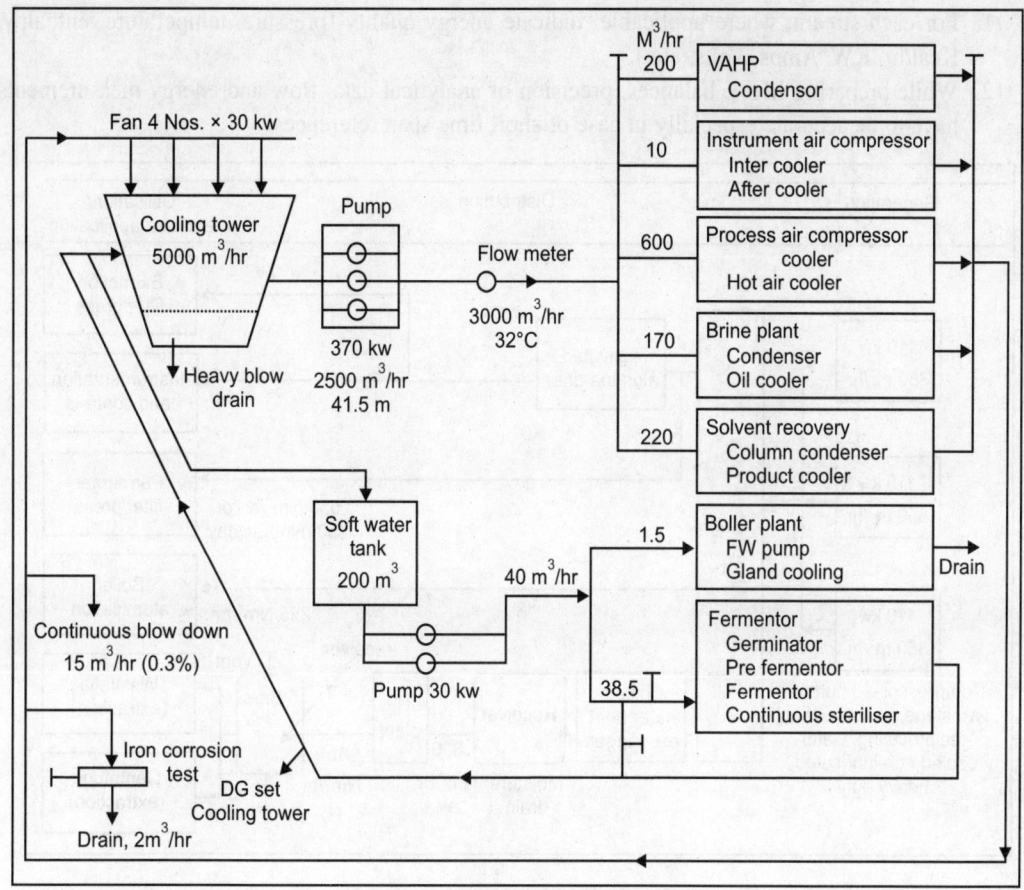

Fig. 5.9. Cooling tower water system.

Guidelines for M & E balance

1. For a complex production stream, it is better to first draft the overall material and energy balance.
2. While splitting up the total system, choose, simple discrete sub-systems. The process flow diagram could be useful here.
3. Choose the material and energy balance envelope such that, the number of streams entering and leaving, is the smallest possible.
4. Always choose recycle streams (material and energy) within the envelope.
5. The measurement units may include, time factor or production linkages.
6. Consider a full batch as the reference in case of batch operations.
7. It is important to include start-up and cleaning operation consumptions [of material and energy resources (M & E)].
8. Calculate the gas volumes at standard conditions.
9. In case of shutdown losses, averaging over long periods may be necessary.
10. Highlight losses and emissions (M&E) at part load operations if prevalent.

11. For each stream, where applicable, indicate energy quality (pressure, temperature, enthalpy, Kcal/hr, KW, Amps, Volts, etc.).

12. While preparing M&E balances, precision of analytical data, flow and energy measurements have to be accurate especially in case of short time span references.

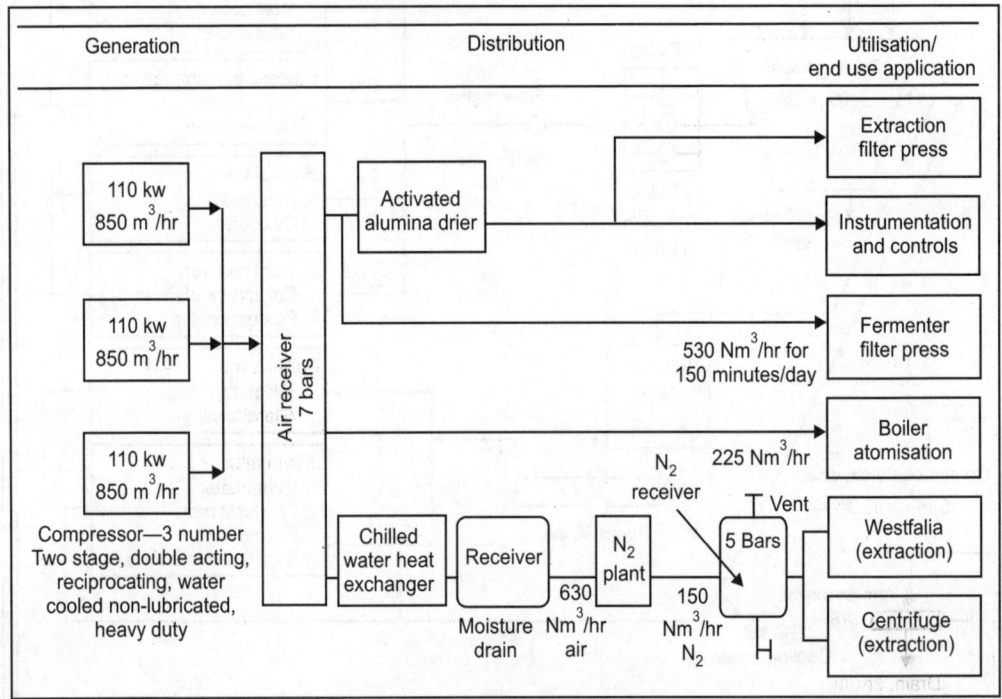

Fig. 5.10. Instrument air system.

The material and energy (M&E) balances along the above guidelines, are required to be developed at the various levels.

1. Overall M&E balance: This involves the input and output streams for complete plant.

2. Section wise M&E balances: In the sequence of process flow, material and energy balances are required to be made for each section/department/cost centres. This would help to prioritise focus areas for efficiency improvement.

3. Equipment-wise M&E balances: M&E balances, for key equipment would help assess performance of equipment, which would in turn help identify and quantify energy and material avoidable losses.

Energy and mass balance calculation procedure

The Energy and Mass balance is a calculation procedure that basically checks if directly or indirectly measured energy and mass flows are in agreement with the energy and mass conservation principles.

This balance is of the utmost importance and is an indispensable tool for a clear understanding of the energy and mass situation achieved in practice.

In order to use it correctly, the following procedure should be used:

1. Clearly identify the problem to be studied.

2. Define a boundary that encloses the entire system or sub-system to be analysed. Entering and leaving mass and energy flows must be measured at the boundary.
3. The boundary must be chosen in such a way that:
    (a) All relevant flows must cross it, all non-relevant flows being within the boundary.
    (b) Measurements at the boundary must be possible in an easy and accurate manner.
4. Select an appropriate test period depending on the type of process and product.
5. Carry out the measurements.
6. Calculate the energy and mass flow.
7. Verify an energy and mass balance. If the balances are outside acceptable limits, then repeat the measurements.
8. The energy release or use in endothermic and exothermic processes should be taken into consideration in the energy balance.

Example/formula

1. Energy supplied by combustion: Q = Fuel consumed × Gross calorific value.
2. Energy supplied by electricity: Q = kWh × 860 kCals.
    Where, Q = thermal energy flow rate produced by electricity (kCals/hr).
3. Continuity equation:

$$\frac{A_1 V_1}{v_1} = \frac{A_2 V_2}{v_2}$$

where, V_1 and V_2 are the velocity in m/s , 'v_1' and 'v_2' the specific volume in m^3/kg and 'A' is the cross sectional area of the pipe in m^2.

4. Heat addition/rejection of a fluid = $mC_p \Delta T$
    where, m is the mass in kg, C_p is the specific heat in kCal/kg·C, ΔT is the difference in temperature in k.

Example 5.1. Heat balance in a boiler.

A heat balance is an attempt to balance the total energy entering a system (e.g. boiler) against that leaving the system in different forms. The Fig. 5.11 illustrates the heat balance and different losses occurring while generating steam.

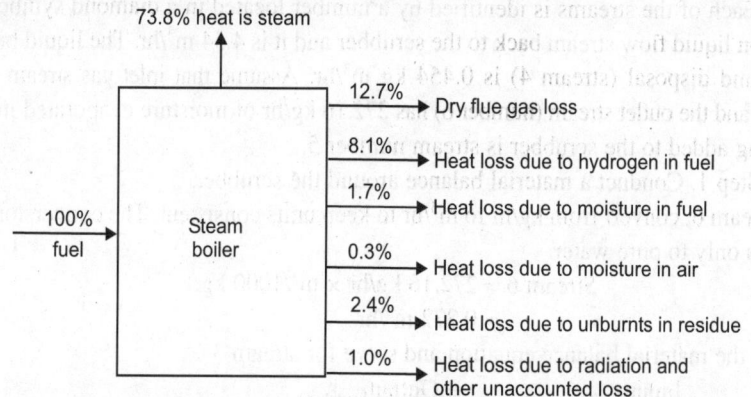

Fig. 5.11. Illustrates the heat balance and different losses occurring while generating steam.

Example 5.2. Mass balance in a cement plant.

The cement process involves gas, liquid and solid flows with heat and mass transfer, combustion of fuel, reactions of clinker compounds and undesired chemical reactions that include sulphur, chlorine, and Alkalies. A typical mass balance is shown in the Fig. 5.12.

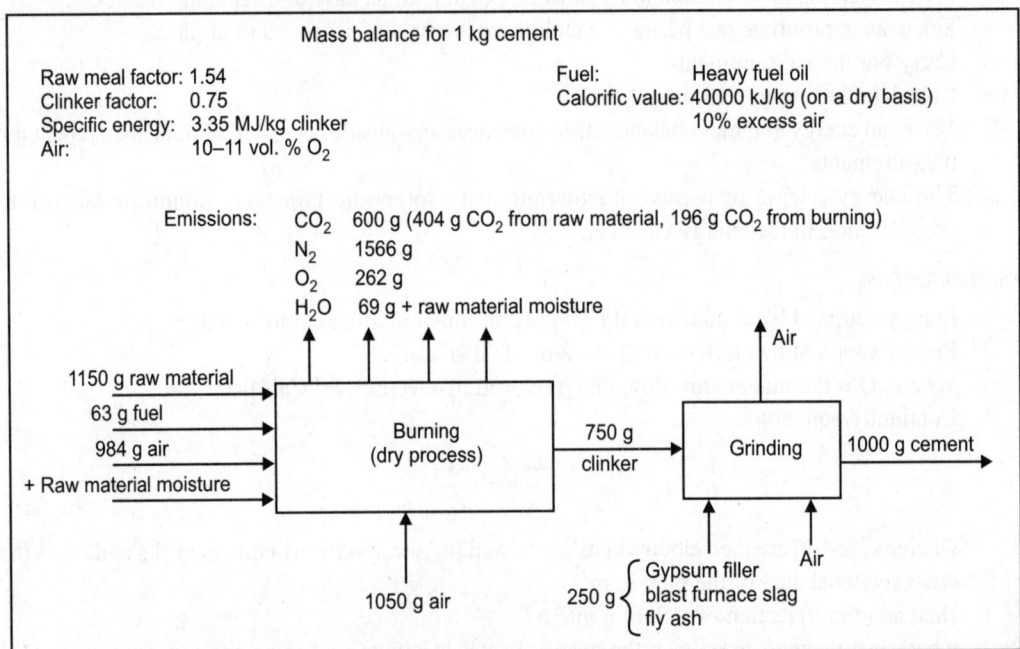

Mass balance for 1 kg cement

Raw meal factor: 1.54
Clinker factor: 0.75
Specific energy: 3.35 MJ/kg clinker
Air: 10–11 vol. % O_2

Fuel: Heavy fuel oil
Calorific value: 40000 kJ/kg (on a dry basis)
10% excess air

Emissions: CO_2 600 g (404 g CO_2 from raw material, 196 g CO_2 from burning)
N_2 1566 g
O_2 262 g
H_2O 69 g + raw material moisture

Air

1150 g raw material
63 g fuel
984 g air
+ Raw material moisture

Burning
(dry process)

750 g
clinker

Grinding

1000 g cement

1050 g air

250 g { Gypsum filler
blast furnace slag
fly ash

Air

Fig. 5.12. A typical mass balance.

Example 5.3. Material requirement for process operations.

A scrubber is used to remove the fine material or dust from the inlet gas stream with a spray of liquid (typically water) so that outlet gas stream meets the required process or emission standards.

How much water must be continually added to wet scrubber shown in Fig. 5.13 in order to keep the unit running? Each of the streams is identified by a number located in a diamond symbol. Stream 1 is the recirculation liquid flow stream back to the scrubber and it is 4.54 m³/hr. The liquid being withdraw for treatment and disposal (stream 4) is 0.454 kg m³/hr. Assume that inlet gas stream (number 2) is completely dry and the outlet stream (number 6) has 272.16 kg/hr of moisture evaporated in the scrubber. The water being added to the scrubber is stream number 5.

Solution: Step 1. Conduct a material balance around the scrubber.

1. For stream 6, convert from kg/hr to m³/hr to keep units consistent. The conversion factor below applies only to pure water.

$$\text{Stream } 6 = 272.16 \text{ kg/hr} \times \text{m}^3/1000 \text{ kg.}$$
$$= 0.272 \text{ m}^3/\text{hr.}$$

2. Set up the material balance equation and solve for stream 3.

$$\text{Input}_{\text{Scrubber}} = \text{Output}_{\text{Scrubber}}.$$
$$\text{Stream 1 + Stream 2} = \text{Stream 3 + Stream 6.}$$

$$4.54 \text{ m}^3/\text{hr} + 0 \qquad = y \text{ m}^3/\text{hr} + 0.272 \text{ m}^3/\text{hr}.$$
$$\text{Stream 3} \qquad = y \text{ m}^3/\text{hr} = 4.27 \text{ m}^3/\text{hr}.$$

Step 2. Conduct a material balance around the recirculation tank. Solve for Stream 5.

$$\text{Input}_{\text{Tank}} \qquad = \text{Output}_{\text{Tank}}.$$
$$\text{Stream 3} + \text{Stream 5} \qquad = \text{Stream 1} + \text{Stream 4}.$$
$$4.25 \text{ m}^3/\text{hr} + x \text{ m}^3/\text{hr} \qquad = 4.54 \text{ m}^3/\text{hr} + 0.454 \text{ m}^3/\text{hr}.$$
$$\text{Stream} \qquad 5 = x \text{ m}^3/\text{hr} \qquad = 5 \text{ m}^3/\text{hr} - 4.27 \text{ m}^3/\text{hr}.$$
$$= 0.73 \text{ m}^3/\text{hr}.$$

If it is to calculate only the makeup water at 5,

$$\text{Stream 5} \qquad = \text{Stream 4} + \text{Stream 6}.$$
$$= 0.454 + 0.272$$
$$= 0.73 \text{ m}^3/\text{hr}.$$

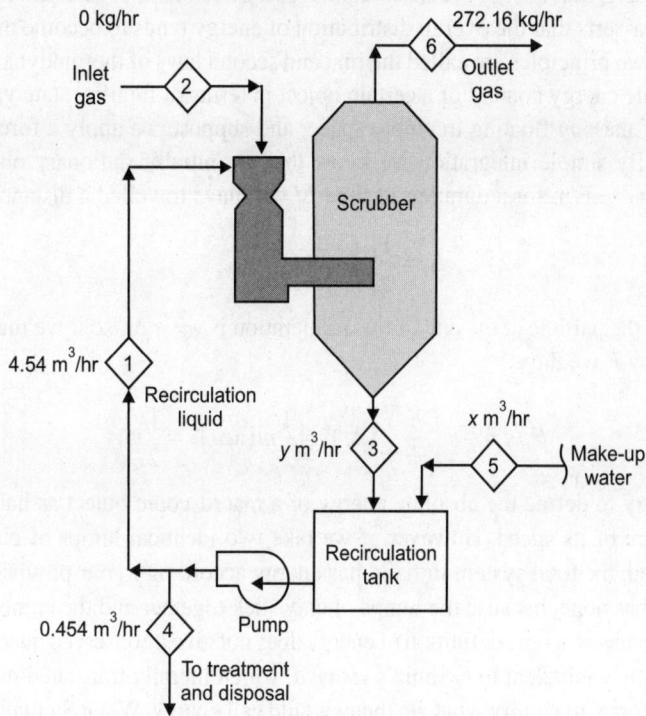

Fig. 5.13. Example of material balance.

One of the key steps in solving Example 5.3 was drawing a simple sketch of the system. This is absolutely necessary so that it is possible to conduct the material balances. Drawings are a valuable first step when solving a wide variety of problems, even ones that appears simple.

The drawing is a very useful way to summarise what we know and what we need to know. It helps visualise the solution. If the problem involves dimensional quantities (such as stream flow quantities), the dimensions should be included on the sketch. They serve as reminders of the need to convert the data into consistent units.

ENERGY, ENTROPY AND ENTHALPY

In the mechanical sense, work was originally defined in terms of lifting a weight to a certain height. The quantity of work was defined as the product of the weight and the height. This definition was then generalised, so that work was considered to be done whenever any kind of force is exerted through some distance. The quantity of work is the force multiplied by the distance. When two physical systems interact, one of them may do work on the other. We find it convenient to assign to each physical system a quantity called energy, with the same units as the units of work. Whenever a system does work on its surroundings, we say its energy has been reduced by the amount of work done, and whenever a system has work done on it (by some other system) we say its energy has been increased by that amount of work. By the law of action and re-action, all work that is done by one system is done on another system. It follows that the total amount of energy is conserved. (Notice that we have not established the absolute value of energy, we have merely discussed changes in the energy levels.)

Classical thermodynamics is founded on two principles, both of which involve the concept of energy. The first principle asserts that energy is conserved, i.e. energy can neither be created nor destroyed, and the second principle asserts that the overall distribution of energy tends to become more uniform, never less uniform. These two principles are called the first and second laws of thermodynamics. In attempting to express the absolute energy content of a certain object in terms of familiar state variables, consider a stationary particle of mass m floating in empty space, and suppose we apply a force F to this particle over a distance Δs. By simple integration we know that an initially stationary object subjected to a constant acceleration $a = F/m$ for a duration of time Δt will have travelled a distance:

$$\Delta s = \frac{1}{2}\left(\frac{F}{m}\right)(\Delta t)^2$$

The velocity v of the particle at the end of the acceleration is $v = a\,\Delta t$, so if we multiply both sides of the above equation by F we have:

$$F\Delta s = \frac{1}{2}m\left(\frac{F}{m}\right)^2(\Delta t)^2 = \frac{1}{2}m(a\Delta t)^2 = \frac{1}{2}mv^2$$

Thus, we might try to define the absolute energy of a macroscopic object as half the product of its mass times the square of its speed. However, if we take two identical lumps of clay and throw them together at high speed, the total system initially has energy according to our provisional definition, but after the collision it has none, because the lumps of clay stick together and the combined lump has zero speed. Therefore, this macroscopic definition of energy does not give a conserved quantity. This definition of energy is essentially equivalent to Leibniz's *vis viva*, which literally translated means 'living force', but using the word 'force' to signify what we today would call energy. When Samuel Clarke pointed out that this quantity is not conserved in such collisions, Leibniz replied.

Here we recognise that in order for energy to be conserved we must consider not only the macroscopic kinetic energies of aggregate bodies, but also the microscopic kinetic energies of their constituent particles. The latter is usually regarded as the heat content of the aggregate body. Hence our concept of energy— if energy is to be conserved—must include not only the mechanical kinetic energies of aggregate bodies but also the internal heats of those bodies.

However, even taking internal heat of massive objects into account, we can still find processes in which the quantity of energy seems not to be conserved. For example, a satellite in an elliptical orbit around a gravitating body moves more rapidly when it is near the gravitating body than when it is far from that body, so it's kinetic energy changes significantly (while it is internal heat content is not significantly altered). This shows that, to maintain the principle of energy conservation, we must include gravitational and other forms of potential in our definition of energy. (This relates to the original conception of work, which was based on raising objects in a gravitational field.) Likewise when we discover that material bodies can lose energy by emitting electromagnetic radiation, we must expand our definition of energy to include electromagnetic waves. This illustrates how we use the principle of energy conservation to define the concept of 'energy'. We classify and quantify phenomena in whatever way is necessary to ensure that energy is conserved. (The great merit of the concept of energy is that the classifications and quantifications to which it leads are extremely useful, and provide a very economical and unified way of formulating physical laws.)

Once we have developed our (provisional) concept of energy, we quickly discover that knowledge of the total quantity of energy in a given system is not sufficient to fully characterise that system. It is also important to specify how the energy is distributed among the different parts of the systems. For example, consider a system consisting of two identical blocks of metal sitting next to each other in an isolated container. If the blocks have the same heat content they will have the same temperature, and the system will be in equilibrium and will not change its condition as time passes. However, if the same total amount of heat energy is distributed asymmetrically, so one block is hotter than the other, the system will not be in equilibrium. In this case, heat will tend to flow from the hot to the cold block, so the condition of the system will change as time passes. Eventually it will approach equilibrium, once enough heat has been transferred to equalise the temperatures. Thus in order to know how a system will change — and even to assess the potential for such change — we need to know not only the total amount of energy in the system, but also how that energy is distributed.

The example of two blocks also illustrates the intuitive fact that the distribution of energy in an isolated system tends to become more uniform as time passes, never less uniform. This is essentially the second law of thermodynamics. Given two blocks in thermal equilibrium (i.e. at the same temperature), we do not expect heat to flow preferentially from one to the other such that one heats up and the other cools down. This would be like a stone rolling uphill.

To quantify the tendency for energy to flow in such a way as to make the distribution more uniform, we need to quantify the notion of 'uniformity'. Historically this was first done in terms of the macroscopic properties of gases, liquids, and solids. We seek a property, which we will call entropy, that is a measure of the uniformity of the distribution of energy, and we would like this property to be such that the entropy of a system is equal to the sum of the entropies of the individual parts of the system. For example, with our two metal blocks we would like to be able to assign values of entropy to each individual block, and then have the total entropy of the system equal to the sum of those two values.

Let s_1 and s_2 denote the entropies of block 1 and block 2 respectively, and let T_1 and T_2 denote the temperatures of the blocks. If a small quantity δq of heat flows from block 1 to block 2, how do the entropies of the blocks change? Clearly the change in entropy cannot be just a multiple of the energy, because the total energy is always conserved, so entropy would always be conserved as well. We want entropy to increase as the uniformity of the energy distribution increases. To accurately represent uniformity, a given amount of heat energy ought to represent more entropy at low temperature than it does at high temperature. Therefore, it is reasonable to weight the changes in energy by the inverse of the temperature. In other words, for a block of temperature T, we define the change in entropy ds

resulting from the addition of a small amount of energy δq to be $\delta q/T$. (We assume δq is small enough that it does not significantly change the temperature of the block.) It follows that if a small amount of heat δq flows from block 1 to block 2, the entropy of block 1 will be decreased by $\delta q/T_1$, and the entropy of block 2 will be increased by $\delta q/T_2$, so the net change in entropy of the whole system is:

$$ds = \delta q \left(\frac{1}{T_2} - \frac{1}{T_1} \right) = \delta q \left(\frac{T_1 - T_2}{T_1 T_2} \right)$$

Thus the net change in entropy is positive if and only if the temperature T_1 of the heat source is greater than the temperature T_2 of the heat sink.

Of course, we might have defined ds corresponding to the addition of a small amount of energy δq in some other way, such as $-T\delta q$. Then the net change in entropy for our example would have been $\delta q(T_1 - T_2)$, which again is positive if and only if T_1 is greater than T_2. However, defining entropy to be a negative value for the addition of heat seems rather incongruous. Also, recognizing that changes in the temperature of a macroscopic object are roughly proportional to changes in its energy, we could conceptually replace T with q (and δq with dq), so our two candidate expressions for the differential entropy are $ds = dq/q$ and $ds = -qdq$. Notice that, up to an additive constant, the first implies $s = \log(q)$ whereas the second would imply $s = -q^2$. When defined in the context of statistical thermodynamics we find that entropy is given by $s = k \log(W)$ where, W signifies the number of microstates for the given macrostate. Thus, up to an exponent, we can roughly equate the heat content of an object with the number of microstates.

Since entropy is a thermodynamic state property, its value depends only on the state of a system, not on the history of the system. Therefore, to determine the change in entropy of a system from one state to another, it is sufficient to evaluate the change for a reversible process between those two states; the change in entropy for any other process connecting the same two states will be the same as the change for the reversible process. (Note that this 'conservatism' applies only to the entropy change between two states, not to the amount of heat flow associated with the process.) For large transfers of heat, such that the temperatures of the objects are significantly altered, we need only imagine a reversible 'quasi-equilibrium' process of extracting heat from the hotter object and then another reversible process of adding heat to the colder object, and integrate the quantity $\delta q/T$ in each case to give the total changes in entropy.

Another useful state variable is enthalpy, defined as the sum of the internal energy and the product of pressure and volume. In other words, $H = U + PV$. The justification for defining this variable is really only a matter of convenience, because we often find that the sum $U + PV$ occurs in thermodynamic equations. This is not surprising, because the work done by a quantity of gas depends on the product of pressure times volume. When a gas expands quasi-statically at constant pressure, the incremental work δW done on the boundary is PdV, so from the energy equation $dU = \delta Q - \delta W$ we have $\delta Q = dU + PdV$. Noting that, at constant pressure, $dH = dU + PdV$, it follows that $\delta Q = dH$ for this process. This explains why enthalpy is often a convenient state variable, especially in open systems. Obviously enthalpy has units of energy, but it does not necessarily have a direct physical interpretation as a quantity of heat. In other words, enthalpy is not any specific form of energy, it is just a defined variable that often simplifies the calculations in the solution of practical thermodynamic problems.

ADIABATIC PROCESS

In thermodynamics, an adiabatic process or an isocaloric process is a thermodynamic process in which no heat is transferred to or from the working fluid. The term 'adiabatic' literally means impassable, coming from the Greek roots; this etymology corresponds here to an absence of heat transfer. Conversely, a process that involves heat transfer (addition or loss of heat to the surroundings) is generally called diabatic. Although the two terms can often be interchanged, adiabatic processes may be considered a subset of isocaloric processes; the remaining complement subset of isocaloric processes being processes where net heat transfer does not diverge regionally such as in an idealised case with mediums of infinite thermal conductivity or non-existent thermal capacity.

In an adiabatic irreversible process, $dQ = 0$ is not equal to TdS ($TdS > 0$). $dQ = TdS = 0$ holds for reversible processes only. For example, an adiabatic boundary is a boundary that is impermeable to heat transfer and the system is said to be adiabatically (or thermally) insulated; an insulated wall approximates an adiabatic boundary. Another example is the adiabatic flame temperature, which is the temperature that would be achieved by a flame in the absence of heat loss to the surroundings. An adiabatic process that is reversible is also called an isentropic process. Additionally, an adiabatic process that is irreversible and extracts no work is in an isenthalpic process, such as viscous drag, progressing towards a nonnegative change in entropy.

One opposite extreme — allowing heat transfer with the surroundings, causing the temperature to remain constant — is known as an isothermal process. Since temperature is thermodynamically conjugate to entropy, the isothermal process is conjugate to the adiabatic process for reversible transformations.

A transformation of a thermodynamic system can be considered adiabatic when it is quick enough that no significant heat is transferred between the system and the outside. At the opposite extreme, a transformation of a thermodynamic system can be considered isothermal if it is slow enough so that the system's temperature remains constant by heat exchange with the outside.

Adiabatic Heating and Cooling

Adiabatic changes in temperature occur due to changes in pressure of a gas while not adding or subtracting any heat. Adiabatic heating occurs when the pressure of a gas is increased from work done on it by its surroundings, i.e. a piston. Diesel engines rely on adiabatic heating during their compression stroke to elevate the temperature sufficiently to ignite the fuel.

Adiabatic heating also occurs in the earth's atmosphere when an air mass descends, for example, in a katabatic wind or Foehn wind flowing downhill.

Adiabatic cooling occurs when the pressure of a substance is decreased as it does work on its surroundings. Adiabatic cooling does not have to involve a fluid. One technique used to reach very low temperatures (thousandths and even millionths of a degree above absolute zero) is adiabatic demagnetisation, where the change in magnetic field on a magnetic material is used to provide adiabatic cooling. Adiabatic cooling also occurs in the earth's atmosphere with orographic lifting and lee waves, and this can form pileus or lenticular clouds if the air is cooled below the dew point. Rising magma also undergoes adiabatic cooling before eruption. Such temperature changes can be quantified using the ideal gas law or the hydrostatic equation for atmospheric processes. It should be noted that no process is truly adiabatic. Many processes are close to adiabatic and can be easily approximated by using an adiabatic assumption, but there is always some heat loss; as no perfect insulators exist.

Ideal Gas (Reversible Case)

The mathematical equation for an ideal fluid undergoing a reversible (i.e. no entropy generation) adiabatic process is:

$$PV^\gamma = \text{Constant}$$

where, P is pressure, V is specific or molar volume, and

$$\gamma = \frac{C_P}{C_V} = \frac{\alpha+1}{\alpha},$$

C_P being the specific heat for constant pressure, C_V being the specific heat for constant volume, γ is the adiabatic index, and α is the number of degrees of freedom divided by 2 (3/2 for monatomic gas, 5/2 for diatomic gas) (Fig. 5.14).

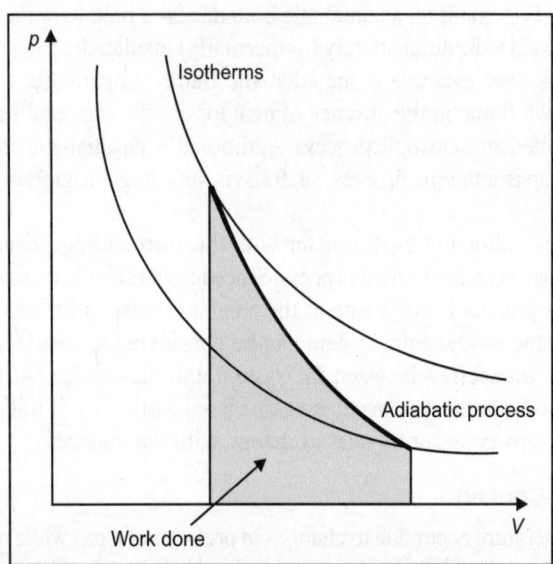

Fig. 5.14. For a simple substance, during an adiabatic process in which the volume increases, the internal energy of the working substance must decrease.

For a monatomic ideal gas $\gamma = 5/3$ and for a diatomic gas (such as nitrogen and oxygen, the main components of air) $\gamma = 7/5$. Note that the above formula is only applicable to classical ideal gases and not Bose–Einstein or Fermi gases.

For reversible adiabatic processes, it is also true that

$$P^{\gamma-1}T^{-\gamma} = \text{constant}$$

$$VT^{\alpha} = \text{constant}$$

where, T is an absolute temperature.

This can also be written as:

$$TV^{\gamma-1} = \text{constant}.$$

Derivation of continuous formula

The definition of an adiabatic process is that heat transfer to the system is zero, $Q = 0$. Then, according to the first law of thermodynamics,

$$dU + \delta W = \delta Q = 0, \qquad \qquad \text{... (5.1)}$$

where, dU is the change in the internal energy of the system and δW is work done by the system. Any work (δW) done must be done at the expense of internal energy U, since no heat δQ is being supplied from the surroundings. Pressure-volume work δW done by the system is defined as:

$$\delta W = P\, dV. \qquad \qquad \text{... (5.2)}$$

However, P does not remain constant during an adiabatic process but instead changes along with V. It is desired to know how the values of dP and dV relate to each other as the adiabatic process proceeds. For an ideal gas the internal energy is given by:

$$U = \alpha n\, RT, \qquad \qquad \dots (5.3)$$

where, R is the universal gas constant and n is the number of moles in the system (a constant).

Differentiating Eq. 5.3 and use of the ideal gas law, $PV = nRT$, yields

$$dU = \alpha n\, RT = \alpha\, d\,(PV) = \alpha\,(P\, dV + V\, dP). \qquad \dots (5.4)$$

Equation 5.4 is often expressed as $dU = nCV\, dT$ because $C_V = \alpha R$.

Now substitute Eqs 5.2 and 5.4 into Eq. 5.1 to obtain:

$$-P\, dV = \alpha P\, dV + \alpha V\, dP,$$

simplify:

$$-(\alpha + 1)\, P\, dV = \alpha V\, dP,$$

and divide both sides by PV:

$$-(\alpha+1)\frac{dV}{V} = \alpha\frac{dP}{P}.$$

After integrating the left and right sides from V_0 to V and from P_0 to P and changing the sides respectively,

$$\ln\left(\frac{P}{P_0}\right) = -\frac{\alpha+1}{\alpha}\ln\left(\frac{V}{V_0}\right).$$

Exponentiate both sides,

$$\left(\frac{P}{P_0}\right) = \left(\frac{V}{V_0}\right)^{-\frac{\alpha+1}{\alpha}}$$

and eliminate the negative sign to obtain:

$$\left(\frac{P}{P_0}\right) = \left(\frac{V_0}{V}\right)^{\frac{\alpha+1}{\alpha}}.$$

Therefore,

$$\left(\frac{P}{P_0}\right)\left(\frac{V}{V_0}\right)^{\frac{\alpha+1}{\alpha}} = 1$$

and

$$PV^{\frac{\alpha+1}{\alpha}} = P_0V_0^{\frac{\alpha+1}{\alpha}} = PV^{\gamma} = \text{Constant}.$$

Derivation of discrete formula

The change in internal energy of a system, measured from Eq. 5.5 to Eq. 5.6, is equal to:

$$\Delta U = \alpha Rn_2T_2 - \alpha Rn_1T_1 = \alpha R(n_2T_2 - n_1T_1) \qquad \dots (5.5)$$

At the same time, the work done by the pressure-volume changes as a result from this process, is equal to:

$$W = \int_{V_1}^{V_2} P\, dV$$

... (5.6)

Since we require the process to be adiabatic, the following equation needs to be true:

$$\Delta U + W = 0$$

... (5.7)

By the previous derivation,

$$PV^{\gamma} = \text{Constant} = P_1 V_1^{\gamma}$$

... (5.8)

Rearranging Eq. 5.8 gives:

$$P = P_1 \left(\frac{V_1}{V} \right)^{\gamma}$$

Substituting this into Eq. 5.6 gives:

$$W = \int_{V_1}^{V_2} P_1 \left(\frac{V_1}{V} \right)^{\gamma} dV$$

Integrating,

$$W = P_1 V_1^{\gamma} \frac{V_2^{1-\gamma} - V_1^{1-\gamma}}{1-\gamma}$$

Substituting,

$$\gamma = \frac{\alpha+1}{\alpha},$$

$$W = \alpha P_1 V_1^{\gamma} \left(V_2^{1-\gamma} - V_1^{1-\gamma} \right)$$

Rearranging,

$$W = \alpha P_1 V_1 \left(\left(\frac{V_2}{V_1} \right)^{1-\gamma} - 1 \right)$$

Using the ideal gas law and assuming a constant molar quantity (as often happens in practical cases),

$$W = -\alpha n R T_1 \left(\left(\frac{V_2}{V_1} \right)^{1-\gamma} - 1 \right)$$

By the continuous formula,

$$\left(\frac{P_2}{P_1} \right) = \left(\frac{V_2}{V_1} \right)^{-\gamma}$$

or,

$$\left(\frac{P_2}{P_1} \right)^{\frac{-1}{\gamma}} = \frac{V_2}{V_1}$$

Substituting into the previous expression for W,

$$W = -\alpha nRT_1\left(\left(\frac{P_2}{P_1}\right)^{\frac{\gamma-1}{\gamma}} - 1\right)$$

Substituting this expression and Eq. 5.5 in Eq. 5.7 gives:

$$\alpha nR(T_2 - T_1)\alpha nRT_1\left(\left(\frac{P_2}{P_1}\right)^{\frac{\gamma-1}{\gamma}} - 1\right)$$

Simplifying,

$$T_2 - T_1 = T_1\left(\left(\frac{P_2}{P_1}\right)^{\frac{\gamma-1}{\gamma}} - 1\right)$$

Simplifying,

$$\frac{T_2}{T_1} - 1 = \left(\frac{P_2}{P_1}\right)^{\frac{\gamma-1}{\gamma}} - 1$$

Simplifying,

$$T_2 - T_1\left(\frac{P_2}{P_1}\right)^{\frac{\gamma-1}{\gamma}}.$$

Graphing Adiabats

An adiabat is a curve of constant entropy on the P–V diagram. Properties of adiabats on a P–V diagram are:

1. Every adiabat asymptotically approaches both the V axis and the P axis (just like isotherms).
2. Each adiabat intersects each isotherm exactly once.
3. An adiabat looks similar to an isotherm, except that during an expansion, an adiabat loses more pressure than an isotherm, so it has a steeper inclination (more vertical).
4. If isotherms are concave towards the 'north-east' direction (45°), then adiabats are concave towards the 'east north-east' (31°).
5. If adiabats and isotherms are graphed severally at regular changes of entropy and temperature, respectively (like altitude on a contour map), then as the eye moves towards the axes (towards the south-west), it sees the density of isotherms stay constant, but it sees the density of adiabats grow. The exception is very near absolute zero, where the density of adiabats drops sharply and they become rare.

The following Fig. 5.15 is a P–V diagram with a superposition of adiabats and isotherms:

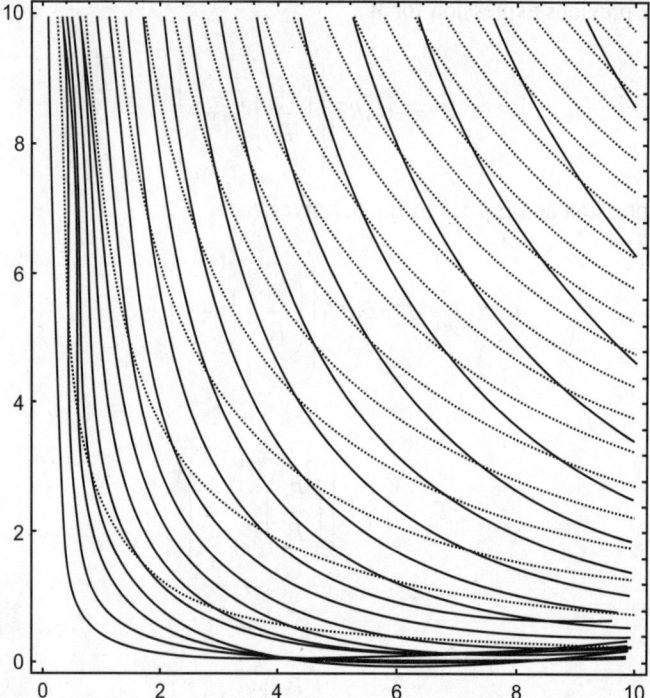

Fig. 5.15. *P–V* diagram with a superposition of adiabats and isotherms.

The isotherms are the red curves and the adiabats are the black curves. The adiabats are isentropic. Volume is the horizontal axis and pressure is the vertical axis.

Adiabatic reactor analysis for methanol synthesis

An important industrial reaction is the combination of carbon monoxide with hydrogen to produce methanol. Methanol is quite useful for a variety of chemical synthesis reactions, including the transesterification of triglycerides in vegetable oils for biodiesel production. The gaseous mixture of carbon monoxide and hydrogen can be used to synthesise a wide array of hydrocarbons, including synthetic fuels, and is therefore often referred to as 'syn-gas'. Syngas can be obtained from coal, as discussed in this chapter by Octave Levenspiel.

The overall reaction for methanol synthesis from syngas is written as:

$$CO + 2H_2 \rightarrow CH_3OH$$

And can be approximated as an elementary reaction, such that the rate expression (assuming irreversible reaction, as written above) is:

$$-r_{CO} = -\frac{1}{2}r_{H_2} = r_{CH_3OH} = k \cdot C_{CO} \cdot C_{H_2}^2$$

where,

$$k = k_o \exp\left[\frac{-E_A}{RT}\right], \ k_o = 5.7 \times 10^{10} \ L^2. \ mol^{-2}. \ min^{-1}, \ E_A = 100.5 \ kJ/mol.$$

Energy balance and the conversion-temperature relationship

The thermodynamics of reaction species are:

Species	$Cp(J/mol.K)$	$H_f(kJ/mol)$ at 298K
CO	29	−110.5
H_2	28.8	0
CH_3OH	43	−201.2

We can then calculate the change in thermodynamic properties (enthalpy and heat capacity) upon reaction, using stoichiometry.

$$\Delta Cp_{rxn} = 1 \times 43 - 2 \times 28.8 - 1 \times 29 = -43.6 \text{ J/mol.K}$$

$$\Delta H_{rxn} = -201.2 - 2 \times 0 - 1 \times (-110.5) = -90.7 \text{ kJ/mol.K}$$

For an adiabatic reactor (either a CSTR or PFR w/o significant heat dispersion), the energy balance yields the following temperature vs. conversion dependency,

$$T = \frac{[-\Delta H_{rxn}]X + \sum \theta_i Cp_i T_0 + X \cdot \Delta Cp \cdot T_R}{\sum \theta_i Cp_i + X \cdot \Delta Cp}$$

Mass balance and rate expression

We can write a general stoichiometric table for this reaction system, accounting for the presence of inert diluent, I.

Species	Initial	Change	Final
CO	F_{CO}	$-F_{CO} \cdot X$	$F_{CO} \cdot (1-X)$
H_2	$\theta_{H_2} F_{CO}$	$-2F_{CO} \cdot X$	$F_{CO} \cdot \left(\theta_{H_2} - 2X\right)$
CH_3OH	$\theta_{CH_3OH} F_{CO}$	$+F_{CO} \cdot X$	$F_{CO} \cdot \left(\theta_{CH_3OH} + X\right)$
Inert	$\theta_I F_{CO}$	0	$\theta_I F_{CO}$
Total	$F_{CO} \cdot \left(1 + \theta_{H_2} + \theta_{CH_3OH} + \theta_I\right)$	$-2F_{CO} \cdot X$	$F_{CO} \cdot \left(1 + \theta_{H_2} + \theta_{CH_3OH} + \theta_I - 2X\right)$

If we assume that everything is ideal gas, then concentration = moles/volume, and accounting for changing volume with temperature and conversion, (Pressure is constant, or pressure drop defined by momentum balance, e.g. Ergun equation),

$$C_i = \frac{N_i P}{N_T RT} = \left(\frac{N_i}{N_T}\right) \cdot \frac{P}{RT}$$

Our concentrations for each species can then be calculated from stoichiometric table, in terms of conversion, pressure and temperature:

$$C_{CO} = \left(\frac{1-X}{1 + \theta_{H_2} + \theta_{CH_3OH} + \theta_I - 2X}\right) \cdot \frac{P}{RT}$$

$$C_{H_2} = \left(\frac{\theta_{H_2} - 2X}{1 + \theta_{H_2} + \theta_{CH_3OH} + \theta_I - 2X} \right) \cdot \frac{P}{RT}$$

$$C_{CH_3OH} = \left(\frac{\theta_{CH_3OH} + X}{1 + \theta_{H_2} + \theta_{CH_3OH} + \theta_I - 2X} \right) \cdot \frac{P}{RT}$$

We can substitute these terms into the rate expression, as follows:

$$r = k_o \left(\frac{P}{RT} \right)^3 \frac{(1-X)(\theta_{H_2} - 2X)^2}{(1 + \theta_{H_2} + \theta_{CH_3OH} + \theta_I - 2X)^3}$$

Inlet conditions

Feed temperature	$T_o = 25°C$ or 298.15 Kelvin
Feed pressure	$P_o = 1$ atm
Molar feed rate	$F_{TO} = 1$ mol/min
Stoichiometric feed, no diluent	$\theta_{H_2} = 2,\ \theta_{CH_3OH} = 0,\ \theta_I = 0$

We further assume that there is no pressure drop associated with gas flow through the continuous reactor, i.e. $P = P_o$.

Calculation 1. Solve for 5 per cent conversion (X = 0.05)

Using Equation from Fogler,

$$T = \frac{(90,700)\cdot(0.05) + (29 + 2\cdot 28.8)\cdot T_0 + (0.05)\cdot(-14.8)\cdot(298.15)}{(29 + 2\cdot 28.8) + (0.05)\cdot(-14.8)} = 350.97 \text{ Kelvin}$$

For our rate expression, substituting $T = 350.97$ and $X = 0.05$,

$$r = 5.7\times10^{10} \cdot \left(\frac{1}{0.0821\cdot 350.97} \right)^3 \frac{(1-0.05)(2 - 2\cdot 0.05)^2}{(1 + 2 - 2\cdot 0.05)^3} \cdot \exp\left[\frac{100,500}{8.314\cdot 350.97} \right]$$

$$= 2.8\times10^{-11} \text{ mol. min}^{-1}.L^{-1}$$

For obtaining a Levenspiel plot (for sizing either a CSTR or PFR), we want to calculate $\dfrac{F_{CO}}{-r_{CO}}$,

$$\frac{F_{CO}}{-r_{CO}} = \frac{\left(\dfrac{1 \text{ mol/min}}{1 + \theta_{H_2}} \right)}{2.8\times10^{-11}} = 1.2\times10^{10} \text{ litres.}$$

Calculation 2: Solve for 10 per cent conversion (X = 0.10)

$$T = \frac{(90,700)\cdot(0.1) + (29 + 2\cdot 28.8)\cdot T_0 + (0.1)\cdot(-14.8)\cdot(298.15)}{(29 + 2\cdot 28.8) + (0.1)\cdot(-14.8)} = 404.7 \text{ Kelvin}$$

For our rate expression, substituting $T = 404.7$ and $X = 0.1$,

$$r = 5.7 \times 10^{10} \cdot \left(\frac{1}{0.0821 \cdot 404.7}\right)^3 \frac{(1-0.1)(2-2\cdot 0.1)^2}{(1+2-2\cdot 0.1)^3} \cdot \exp\left[\frac{100,500}{8.314 \cdot 404.7}\right]$$

$$= 3.4 \times 10^{-9} \text{ mol. min}^{-1} \cdot l^{-1}$$

For obtaining a Levenspiel plot (for sizing either a CSTR or PFR), we want to calculate $\dfrac{F_{CO}}{-r_{CO}}$,

$$\frac{F_{CO}}{-r_{CO}} = \frac{\left(\dfrac{1 \text{ mol/min}}{1+\theta_{H_2}}\right)}{2.4 \times 10^{-9}} = 9.8 \times 10^7 \text{ litres.}$$

Calculation 3: Solve for 50 per cent conversion (X = 0.50)
Using Eq. 8.30 (page 487, 4th edition 2009) from Fogler,

$$T = \frac{(90,700)\cdot(0.5)+(29+2\cdot 28.8)\cdot T_0 +(0.5)\cdot(-14.8)\cdot(298.15)}{(29+2\cdot 28.8)+(0.5)\cdot(-14.8)} = 870.8 \text{ Kelvin}$$

For our rate expression, substituting $T = 870.8$ and $X = 0.50$,

$$r = 5.7 \times 10^{10} \cdot \left(\frac{1}{0.0821 \cdot 870.8}\right)^3 \frac{(1-0.5)(2-2\cdot 0.5)^2}{(1+2-2\cdot 0.5)^3} \cdot \exp\left[\frac{100,500}{8.314 \cdot 870.8}\right]$$

$$= 9.1 \times 10^{-3} \text{ mol. min}^{-1} \cdot l^{-1}$$

For obtaining a Levenspiel plot (for sizing either a CSTR or PFR), we want to calculate $\dfrac{F_{CO}}{-r_{CO}}$,

$$\frac{F_{CO}}{-r_{CO}} = \frac{\left(\dfrac{1 \text{ mol/min.}}{1+\theta_{H_2}}\right)}{9.1 \times 10^{-3}} = 36.6 \text{ litres.}$$

Generate data for Levenspiel plot

X	T	k	$-r_{CO}$	$F_{ao}/-r_{CO}$
0.05	351.0	6.28×10^{-5}	2.82×10^{-11}	1.18×10^{10}
0.10	404.7	6.08×10^{-3}	3.43×10^{-9}	9.73×10^7
0.15	459.4	2.13×10^{-1}	1.18×10^{-7}	2.83×10^6
0.20	515.0	3.65×10^0	1.83×10^{-6}	1.82×10^5
0.25	571.7	3.74×10^1	1.63×10^{-5}	2.05×10^4
0.30	629.3	2.59×10^2	9.60×10^{-5}	3.47×10^3
0.35	688.0	1.34×10^3	4.14×10^{-4}	8.04×10^2

(Contd ...)

X	T	k	$-r_{CO}$	$F_{ao}/-r_{CO}$
0.40	747.8	5.44×10^3	1.30×10^{-3}	2.38×10^2
0.45	808.7	1.84×10^4	3.88×10^{-3}	8.60×10^1
0.50	870.5	5.33×10^4	9.12×10^{-3}	3.65×10^1
0.55	934.0	1.36×10^5	1.87×10^{-2}	1.79×10^1
0.60	998.4	3.14×10^5	3.38×10^{-2}	9.85×10^0
0.65	1064.0	6.64×10^5	5.49×10^{-2}	6.08×10^0
0.70	1130.9	1.30×10^6	7.99×10^{-2}	4.17×10^0
0.75	1199.1	2.39×10^6	1.04×10^{-1}	3.20×10^0
0.80	1268.7	4.15×10^6	1.20×10^{-1}	2.78×10^0
0.85	1339.7	6.87×10^6	1.17×10^{-1}	2.85×10^0
0.90	1412.1	1.09×10^7	8.76×10^{-2}	3.81×10^0
0.95	1486.0	1.67×10^7	3.61×10^{-2}	9.23×10^0
0.96	1501.0	1.81×10^7	2.55×10^{-2}	1.31×10^1
0.97	1516.0	1.96×10^7	1.58×10^{-2}	2.11×10^1
0.98	1531.0	2.12×10^7	7.75×10^{-3}	4.30×10^1
0.99	1546.2	2.29×10^7	2.13×10^{-3}	1.56×10^2

We can then plot this data to see how to best perform this reaction:

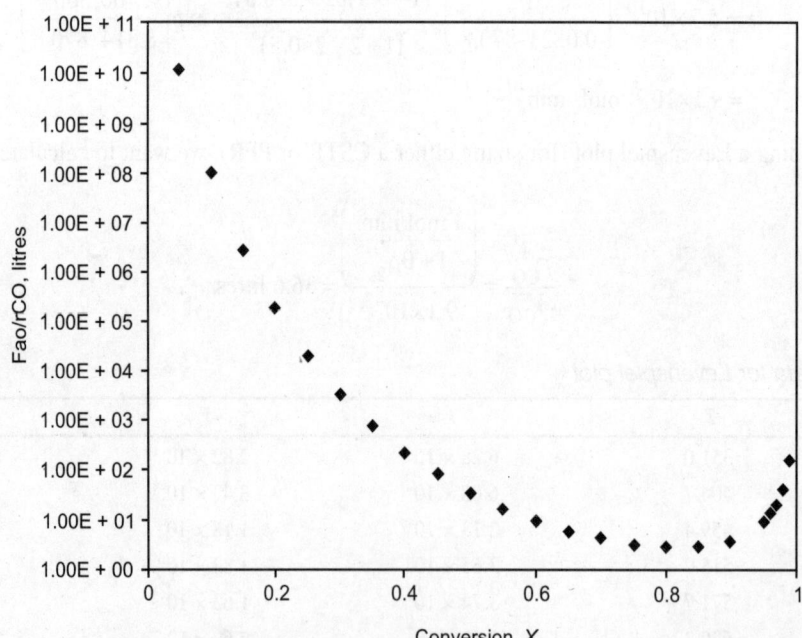

We can see from the plot that a CSTR will get us to a conversion of ~85 per cent after that we would prefer a PFR to keep reactor volume to a minimum.

Caveat

Sometimes people write for a gas-phase reaction the rate expression in terms of partial pressures:

$$r' = k' \cdot P_{CO} \cdot P_{H_2}^2$$

Our true form of the rate expression, which is in terms of concentrations, is:

$$r = k \cdot C_{CO} \cdot C_{H_2}^2 = k \cdot \left(\frac{1}{RT}\right)^3 P_{CO} \cdot P_{H_2}^2$$

Comparing the two, we see that:

$$k' = k\left(\frac{1}{RT}\right)^3 = k_o \left(\frac{1}{RT}\right)^3 \exp\left[\frac{-E_A}{RT}\right]$$

Which when linearised does not fit an Arrhenius relationship.

Chapter 6

Fundamentals of Catalytic Processes

INTRODUCTION

Catalysis is the change in rate of a chemical reaction due to the participation of a substance called a catalyst. Unlike other reagents that participate in the chemical reaction, a catalyst is not consumed by the reaction itself. A catalyst may participate in multiple chemical transformations. Catalysts that speed the reaction are called positive catalysts. Catalysts that slow the reaction are called negative catalysts, or inhibitors. Substances that increase the activity of catalysts are called promoters, and substances that deactivate catalysts are called catalytic poisons.

The general feature of catalysis is that the catalytic reaction has a lower rate-limiting free energy change to the transition state than the corresponding uncatalysed reaction, resulting in a larger reaction rate at the same temperature. However, the mechanistic origin of catalysis is complex. Catalysts may affect the reaction environment favourably, or bind to the reagents to polarise bonds, e.g. acid catalysts for reactions of carbonyl compounds, form specific intermediates that are not produced naturally, such as osmate esters in osmium tetroxide-catalysed dihydroxylation of alkenes, or cause lysis of reagents to reactive forms, such as atomic hydrogen in catalytic hydrogenation.

Kinetically, catalytic reactions behave like typical chemical reactions, i.e. the reaction rate depends on the frequency of contact of the reactants in the rate-determining step. Usually, the catalyst participates in this slow step, and rates are limited by amount of catalyst. In heterogeneous catalysis, the diffusion of reagents to the surface and diffusion of products from the surface can be rate determining. Analogous events associated with substrate binding and product dissociation apply to homogeneous catalysts.

Although catalysts are not consumed by the reaction itself, they may be inhibited, deactivated or destroyed by secondary processes. In heterogeneous catalysis, typical secondary processes include coking where the catalyst becomes covered by polymeric side products. Additionally, heterogeneous catalysts can dissolve into the solution in a solid-liquid system or evaporate in a solid-gas system.

The production of most industrially important chemicals involves catalysis. Similarly, most biochemically significant processes are catalysed. Research into catalysis is a major field in applied science and involves many areas of chemistry, notably in organometallic chemistry and materials science. Catalysis is relevant to many aspects of environmental science, e.g. the catalytic converter in automobiles and the dynamics of the ozone hole. Catalytic reactions are preferred in environmentally friendly green chemistry due to the reduced amount of waste generated, as opposed to stoichiometric reactions in which all reactants are consumed and more side products are formed. The most common catalyst is the proton (H^+). Many transition metals and transition metal complexes are used in catalysis as well. Catalysts called enzymes are important in biology.

A catalyst works by providing an alternative reaction pathway to the reaction product. The rate of the reaction is increased as this alternative route has a lower activation energy than the reaction route not mediated by the catalyst. The disproportionation of hydrogen peroxide to give water and oxygen is a reaction that is strongly affected by catalysts:

$$2H_2O_2 \rightarrow 2H_2O + O_2$$

This reaction is favoured in the sense that reaction products are more stable than the starting material, however the uncatalysed reaction is slow. The decomposition of hydrogen peroxide is in fact so slow that hydrogen peroxide solutions are commercially available. Upon the addition of a small amount of manganese dioxide, the hydrogen peroxide rapidly reacts according to the above equation. This effect is readily seen by the effervescence of oxygen. The manganese dioxide may be recovered unchanged, and reused indefinitely, and thus is not consumed in the reaction. Accordingly, manganese dioxide catalyses this reaction.

GENERAL PRINCIPLES OF CATALYSIS

Typical Mechanism

Catalysts generally react with one or more reactants to form intermediates that subsequently give the final reaction product, in the process regenerating the catalyst. The following is a typical reaction scheme, where C represents the catalyst, X and Y are reactants, and Z is the product of the reaction of X and Y:

$$X + C \rightarrow XC \qquad \text{... (6.1)}$$
$$Y + XC \rightarrow XYC \qquad \text{... (6.2)}$$
$$XYC \rightarrow CZ \qquad \text{...(6.3)}$$
$$CZ \rightarrow C + Z \qquad \text{...(6.4)}$$

Although the catalyst is consumed by Eq. 6.1, it is subsequently produced by Eq. 6.4, so for the overall reaction:

$$X + Y \rightarrow Z$$

As a catalyst is regenerated in a reaction, often only small amounts are needed to increase the rate of the reaction. In practice, however, catalysts are sometimes consumed in secondary processes.

As an example of this process, in 2008 Danish researchers first revealed the sequence of events when oxygen and hydrogen combine on the surface of titanium dioxide (TiO_2, or titania) to produce water. With a time-lapse series of scanning tunnelling microscopy images, they determined the molecules undergo adsorption, dissociation and diffusion before reacting. The intermediate reaction states were: HO_2, H_2O_2, then H_3O_2 and the final reaction product (water molecule dimers), after which the water molecule desorbs from the catalyst surface.

Catalysis and reaction energetics

Catalysts work by providing an (alternative) mechanism involving a different transition state and lower activation energy. Consequently, more molecular collisions have the energy needed to reach the transition state. Hence, catalysts can enable reactions that would otherwise be blocked or slowed by a kinetic barrier. The catalyst may increase reaction rate or selectivity, or enable the reaction at lower temperatures. This effect can be illustrated with a Boltzmann distribution and energy profile diagram (Fig. 6.1).

Catalysts do not change the extent of a reaction: they have no effect on the chemical equilibrium of a reaction because the rate of both the forward and the reverse reaction are both affected. The fact that a catalyst does not change the equilibrium is a consequence of the second law of thermodynamics. Suppose there was such a catalyst that shifted an equilibrium. Introducing the catalyst to the system would result in reaction to move to the new equilibrium, producing energy. Production of energy is a necessary result since reactions are spontaneous if and only if Gibbs free energy is produced, and if there is no energy barrier, there is no need for a catalyst. Then, removing the catalyst would also result in reaction, producing energy, i.e. the addition and its reverse process, removal, would both produce energy. Thus, a catalyst that could change the equilibrium would be a perpetual motion machine, a contradiction to the laws of thermodynamics.

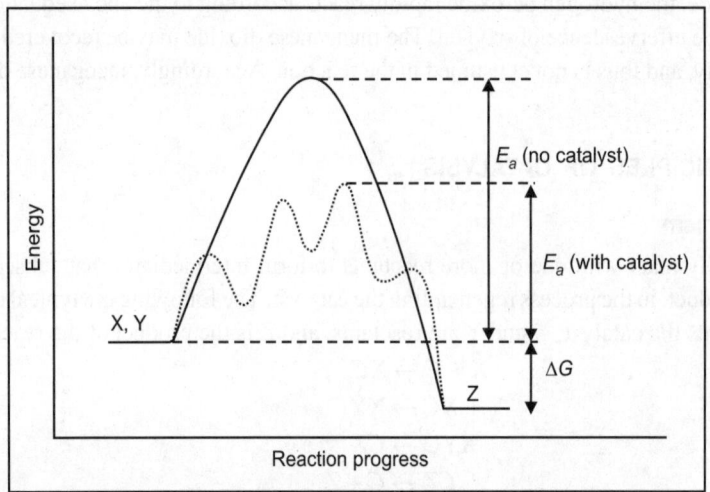

Fig. 6.1. Generic potential energy diagram showing the effect of a catalyst in a hypothetical exothermic chemical reaction X + Y to give Z. The presence of the catalyst opens a different reaction pathway (shown in dotted line) with a lower activation energy. The final result and the overall thermodynamics are the same.

If a catalyst does change the equilibrium, then it must be consumed as the reaction proceeds, and thus it is also a reactant. Illustrative is the base-catalysed hydrolysis of esters, where the produced carboxylic acid immediately reacts with the base catalyst and thus the reaction equilibrium is shifted towards hydrolysis.

The SI derived unit for measuring the catalytic activity of a catalyst is the katal, which is moles per second. The activity of a catalyst can also be described by the turn over number (or TON) and the catalytic efficiency by the turn over frequency (TOF). The biochemical equivalent is the enzyme unit.

The catalyst stabilises the transition state more than it stabilises the starting material. It decreases the kinetic barrier by decreasing the difference in energy between starting material and transition state.

Typical catalytic materials

The chemical nature of catalysts is as diverse as catalysis itself, although some generalisations can be made. Proton acids are probably the most widely used catalysts, especially for the many reactions involving water, including hydrolysis and its reverse. Multifunctional solids often are catalytically active, e.g. zeolites, alumina and certain forms of graphitic carbon. Transition metals are often used to catalyse redox reactions (oxidation, hydrogenation). Many catalytic processes, especially those involving hydrogen, require platinum metals. Some so-called catalysts are really precatalysts. Precatalysts convert to catalysts in the reaction. For example, Wilkinson's catalyst $RhCl(PPh_3)_3$ loses one triphenylphosphine ligand before entering the true catalytic cycle. Precatalysts are easier to store but are easily activated *in situ*. Because of this preactivation step, many catalytic reactions involve an induction period.

Chemical species that improve catalytic activity are called cocatalysts or promotors in cooperative catalysis.

Types of Catalysis

Catalysts can be either heterogeneous or homogeneous, depending on whether a catalyst exists in the same phase as the substrate. Biocatalysts are often seen as a separate group.

Heterogeneous catalysts

Heterogeneous catalysts act in a different phase than the reactants. Most heterogeneous catalysts are solids that act on substrates in a liquid or gaseous reaction mixture. Diverse mechanisms for reactions on surfaces are known, depending on how the adsorption takes place (Langmuir-Hinshelwood, Eley-Rideal, and Mars-van Krevelen). The total surface area of solid has an important effect on the reaction rate. The smaller the catalyst particle size, the larger the surface area for a given mass of particles.

For example, in the Haber process, finely divided iron serves as a catalyst for the synthesis of ammonia from nitrogen and hydrogen. The reacting gases adsorb onto 'active sites' on the iron particles. Once adsorbed, the bonds within the reacting molecules are weakened, and new bonds between the resulting fragments form in part due to their close proximity. In this way the particularly strong triple bond in nitrogen is weakened and the hydrogen and nitrogen atoms combine faster than would be the case in the gas phase, so the rate of reaction increases.

Heterogeneous catalysts are typically 'supported', which means that the catalyst is dispersed on a second material that enhances the effectiveness or minimises their cost. Sometimes the support is merely a surface on which the catalyst is spread to increase the surface area. More often, the support and the catalyst interact, affecting the catalytic reaction. Supports are porous materials with a high surface area, most commonly activated alumina or activated carbon, but special applications silicon dioxide, titanium dioxide, calcium carbonate or barium sulphate can be found.

Homogeneous catalysts

Homogeneous catalysts function in the same phase as the reactants, but the mechanistic principles invoked in heterogeneous catalysis are generally applicable. Typically homogeneous catalysts are dissolved in a solvent with the substrates. One example of homogeneous catalysis involves the influence of H^+ on the esterification of esters, e.g. methyl acetate from acetic acid and methanol. For inorganic chemists, homogeneous catalysis is often synonymous with organometallic catalysts.

Electrocatalysts

In the context of electrochemistry, specifically in fuel cell engineering, various metal-containing catalysts are used to enhance the rates of the half reactions that comprise the fuel cell. One common type of fuel cell electrocatalyst is based upon nanoparticles of platinum that are supported on slightly larger carbon particles.

When this platinum electrocatalyst is in contact with one of the electrodes in a fuel cell, it increases the rate of oxygen reduction to water (or hydroxide or hydrogen peroxide).

Organocatalysis

Whereas transition metals sometimes attract most of the attention in the study of catalysis, organic molecules without metals can also possess catalytic properties. Typically, organic catalysts require a higher loading (or amount of catalyst per unit amount of reactant) than transition metal-based catalysts, but these catalysts are usually commercially available in bulk, helping to reduce costs. In the early 2000s, organocatalysts were considered 'new generation' and are competitive to traditional metal-containing catalysts. Enzymatic reactions operate via the principles of organic catalysis.

Significance of Catalysis

Estimates are that 90 per cent of all commercially produced chemical products involve catalysts at some stage in the process of their manufacture. Catalysis is so pervasive that subareas are not readily classified. Some areas of particular concentration are surveyed below.

Energy processing

Petroleum refining makes intensive use of catalysis for alkylation, catalytic cracking (breaking long-chain hydrocarbons into smaller pieces), naphtha reforming and steam reforming (conversion of hydrocarbons into synthesis gas). Even the exhaust from the burning of fossil fuels is treated via catalysis: Catalytic converters, typically composed of platinum and rhodium, break down some of the more harmful by-products of automobile exhaust.

$$2CO + 2NO \rightarrow 2CO_2 + N_2$$

With regard to synthetic fuels, an old but still important process is the Fischer-Tropsch synthesis of hydrocarbons from synthesis gas, which itself is processed via water-gas shift reactions, catalysed by iron. Biodiesel and related biofuels require processing via both inorganic and biocatalysts. Fuel cells rely on catalysts for both the anodic and cathodic reactions.

Bulk chemicals

Some of the largest-scale chemicals are produced via catalytic oxidation, often using oxygen. Examples include nitric acid (from ammonia), sulphuric acid (from sulphur dioxide to sulphur trioxide by the chamber process), terephthalic acid from *p*-xylene, and acrylonitrile from propane and ammonia. Many other chemical products are generated by large-scale reduction, often via hydrogenation. The largest-scale example is ammonia, which is prepared via the Haber process from nitrogen. Methanol is prepared from carbon monoxide. Bulk polymers derived from ethylene and propylene are often prepared via Ziegler-Natta catalysis. Polyesters, polyamides, and isocyanates are derived via acid-base catalysis. Most carbonylation processes require metal catalysts, examples include the Monsanto acetic acid process and hydroformylation.

Fine chemicals

Many fine chemicals are prepared via catalysis; methods include those of heavy industry as well as more specialised processes that would be prohibitively expensive on a large scale. Examples include olefin metathesis using Grubbs' catalyst, the Heck reaction, and Friedel-Crafts reactions. Because most bioactive compounds are chiral, many pharmaceuticals are produced by enantioselective catalysis.

Food processing

One of the most obvious applications of catalysis is the hydrogenation (reaction with hydrogen gas) of fats using nickel catalyst to produce margarine. Many other foodstuffs are prepared via biocatalysis.

Biology

In nature, enzymes are catalysts in metabolism and catabolism. Most biocatalysts are protein-based, i.e. enzymes, but other classes of biomolecules also exhibit catalytic properties including ribozymes, and synthetic deoxyribozymes.

Biocatalysts can be thought of as intermediate between homogenous and heterogeneous catalysts, although strictly speaking soluble enzymes are homogeneous catalysts and membrane-bound enzymes are heterogeneous. Several factors affect the activity of enzymes (and other catalysts) including temperature, pH, concentration of enzyme, substrate, and products. A particularly important reagent in enzymatic reactions is water, which is the product of many bond-forming reactions and a reactant in many bond-breaking processes. Enzymes are employed to prepare many commodity chemicals including high-fructose corn syrup and acrylamide.

In the environment

Catalysis impacts the environment by increasing the efficiency of industrial processes, but catalysis also plays a direct role in the environment. A notable example is the catalytic role of chlorine free radicals in the breakdown of ozone. These radicals are formed by the action of ultraviolet radiation on chlorofluorocarbons (CFCs).

$$Cl^{\cdot} + O_3 \rightarrow ClO^{\cdot} + O_2$$
$$ClO^{\cdot} + O^{\cdot} \rightarrow Cl^{\cdot} + O_2$$

Inhibitors, Poisons and Promoters

Substances that reduce the action of catalysts are called catalyst inhibitors if reversible, and catalyst poisons if irreversible. Promoters are substances that increase the catalytic activity, particularly when not being catalysts unto themselves.

The inhibitor may modify selectivity in addition to rate. For instance, in the reduction of ethane to ethene, the catalyst is palladium (Pd) partly 'poisoned' with lead(II) acetate [Pb(CH$_3$COO)$_2$]. Without the deactivation of the catalyst, the ethene produced will be further reduced to ethane.

The inhibitor can produce this effect by e.g. selectively poisoning only certain types of active sites. Another mechanism is the modification of surface geometry. For instance, in hydrogenation operations, large planes of metal surface function as sites of hydrogenolysis catalysis while sites catalysing hydrogenation of unsaturates are smaller. Thus, a poison that covers surface randomly will tend to reduce the number of uncontaminated large planes but leave proportionally more smaller sites free, thus changing the hydrogenation vs. hydrogenolysis selectivity. Many other mechanisms are also possible.

Promoters can cover up surface to prevent production of a mat of coke, or even actively remove such material (e.g. rhenium on platinum in platforming). They can aid the dispersion of the catalytic material or bind to reagents.

FISCHER–TROPSCH PROCESS

The Fischer–Tropsch process (or Fischer–Tropsch Synthesis) is a set of chemical reactions that convert a mixture of carbon monoxide and hydrogen into liquid hydrocarbons. The process, a key component of gas to liquids technology, produces a petroleum substitute, typically from coal, natural gas or biomass for use as synthetic lubrication oil and as synthetic fuel. The F-T process has received intermittent attention as a source of low-sulphur diesel fuel and to address the supply or cost of petroleum-derived hydrocarbons.

Process Chemistry

The Fischer–Tropsch process involves a series of chemical reactions that lead to a variety of hydrocarbons. Useful reactions give alkanes:

$$(2n+1)\ H_2 + nCO \rightarrow C_nH_{(2n+2)} + nH_2O$$

where, n is a positive integer. The formation of methane ($n = 1$) is generally unwanted. Most of the alkanes produced tend to be straight-chain alkanes, although some branched alkanes are also formed. In addition to alkane formation, competing reactions result in the formation of alkenes, as well as alcohols and other oxygenated hydrocarbons. Usually, only relatively small quantities of these non-alkane products are formed, although catalysts favouring some of these products have been developed.

Other reactions relevant to the F-T process

Several reactions are required to obtain the gaseous reactants required for F-T catalysis. First, reactant gases entering a F-T reactor must first be desulphurised to protect the catalysts that are readily poisoned. The other major class of reactions are employed to adjust the H_2:CO ratio:

1. Water gas shift reaction provides a source of hydrogen:

$$H_2O + CO \rightarrow H_2 + CO_2$$

2. For F-T plants that start with methane, another important reaction is steam reforming, which converts the methane into CO and H_2:

$$H_2O + CH_4 \rightarrow CO + 3H_2$$

Chemical mechanisms

The conversion of CO to alkanes involves net hydrogenation of CO, the hydrogenolysis of C-O bonds, and the formation of C-C bonds. Such reactions are assumed to proceed via initial formation of surface-bound metal carbonyls. The CO ligand is speculated to undergo dissociation, possibly into oxide and carbide ligands. Other potential intermediates are various C-1 fragments including formyl (CHO), hydroxycarbene (HCOH), hydroxymethyl (CH_2OH), methyl (CH_3), methylene (CH_2), methylidyne (CH), and hydroxymethylidyne (COH). Furthermore, and critical to the production of liquid fuels, are reactions that form C-C bonds, such as migratory insertion. Many related stoichiometric reactions have been simulated on discrete metal clusters, but homogeneous F-T catalysts are poorly developed and of no commercial importance.

Process conditions

Generally, the Fischer–Tropsch process is operated in the temperature range of 150°–300°C (302°–572°F). Higher temperatures lead to faster reactions and higher conversion rates but also tend to favour methane production. As a result, the temperature is usually maintained at the low to middle part of the range. Increasing the pressure leads to higher conversion rates and also favours formation of long-chained alkanes both of which are desirable. Typical pressures range from one to several tens of atmospheres. Even higher pressures would be favourable, but the benefits may not justify the additional costs of high-pressure equipment.

A variety of synthesis gas compositions can be used. For cobalt-based catalysts the optimal H_2:CO ratio is around 1.8–2.1. Iron-based catalysts promote the water-gas-shift reaction and thus can tolerate significantly lower ratios. This reactivity can be important for synthesis gas derived from coal or biomass, which tend to have relatively low H_2:CO ratios (<1).

Product distribution

In general the product distribution of hydrocarbons formed during the Fischer–Tropsch process follows an Anderson-Schulz-Flory distribution, which can be expressed as:

$$W_n/n = (1-\alpha)^2\alpha^{n-1}$$

where, W_n is the weight fraction of hydrocarbon molecules containing n carbon atoms, α is the chain growth probability or the probability that a molecule will continue reacting to form a longer chain. In general, α is largely determined by the catalyst and the specific process conditions.

Examination of the above equation reveals that methane will always be the largest single product; however by increasing α close to one, the total amount of methane formed can be minimised compared to the sum of all of the various long-chained products. Increasing α increases the formation of long-chained hydrocarbons. The very long-chained hydrocarbons are waxes, which are solid at room temperature. Therefore, for production of liquid transportation fuels it may be necessary to crack some of the Fischer-Tropsch products. In order to avoid this, some researchers have proposed using zeolites or other catalyst substrates with fixed sized pores that can restrict the formation of hydrocarbons longer than some characteristic size (usually $n < 10$). This way they can drive the reaction so as to minimise methane formation without producing lots of long-chained hydrocarbons. Such efforts have met with only limited success.

Fischer–Tropsch catalysts

A variety of catalysts can be used for the Fischer–Tropsch process, but the most common are the transition metals cobalt, iron, and ruthenium. Nickel can also be used, but tends to favour methane formation (methanation).

Cobalt seems to be the most active catalyst, although iron may be more suitable for low-hydrogen-content synthesis gases such as those derived from coal due to its promotion of the water-gas-shift reaction. In addition to the active metal the catalysts typically contain a number of 'promoters', including potassium and copper. Catalysts are supported on high-surface-area binders/supports such as silica, alumina or zeolites. Cobalt catalysts are more active for Fischer–Tropsch synthesis when the feedstock is natural gas. Natural gas has a high hydrogen to carbon ratio, so the water-gas-shift is not needed for cobalt catalysts. Iron catalysts are preferred for lower quality feedstocks such as coal or biomass.

Unlike the other metals used for this process (Co, Ni, Ru) which remain in the metallic state during synthesis, iron catalysts tend to form a number of phases, including various oxides and carbides during the reaction. Control of these phase transformations can be important in maintaining catalytic activity and preventing breakdown of the catalyst particles.

Fischer-Tropsch catalysts are notoriously sensitive to poisoning by sulphur-containing compounds. The sensitivity of the catalyst to sulphur is greater for cobalt-based catalysts than for their iron counterparts.

Alkylation

Alkylation is the transfer of an alkyl group from one molecule to another. The alkyl group may be transferred as an alkyl carbocation, a free radical, a carbanion or a carbene (or their equivalents). Alkylating agents are widely used in chemistry because the alkyl group is probably the most common group encountered in organic molecules. Many biological target molecules or their synthetic precursors are composed of an alkyl chain with specific functional groups in a specific order. Selective alkylation or adding parts to the chain with the desired functional groups, is used, especially if there is no commonly available biological precursor. Alkylation with only one carbon is termed methylation (Fig. 6.2).

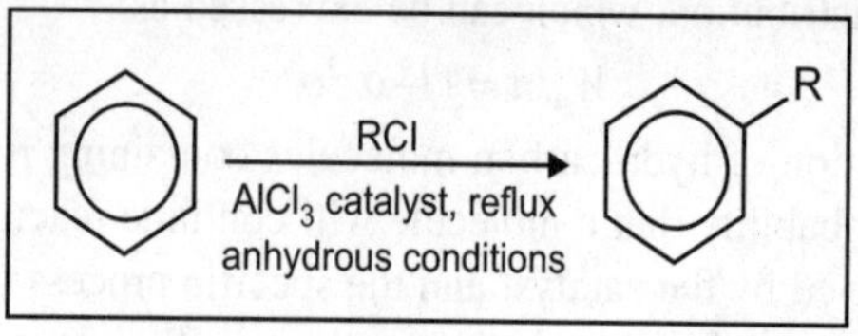

Fig. 6.2. Benzene Friedel–Crafts alkylation.

In oil refining contexts, alkylation refers to a particular alkylation of isobutane with olefins. It is a major aspect of the upgrading of petroleum. In medicine, alkylation of DNA is used in chemotherapy to damage the DNA of cancer cells. Alkylation is accomplished with the class of drugs called alkylating antineoplastic agents.

Friedel–Crafts alkylation

Friedel–Crafts alkylation involves the alkylation of an aromatic ring with an alkyl halide using a strong Lewis acid catalyst. With anhydrous ferric chloride as a catalyst, the alkyl group attaches at the former site of the chloride ion. The general mechanism is shown below.

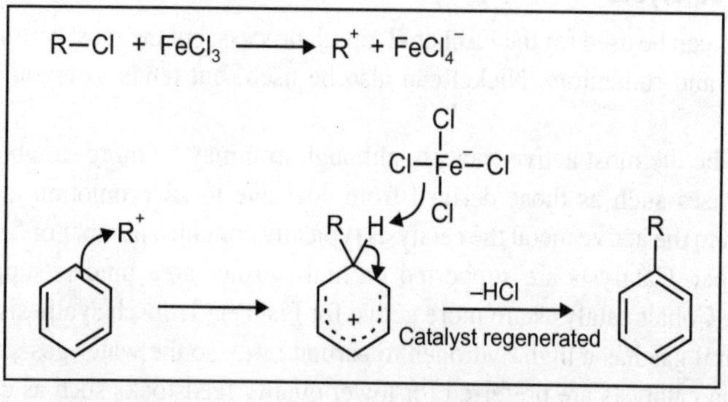

This reaction has one big disadvantage, namely that the product is more nucleophilic than the reactant due to the electron donating alkyl-chain. Therefore, another hydrogen is substituted with an alkyl-chain, which leads to overalkylation of the molecule. Also, if the chlorine is not on a tertiary carbon, carbocation rearrangement reaction will occur. This is due to the relative stability of the tertiary carbocation over the secondary and primary carbocations. Steric hindrance can be exploited to limit the number of alkylations, as in the t-butylation of 1,4-dimethoxybenzene.

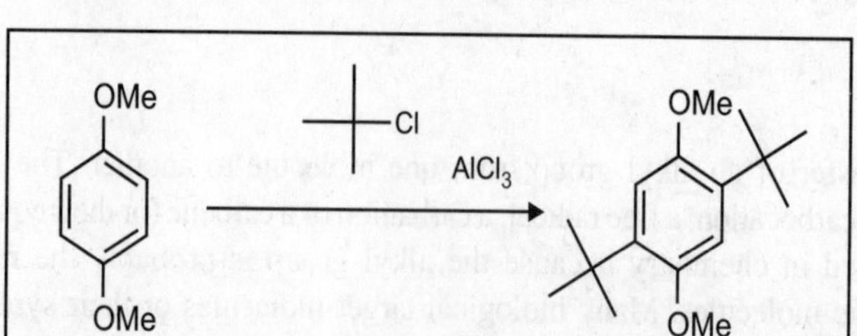

Alkylations are not limited to alkyl halides: Friedel–Crafts reactions are possible with any carbo-cationic intermediate such as those derived from alkenes and a protic acid, Lewis acid, enones, and epoxides. An example is the synthesis of neophyl chloride from benzene and methallyl chloride:

$$H_2C = C(CH_3)_2CH_2Cl + C_6H_6 \rightarrow C_6H_5C(CH_3)_2CH_2Cl$$

In one study the electrophile is a bromonium ion derived from an alkene and NBS:

In this reaction samarium(III) triflate is believed to activate the NBS halogen donor in halonium ion formation.

Friedel–Crafts dealkylation

Friedel–Crafts alkylation is a reversible reaction. In a reversed Friedel–Crafts reaction or Friedel–Crafts dealkylation, alkyl groups can be removed in the presence of protons and a Lewis acid.

For example, in a multiple addition of ethyl bromide to benzene, *ortho* and *para* substitution is expected after the first monosubstitution step because an alkyl group is an activating group. However, the actual reaction product is 1,3,5-triethylbenzene with all alkyl groups as a *meta* substituent. Thermodynamic reaction control makes sure that thermodynamically favoured *meta* substitution with steric hindrance minimised takes prevalence over less favourable *ortho* and *para* substitution by chemical equilibration.

The ultimate reaction product is thus the result of a series of alkylations and dealkylations.

Friedel–Crafts acylation

Friedel–Crafts acylation is the acylation of aromatic rings with an acyl chloride using a strong Lewis acid catalyst. Friedel–Crafts acylation is also possible with acid anhydrides. Reaction conditions are similar to the Friedel–Crafts alkylation mentioned above. This reaction has several advantages over the alkylation reaction.

Due to the electron-withdrawing effect of the carbonyl group, the ketone product is always less reactive than the original molecule, so multiple acylations do not occur. Also, there are no carbocation rearrangements, as the carbonium ion is stabilised by a resonance structure in which the positive charge is on the oxygen.

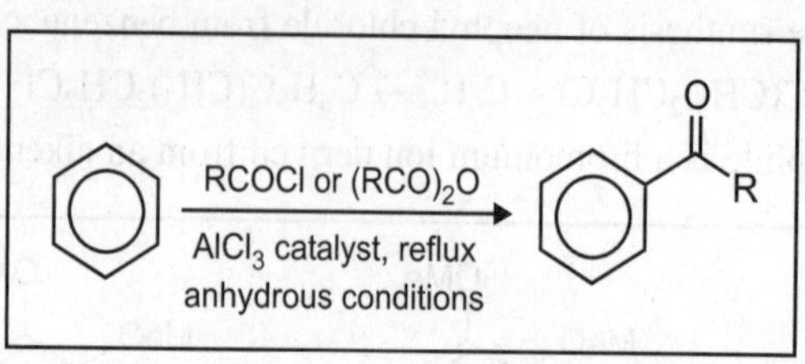

The viability of the Friedel–Crafts acylation depends on the stability of the acyl chloride reagent. Formyl chloride, for example, is too unstable to be isolated. Thus, synthesis of benzaldehyde via the Friedel–Crafts pathway requires that formyl chloride be synthesised *in situ*.

This is accomplished via the Gattermann–Koch reaction, accomplished by reacting benzene with carbon monoxide and hydrogen chloride under high pressure, catalysed by a mixture of aluminium chloride and cuprous chloride.

Reaction mechanism

In a simple mechanistic view, the first step consists of dissociation of a chlorine atom to form an acyl cation:

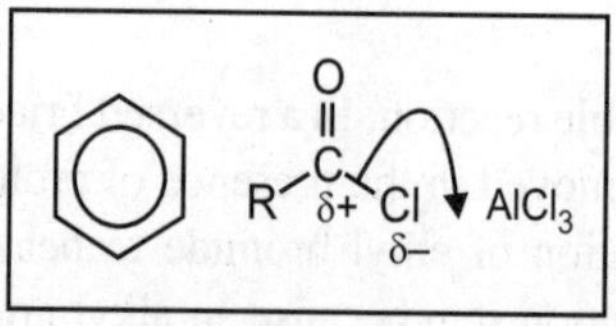

This is followed by nucleophilic attack of the arene toward the acyl group:

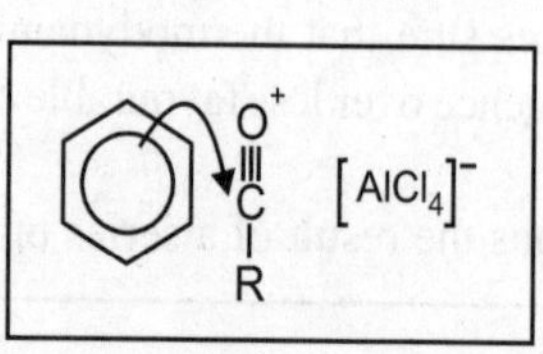

Finally, a chlorine atom reacts to form HCl, and the $AlCl_3$ catalyst is regenerated:

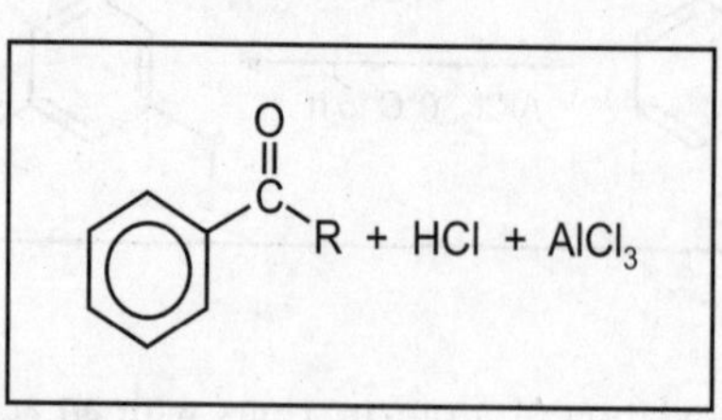

Friedel–Crafts hydroxyalkylation

Arenes react with certain aldehydes and ketones to form the hydroxyalkylated product for example in the reaction of the mesityl derivative of glyoxal with benzene to form a benzoin with an alcohol rather than a carbonyl group:

Scope and variations

This reaction is related to several classic named reactions:

1. The acylated reaction product can be converted into the alkylated product via a Clemmensen reduction.
2. The Gattermann-Koch reaction can be used to synthesise benzaldehyde from benzene.
3. The Gatterman reaction describes arene reactions with hydrocyanic acid.
4. The Houben-Hoesch reaction describes arene reactions with nitriles.
5. A reaction modification with an aromatic phenyl ester as a reactant is called the Fries rearrangement.
6. In the Scholl reaction two arenes couple directly (sometimes called Friedel–Crafts arylation).
7. In the Zincke-Suhl reaction p-cresol is alkylated to a cyclohexadienone with tetrachloromethane.
8. In the Blanc chloromethylation a chloromethyl group is added to an arene with formaldehyde, hydrochloric acid and zinc chloride.

Isomerisation

In chemistry isomerisation is the process by which one molecule is transformed into another molecule which has exactly the same atoms, but the atoms are rearranged e.g. A-B-C → B-A-C (these related molecules are known as isomers). In some molecules and under some conditions, isomerisation occurs spontaneously. Many isomers are equal or roughly equal in bond energy, and so exist in roughly equal amounts, provided that they can interconvert relatively freely, that is the energy barrier between the two isomers is not too high. When the isomerisation occurs intramolecularly it is considered a rearrangement reaction.

Instances of isomerisation

1. Isomerisations in hydrocarbon cracking. This is usually employed in organic chemistry, where fuels, such as pentane, a straight-chain isomer, are heated in the presence of a platinum catalyst. The resulting mixture of straight and branched-chain isomers than have to be separated. An industrial process is also the isomerisation of n-butane into isobutane.

Pentane   2-Methylbutane   2,2-Dimethylpropane

2. *Trans-cis* isomerism: In certain compounds an interconversion of *cis* and *trans* isomers can be observed, for instance, with maleic acid and with azobenzene often by photoisomerisation. An example is the photochemical conversion of the *trans* isomer to the *cis* isomer of resveratrol.

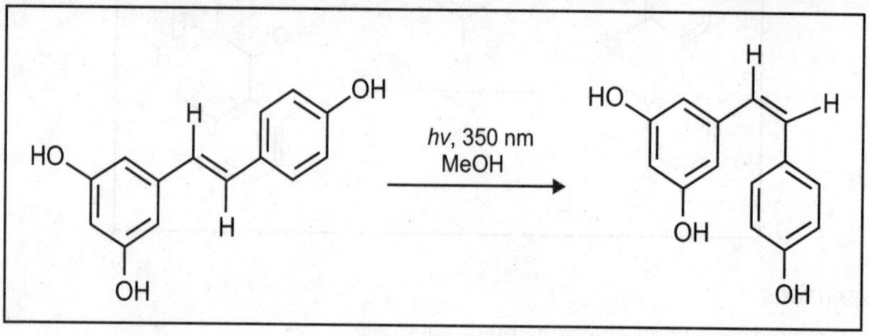

3. Aldose-ketose isomerism in biochemistry.
4. Isomerisations between conformational isomers. These take place without an actual rearrangement for instance incoversion of two cyclohexane conformations.
5. Fluxional molecules display rapid intercoversion of isomers, e.g. bullvalene.
6. Valence isomerisation: The isomerisation of molecules which involve structural changes resulting only from a relocation of single and double bonds. If a dynamic equilibrium is established between the two isomers it is also referred to as valence tautomerism.

Hydrogenation

Hydrogenation, a form of chemical reduction, is a chemical reaction between molecular hydrogen (H_2) and another compound or element, usually in the presence of a catalyst. The process is commonly employed to reduce or saturate organic compounds. Hydrogenation typically constitutes the addition of pairs of hydrogen atoms to a molecule, generally an alkene. Catalysts are required for the reaction to be usable; non-catalytic hydrogenation takes place only at very high temperatures. Hydrogen adds to double and triple bonds in hydrocarbons.

Because of the importance of hydrogen, many related reactions have been developed for its use. Most hydrogenations use gaseous hydrogen (H_2), but some involve the alternative sources of hydrogen, not H_2: these processes are called transfer hydrogenations. The reverse reaction, removal of hydrogen from a molecule, is called dehydrogenation. A reaction where bonds are broken while hydrogen is added is called hydrogenolysis, a reaction that may occur to carbon–carbon and carbon–heteroatom (O, N, X) bonds. Hydrogenation differs from protonation or hydride addition: in hydrogenation, the products have the same charge as the reactants. Figure 6.3 is an illustrative example of a hydrogenation reaction is the addition of hydrogen to maleic acid to succinic acid depicted on the right.

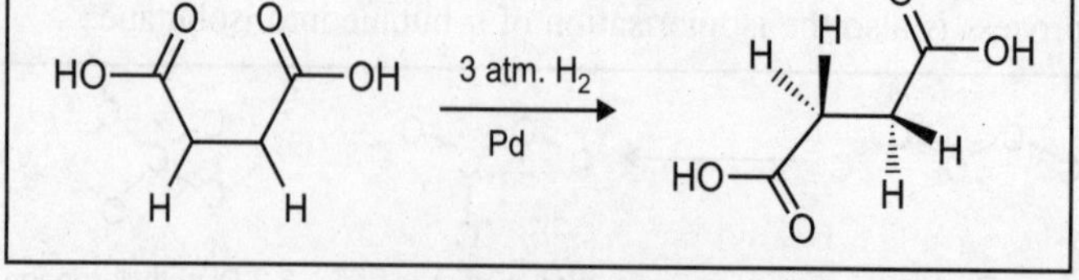

Fig. 6.3. Example of a hydrogenation reaction.

Numerous important applications are found in the petrochemical, pharmaceutical and food industries. Hydrogenation of unsaturated fats produces saturated fats and, in some cases, *trans* fats.

Process

Hydrogenation has three components, the unsaturated substrate, the hydrogen (or hydrogen source) and, invariably, a catalyst. The reaction is carried out at different temperatures and pressures depending upon the substrate and the activity of the catalyst.

Substrate

The addition of H_2 to an alkene affords an alkane in the protypical reaction:

$$RCH = CH_2 + H_2 \rightarrow RCH_2CH_3 \text{ (R = alkyl, aryl)}$$

Hydrogenation is sensitive to steric hindrance explaining the selectivity for reaction with the exocyclic double bond but not the internal double bond. An important characteristic of alkene and alkyne hydrogenations, both the homogeneously and heterogeneously catalysed versions, is that hydrogen addition occurs with 'syn addition', with hydrogen entering from the least hindered side.

Dehydrogenation

Dehydrogenation is a chemical reaction that involves the elimination of hydrogen (H_2). It is the reverse process of hydrogenation. Dehydrogenation reactions may be either large scale industrial processes or smaller scale laboratory procedures. There are a variety of classes of dehydrogenations:

1. Aromatisation: Six-membered alicyclic rings can be aromatised in the presence of hydrogenation catalysts, the elements sulphur and selenium or quinones (such as DDQ).
2. Oxidation: The conversion of alcohols to ketones or aldehydes can be effected by metal catalysts such as copper chromite. In the Oppenauer oxidation, hydrogen is transferred from one alcohol to another to bring about the oxidation.
3. Dehydrogenation of amines: Amines can be converted to nitriles using a variety of reagents, such as IF_5.
4. Dehydrogenation of paraffins and olefins: Paraffins like *n*-pentane and isopentane can be converted to pentene and isoprene using chromium (III) oxide as a catalyst at 500°C.

Dehydrogenation converts saturated fats to unsaturated fats. Enzymes that catalyse dehydrogenation are called dehydrogenases.

Adsorption and Isotherms

Adsorption

Adsorption is the process of attraction of atoms or molecules (generically known as 'monomers') from an adjacent gas or liquid to an exposed solid surface. Such attraction forces (adhesion or cohesion) align the monomers into layers (films) onto the existent surface.

The deposition may be driven by:

1. Long range weak forces among atomic or molecular electric multipoles ('van der Waals', among induced and/or fluctuating dipoles or 'London', dipole-quadrupole, etc.) will initiate the initial attraction.
2. Short range strong ionic or metallic forces may finalise the setting of new layers onto the solid surface (without generating new chemical species)—as salt deposits (crystalline growth) from supersaturated solutions or as metal vapour deposition onto metallic surfaces.

Covalent forces at solid surfaces will always create new chemical species; the formation of covalent solids involves energy transfers that penetrate deep into the bulk, far beyond the surfaces—these are valence electron rearrangements (phase transitions) at the whole scale of the involved bulk.

Energetically the adsorbed species loses its internal kinetic energy (of thermal agitation) which it transfers into the solid surface to which it is being adsorbed—so the adsorption rates increase with an increasing temperature gap between a warmer monomer gas/liquid and a cooler surface.

Adsorption is present in many natural physical, biological, and chemical systems, and is widely used in industrial applications such as activated charcoal, capturing and using waste heat to provide cold water for air conditioning and other process requirements (adsorption refrigerators), synthetic resins, increase storage capacity of carbide-derived carbons for tunable nanoporous carbon, and water purification.

Adsorption, ion exchange, and chromatography are sorption processes in which certain adsorbates are selectively transferred from the fluid phase to the surface of insoluble, rigid particles suspended in a vessel or packed in a column.

Isotherms

Adsorption is usually described through isotherms, that is, the amount of adsorbate on the adsorbent as a function of its pressure (if gas) or concentration (if liquid) at constant temperature. The quantity adsorbed is nearly always normalised by the mass of the adsorbent to allow comparison of different materials.

Freundlich

The first mathematical fit to an isotherm was published by Freundlich and Küster and is a purely empirical formula for gaseous adsorbates,

$$\frac{x}{m} = kP^{1/n}$$

where, x is the quantity adsorbed, m is the mass of the adsorbent, P is the pressure of adsorbate and k and n are empirical constants for each adsorbent-adsorbate pair at a given temperature. The function has an asymptotic maximum as pressure increases without bound. As the temperature increases, the constants k and n change to reflect the empirical observation that the quantity adsorbed rises more slowly and higher pressures are required to saturate the surface.

Langmuir

In 1916, Irving Langmuir published a new model isotherm for gases adsorbed on solids, which retained his name. It is a semi-empirical isotherm derived from a proposed kinetic mechanism. It is based on four assumptions:

1. The surface of the adsorbent is uniform, that is, all the adsorption sites are equivalent.
2. Adsorbed molecules do not interact.
3. All adsorption occurs through the same mechanism.
4. At the maximum adsorption, only a monolayer is formed: molecules of adsorbate do not deposit on other, already adsorbed, molecules of adsorbate, only on the free surface of the adsorbent.

These four assumptions are seldom all true: there are always imperfections on the surface, adsorbed molecules are not necessarily inert, and the mechanism is clearly not the same for the very first molecules to adsorb as for the last.

The fourth condition is the most troublesome, as frequently more molecules will adsorb on the monolayer; this problem is addressed by the BET isotherm for relatively flat (non-microporous) surfaces. The Langmuir isotherm is nonetheless the first choice for most models of adsorption, and has many applications in surface kinetics (usually called Langmuir–Hinshelwood kinetics) and thermodynamics.

Langmuir suggested that adsorption takes place through this mechanism: $A_g + S \rightleftharpoons AS$, where A is a gas molecule and S is an adsorption site. The direct and inverse rate constants are k and k_{-1}. If we define surface coverage, θ, as the fraction of the adsorption sites occupied, in the equilibrium we have

$$K = \frac{k}{k_{-1}} = \frac{\theta}{(1-\theta)P} \quad \text{or } \theta = \frac{KP}{1+KP}$$

where, P is the partial pressure of the gas or the molar concentration of the solution. For very low pressures $\theta \approx KP$ and for high pressures $\theta \approx 1$.

θ is difficult to measure experimentally; usually, the adsorbate is a gas and the quantity adsorbed is given in moles, grams or gas volumes at standard temperature and pressure (STP) per gram of adsorbent. If we call v_{mon} the STP volume of adsorbate required to form a monolayer on the adsorbent (per gram of adsorbent), $\theta = \dfrac{v}{v_{mon}}$ and we obtain an expression for a straight line:

$$\frac{1}{v} = \frac{1}{Kv_{mon}}\frac{1}{P} + \frac{1}{v_{mon}}$$

Through its slope and y-intercept we can obtain v_{mon} and K, which are constants for each adsorbent/adsorbate pair at a given temperature. v_{mon} is related to the number of adsorption sites through the ideal gas law. If we assume that the number of sites is just the whole area of the solid divided into the cross section of the adsorbate molecules, we can easily calculate the surface area of the adsorbent. The surface area of an adsorbent depends on its structure; the more pores it has, the greater the area, which has a big influence on reactions on surfaces.

If more than one gas adsorbs on the surface, we define θ_E as the fraction of empty sites and we have

$$\theta_E = \frac{1}{1 + \displaystyle\sum_{i=1}^{n} K_i P_i}$$

and

$$\theta_j = \frac{K_j P_j}{1 + \displaystyle\sum_{i=1}^{n} K_i P_i}$$

where, i is each one of the gases that adsorb.

BET

Often molecules do form multilayers, that is, some are adsorbed on already adsorbed molecules and the Langmuir isotherm is not valid. In 1938 Stephen Brunauer, Paul Emmett, and Edward Teller developed a model isotherm that takes that possibility into account. Their theory is called BET theory, after the initials in their last names.

They modified Langmuir's mechanism as follows:

$$A_{(g)} + S \rightleftharpoons AS$$

$$A_{(g)} + AS \rightleftharpoons A_2S$$

$$A_{(g)} + A_2S \rightleftharpoons A_3S \text{ and so on}$$

The derivation of the formula is more complicated than Langmuir's. We obtain:

$$\frac{x}{v(1-x)} = \frac{1}{v_{mon}c} + \frac{x(c-1)}{v_{mon}c}$$

x is the pressure divided by the vapour pressure for the adsorbate at that temperature (usually denoted P/P^0), v is the STP volume of adsorbed adsorbate, v_{mon} is the STP volume of the amount of adsorbate required to form a monolayer and c is the equilibrium constant K we used in Langmuir isotherm multiplied by the vapour pressure of the adsorbate. The key assumption used in deriving the BET equation that the successive heats of adsorption for all layers except the first are equal to the heat of condensation of the adsorbate.

The Langmuir isotherm is usually better for chemisorption and the BET isotherm works better for physisorption for non-microporous surfaces (Fig. 6.4).

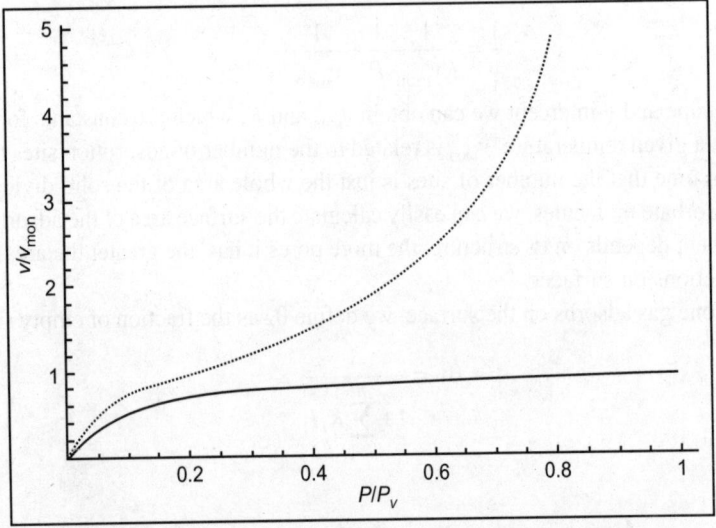

Fig. 6.4. Langmuir isotherm (black) and BET isotherm (dotted).

Kisliuk

In other instances, molecular interactions between gas molecules previously adsorbed on a solid surface form significant interactions with gas molecules in the gaseous phase. Hence, adsorption of gas molecules to the surface is more likely to occur around gas molecules that are already present on the solid surface, rendering the Langmuir adsorption isotherm ineffective for the purposes of modelling. This effect was studied in a system where nitrogen was the adsorbate and tungsten was the adsorbent by Paul Kisliuk. To compensate for the increased probability of adsorption occurring around molecules present on the substrate surface, Kisliuk developed the precursor state theory, whereby molecules would enter a precursor state at the interface between the solid adsorbent and adsorbate in the gaseous phase. From here, adsorbate molecules would either adsorb to the adsorbent or desorb into the gaseous phase. The probability of adsorption occurring from the precursor state is dependent on the adsorbate's proximity to other adsorbate molecules that have already been adsorbed. If the adsorbate molecule in the precursor state is in close proximity to an adsorbate molecule which has already formed on the surface, it has a sticking probability reflected by the size of the S_E constant and will either be adsorbed from the precursor state at a rate of k_{EC} or will desorb into the gaseous phase at a rate of k_{ES}. If an adsorbate molecule enters the precursor state at a location that is remote from any other previously adsorbed adsorbate molecules, the sticking probability is reflected by the size of the S_D constant (Fig. 6.5).

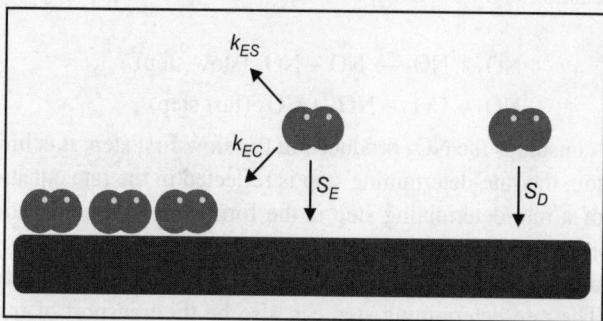

Fig. 6.5. Two adsorbate nitrogen molecules adsorbing onto a tungsten adsorbent from the precursor state around an island of previously adsorbed adsobate (left) and via random adsorption (right).

These factors were included as part of a single constant termed a 'sticking coefficient', k_E, described below:

$$k_E = \frac{S_E}{k_{ES} \cdot S_D}.$$

As S_D is dictated by factors that are taken into account by the Langmuir model, S_D can be assumed to be the adsorption rate constant. However, the rate constant for the Kisliuk model (R') is different to that of the Langmuir model, as R' is used to represent the impact of diffusion on monolayer formation and is proportional to the square root of the system's diffusion coefficient. The Kisliuk adsorption isotherm is written as follows, where, $\theta_{(t)}$ is fractional coverage of the adsorbent with adsorbate, and t is immersion time:

$$\frac{d\theta(t)}{dt} = R'(1-\theta)(1+k_E\theta).$$

Solving for $\theta_{(t)}$ yields:

$$\theta_{(t)} = \frac{1-e^{-R'(1+k_E)t}}{1+k_E e^{-R'(1+k_E)t}}.$$

Rate-determining Step

The rate-determining step (RDS) is a chemistry term for the slowest step in a chemical reaction. The rate-determining step is often compared to the neck of a funnel; the rate at which water flows through the funnel is determined by the width of the neck, not by the speed at which water is poured in. In similar manner, the rate of reaction depends on the rate of the slowest step. However it is not clear what 'slowest step' means. For example does it refer to the step with the smallest rate constant, if the reaction is reversible, does it refer to the ratio of the forward and reverse rate constants or does it simply refer to the step that has the smallest flux? If the later then at steady state all steps carry the same flux and therefore there is no slowest step. In metabolic pathways the rate limiting step is more well defined and a measure, the flux control coefficient is given to a step to signify how rate limiting the step is. Theory and experiment also suggests that there is no single rate limiting step but a range of rate limitingness across the entire reaction network.

For example, the reaction $NO_{2(g)} + CO_{(g)} \rightarrow NO_{(g)} + CO_{2(g)}$ can be thought of as occurring in two elementary steps:

$$NO_2 + NO_2 \rightarrow NO + NO_3 \text{ (slow step)}$$
$$NO_3 + CO \rightarrow NO_2 + CO_2 \text{ (fast step)}$$

As the second step consumes the NO_3 produced in the slow first step, it is limited by the rate of the first step. For this reason, the rate-determining step is reflected in the rate equation of a reaction.

Another example of a rate-determining step is the formation of a carbocation from a haloalkane during the S_N1 reaction of tertiary haloalkanes with sodium hydroxide.

In the previous examples, the rate determining step was one of the sequential chemical reactions leading to a product. The rate-determining step can also be the transport of reactants to interact and form the product. This case is referred to as diffusion control and, in general, occurs where the formation of product from the activated complex is very rapid and thus the supply of reactants and their interaction is rate determining. The concept of the rate-determining step is very important to the optimisation and understanding of many chemical processes such as catalysis and combustion. In a reaction coordinate, the transition state with the highest energy is the rate-determining step of a given reaction.

CATALYTIC REFORMING

Catalytic reforming is a chemical process used to convert petroleum refinery naphthas, typically having low octane ratings, into high-octane liquid products called reformates which are components of high-octane gasoline (also known as petrol). Basically, the process rearranges or restructures the hydrocarbon molecules in the naphtha feedstocks as well as breaking some of the molecules into smaller molecules. The overall effect is that the product reformate contains hydrocarbons with more complex molecular shapes having higher octane values than the hydrocarbons in the naphtha feedstock. In so doing, the process separates hydrogen atoms from the hydrocarbon molecules and produces very significant amounts of by-product hydrogen gas for use in a number of the other processes involved in a modern petroleum refinery. Other by-products are small amounts of methane, ethane, propane and butanes.

This process is quite different from and not to be confused with the catalytic steam reforming process used industrially to produce various products such as hydrogen, ammonia and methanol from natural gas, naphtha or other petroleum-derived feedstocks. Nor is this process to be confused with various other catalytic reforming processes that use methanol or biomass-derived feedstocks to produce hydrogen for fuel cells or other uses.

Chemistry

Before describing the reaction chemistry of the catalytic reforming process as used in petroleum refineries, the typical naphthas used as catalytic reforming feedstocks will be discussed.

Typical naphtha feedstocks

A petroleum refinery includes many unit operations and unit processes. The first unit operation in a refinery is the continuous distillation of the petroleum crude oil being refined. The overhead liquid distillate is called naphtha and will become a major component of the refinery's gasoline (petrol) product after it is further processed through a catalytic hydrodesulphuriser to remove sulphur-containing hydrocarbons and a catalytic reformer to reform its hydrocarbon molecules into more complex molecules with a higher octane rating value.

The naphtha is a mixture of very many different hydrocarbon compounds. It has an initial boiling point of about 35°C and a final boiling point of about 200°C, and it contains paraffin, naphthene (cyclic paraffins) and aromatic hydrocarbons ranging from those containing 4 carbon atoms to those containing about 10 or 11 carbon atoms.

The naphtha from the crude oil distillation is often further distilled to produce a 'light' naphtha containing most (but not all) of the hydrocarbons with 6 or less carbon atoms and a 'heavy' naphtha containing most (but not all) of the hydrocarbons with more than 6 carbon atoms. The heavy naphtha has an initial boiling point of about 140° to 150°C and a final boiling point of about 190° to 205°C. The naphthas derived from the distillation of crude oils are referred to as 'straight-run' naphthas.

It is the straight-run heavy naphtha that is usually processed in a catalytic reformer because the light naphtha has molecules with 6 or less carbon atoms which, when reformed, tend to crack into butane and lower molecular weight hydrocarbons which are not useful as high-octane gasoline blending components. Also, the molecules with 6 carbon atoms tend to form aromatics which is undesirable because governmental environmental regulations in a number of countries limit the amount of aromatics (most particularly benzene) that gasoline may contain.

It should be noted that there are a great many petroleum crude oil sources worldwide and each crude oil has its own unique composition or 'assay'. Also, not all refineries process the same crude oils and each refinery produces its own straight-run naphthas with their own unique initial and final boiling points. In other words, naphtha is a generic term rather than a specific term.

Some refinery naphthas include olefinic hydrocarbons, such as naphthas derived from the fluid catalytic cracking and coking processes used in many refineries. Some refineries may also desulphurise and catalytically reform those naphthas. However, for the most part, catalytic reforming is mainly used on the straight-run heavy naphthas, such as those in the above table, derived from the distillation of crude oils.

Reaction chemistry

There are a good many chemical reactions that occur in the catalytic reforming process, all of which occur in the presence of a catalyst and a high partial pressure of hydrogen. Depending upon the type or version of catalytic reforming used as well as the desired reaction severity, the reaction conditions range from temperatures of about 495° to 525°C and from pressures of about 5 to 45 atm.

The commonly used catalytic reforming catalysts contain noble metals such as platinum and/or rhenium, which are very susceptible to poisoning by sulphur and nitrogen compounds. Therefore, the naphtha feedstock to a catalytic reformer is always preprocessed in a hydrodesulphurisation unit which removes both the sulphur and the nitrogen compounds.

The four major catalytic reforming reactions are:

1. The dehydrogenation of naphthenes to convert them into aromatics as exemplified in the conversion methylcyclohexane (a naphthene) to toluene (an aromatic), as shown below:

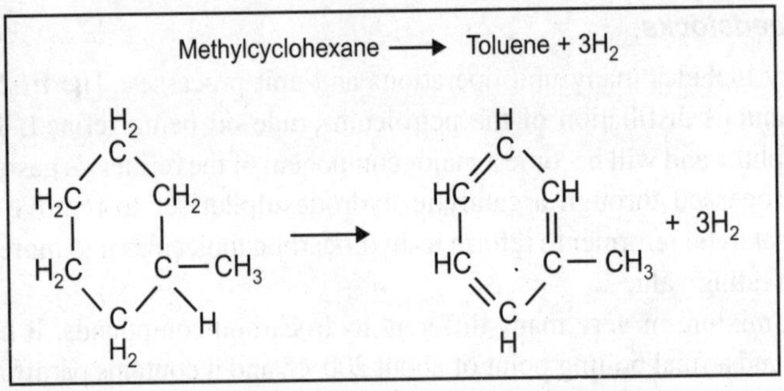

Methylcyclohexane $\longrightarrow$ Toluene + 3H$_2$

2. The isomerisation of normal paraffins to isoparaffins as exemplified in the conversion of normal octane to 2,5-dimethylhexane (an isoparaffin), as shown below:

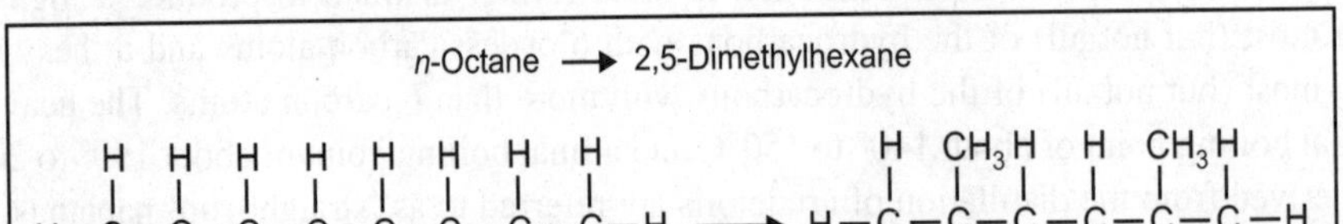

n-Octane $\longrightarrow$ 2,5-Dimethylhexane

3. The dehydrogenation and aromatisation of paraffins to aromatics (commonly called dehydrocyclisation) as exemplified in the conversion of normal heptane to toluene, as shown below:

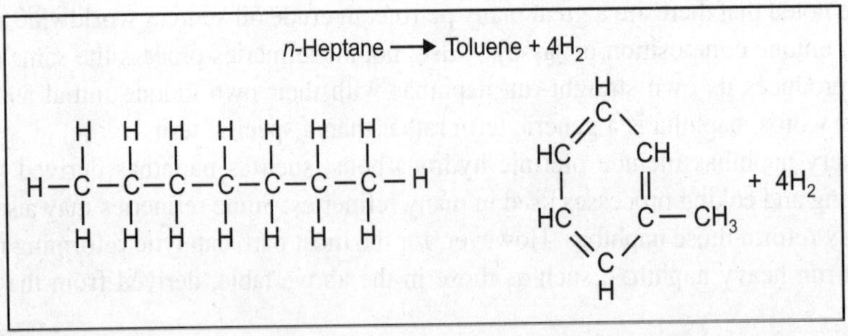

n-Heptane $\longrightarrow$ Toluene + 4H$_2$

4. The hydrocracking of paraffins into smaller molecules as exemplified by the cracking of normal heptane into isopentane and ethane, as shown below:

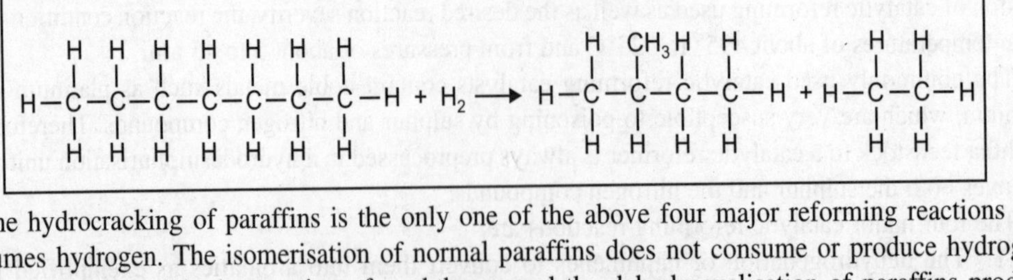

n-Heptane + H$_2$ $\longrightarrow$ Isopentane + Ethane

The hydrocracking of paraffins is the only one of the above four major reforming reactions that consumes hydrogen. The isomerisation of normal paraffins does not consume or produce hydrogen. However, both the dehydrogenation of naphthenes and the dehydrocyclisation of paraffins produce hydrogen. The overall net production of hydrogen in the catalytic reforming of petroleum naphthas ranges from about 50 to 200 cubic metres of hydrogen gas (at 0°C and 1 atm.) per cubic metre of liquid naphtha feedstock. In the United States customary units, that is equivalent to 300 to 1200 cubic feet of hydrogen gas (at 60°F and 1 atm.) per barrel of liquid naphtha feedstock. In many petroleum refineries, the net hydrogen produced in catalytic reforming supplies a significant part of the hydrogen used elsewhere in the refinery (for example, in hydrodesulphurisation processes). The hydrogen is also necessary in order to hydrogenolyse any polymers that form on the catalyst.

Process Description

The most commonly used type of catalytic reforming unit has three reactors, each with a fixed bed of catalyst, and all of the catalyst is regenerated *in situ* during routine catalyst regeneration shutdowns which occur approximately once each 6 to 24 months. Such a unit is referred to as a semi-regenerative catalytic reformer (SRR). Some catalytic reforming units have an extra *spare* or *swing* reactor and each reactor can be individually isolated so that any one reactor can be undergoing *in situ* regeneration while the other reactors are in operation. When that reactor is regenerated, it replaces another reactor which, in turn, is isolated so that it can then be regenerated. Such units, referred to as cyclic catalytic reformers, are not very common. Cyclic catalytic reformers serve to extend the period between required shutdowns.

The latest and most modern type of catalytic reformers are called continuous catalyst regeneration reformers (CCR). Such units are characterised by continuous *in situ* regeneration of part of the catalyst in a special regenerator, and by continuous addition of the regenerated catalyst to the operating reactors. Many of the earliest catalytic reforming units were non-regenerative in that they did not perform *in situ* catalyst regeneration.

Instead, when needed, the aged catalyst was replaced by fresh catalyst and the aged catalyst was shipped to catalyst manufacturer's to be either regenerated or to recover the platinum content of the aged catalyst. Very few, if any, catalytic reformers currently in operation are non-regenerative. The process flow diagram (Fig. 6.6) shows a typical semi-regenerative catalytic reforming unit.

The liquid feed (at the bottom left in the diagram) is pumped up to the reaction pressure (5 to 45 atm.) and is joined by a stream of hydrogen-rich recycle gas. The resulting liquid-gas mixture is preheated by flowing through a heat exchanger. The preheated feed mixture is then totally vapourised and heated to the reaction temperature (495° to 520°C) before the vapourised reactants enter the first reactor. As the vapourised reactants flow through the fixed bed of catalyst in the reactor, the major reaction is the dehydrogenation of naphthenes to aromatics (as described earlier herein) which is highly endothermic and results in a large temperature decrease between the inlet and outlet of the reactor. To maintain the required reaction temperature and the rate of reaction, the vapourised stream is reheated in the second fired heater before it flows through the second reactor. The temperature again decreases across the second reactor and the vapourised stream must again be reheated in the third fired heater before it flows through the third reactor.

As the vapourised stream proceeds through the three reactors, the reaction rates decrease and the reactors therefore become larger. At the same time, the amount of reheat required between the reactors becomes smaller. Usually, three reactors are all that is required to provide the desired performance of the catalytic reforming unit.

Some installations use three separate fired heaters as shown in the schematic diagram and some installations use a single fired heater with three separate heating coils.

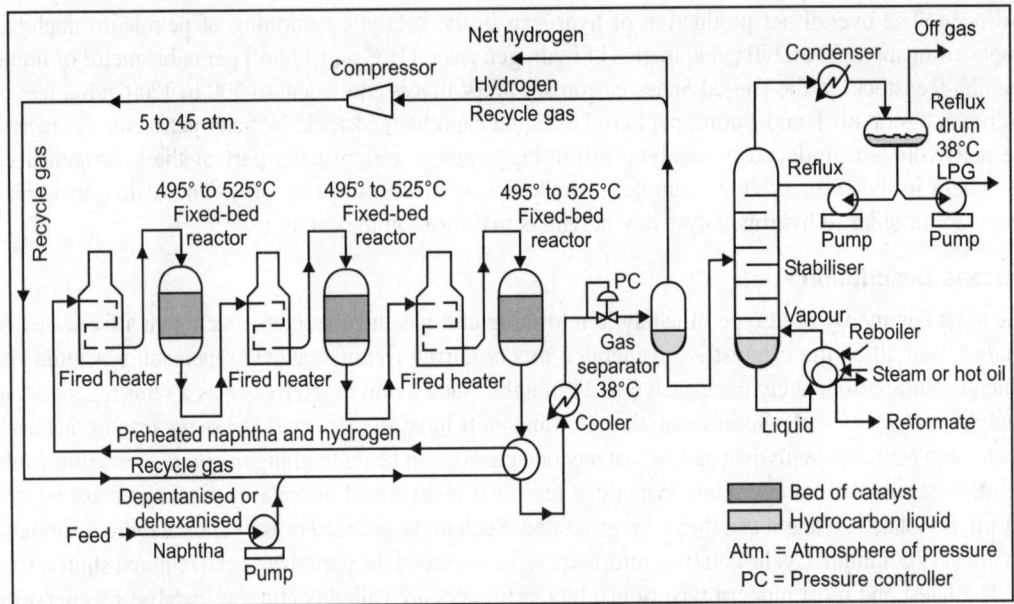

Fig. 6.6. Schematic diagram of a typical semi-regenerative catalytic reformer unit in a petroleum refinery.

The hot reaction products from the third reactor are partially cooled by flowing through the heat exchanger where the feed to the first reactor is preheated and then flow through a water-cooled heat exchanger before flowing through the pressure controller (PC) into the gas separator.

Most of the hydrogen-rich gas from the gas separator vessel returns to the suction of the recycle hydrogen gas compressor and the net production of hydrogen-rich gas from the reforming reactions is exported for use in the other refinery processes that consume hydrogen (such as hydrodesulphurisation units and/or a hydrocracker unit).

The liquid from the gas separator vessel is routed into a fractionating column commonly called a stabiliser. The overhead offgas product from the stabiliser contains the by-product methane, ethane, propane and butane gases produced by the hydrocracking reactions as explained in the above discussion of the reaction chemistry of a catalytic reformer, and it may also contain some small amount of hydrogen. That offgas is routed to the refinery's central gas processing plant for removal and recovery of propane and butane. The residual gas after such processing becomes part of the refinery's fuel gas system.

The bottoms product from the stabiliser is the high-octane liquid reformate that will become a component of the refinery's product gasoline.

Catalysts and Mechanisms

Most catalytic reforming catalysts contain platinum or rhenium on a silica or silica-alumina support base, and some contain both platinum and rhenium. Fresh catalyst is chlorided (chlorinated) prior to use. The noble metals (platinum and rhenium) are considered to be catalytic sites for the dehydrogenation reactions and the chlorinated alumina provides the acid sites needed for isomerisation, cyclisation and hydrocracking reactions. The activity (i.e. effectiveness) of the catalyst in a semi-regenerative catalytic reformer is reduced over time during operation by carbonaceous coke deposition and chloride loss. The activity of the catalyst can be periodically regenerated or restored by *in situ* high temperature oxidation of the coke followed by chlorination. As stated earlier herein, semi-regenerative catalytic reformers are regenerated about once per 6 to 24 months.

Normally, the catalyst can be regenerated perhaps 3 or 4 times before it must be returned to the manufacturer for reclamation of the valuable platinum and/or rhenium content.

REACTION ENGINEERING IN MICROELECTRONIC FABRICATION

Microfabrication or micromanufacturing are the terms to describe processes of fabrication of miniature structures, of micrometer sizes and smaller. Practical advances in microelectromechanical systems (MEMS) and other nanotechnology, where the technologies from IC fabrication are being reused, adapted or extended have led to the extension of the scope and techniques of microfabrication.

The major concepts and principles of micromanufacturing are laser technology, microlithography, micromechatronics, micromachining and microfinishing (nanofinishing).

Microfabrication technologies originate from the microelectronics industry, and the devices are usually made on silicon wafers even though glass, plastics and many other substrate are in use. Micromachining, semiconductor processing, microelectronic fabrication, semiconductor fabrication, MEMS fabrication and integrated circuit technology are terms used instead of microfabrication, but microfabrication is the broad general term.

Microfabrication is actually a collection of technologies which are utilised in making microdevices. Some of them have very old origins, not connected to manufacturing, like lithography or etching.

To fabricate a microdevice, many processes must be performed, one after the other, many times repeatedly. These processes typically include depositing a film, patterning the film with the desired micro features, and removing (or etching) portions of the film. For example, in memory chip fabrication there are some 30 lithography steps, 10 oxidation steps, 20 etching steps, 10 doping steps, and many others are performed. The complexity of microfabrication processes can be described by their mask count. This is the number of different pattern layers that constitute the final device. Modern microprocessors are made with 30 masks while a few masks suffice for a microfluidic device or a laser diode. Microfabrication resembles multiple exposure photography, with many patterns aligned to each other to create the final structure.

Microfabricated devices are not generally freestanding devices but are usually formed over or in a thicker support substrate. For electronic applications, semiconducting substrates such as silicon wafers can be used. For optical devices or flat panel displays, transparent substrates such as glass or quartz are common. The substrate enables easy handling of the micro device through the many fabrication steps. Often many individual devices are made together on one substrate and then singulated into separated devices toward the end of fabrication.

Microfabricated devices are typically constructed using one or more thin films. The purpose of these thin films depends on the type of device. Electronic devices may have thin films which are conductors (metals), insulators (dielectrics) or semiconductors. Optical devices may have films which are reflective, transparent, light guiding or scattering.

Films may also have a chemical or mechanical purpose as well as for MEMS applications. Examples of deposition techniques include:

1. Thermal oxidation.
2. Chemical vapour deposition (CVD).
3. Physical vapour deposition (PVD).
4. Epitaxy.

It is often desirable to pattern a film into distinct features or to form openings (or vias) in some of the layers. These features are on the micrometer or nanometer scale and the patterning technology is what defines microfabrication. The patterning technique typically uses a 'mask' to define portions of the film which will be removed.

Examples of patterning techniques include:

1. Photolithography.
2. Shadow masking.

Etching is the removal of some portion of the thin film or substrate. The substrate is exposed to an etching (such as an acid or plasma) which chemically or physically attacks the film until it is removed. Etching techniques include:

1. Dry etching (Plasma etching) such as reactive-ion etching (RIE) or deep reactive-ion etching (DRIE).
2. Wet etching or chemical etching.

A wide variety of other processes for cleaning, planarising or modifying the chemical properties of the microfabricated devices can also be performed. Some examples include:

1. Doping by either thermal diffusion or ion implantation.
2. Chemical-mechanical planarisation (CMP).
3. Wafer cleaning, also known as 'surface preparation'.
4. Wire bonding.

Microfabrication is carried out in cleanrooms, where air has been filtered of particle contamination and temperature, humidity, vibrations and electrical disturbances are under stringent control. Smoke, dust, bacteria and cells are micrometers in size, and their presence will destroy the functionality of a microfabricated device.

Etching

Etching is the process of using strong acid or mordant to cut into the unprotected parts of a metal surface to create a design in intaglio in the metal (the original process—in modern manufacturing other chemicals may be used on other types of material). As an intaglio method of printmaking it is, along with engraving, the most important technique for old master prints, and remains widely used today.

In pure etching, a metal (usually copper, zinc or steel) plate is covered with a waxy ground which is resistant to acid. The artist then scratches off the ground with a pointed etching needle where he wants a line to appear in the finished piece, so exposing the bare metal. The échoppe, a tool with a slanted oval section is also used for 'swelling' lines.

The plate is then dipped in a bath of acid, technically called the mordant or etchant or has acid washed over it. The acid 'bites' into the metal, where it is exposed, leaving behind lines sunk into the plate. The remaining ground is then cleaned off the plate. The plate is inked all over, and then the ink wiped off the surface, leaving only the ink in the etched lines.

The plate is then put through a high-pressure printing press together with a sheet of paper (often moistened to soften it). The paper picks up the ink from the etched lines, making a print. The process can be repeated many times; typically several hundred impressions (copies) could be printed before the plate shows much sign of wear. The work on the plate can also be added to by repeating the whole process; this creates an etching which exists in more than one state. Etching has often been combined with other intaglio techniques such as engraving (e.g. Rembrandt) or aquatint (e.g. Goya).

Modern technique in etching

A waxy acid-resist, known as a ground, is applied to a metal plate, most often copper or zinc but steel plate is another medium with different qualities. There are two common types of ground: hard ground and soft ground.

Hard ground can be applied in two ways. Solid hard ground comes in a hard waxy block. To apply hard ground of this variety, the plate to be etched is placed upon a hotplate (set at 70°C), a kind of metal worktop that is heated up. The plate heats up and the ground is applied by hand, melting onto the plate as it is applied. The ground is spread over the plate as evenly as possible using a roller. Once applied the etching plate is removed from the hotplate and allowed to cool which hardens the ground.

After the ground has hardened the artist 'smokes' the plate, classically with 3 beeswax tapers, applying the flame to the plate to darken the ground and make it easier to see what parts of the plate are exposed. Smoking not only darkens the plate but adds a small amount of wax. Afterwards the artist uses a sharp tool to scratch into the ground, exposing the metal.

The second way to apply hard ground is by liquid hard ground. This comes in a can and is applied with a brush upon the plate to be etched. Exposed to air the hard ground will harden. Some printmakers use oil/tar based asphaltum or bitumen as hard ground, although often bitumen is used to protect steel plates from rust and copper plates from ageing.

Soft ground also comes in liquid form and is allowed to dry but it does not dry hard like hard ground and is impressionable. After the soft ground has dried the printmaker may apply materials such as leaves, objects, hand prints and so on which will penetrate the soft ground and expose the plate underneath.

The ground can also be applied in a fine mist, using powdered rosin or spraypaint. This process is called aquatint, and allows for the creation of tones, shadows, and solid areas of colour.

The design is then drawn (in reverse) with an etching-needle or échoppe. An 'echoppe' point can be made from an ordinary tempered steel etching needle, by grinding the point back on a carborundum stone, at a 45°–60° angle. The 'echoppe' works on the same principle that makes a fountain pen's line more attractive than a ballpoint's: The slight swelling variation caused by the natural movement of the hand 'warms up' the line, and although hardly noticeable in any individual line, has a very attractive overall effect on the finished plate. It can be drawn with in the same way as an ordinary needle.

The plate is then completely submerged in an acid that eats away at the exposed metal. Ferric chloride may be used for etching copper or zinc plates, whereas nitric acid may be used for etching zinc or steel plates. Typical solutions are 2 parts $FeCl_3$ to 2 parts water and 1 part nitric to 3 parts water. The strength of the acid determines the speed of the etching process.

1. The etching process is known as biting.
2. The waxy resist prevents the acid from biting the parts of the plate which have been covered.
3. The longer the plate remains in the acid the deeper the 'bites' become.

During the etching process the printmaker uses a bird feather or similar item to wave away bubbles and detritus produced by the dissolving process, from the surface of the plate, or the plate may be periodically lifted from the acid bath. If a bubble is allowed to remain on the plate then it will stop the acid biting into the plate where the bubble touches it. Zinc produces more bubbles much more rapidly than copper and steel and some artists use this to produce interesting round bubble-like circles within their prints for a Milky Way effect.

The detritus is powdery dissolved metal that fills the etched grooves and can also block the acid from biting evenly into the exposed plate surfaces. Another way to remove detritus from a plate is to place the plate to be etched face down within the acid upon plasticine balls or marbles, although the drawback of this technique is the exposure to bubbles and the inability to remove them readily.

For aquatinting a printmaker will often use a test strip of metal about a centimetre to three centimetres wide. The strip will be dipped into the acid for a specific number of minutes or seconds. The metal strip will then be removed and the acid washed off with water. Part of the strip will be covered in ground and then the strip is redipped into the acid and the process repeated. The ground will then be removed from the strip and the strip inked up and printed. This will show the printmaker the different degrees or depths of the etch, and therefore the strength of the ink colour, based upon how long the plate is left in the acid. The plate is removed from the acid and washed over with water to remove the acid. The ground is removed with a solvent such as turpentine. Turpentine is often removed from the plate using methylated spirits since turpentine is greasy and can affect the application of ink and the printing of the plate.

Spit-biting is a process whereby the printmaker will apply acid to a plate with a brush in certain areas of the plate. The plate may be aquatinted for this purpose or exposed directly to the acid. The process is known as spit-biting due to the use of saliva once used as a medium to dilute the acid, although gum arabic or water are now commonly used.

A piece of matte board, a plastic 'card', or a wad of cloth is often used to push the ink into the incised lines. The surface is wiped clean with a piece of stiff fabric known as tarlatan and then either wiped with newsprint paper; some printmakers prefer to use the blade part of their hand or palm at the base of their thumb. The wiping leaves ink in the incisions. You may also use a folded piece of organza silk to do the final wipe. If copper or zinc plates are used plate surface is left very clean and therefore white in the print. If steel plate is used then the plate's natural tooth gives the print a grey background similar to the effects of aquatinting. As a result steel plates do not need aquatinting as gradual exposure of the plate via successive dips into acid will produce the same result. A damp piece of paper is placed over the plate and it is run through the press.

Nontoxic etching

Growing concerns about the health effects of acids and solvents led to the development of less toxic etching methods in the late 20th century. An early innovation was the use of floor wax as a hard ground for coating the plate. Others, such as printmakers Mark Zaffron and Keith Howard, developed systems using acrylic polymers as a ground and ferric chloride for etching. The polymers are removed with sodium carbonate (washing soda) solution, rather than solvents. When used for etching, ferric chloride does not produce a corrosive gas, as acids do, thus eliminating another danger of traditional etching.

The traditional aquatint, which uses either powdered rosin or enamel spray paint, is replaced with an airbrush application of the acrylic polymer hard ground. Again, no solvents are needed beyond the soda-ash solution, though a ventilation hood is needed due to acrylic particulates from the air brush spray. The traditional soft ground, requiring solvents for removal from the plate, is replaced with water-based relief printing ink. The ink receives impressions like traditional soft ground, resists the ferric chloride etchant, yet can be cleaned up with warm water and either soda ash solution or ammonia.

Anodic etching has been used in industrial processes for over a century. The etching power is a source of direct current. The item to be etched (anode) is connected to its positive pole. A receiver plate (cathode) is connected to its negative pole. Both, spaced slightly apart, are immersed in a suitable aqueous solution of a suitable electrolyte. The current pushes the metal out from the anode into solution and deposits it as metal on the cathode.

Shortly before 1990, two groups working independently developed different ways of applying it to creating intaglio printing plates. In the patented Electroetch system, invented by Marion and Omri Behr, in contrast to certain non toxic etching methods, an etched plate can be reworked as often as the artist desires.

The system uses voltages below 2 volts which exposes the uneven metal crystals in the etched areas resulting in superior ink retention and printed image appearance of quality equivalent to traditional acid methods. With polarity reversed the low voltage provides a simpler method of making mezzotint plates as well as the 'steel facing' copper plates.

Photo-etching

Light sensitive polymer plates allow for photorealistic etchings. A photosensitive coating is applied to the plate by either the plate supplier or the artist. Light is projected onto the plate as a negative image to expose it. Photopolymer plates are either washed in hot water or under other chemicals according to the plate manufacturers' instructions. Areas of the photo-etch image may be stopped-out before etching to exclude them from the final image on the plate, or removed or lightened by scraping and burnishing once the plate has been etched.

Once the photo-etching process is complete, the plate can be worked further as a normal intaglio plate, using drypoint, further etching, engraving, etc. The final result is an intaglio plate which is printed like any other.

Types of metal plates

Copper was always the traditional metal, and is still preferred, for etching, as it bites evenly, holds texture well, and does not distort the colour of the ink when wiped. Zinc is cheaper than copper, so preferable for beginners, but it does not bite as cleanly as copper, and it alters some colours of ink. Steel is growing in popularity as an etching substrate. Prices of copper and zinc have steered steel to an acceptable alternative. The line quality of steel is less fine than copper but finer than zinc. Steel has a natural and rich aquatint.

Industrial uses

Etching is also used in the manufacturing of printed circuit boards and semiconductor devices, on glass, and in the preparation of metallic specimens for microscopic observation.

Chemical Vapour Deposition

Chemical vapour deposition (CVD) is a chemical process used to produce high-purity, high-performance solid materials. The process is often used in the semiconductor industry to produce thin films. In a typical CVD process, the wafer (substrate) is exposed to one or more volatile precursors, which react and/or decompose on the substrate surface to produce the desired deposit. Frequently, volatile by-products are also produced, which are removed by gas flow through the reaction chamber.

Microfabrication processes widely use CVD to deposit materials in various forms, including: monocrystalline, polycrystalline, amorphous, and epitaxial. These materials include: silicon, carbon fibre, carbon nanofibres, filaments, carbon nanotubes, SiO_2, silicon-germanium, tungsten, silicon carbide, silicon nitride, silicon oxynitride, titanium nitride, and various high-k dielectrics. The CVD process is also used to produce synthetic diamonds.

Types of chemical vapour deposition

A number of forms of CVD are in wide use and are frequently referenced in the literature. These processes differ in the means by which chemical reactions are initiated (e.g. activation process) and process conditions.

1. Classified by operating pressure:
   (a) Atmospheric pressure CVD (APCVD): CVD processes at atmospheric pressure.
   (b) Low-pressure CVD (LPCVD): CVD processes at subatmospheric pressures. Reduced pressures tend to reduce unwanted gas-phase reactions and improve film uniformity across the wafer. Most modern CVD processes are either LPCVD or UHVCVD.
   (c) Ultrahigh vacuum CVD (UHVCVD): CVD processes at a very low pressure, typically below 10^{-6} Pa ($\sim 10^{-8}$ torr). Note that in other fields, a lower division between high and ultra-high vacuum is common, often 10^{-7} Pa.
2. Classified by physical characteristics of vapour:
   (a) Aerosol assisted CVD (AACVD): A CVD process in which the precursors are transported to the substrate by means of a liquid/gas aerosol, which can be generated ultrasonically. This technique is suitable for use with nonvolatile precursors.
   (b) Direct liquid injection CVD (DLICVD): A CVD process in which the precursors are in liquid form (liquid or solid dissolved in a convenient solvent). Liquid solutions are injected in a vapourisation chamber towards injectors (typically car injectors). Then the precursor vapours are transported to the substrate as in classical CVD process. This technique is suitable for use on liquid or solid precursors. High growth rates can be reached using this technique.

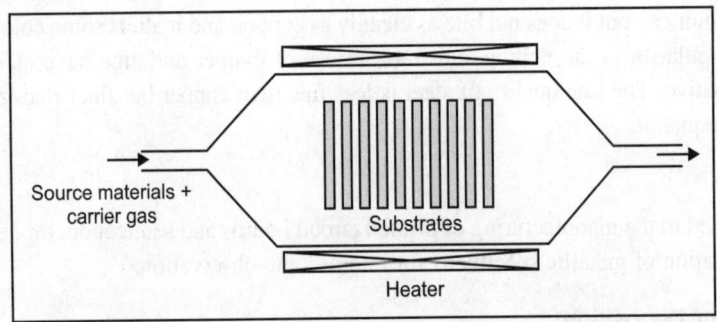

Fig. 6.7. Hot-wall thermal CVD (batch operation type).

3. Plasma methods:
   (a) Microwave plasma-assisted CVD (MPCVD).
   (b) Plasma-Enhanced CVD (PECVD): CVD processes that utilise plasma to enhance chemical reaction rates of the precursors. PECVD processing allows deposition at lower temperatures, which is often critical in the manufacture of semiconductors.
   (c) Remote plasma-enhanced CVD (RPECVD): Similar to PECVD except that the wafer substrate is not directly in the plasma discharge region. Removing the wafer from the plasma region allows processing temperatures down to room temperature.
4. Atomic layer CVD (ALCVD): Deposits successive layers of different substances to produce layered, crystalline films.

5. Combustion chemical vapour deposition (CCVD): nGimat's proprietary combustion chemical vapour deposition process is an open-atmosphere, flame-based technique for depositing high-quality thin films and nanomaterials.
6. Hot wire CVD (HWCVD): Also known as catalytic CVD (Cat-CVD) or hot filament CVD (HFCVD). Uses a hot filament to chemically decompose the source gases.
7. Metalorganic chemical vapour deposition (MOCVD): CVD processes based on metalorganic precursors.
8. Hybrid physical-chemical vapour deposition (HPCVD): Vapour deposition processes that involve both chemical decomposition of precursor gas and vapourisation of a solid source.
9. Rapid thermal CVD (RTCVD): CVD processes that use heating lamps or other methods to rapidly heat the wafer substrate. Heating only the substrate rather than the gas or chamber walls helps reduce unwanted gas phase reactions that can lead to particle formation.
10. Vapour phase epitaxy (VPE).

Figure 6.8 shows the plasma assisted CVD.

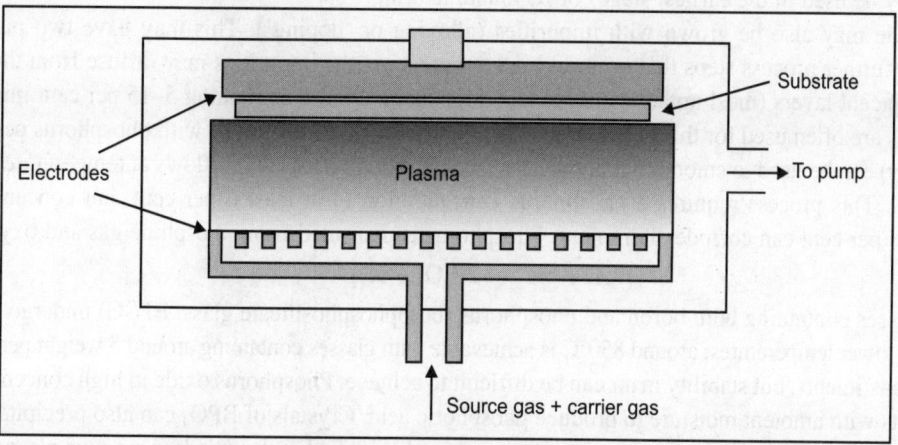

Fig. 6.8. Plasma assisted CVD.

Substances commonly deposited for ICs

This section discusses the CVD processes often used for integrated circuits (ICs). Particular materials are deposited best under particular conditions.

Polysilicon

Polycrystalline silicon is deposited from silane (SiH_4), using the following reaction:

$$SiH_4 \rightarrow Si + 2H_2$$

This reaction is usually performed in LPCVD systems, with either pure silane feedstock or a solution of silane with 70–80 per cent nitrogen. Temperatures between 600° and 650°C and pressures between 25 and 150 Pa yield a growth rate between 10 and 20 nm per minute. An alternative process uses a hydrogen-based solution. The hydrogen reduces the growth rate, but the temperature is raised to 850°C or even 1050°C to compensate.

Polysilicon may be grown directly with doping, if gases such as phosphine, arsine or diborane are added to the CVD chamber. Diborane increases the growth rate, but arsine and phosphine decrease it.

Silicon dioxide

Silicon dioxide (usually called simply 'oxide' in the semiconductor industry) may be deposited by several different processes. Common source gases include silane and oxygen, dichlorosilane ($SiCl_2H_2$) and nitrous oxide (N_2O), or tetraethylorthosilicate [TEOS; $Si(OC_2H_5)_4$]. The reactions are as follows:

$$SiH_4 + O_2 \rightarrow SiO_2 + 2H_2$$

$$SiCl_2H_2 + 2\ N_2O \rightarrow SiO_2 + 2N_2 + 2HCl$$

$$Si(OC_2H_5)_4 \rightarrow SiO_2 + \text{by-products}$$

The choice of source gas depends on the thermal stability of the substrate; for instance, aluminium is sensitive to high temperature. Silane deposits between 300° and 500°C, dichlorosilane at around 900°C, and TEOS between 650° and 750°C, resulting in a layer of low-temperature oxide (LTO). However, silane produces a lower-quality oxide than the other methods (lower dielectric strength, for instance), and it deposits nonconformally. Any of these reactions may be used in LPCVD, but the silane reaction is also done in APCVD. CVD oxide invariably has lower quality than thermal oxide, but thermal oxidation can only be used in the earliest stages of IC manufacturing.

Oxide may also be grown with impurities (alloying or 'doping'). This may have two purposes. During further process steps that occur at high temperature, the impurities may diffuse from the oxide into adjacent layers (most notably silicon) and dope them. Oxides containing 5–15 per cent impurities by mass are often used for this purpose. In addition, silicon dioxide alloyed with phosphorus pentoxide (P-glass) can be used to smooth out uneven surfaces. P-glass softens and reflows at temperatures above 1000°C. This process requires a phosphorus concentration of at least 6 per cent, but concentrations above 8 per cent can corrode aluminium. Phosphorus is deposited from phosphine gas and oxygen:

$$4PH_3 + 5O_2 \rightarrow 2P_2O_5 + 6H_2$$

Glasses containing both boron and phosphorus (borophosphosilicate glass, BPSG) undergo viscous flow at lower temperatures; around 850°C is achievable with glasses containing around 5 weight per cent of both constituents, but stability in air can be difficult to achieve. Phosphorus oxide in high concentrations interacts with ambient moisture to produce phosphoric acid. Crystals of BPO_4 can also precipitate from the flowing glass on cooling; these crystals are not readily etched in the standard reactive plasmas used to pattern oxides, and will result in circuit defects in integrated circuit manufacturing.

Besides these intentional impurities, CVD oxide may contain by-products of the deposition process. TEOS produces a relatively pure oxide, whereas silane introduces hydrogen impurities, and dichlorosilane introduces chlorine.

Lower temperature deposition of silicon dioxide and doped glasses from TEOS using ozone rather than oxygen has also been explored (350° to 500°C). Ozone glasses have excellent conformality but tend to be hygroscopic—that is, they absorb water from the air due to the incorporation of silanol (Si-OH) in the glass. Infrared spectroscopy and mechanical strain as a function of temperature are valuable diagnostic tools for diagnosing such problems.

Silicon nitride

Silicon nitride is often used as an insulator and chemical barrier in manufacturing ICs. The following two reactions deposit nitride from the gas phase:

$$3SiH_4 + 4NH_3 \rightarrow Si_3N_4 + 12H_2$$

$$3\ SiCl_2H_2 + 4NH_3 \rightarrow Si_3N_4 + 6HCl + 6H_2$$

Silicon nitride deposited by LPCVD contains up to 8 per cent hydrogen. It also experiences strong tensile stress, which may crack films thicker than 200 nm. However, it has higher resistivity and dielectric strength than most insulators commonly available in microfabrication (10^{16} Ω·cm and 10 MV/cm, respectively).

Another two reactions may be used in plasma to deposit SiNH:

$$2SiH_4 + N_2 \rightarrow 2SiNH + 3H_2$$
$$SiH_4 + NH_3 \rightarrow SiNH + 3H_2$$

These films have much less tensile stress, but worse electrical properties (resistivity 10^6 to 10^{15} Ω·cm, and dielectric strength 1 to 5 MV/cm).

Metals

Some metals (notably aluminium and copper) are seldom or never deposited by CVD. As of 2002, a commercially, cost effective, viable CVD process for copper did not exist, though copper formate, copper(hfac)2, Cu(II) ethyl acetoacetate, and other precursors have been used. Copper deposition of the metal has been done mostly by electroplating, in order to reduce the cost. Aluminium can be deposited from tri-isobutyl aluminium (TIBAL), triethyl/methyl aluminium (TEA, TMA), or dimethylaluminium hydride (DMAH), but physical vapour deposition methods are usually preferred.

However, CVD processes for molybdenum, tantalum, titanium, nickel, and tungsten are widely used. These metals can form useful silicides when deposited onto silicon. Mo, Ta and Ti are deposited by LPCVD, from their pentachlorides. Nickel, molybdenum, and tungsten can be deposited at low temperatures from their carbonyl precursors. In general, for an arbitrary metal M, the reaction is as follows:

$$2MCl_5 + 5H_2 \rightarrow 2M + 10HCl$$

The usual source for tungsten is tungsten hexafluoride, which may be deposited in two ways:

$$WF_6 \rightarrow W + 3F_2$$
$$WF_6 + 3H_2 \rightarrow W + 6HF.$$

CATALYST DEACTIVATION

Catalyst deactivation, the loss of catalytic activity and/or selectivity over time, is of crucial importance in three-way catalysis where catalytic materials are exposed to high temperatures under fluctuating conditions. In the literature, there are several definitions for deactivation, but none of them, is exclusively broad enough as a definition of the deactivation of a three-way catalyst.

The catalyst deactivation is defined as a phenomenon in which the structure and state of the catalyst change, leading to the loss of active sites on the catalyst's surface and thus causing a decrease in the catalyst's performance.

Catalyst deactivation is a result of a number of unwanted chemical and physical changes. The causes of deactivation are classically divided to three categories: chemical, thermal and mechanical. In this section, mechanisms of thermal and chemical deactivation are mostly considered.

Mechanical deactivation as a result of physical breakage, attrition or crushing is also an important deactivation phenomenon. However, for the current catalytic converters, deactivation during the normal vehicle operation is typically a result of chemical and thermal mechanisms, rather than fouling and mechanical factors.

Deactivation of a catalyst is usually an inevitable and slow phenomenon. Despite its inevitable nature, some immediate consequences of deactivation may at least be partly avoided, or even reversed. Deactivation of a three-way catalyst can result from various processes (deactivation mechanisms) as summarised in Table 6.1. The three major categories of deactivation mechanisms are sintering, poisoning, and coke formation or fouling. They may occur separately or in combination, but the net effect is always the removal of active sites from the catalytic surface. Some other deactivation mechanisms, such as pore blockage, volatilisation of active component, destruction of the active surface and incorporation of the active component into the washcoat in an inactive form can also cause decline in the catalyst's activity.

Table 6.1. A summary of the deactivation mechanisms of three-way catalysts.

Thermal	Chemical	Fouling	Mechanical
Sintering	Poisoning: irreversible adsorption or reaction on/with the surface	Coke formation (carbon deposits)	Thermal shock
Alloying	Inhibition: competive reversible adsorption of poison precursors		Attrition
Support changes	Poison-induced reconstruction of catalytic surfaces		Physical breakage
Precious metal-base metal interactions	Physical/chemical blockage of support pore structure		
Metal/metal oxide- support interactions			
Oxidation			
Precious metal surface orientation			
Metal volatilisation			

Deactivation of a three-way catalyst is a complex phenomenon since the purification performance of a three-way catalytic converter is affected by many factors, such as changes in the exhaust gas velocity and composition, temperature, precious metal loading and the catalyst's age. Three-way catalysts are designed to withstand momentarily high operation temperatures, but the long-lasting exposure to high thermal loading increases the risk of thermal deactivation.

Thermal or thermochemical degradation, is probably the main cause for the deactivation of automotive exhaust gas catalysts. Three-way catalysts are known to loose their activity, especially under oxidising conditions at temperatures higher than 900°C. Exposure to high operation temperatures enhances the reduction of the alumina surface area and sintering of the precious metals, resulting in a loss of effective catalytic area.

The thermal degradation of three-way catalysts is caused not only by high temperature but also by sudden temperature changes in the catalytic converter. Catalysts may also be poisoned in the presence of some pollutants, such as sulphur or phosphorus. These components contaminate the washcoat and precious metals and reduce the active catalytic area by blocking the active sites. On the other hand, deactivation by fouling or coke formation is not regarded as a major problem in the current high-temperature catalytic purification systems.

Catalyst Deactivation Mechanisms

Deactivation by thermal degradation and sintering

Temperature has become an increasingly important factor for deactivation of three-way catalysts due to the fact that the converter is installed near the engine to confirm the efficient purification of hydrocarbons. Most of the emissions are formed during cold start, and during the low temperature operation of the catalyst, as mentioned earlier. Since the pre-converter is installed near the engine, the temperature inside the converter is higher than in the main converter due to higher temperature of exhaust gas. Thus, catalytic materials have to work even at temperatures higher than 1000°C. Thermal degradation of a three-way catalyst begins in the temperature area of 800°–900°C, or even at lower temperatures, depending on the materials used. It is a physical process leading to a catalyst deactivation at high temperatures because of the loss of catalytic surface area due to crystal growth of the catalytic phase, the loss of washcoat area due to a collapse of pore structure, and/or chemical transformations of catalytic phases to non-catalytic phases. The first two processes are typically referred to as sintering, and the third as the solid-solid phase transition at high temperatures.

In Fig. 6.9, the deactivation mechanisms of a three-way catalyst are illustrated. These mechanisms are examined in detail in the following paragraphs and references to Fig. 6.9 are given concurrently. Sintering, as illustrated in Figs 6.9(c) and 6.9(d), is the loss of catalyst's active surface due to crystal growth of either the bulk material or the active phase. In the case of supported metal catalysts, reduction of the active surface area is provoked via agglomeration and coalescence of small metal crystallites into larger ones.

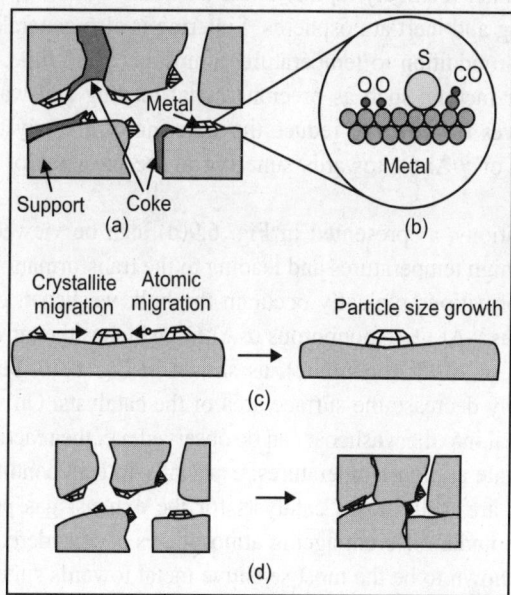

Fig. 6.9. Deactivation mechanisms: (a) coke formation, (b) poisoning, (c) sintering of the active metal particles, and (d) sintering and solid-solid phase transitions of the washcoat and encapsulation of active metal particles.

Two different models have been proposed for sintering, i.e. the atomic migration and the crystallite migration models. As such, sintering occurs either due to metal atoms migrating from one crystallite to another via the surface or gas phase by diminishing small crystallites in size and increasing the larger ones (atomic migration model).

Or sintering can occur via migration of the crystallites along the surface, followed by collision and coalescence of two crystallites (crystallite migration model). Figure 6.9(c) presents a schematic representation of atomic migration and crystallite migration models.

As mentioned earlier, sintering on supported metal catalysts involves complex physical and chemical phenomena that make the understanding of the mechanistic aspects of sintering difficult. Experimental observations have shown that sintering is strongly temperature-dependent, but is also affected by the surrounding gas atmosphere. The rate of sintering increases exponentially with temperature and, for example, the sintering of precious metals becomes significant above 600°C. The underlying mechanism of sintering of small metal particles is the surface diffusion or at higher temperatures, the mobility of larger agglomerates. The so-called Hüttig and Tamman temperatures indicate the temperature at which sintering starts. The following semi-empirical relations for Hüttig and Tamman temperatures are more commonly used:

$$T_{Hüttig} = 0.3\ T_{melting} \qquad \qquad \text{... (6.5)}$$

$$T_{Tamman} = 0.5\ T_{melting} \qquad \qquad \text{... (6.6)}$$

Temperature at which the solid phase becomes mobile depends on several factors such as texture, size and morphology. For instance, highly porous γ-alumina is much more sensitive to sintering than nonporous α-alumina.

Sintering processes at high temperatures are also affected by atmosphere, as expressed earlier. Supported metal catalysts sinter relatively rapidly under an oxidising atmosphere, however the process is more slow under reducing and inert atmospheres. Sintering is also generally accelerated, e.g. in the presence of water vapour. In addition to temperature, atmosphere and time, the sintering rate is also dependent on several other factors, such as precious metal loading and washcoat composition. The presence of specific additives is known to reduce the sintering of a catalyst: BaO, CeO_2, La_2O_3 and ZrO_2 improve the stability of γ-Al_2O_3 towards sintering in the presence of high H_2O content in the exhaust gas.

Solid-solid phase transitions, as presented in Fig. 6.9(d), can be viewed as an extreme form of sintering occurring at very high temperatures and leading to the transformation of one crystalline phase into another. Phase transformations typically occur in the bulk washcoat, e.g. aluminium oxide has many phases from the porous γ-Al_2O_3 to nonporous α-Al_2O_3, which is the most stable phase of alumina. The phase transformations of Al_2O_3 (boehmite), as shown in Fig. 6.10, become significant at high temperatures and remarkably decrease the surface area of the catalysts. On supported metal catalysts, the incorporation of the metal into the washcoat can be observed, i.e. the reaction of Rh_2O_3 with alumina to form inactive Rh-aluminate at high-temperatures, especially in lean conditions.

Active precious metals are well-known catalysts for the exhaust gas purification. The sintering behaviour of Pt, Pd and Rh under different ageing atmospheres is considered in Table 6.2. Among the active metals, rhodium is known to be the most sensitive metal towards sintering at high temperatures under the exhaust gas conditions.

This leads to poor activity, especially in the reduction of NO_x. The use of a bimetallic catalyst, such as Pd-Rh or Pt-Rh, gives a better catalytic activity at high temperatures. The operation conditions (rich or lean) also affect the sintering of active metals; for example, ageing atmosphere and the oscillation between oxidising and reducing atmospheres can accelerate deactivation.

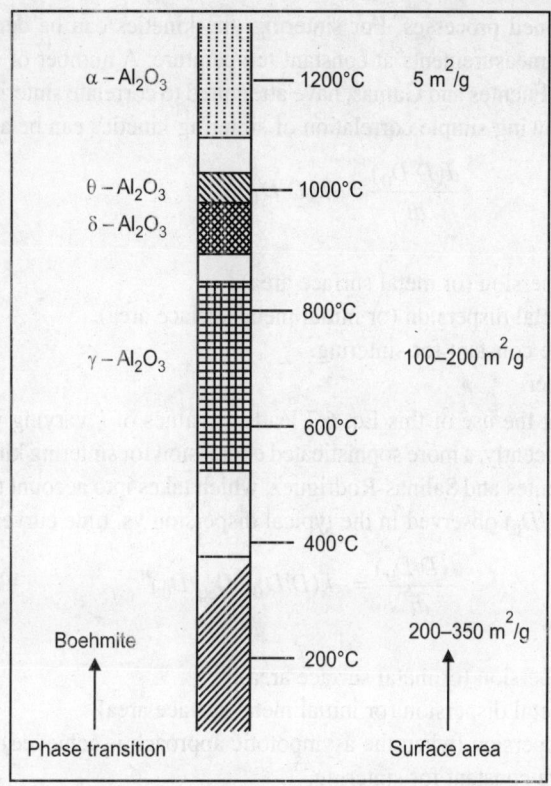

Fig. 6.10. Phase transitions and surface areas of Al_2O_3 (boehmite) as a function of temperature.

Table 6.2. Effect of ageing atmosphere on the sintering behaviour (particle size) of Pt, Pd and Rh (washcoat La_2O_3-doped Al_2O_3, precious metal content 0.14 wt-%).

Particle size (Å)	Pt	Pd	Rh
N$_2$/1100°C	210	970	140
Exhaust gas/1100°C	780	680	880
Air/1100°C	970	n.d.	n.d.

n.d. = not determined.

Redispersion is an opposite process to sintering. During redispersion, many complex phenomena take place, the particle sizes decrease and surface areas increase. In particular, the interaction between oxygen and precious metals may lead to the formation of species that are mobile on the surface and reverse the process of agglomeration. Sintering is normally physical rather than chemical in nature and, therefore, the magnitudes of thermal activation are quite different. Typical activation energies for sintering may be twice or even three times lower than those associated with the chemical processes in poisoning or coke formation. Furthermore, ageing time is important because it correlates both with sintering and redispersion.

The kinetics of a catalyst deactivation is a function of temperature, time, pressure and the concentrations of different substances. The change in catalytic activity can be an effect of one or several of the previously mentioned processes. For sintering, the kinetics can be derived from active metal surface area versus time measurements at constant temperature. A number of researchers, Wanke and Flynn, Bartholomew and Fuentes and Gamas, have attempted to correlate sintering kinetics of supported metal catalysts. The following simple correlation of sintering kinetics can be applied:

$$\frac{d(D/D_0)}{dt} = -k(D/D_0)^n \qquad \text{... (6.7)}$$

where,

> D is the metal dispersion (or metal surface area).
> D_0 is the initial metal dispersion (or initial metal surface area).
> k is the kinetic rate constant for sintering.
> n the sintering order.

It has been found that the use of this Eq. 6.7 leads to values of k varying with sintering time, and hence with dispersion. Recently, a more sophisticated expression for sintering kinetics has been proposed by Bartholomew and Fuentes and Salinas-Rodriguez, which takes into account the asymptotic approach (by adding the term $-D_{eq}/D_0$) observed in the typical dispersion vs. time curves:

$$\frac{d(D/D_0)}{dt} = -k(D/D_0 - D_{eq}/D_0)^n \qquad \text{... (6.8)}$$

where,

> D is the metal dispersion (or metal surface area).
> D_0 is the initial metal dispersion (or initial metal surface area).
> D_{eq} is the final dispersion (when the asympototic approach is achieved).
> k is the kinetic rate constant for sintering.
> n the sintering order.

Equation 6.8 can be applied in a quantitative comparison regarding the effect of temperature, time and atmosphere on the sintering rate of supported metal catalysts.

Deactivation by poisoning

The activity of a three-way catalyst reduces gradually when the unwanted, harmful components of fuels and lubricants or other impurities, are accumulated on the catalyst's surface and slowly poison the catalyst. Poisoning is defined as a loss of catalytic activity due to the chemisorption of impurities on the active sites of the catalyst. Usually, a distinction is made between poisons and inhibitors. Poisons are substances that interact very strongly and irreversibly with the active sites, whereas the adsorption of inhibitors on the catalyst surface is weak and reversible. In the latter case, the catalytic activity can be at least partly restored by regeneration. This irreversible/reversible or permanent/temporary nature of deactivation and the regeneration possibility of a catalyst are the main differences between poisoning and inhibition. However, the distinction between permanent and temporary poisoning is not always so clear, since strong poisons at low temperatures may be less harmful in high-temperature applications. Catalyst poisons can also be classified as selective or nonselective. The description of a poison as selective or nonselective is related to the nature of the surface and the degree of interaction of the poison with the surface. A poison can also be selective in one reaction, but not in another.

Poisoning of a three-way catalyst as a result of the accumulation of impurities on the active sites [Fig. 6.9(b)] is typically a slow and irreversible phenomenon. The accumulation of poisons on the active sites blocks the access of reactants to these active sites. As a result of poisoning, the catalytic activity may be decreased without affecting the selectivity, but often selectivity is also changed since some of the active sites are deactivated while others are practically unaffected. In some cases, depending on the adsorbed poison, the poisoned catalyst can be regenerated and its activity can be at least partly restored. However, the poisoned three-way catalyst can hardly be regenerated and, therefore, the best method to reduce poisoning is to decrease the amount of poisons in the fuel and lubrication oils to more acceptable levels.

Catalytic converters are poisoned by the impurities in fuel and lubrication oil or by shavings from the exhaust tailpipe. Even the low levels of impurities are enough to cover the active sites and decrease the performance of a catalytic converter. It follows that the analysis of poisoned catalysts may be complicated since the content of poison of a fully deactivated catalyst can be as low as 0.1 wt-% or even less. Lead (Pb), sulphur (S), phosphorus (P), zinc (Zn), calcium (Ca), and magnesium (Mg) compounds are typical catalyst poisons. Earlier, the effects of lead (Pb) had been studied extremely carefully. The catalytic converters were known to loose already their effectiveness after 10 refills with leaded gasoline. Nowadays, mostly due to the use of unleaded or low Pb concentrations in gasoline, the role of lead as a catalyst poison is far less significant than in the past.

Fuel (gasoline) contains sulphur in small amounts. New requirements for a low sulphur content (<50 ppm) in the gasoline fuel are introduced together with Euro IV (Directive 98/70/EC). Sulphur clearly affects, and often very quickly, the efficiency and oxygen storage capacity (OSC) of the catalyst. Sulphur poisoning can lead to the formation of new inactive compounds on the catalyst's surface and also to the morphological changes in the catalyst. Fast poisoning by sulphur can be to some extent reversible and the poisoned catalyst can be regenerated. Beck and Sommers have shown that the impact of sulphur on vehicle-aged catalysts was irreversible at temperatures below 650°C, but the original activity could be restored at higher temperatures. However, it should be noted that although the purification efficiency is recovered, the oxygen storage capacity is not. During the combustion processes in the engine, fuel sulphur oxidises to SO_2 and SO_3. These compounds adsorb on the precious metal sites on the catalyst's surface at low temperatures (below 300°C) and react with alumina to form aluminium sulphates that reduce the active surface of washcoat and deactivate the catalyst. The air-to-fuel ratio also affects the behaviour of sulphur. In lean conditions, SO_2 is stored on cerium, and in rich conditions both SO_2 and SO_3 are reduced to form hydrogen sulphide (H_2S). Three-way catalysts are known to loose their activity more in oxidising (lean) than in reducing (rich) conditions in the presence of sulphur compounds. The problem of sulphur poisoning also appears to be more significant at low temperatures, whereas at elevated temperatures (T>1000°C), the adsorption of sulphur species is almost absent.

Phosphorus (P), zinc (Zn), calcium (Ca), and magnesium (Mg) compounds are typical impurities in the lubrication oils. These substances and/or their compounds accumulate on the catalyst's surface and they can be regarded as notable as fuel poisons. The considerable amounts of phosphorus, zinc, calcium, and/or magnesium are normally observed on the surface of an aged catalyst after years of driving. Several studies of the deactivation of a three-way catalyst by phosphorus, calcium and zinc compounds have been published.

Zinc dialkyldithiophosphate (ZDP), a typical oil additive, is a common source of phosphorus and zinc. Several studies have shown that the individual effects of P and Zn on deactivation are small compared to that of the combined effect of P and Zn. At low exhaust temperatures in particular the formation of zinc pyrophosphate ($Zn_2P_2O_7$) decreases the catalytic activity. The phosphorus contamination is observed either as an overlayer of Zn, Ca and Mg phosphates ($M_3(PO_4)_2$, M = Zn, Ca or Mg), or as aluminium phosphate ($AlPO_4$) within the washcoat. Recently, cerium has also been observed to form cerium phosphates, $CePO_4$ and/or $Ce(PO_3)_3$. Phosphates form a film layer on the catalyst surface that covers the precious metals in the porous washcoat and prevents contact between the catalyst and the surrounding gas atmosphere.

The poisoning of a catalyst is clearly dependent on the phosphorus level in the lubrication oil. The use of calcium or magnesium containing oil additives can decrease the harmful effects of phosphorus. Calcium and magnesium sulphonates form Ca and Mg phosphates and thus prevent the accumulation of phosphorus on the catalyst's surface. Similar observations have also been made in the case of zinc compounds. The largest contaminant levels are typically observed in the front edge of the catalyst. Experimental observations have also shown that even small amounts of these compounds are high enough to decrease the performance of a catalytic converter.

Active metal catalysts are preferred in the controlling of the exhaust gas emissions, because they are less liable to sulphur poisoning than metal oxide catalysts, as reported above. Precious metals have different types of resistance against poisoning. Palladium is more sensitive than platinum and rhodium to chemical deactivation, in particular to poisoning by sulphur and lead. Currently, the use of Pd catalysts is possible because of the rapid decrease in fuel lead content, as discussed previously. The additives used and the chemical composition of the washcoat has an effect on the sulphur behaviour in the catalyst. In particular additives, which play a significant role in Pd-only catalysts.

Driving conditions also affect the catalyst's chemical deactivation. Especially in Nordic countries, where the cold weather and urban driving keep the catalyst's temperature low during a long time period. This accelerates the catalyst's chemical ageing, because the unburned soot and particles adsorb on the active material. The stability against thermal and chemical deactivation can be improved by a proper choice of the catalyst material. In addition, the placement of the active material in separate washcoat layers improves the durability.

Other relevant mechanisms of deactivation

There are other essential forms of the deactivation of three-way catalysts. For example, pore blockage, encapsulation of metal particles, volatilisation of active compounds, fouling or coke formation and metal-metal or metal-washcoat interactions, which will be briefly discussed below.

According to Graham, high temperature ageing may result in deep encapsulation of sintered precious metal particles [Fig. 6.9(d)] as the surface area of the washcoat decreases. This is a serious type of deactivation because of its permanent nature. The encapsulated metal particles cannot participate in catalysis since they are inaccessible to gas phase molecules. Furthermore, support can interact with the metal catalyst also by the support-induced changes observed in the metal particle morphology, by the formation of specific active sites on the metal-support interface and by the charging of metal particles.

Fouling covers all phenomena where the surface is covered with a deposit, e.g. with combustion residues such as soot or with mechanical wear. Coke formation is the most widely known form of fouling (it is even used as a synonym for fouling). Coke formation is not very clearly defined. There are probably as many mechanisms of coke formation as there are reactions and catalysts where this phenomenon is encountered. During the coke formation, carbonaceous residues cover the active surface sites [Fig. 6.9(a)], and decrease the active surface area. First, this blocks out the active compounds to reach the surface sites, and second, the amount of coke might be so large that carbon deposits block the internal pores in the catalyst. In many cases, hydrocarbons and aromatic materials are primarily responsible for coke formation. Among these other deactivation mechanisms, pore blocking is probably one of the most important mechanisms. Pore blocking is often connected to coke formation, and when the amount of coke is high on the catalyst's surface, it may be possible for the coke itself to block off the pore structure.

At high temperatures, catalysts may suffer from the loss of active phase through volatilisation. Metal loss through direct volatilisation is generally an insignificant route of the catalyst deactivation. By contrast, metal loss through the formation of volatile compounds is important over a wide range of conditions. Large amounts of catalytic materials can be transported to either substrate where they can react, or into the gas phase where they are lost in the effluent gas stream. High volatility limits the selection of otherwise useful catalytic materials, e.g. the oxides of Pt, Pd and Rh formed during the reaction cycles are not as volatile as the other noble metal oxides, such as RuO_2, OsO_4 and Ir_2O_3.

The thermodynamics of volatilisation and thermodynamic equilibrium calculations are useful in the evaluation of the volatility of metals and metal oxides in order to assess which materials are stable over long periods at high temperatures. Thermodynamic equilibrium calculations of the oxidation/reduction behaviour of palladium have shown that phase stability in a Pd/PdO system changes as a function of temperature and oxygen partial pressure. The lower the pressure and the higher the temperature are, the more likely is the Pd phase. In Table 6.3, vapour pressures of Pt, Pd and Rh as metals and metal oxides are given at a temperature of 800°C in air. The vapour pressure increases with temperature, and it is also strongly dependent on the composition of the surrounding atmosphere, i.e. Pd is volatile at temperatures around 850°C and above, depending on the surrounding environment. As can be seen in Table 6.3, the orders of vapour pressures of active metals and their oxides are as follows:

Metals: Pd > > Pt > Rh

Oxides: Pt > Rh >> Pd

Hence the vapour pressure of metallic Pd is clearly several magnitudes higher than the vapour pressures of Pt and Rh, while as oxides, the situation is the reverse.

Table 6.3. Vapour pressures (torr) of Pt, Pd and Rh at 800°C in air.

	Pt	Pd	Rh
Metal	9.1×10^{-17}	1.2×10^{-9}	2.9×10^{-17}
Oxide	1.2×10^{-5}	Negligible	5.8×10^{-6}

As an example, in Figs 6.11(a) and 6.11(b), thermodynamics equilibrium curves of Pd and Rh are presented respectively. According to the thermodynamics, Pd is easily oxidised at room temperature to PdO and it reduces to metallic Pd in the temperature range of 500°–1200°C. The formation and decomposition of PdO occurs as follows:

$$2PdO \rightleftharpoons 2Pd + O_2 \quad\quad\quad ... (6.9)$$

The most stable oxidation state of Pd is +2 and the formation of PdO is kinetically restricted at low temperatures. As shown in Fig. 6.11(a), the metallic Pd is totally volatilised in a 5 per cent O_2/N_2 atmosphere at a temperature of 1400°C, and the increased amount of oxygen in the gas phase moves the reduction curve of Pd to higher temperatures. According to Farrauto, two kinds of palladium oxides, PdO_x-Pd and PdO, and metallic Pd have been observed on the surface supported on γ-alumina. PdO supported on pure alumina is known to decompose in two steps to metallic palladium in air at a temperature above 800°C. Instead, the reoxidation of metallic Pd to PdO and PdO_x species during the cooling process is very slow at temperatures 550°–650°C, a temperature range at which PdO is the thermodynamically favoured phase.

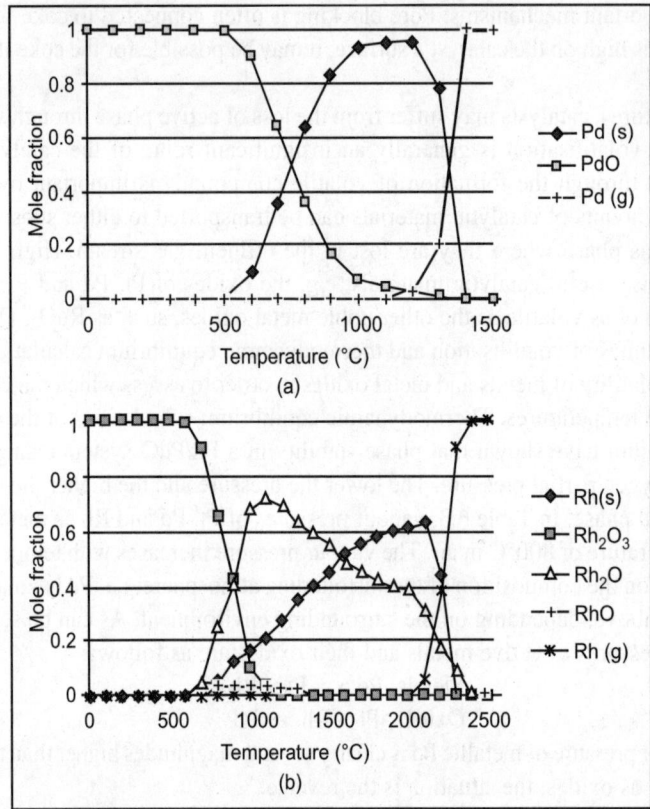

Fig. 6.11. Thermodynamic equilibrium calculations of volatilisation of (a) Pd and (b) Rh in a 5 per cent O_2/N_2 atmosphere (lean).

Temperatures above 800°C convert all PdO_x and PdO to metallic Pd and subsequent cooling again leads to redispersed PdO/Al_2O_3 and PdO_x-Pd/Al_2O_3 phases. Therefore, there is a window of a few hundred degrees in which the catalyst could be in the form of Pd metal or PdO. This hysteresis-like behaviour is strongly dependent on the surroundings of Pd/PdO phases, especially the chemistry of washcoat material and stabilisers.

Studies of the stabilisation of rhodium oxide phases supported on γ-alumina have shown several thermodynamically stable bulk rhodium oxide phases within the ageing temperatures of 500°–1050°C. According to Weng-Sieh, ageing in air below 650°C results in the formation of highly dispersed rhodium oxide – RhO_2, and above 650°C large particles of Rh_2O_3 are observed together with smaller particles of RhO_2. The observed low thermal stability and catalytic activity of rhodium under oxidising conditions has been attributed to the interaction of Rh with the alumina support and the diffusion of rhodium into the bulk of alumina at high temperatures. However, the nature of rhodium oxides formed during the ageings in air is still unclear and it does not necessarily coincide with that expected on the basis of bulk-phase thermodynamics. In Fig. 6.11(b), the thermodynamical equilibrium curves of Rh are presented in a 5 per cent O_2/N_2 atmosphere. At room temperature, the most stable oxidation state of Rh is +3, and Rh has many oxidation states, as can be seen in Fig. 6.11(b). The oxidation/reduction of Rh occurs as follows:

$$2Rh_2O_3 \rightleftharpoons 2Rh_2O + 2O_2 \rightleftharpoons 4Rh + 3O_2 \qquad \text{... (6.10)}$$

At normal operation temperatures of the catalytic converter, Rh is in the form of Rh_2O_3, if the oxidation of Rh is kinetically favoured. The oxygen content in the exhaust gas strongly affects the formation of Rh oxides; the higher the amount of oxygen, the higher the transition temperature is.

SOLVED EXAMPLES

Example 6.1. A catalytic reaction $A \rightarrow 4R$ is being carried out at 3.2 bar and 115°C in a plug flow reactor which contains 0.01 kg of catalyst. Determine the catalyst needed in a packed bed for 3.5 per cent conversion of A to R. For a feed of 2000 moles/hr of pure A.

Solution:

Assume first order rate expression. For first order reaction in the packed bed one has:

$$\frac{W}{F_{A0}} = \int_0^{X_A} \frac{dX_A}{(-r_A)} = \int_0^{X_A} \frac{dX_A}{kC_A} = \frac{1}{kC_A} = \int_0^{X_A} \frac{(1+\varepsilon_A X_A)}{(1-X_A)} dX_A$$

or

$$W = \frac{F_{A0}}{kC_{A0}} \left\{ (1+\varepsilon_A) \ln \frac{1}{1-x_A} - \varepsilon_A x_A \right\}$$

Substituting the values:

$$W = \frac{2000}{96 \times 0.1} \left\{ 4\ln \frac{1}{1.65} - 1.05 \right\} = 140 \text{ kg catalyst.}$$

Example 6.2. The catalytic reaction $A \rightarrow 4R$ was studied in a P.F.R. with various amounts of catalyst and 20 litre/hr of pure A at 3.2 bar and 115°C. The concentration of A in the effluent stream was reported for various run:

Runs	1	2	3	4	5
Catalyst used, kg $\times 10^3$	20	40	80	120	160
CA, mole/litre $\times 10^3$	74	60	44	35	29

   (i) Develop the rate equation for this reaction.
   (ii) Repeat (i) with differential method of analysis.

Solution:

Assume: First order reaction. Performance equation for P.F.R. states:

$$\frac{W}{F_{A0}} = \int_0^{X_A} \frac{dX_A}{-r_A'}$$

or

$$\frac{W}{F_{A0}} = \int_0^{X_A} \frac{dX_A}{kC_A} = \frac{1}{kC_{A0}} \int_0^{X_A} \frac{1+\varepsilon_A X_A}{1-x_A} dX_A$$

$$\frac{kC_{A0}W}{F_{A0}} = (1+\varepsilon_A) \ln \frac{1}{1-X_A} - \varepsilon_A X_A$$

or

$$k \left(\frac{W}{20} \right) = \left(4 \ln \frac{1}{1-X_A} - 3X_A \right)$$

A plot of $4 \ln \dfrac{1}{1-X_A} - 3X_A$ versus $W/20$ is shown below to yield $-r_A' = 95\, C_{AS}$.

Results are also tabulated below:

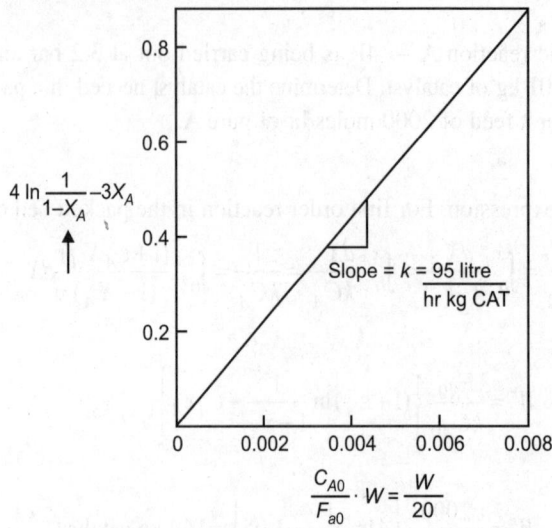

Tabulation based on the measured slopes of plot

$X_A \times 10^3$	$4 \ln 1/1{-}X \times 10^3$	$3X_A \times 10^3$	$4 \ln 1/1 - X_A - 3X_A$	W	$W/20 \times 10^3$
808	337.2	242.4	0.0748	0.02	1
142.9	616.0	428.7	0.1873	0.04	2
241.5	1108.0	724.5	0.3835	0.08	4
317	1526.8	951	0.5758	0.12	6
379	1908	1131	0.771	0.16	8

Tabulation based on the measured slopes of plot

$$4 \ln \frac{1}{1-x_A} - 3x_A \; V_s \frac{W}{20}$$

are given below to predict the rate of reaction at various C_A.

$W \times 10^2$	$W/F_{A0} \times 10^2$	$C_A/C_{A0} \times 10^2$	$X_A \times 10^3$	$-r_A'$ (from plot mentioned)
0	0	100	0	9.3
2	1	74	80.8	–
4	2	60	14.29	5.62
8	4	44	24.15	4.13
12	6	35	31.7	3.34
16	8	29	37.9	2.715

A plot of $-r_A'$ versus C_A is shown below to yield:

$$r_A' = 93\, CA.$$

Results are also tabulated below:

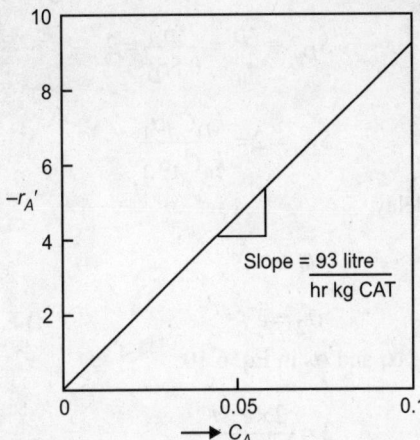

Example 6.3. The following results were obtained on a sample of activated silica granular 4 to 12 mesh size.

Mass of catalyst sample used = 101.5×10^{-3} kg
Volume of helium displaced = 45.1×10^{-6} m³
Volume of mercury displaced = 82.7×10^{-6} m³

Determine the pore volume, density of catalyst and the porosity of the catalyst.

Solution:

$$\text{Pore volume} = \frac{(\text{Volume of } H_g - \text{volume of } H_e) \text{ displaced}}{\text{Mass of catalyst}}$$

$$= \frac{(82.7 - 45.1) \times 10^{-6}}{101.5 \times 10^{-3}} = 0.371 \times 10^{-3} \, \text{m}^3/\text{kg}$$

$$\rho_s = \frac{101.5 \times 10^{-3}}{45.1 \times 10^{-6}} = 2.25 \times 10^{-3} \, \text{kg/m}^3$$

$$\varepsilon_P = \frac{(0.371 \times 10^{-3}) 2.25 \times 10^3}{1 + \frac{0.371 \times 10^{-3}}{2.25 \times 10^3}} = 0.455$$

Example 6.4. The rate of catalyst decay is first order for both reactions $A \underset{k_u}{\overset{k_D}{\rightleftarrows}} \genfrac{}{}{0pt}{}{D}{u}$ with $k_{du} = 5 \times 10^{-5}$

min⁻¹ and $k_{dD} = 10^{-4}$ min. and is independent of reactant concentration. If it is decided to change the catalyst if rate of formation of the undesired product is just one half of the desired product, calculate the time of this change. Assume rate of formation of both D and U be first order with respect to reactant concentration if $k_D = 2$ min⁻¹ and $k_u = 0.2$ min⁻¹. Take U as undesired product.

Solution: The selectivity parameter,

$$S_{Du} = \frac{r_D}{r_u} = \frac{r_D}{0.5 r_D} = 2 \qquad \qquad \text{... (6.9)}$$

or

$$S_{Du} = 2 = \frac{k_D C_A \alpha_1}{k_u C_A \alpha_2} \qquad \qquad \text{... (6.10)}$$

It is known that for first order delay,

$$\alpha_1 = e^{-k_{dD} t} \qquad \qquad \text{... (6.11)}$$

$$\alpha_2 = e^{-k_{du} t} \qquad \qquad \text{... (6.12)}$$

Substituting the values of k_D, k_u, α_1 and α_2 in Eq. 6.10:

$$2 = \frac{2 \times e^{-k_{dD} t}}{0.2 \times e^{-k_{du} t}} \qquad \qquad \text{... (6.13)}$$

or

$$\ln 5 = (k_{dD} - k_{du} t) \qquad \qquad \text{... (6.14)}$$

or

$$t = \frac{\ln 5}{(k_{dD} - k_{du})} = 32{,}189 \text{ min.}$$

$$\approx 22.4 \text{ days.}$$

To check the amount of decay, the desired specific reaction rate is:

$$k_D \alpha_1 = 2e^{-10^{-4} t} = 2e^{-3.22} = 2(0.04) = 0.08 \text{ min}^{-1}.$$

Note: It is clear from above analysis that the catalyst has decayed to the extend that the rate is only 4 per cent of original rate. Consequently, the catalyst should be changed before the rate of formation of undesired product reaches to 20 per cent (say) of the desired product.

Chapter 7

Diffusion and Chemical Kinetics

INTRODUCTION

Diffusion is a time-dependent process constituted by random motion of given entities and causing the statistical distribution of these entities to spread in space. The concept of diffusion is tied to notion of mass transfer, driven by a concentration gradient (Fig. 7.1).

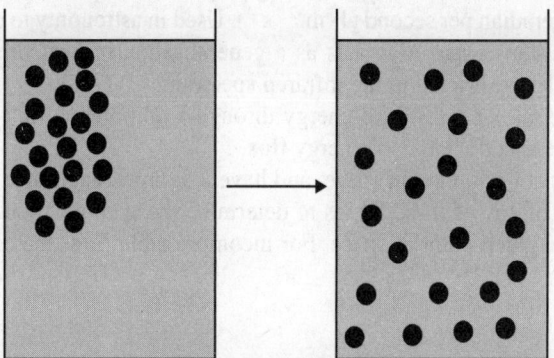

Fig. 7.1. A diffusion process in science. Some particles are dissolved in a glass of water. Initially, the particles are all near one corner of the glass. If the particles all randomly move around ('diffuse') in the water, then the particles will eventually become distributed randomly and uniformly (but diffusion will still continue to occur, just that there will be no net flux).

In molecular diffusion, the moving entities are small molecules. They move at random because they frequently collide. Diffusion is this thermal motion of all (liquid and gas) molecules at temperatures above absolute zero. Diffusion rate is a function of only temperature, and is not affected by concentration. Brownian motion is observed in molecules that are so large that they are not driven by their own thermal energy but by collisions with solvent particles.

While Brownian motion of large molecules is observable under a microscope, small-molecule diffusion can only be probed in carefully controlled experimental conditions. Under normal conditions, molecular diffusion is relevant only on length scales between nanometer and millimeter. On larger length scales, transport in liquids and gases is normally due to another transport phenomenon, convection.

In contrast, heat conduction through solid media is an everyday occurrence (e.g. a metal spoon partly immersed in a hot liquid). This explains why the diffusion of heat was explained mathematically before the diffusion of mass.

DIFFUSION FUNDAMENTALS

Flux Definition and Theorems

Flux is surface bombardment rate. There are many fluxes used in the study of transport phenomena. Each type of flux has its own distinct unit of measurement along with distinct physical constants. Seven of the most common forms of flux from the transport literature are defined as:

1. Momentum flux, the rate of transfer of momentum across a unit area ($N \cdot s \cdot m^{-2} \cdot s^{-1}$) (Newon's law of viscosity).
2. Heat flux, the rate of heat flow across a unit area ($J \cdot m^{-2} \cdot s^{-1}$). (Fourier's law of conduction) (This definition of heat flux fits Maxwell's original definition).
3. Chemical flux, the rate of movement of molecules across a unit area ($mol \cdot m^{-2} \cdot s^{-1}$). (Fick's law of diffusion).
4. Volumetric flux, the rate of volume flow across a unit area ($m^3 \cdot m^{-2} \cdot s^{-1}$). (Darcy's law of groundwater flow).
5. Mass flux, the rate of mass flow across a unit area ($kg \cdot m^{-2} \cdot s^{-1}$). (Either an alternate form of Fick's law that includes the molecular mass, or an alternate form of Darcy's law that includes the density).
6. Radiative flux, the amount of energy moving in the form of photons at a certain distance from the source per steradian per second ($J \cdot m^{-2} \cdot s^{-1}$). Used in astronomy to determine the magnitude and spectral class of a star. Also acts as a generalisation of heat flux, which is equal to the radiative flux when restricted to the infrared spectrum.
7. Energy flux, the rate of transfer of energy through a unit area ($J \cdot m^{-2} \cdot s^{-1}$). The radiative flux and heat flux are specific cases of energy flux.

These fluxes are vectors at each point in space, and have a definite magnitude and direction. Also, one can take the divergence of any of these fluxes to determine the accumulation rate of the quantity in a control volume around a given point in space. For incompressible flow, the divergence of the volume flux is zero.

Chemical Diffusion

Chemical molar flux of a component A in an isothermal, isobaric system is defined in above-mentioned Fick's first law as:

$$J_A = -D_{AB} \nabla c_A$$

where:

D_{AB} = is the diffusion coefficient (m^2/s) of component A diffusing through component B.

c_A = is the concentration (mol/m^3) of species A.

This flux has units of $mol \cdot m^{-2} \cdot s^{-1}$, and fits Maxwell's original definition of flux.

Note: ∇ ('nabla') denotes the del operator.

For dilute gases, kinetic molecular theory relates the diffusion coefficient D to the particle density $n = N/V$, the molecular mass m, the collision cross section σ, and the absolute temperature T by:

$$D = \frac{1}{3} \frac{1}{\sqrt{2}n\sigma} \sqrt{\frac{8kT}{\pi m}}$$

where, the second factor is the mean free path and the square root (with Boltzmann's constant k) is the mean velocity of the particles. In turbulent flows, the transport by eddy motion can be expressed as a grossly increased diffusion coefficient.

Quantum Mechanics

In quantum mechanics, particles of mass m in the state $\psi(r, t)$ have a probability density defined as:

$$\rho = \psi * \psi = |\psi|^2.$$

So the probability of finding a particle in a unit of volume, say d^3x, is:

$$|\psi|^2 \, d^3x.$$

Then the number of particles passing through a perpendicular unit of area per unit time is:

$$J = -i\frac{\hbar}{2m}(\psi * \nabla\Psi - \psi\nabla\Psi *).$$

This is sometimes referred to as the 'flux density'.

FICK'S LAWS OF DIFFUSION

Fick's laws of diffusion describe diffusion and can be used to solve for the diffusion coefficient, D (Fig. 7.2).

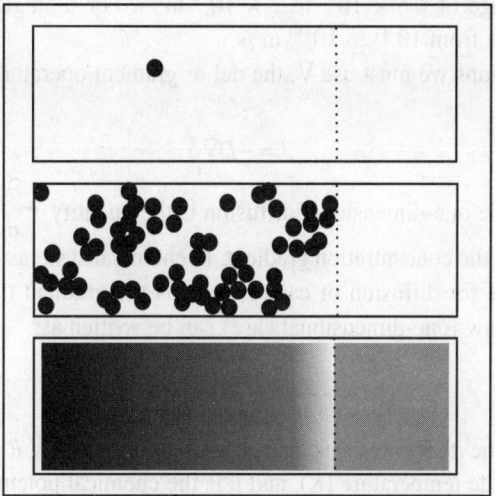

Fig. 7.2. Molecular diffusion from a microscopic and macroscopic point of view. Initially, there are solute molecules on the left side of a barrier (dotted line) and none on the right. The barrier is removed, and the solute diffuses to fill the whole container. Top: A single molecule moves around randomly. Middle: With more molecules, there is a clear trend where the solute fills the container more and more uniformly. Bottom: With an enormous number of solute molecules, all randomness is gone: The solute appears to move smoothly and systematically from high-concentration areas to low-concentration areas. This smooth flow is described by Fick's laws.

Fick's First Law

Fick's first law relates the diffusive flux to the concentration field, by postulating that the flux goes from regions of high concentration to regions of low concentration, with a magnitude that is proportional to the concentration gradient (spatial derivative). In one (spatial) dimension, this is:

$$J = -D\frac{\partial\phi}{\partial x}$$

where,

J = is the diffusion flux in dimensions of [(amount of substance) length^{-2} time^{-1}], example $\dfrac{mol}{m^2 \cdot s}$.

J = measures the amount of substance that will flow through a small area during a small time interval.

D = is the diffusion coefficient or diffusivity in dimensions of [length^2 time^{-1}], example $\left(\dfrac{m^2}{s}\right)$.

ϕ = (for ideal mixtures) is the concentration in dimensions of [(amount of substance) length^{-3}], example $\left(\dfrac{mol}{m^3}\right)$.

x = is the position [length], example m.

D is proportional to the squared velocity of the diffusing particles, which depends on the temperature, viscosity of the fluid and the size of the particles according to the Stokes–Einstein equation. In dilute aqueous solutions the diffusion coefficients of most ions are similar and have values that at room temperature are in the range of 0.6×10^{-9} to 2×10^{-9} m^2/s. For biological molecules the diffusion coefficients normally range from 10^{-11} to 10^{-10} m^2/s.

In two or more dimensions we must use ∇, the del or gradient operator, which generalises the first derivative, obtaining:

$$J = -D\nabla\phi.$$

The driving force for the one-dimensional diffusion is the quantity $-\dfrac{\partial\phi}{\partial x}$ which for ideal mixtures is the concentration gradient. In chemical systems other than ideal solutions or mixtures, the driving force for diffusion of each species is the gradient of chemical potential of this species. Then Fick's first law (one-dimensional case) can be written as:

$$J_i = -\frac{Dc_i}{RT}\frac{\partial\mu_i}{\partial x}$$

where the index i denotes the ith species, c is the concentration (mol/m^3), R is the universal gas constant [$J/(K\ mol)$], T is the absolute temperature (K), and μ is the chemical potential (J/mol).

Fick's Second Law

Fick's second law predicts how diffusion causes the concentration field to change with time:

$$\frac{\partial\phi}{\partial t} = D\frac{\partial^2\phi}{\partial x^2}$$

where,

ϕ = the concentration in dimensions of [(amount of substance) length^{-3}], example $\left(\dfrac{mol}{m^3}\right)$.

t = time [s].

D = the diffusion coefficient in dimensions of [length^2 time^{-1}], example $\left(\dfrac{m^2}{s}\right)$.

x = the position [length], example m.

It can be derived from Fick's First law and the mass balance:

$$\frac{\partial \phi}{\partial t} = -\frac{\partial}{\partial x} J = \frac{\partial}{\partial x}\left(D\frac{\partial}{\partial x}\phi \right)$$

Assuming the diffusion coefficient D to be a constant we can exchange the orders of the differentiating and multiplying by the constant:

$$\frac{\partial}{\partial x}\left(D\frac{\partial}{\partial x}\phi \right) = D\frac{\partial}{\partial x}\frac{\partial}{\partial x}\phi = D\frac{\partial^2 \phi}{\partial x^2}$$

and, thus, receive the form of the Fick's equations as was stated above.

For the case of diffusion in two or more dimensions Fick's Second Law becomes:

$$\frac{\partial \phi}{\partial t} = D\nabla^2 \phi,$$

which is analogous to the heat equation.

If the diffusion coefficient is not a constant, but depends upon the coordinate and/or concentration, Fick's Second Law yields:

$$\frac{\partial \phi}{\partial t} = \nabla \cdot \left(D\nabla \phi \right)$$

An important example is the case where, ϕ is at a steady state, i.e. the concentration does not change by time, so that the left part of the above equation is identically zero. In one dimension with constant D, the solution for the concentration will be a linear change of concentrations along x. In two or more dimensions we obtain:

$$\nabla^2 \phi = 0$$

which is Laplace's equation, the solutions to which are called harmonic functions by mathematicians.

Example solution in one dimension: diffusion length

A simple case of diffusion with time t in one dimension (taken as the x-axis) of a density $n(x,t)$ from a boundary located at position $x = 0$ where the density is maintained at a value $n(0)$ is:

$$n(x,t) = n(0)\, erfc\left(\frac{x}{2\sqrt{Dt}} \right)$$

where $erfc$ is the complementary error function. The length $2\sqrt{Dt}$ is called the diffusion length and provides a measure of how far the density has propagated in the x-direction by diffusion in time t.

As a quick approximation of the error function, the first 2 terms of the Taylor series can be used:

$$n(x,t) = n(0)\left(1 - 2\left(\frac{x}{2\sqrt{Dt\pi}} \right) \right)$$

Applications

Equations based on Fick's law have been commonly used to model transport processes in foods, neurons, biopolymers, pharmaceuticals, porous soils, population dynamics, semiconductor doping process, etc. Theory of all voltammetric methods is based on solutions of Fick's equation. A large amount of experimental research in polymer science and food science has shown that a more general approach is required to describe transport of components in materials undergoing glass transition. In the vicinity of glass transition the flow behaviour becomes 'non-Fickian'.

Biological perspective

The first law gives rise to the following formula:

$$\text{Flux} = -P \cdot A \cdot (c_2 - c_1)$$

in which,

P = is the permeability, an experimentally determined membrane 'conductance' for a given gas at a given temperature.

A = is the surface area over which diffusion is taking place.

$c_2 - c_1$ = is the difference in concentration of the gas across the membrane for the direction of flow (from c_1 to c_2).

Fick's first law is also important in radiation transfer equations. However, in this context it becomes inaccurate when the diffusion constant is low and the radiation becomes limited by the speed of light rather than by the resistance of the material the radiation is flowing through. In this situation, one can use a flux limiter. The exchange rate of a gas across a fluid membrane can be determined by using this law together with Graham's law.

Fick's Flow in Liquids

When two miscible liquids are brought into contact, and diffusion takes place, the macroscopic (or average) concentration evolves following Fick's law. On a mesoscopic scale, that is, between the macroscopic scale described by Fick's law and molecular scale, where molecular random walks take place, fluctuations cannot be neglected. Such situations can be successfully modelled with Landau–Lifshitz fluctuating hydrodynamics. In this theoretical framework, diffusion is due to fluctuations whose dimensions range from the molecular scale to the macroscopic scale.

In particular, fluctuating hydrodynamic equations include a Fick's flow term, with a given diffusion coefficient, along with hydrodynamics equations and stochastic terms describing fluctuations. When calculating the fluctuations with a perturbative approach, the zero order approximation is Fick's law. The first order gives the fluctuations, and it comes out that fluctuations contribute to diffusion. This represents somehow a tautology, since the phenomena described by a lower order approximation is the result of a higher order approximation: this problem is solved only by renormalising fluctuating hydrodynamics equations.

MOLECULAR DIFFUSION

Molecular diffusion, often called simply diffusion, is the thermal motion of all (liquid or gas) particles at temperatures above absolute zero. The rate of this movement is a function of temperature, viscosity of the fluid and the size (mass) of the particles, but is not a function of concentration. Diffusion explains the net flux of molecules from a region of higher concentration to one of lower concentration, but it is important to note that diffusion also occurs when there is no concentration gradient. The result of diffusion is a gradual mixing of material. In a phase with uniform temperature, absent external net forces acting on the particles, the diffusion process will eventually result in complete mixing. Molecular diffusion is typically described mathematically using Fick's laws of diffusion. Diffusion is part of the transport phenomena. Of mass transport mechanisms, molecular diffusion is known as a slower one.

In cell biology, diffusion is a main form of transport for necessary materials such as amino acids within cells. Diffusion of a fluid (anything that moves like a liquid) through a partially permeable membrane is classified as osmosis (Fig. 7.3).

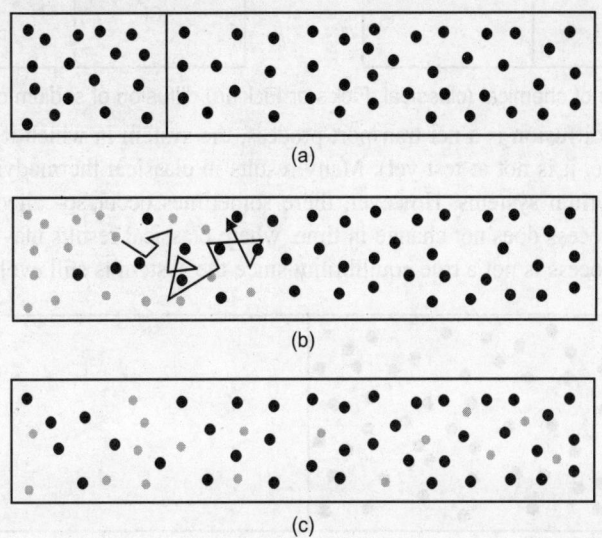

(a)

(b)

(c)

Fig. 7.3. Schematic representation of mixing of two substances by diffusion.

Metabolism and respiration rely in part upon diffusion in addition to bulk or active processes. For example, in the alveoli of mammalian lungs, due to differences in partial pressures across the alveolar-capillary membrane, oxygen diffuses into the blood and carbon dioxide diffuses out. Lungs contain a large surface area to facilitate this gas exchange process.

Fundamentally, two types of diffusion are distinguished:

1. Tracer diffusion: which is a spontaneous mixing of molecules taking place in the absence of concentration (or chemical potential) gradient. This type of diffusion can be followed using isotopic tracers, hence the name. The tracer diffusion is usually assumed to be identical to self-diffusion (assuming no significant isotopic effect). This diffusion can take place under equilibrium (Fig 7.4).

2. Chemical diffusion: occurs in a presence of concentration (or chemical potential) gradient and it results in net transport of mass. This is the process described by the diffusion equation. This diffusion is always a non-equilibrium process, increases the system entropy, and brings the system closer to equilibrium (Fig 7.5).

Fig. 7.4. Self diffusion, exemplified with an isotopic tracer of radioactive isotope ^{22}Na.

The diffusion coefficients for these two types of diffusion are generally different because the diffusion coefficient for chemical diffusion is binary and it includes the effects due to the correlation of the movement of the different diffusing species.

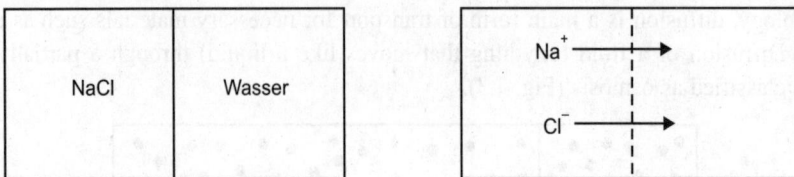

Fig. 7.5. Example of chemical (classical, Fick's or Fickian) diffusion of sodium chloride in water.

Because chemical diffusion is a net transport process, the system in which it takes place is not an equilibrium system (i.e. it is not at rest yet). Many results in classical thermodynamics are not easily applied to non-equilibrium systems. However, there sometimes occur so-called quasi-steady states, where the diffusion process does not change in time, where classical results may locally apply. As the name suggests, this process is not a true equilibrium since the system is still evolving (Fig. 7.6).

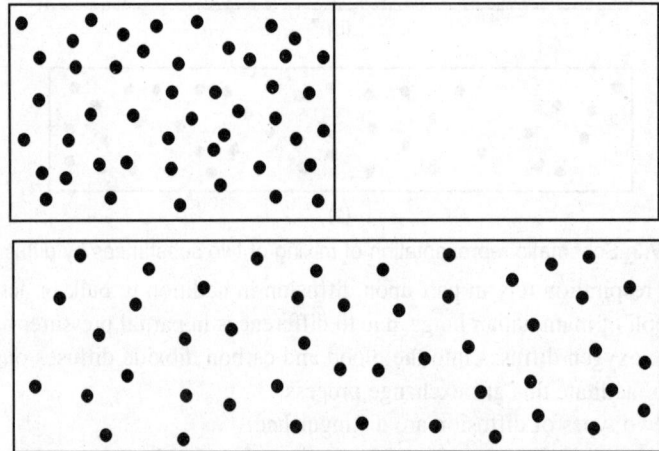

Fig. 7.6. Illustration of low entropy (top) and high entropy (bottom).

Non-equilibrium fluid systems can be successfully modelled with Landau-Lifshitz fluctuating hydrodynamics. In this theoretical framework, diffusion is due to fluctuations whose dimensions range from the molecular scale to the macroscopic scale.

Chemical diffusion increases the entropy of a system, i.e. diffusion is a spontaneous and irreversible process. Particles can spread out by diffusion, but will not spontaneously reorder themselves (absent changes to the system, assuming no creation of new chemical bonds, and absent external forces acting on the particles).

Error Diffusion

Error diffusion is a type of halftoning in which the quantisation residual is distributed to neighbouring pixels that have not yet been processed. Its main use is to convert a multi-level image into a binary image, though it has other applications.

Unlike many other halftoning methods, error diffusion is classified as an area operation, because what the algorithm does at one location influences what happens at other locations.

This means buffering is required, and complicates parallel processing. Point operations, such as ordered dither, do not have these complications.

Error diffusion has the tendency to enhance edges in an image. This can make text in images more readable than in other halftoning techniques.

Equimolar Counter-Diffusion

In equimolar counter-diffusion, the molar fluxes or A and B are equal, but opposite in direction, and the total pressure is constant throughout. Hence we can write:

$$N = N_A + N_B = 0$$

$$J_A = -J_B$$

(Pressure is caused by the collisions of molecules with the container wall. If the pressure is constant at any point in the container, then it must be implied that the number of molecules acting on the wall at any point is also constant. In other words, if certain amount of A has diffuse away, then they are replaced by the same amount of B.)

Under equimolar counter-diffusion, the diffusivity of A in B is the same as the diffusivity of B in A, i.e. $D_{AB} = D_{BA}$.

Fick's law for steady-state equimolar counter-diffusion of ideal gas mixture

Consider 2-component gas mixture (A and B).

Ideal Gas Law: $Pv = nRT$

where P is the total pressure, and n is the total moles of gas.

For component-A: $p_A v = n_A RT$

where p_A is the partial pressure of A and n_A is the moles of A.

Concentration of A:

$$C_A = \left(\frac{n_A}{V}\right) = \left(\frac{p_A}{RT}\right)$$

Differentiating with respect to distance z, we obtain:

$$\left(\frac{dC_A}{dz}\right) = \left(\frac{1}{RT}\right)\left(\frac{dp_A}{dz}\right)$$

Replacing into the original Fick's Law, we have an alternative equation for ideal gas using partial pressure:

$$J_A = -\frac{D_{AB}}{RT}\frac{dp_A}{dz}$$

Under constant total pressure and temperature conditions, the above equation for can be integrated over a diffusional path from z_2 to z_1 to as follows:

$$J_A = -\frac{D_{AB}}{RT}\left(\frac{dp_A}{dz}\right)$$

$$J_A dz = -\left(\frac{D_{AB}}{RT}\right)dp_A$$

Integrate from p_{A1} to p_{A2}, where p_{A1} is the partial pressure of A at point 1 and p_{A2} is the partial pressure of A at point 2:

$$J_A \int_{z_1}^{z_2} dz = -\left(\frac{D_{AB}}{RT}\right) \int_{p_{A1}}^{p_{A2}} dp A$$

$$J_A(z_2 - z_1) = -\left(\frac{D_{AB}}{RT}\right)(p_{A2} - p_{A1})$$

We have, for steady-state equimolar counter-diffusion of ideal gas mixture:

$$J_A = \frac{D_{AB}(p_{A1} - p_{A2})}{RT(z_2 - z_1)}$$

Similarly, the molar flux for component-B can be written as:

$$J_B = \frac{D_{AB}(p_{B1} - p_{B2})}{RT(z_2 - z_1)}.$$

Forced Convection

Forced convection is a mechanism or type of heat transport in which fluid motion is generated by an external source (like a pump, fan, suction device, etc.). It should be considered as one of the main methods of useful heat transfer as significant amounts of heat energy can be transported very efficiently and this mechanism is found very commonly in everyday life, including central heating, air conditioning, steam turbines and in many other machines. Forced convection is often encountered by engineers designing or analysing heat exchangers, pipe flow, and flow over a plate at a different temperature than the stream (the case of a shuttle wing during re-entry, for example). However, in any forced convection situation, some amount of natural convection is always present whenever there are g-forces present (i.e. unless the system is in free fall). When the natural convection is not negligible, such flows are typically referred to as mixed convection.

When analysing potentially mixed convection, a parameter called the Archimedes number (Ar) parametrises the relative strength of free and forced convection. The Archimedes number is the ratio of Grashof number and the square of Reynolds number, which represents the ratio of buoyancy force and inertia force, and which stands in for the contribution of natural convection. When $Ar \gg 1$, natural convection dominates and when $Ar \ll 1$, forced convection dominates.

$$A_r = \frac{Gr}{Re^2}$$

When natural convection isn't a significant factor, mathematical analysis with forced convection theories typically yields accurate results. The parameter of importance in forced convection is the Peclet number, which is the ratio of advection (movement by currents) and diffusion (movement from high to low concentrations) of heat.

$$Pe = \frac{UL}{\alpha}$$

When the Peclet number is much greater than unity, advection dominates diffusion. Similarly, much smaller ratios indicate a higher rate of diffusion relative to advection.

Reaction–Diffusion System

Reaction–diffusion systems are mathematical models which explain how the concentration of one or more substances distributed in space changes under the influence of two processes: local chemical reactions in which the substances are transformed into each other, and diffusion which causes the substances to spread out over a surface in space.

This description implies that reaction–diffusion systems are naturally applied in chemistry. However, the system can also describe dynamical processes of non-chemical nature. Examples are found in biology, geology and physics and ecology. Mathematically, reaction–diffusion systems take the form of semi-linear parabolic partial differential equations. They can be represented in the general form

$$\partial_t q = \underline{D} \nabla^2 q + R(q),$$

where, each component of the vector $q(x, t)$ represents the concentration of one substance, $\underline{D}$ is a diagonal matrix of diffusion coefficients, and R accounts for all local reactions. The solutions of reaction–diffusion equations display a wide range of behaviours, including the formation of travelling waves and wave-like phenomena as well as other self-organised patterns like stripes, hexagons or more intricate structure like dissipative solitons.

One-component reaction–diffusion equations

The most simple reaction–diffusion equation concerning the concentration u of a single substance in one spatial dimension,

$$\partial_t u = D \partial_x^2 u + R(u),$$

is also referred to as the KPP (Kolmogorov–Petrovsky-Piskounov) equation. If the reaction term vanishes, then the equation represents a pure diffusion process. The corresponding equation is Fick's second law. The choice $R(u) = u(1 - u)$ yields Fisher's equation that was originally used to describe the spreading of biological populations, the Newell–Whitehead–Segel equation with $R(u) = u(1 - u^2)$ to describe Rayleigh–Benard convection, the more general Zeldovich equation with $R(u) = u(1 - u)(u - \alpha)$ and $0 < \alpha < 1$ that arises in combustion theory, and its particular degenerate case with $R(u) = u^2 - u^3$ that is sometimes referred to as the Zeldovich equation as well.

The dynamics of one-component systems is subject to certain restrictions as the evolution equation can also be written in the variational form

$$\partial_t u = -\frac{\partial \xi}{\partial u}$$

and therefore describes a permanent decrease of the 'free energy' ξ given by the functional:

$$\xi = \int_{-\infty}^{\infty} \left[\frac{D}{2} (\partial_x u)^2 + V(u) \right] dx$$

with a potential $V(u)$ such that $R(u) = dV(u)/du$.

In systems with more than one stationary homogeneous solution, a typical solution is given by travelling fronts connecting the homogeneous states. These solutions move with constant speed without changing their shape and are of the form $u(x, t) = \hat{u}(\xi)$ with $\xi = x - ct$, where, c is the speed of the travelling wave. Note that while travelling waves are generically stable structures, all non-monotonous stationary solutions (e.g. localised domains composed of a front-antifront pair) are unstable. For $c = 0$, there is a simple proof for this statement: if $u_0(x)$ is a stationary solution and $u = u_0(x) + \tilde{u}(x, t)$ is an infinitesimally perturbed solution, linear stability analysis yields the equation (Fig. 7.7).

$$\partial_t \tilde{u} = D\partial_x^2 \tilde{u} - U(x)\tilde{u}, \quad U(x) = -R'(u)\big|_{u=u_0(x)}.$$

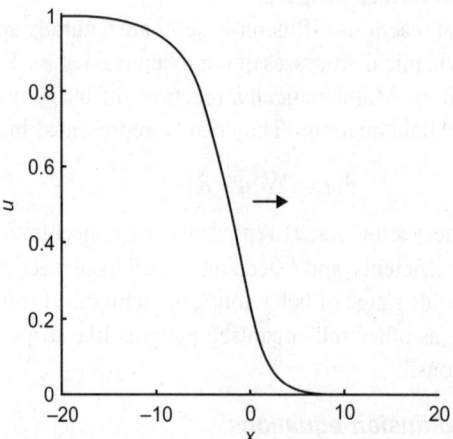

Fig. 7.7. A travelling wave front solution for Fisher's equation.

With the ansatz $\tilde{u} = \psi(x)\exp(-\lambda t)$ we arrive at the eigenvalue problem:

$$\hat{H}\psi = \lambda\psi, \quad \hat{H} = -D\partial_x^2 + U(x),$$

of Schrödinger type where negative eigenvalues result in the instability of the solution. Due to translational invariance $\psi = \partial_x u_0(x)$ is a neutral eigenfunction with the eigenvalue $\lambda = 0$, and all other eigenfunctions can be sorted according to an increasing number of knots with the magnitude of the corresponding real eigenvalue increases monotonically with the number of zeros. The eigenfunction $\psi = \partial_x u_0(x)$ should have at least one zero, and for a non-monotonic stationary solution the corresponding eigenvalue $\lambda = 0$ cannot be the lowest one, thereby implying instability.

To determine the velocity c of a moving front, one may go to a moving coordinate system and look at stationary solutions:

$$D\partial_\xi^2 \hat{u}(\xi) + c\partial_\xi \hat{u}(\xi) + R(\hat{u}(\xi)) = 0.$$

This equation has a nice mechanical analogue as the motion of a mass D with position $\hat{u}$ in the course of the 'time' ξ under the force R with the damping coefficient c which allows for a rather illustrative access to the construction of different types of solutions and the determination of c.

When going from one to more space dimensions, a number of statements from one-dimensional systems can still be applied. Planar or curved wave fronts are typical structures, and a new effect arises as the local velocity of a curved front becomes dependent on the local radius of curvature (this can be seen by going to polar coordinates). This phenomenon leads to the so-called curvature-driven instability.

Two-component reaction–diffusion equations

Two-component systems allow for a much larger range of possible phenomena than their one-component counterparts. An important idea that was first proposed by Alan Turing is that a state that is stable in the local system should become unstable in the presence of diffusion. This idea seems unintuitive at first glance as diffusion is commonly associated with a stabilising effect.

A linear stability analysis however shows that when linearising the general two-component system:

$$\begin{pmatrix} \partial_t u \\ \partial_t v \end{pmatrix} = \begin{pmatrix} Du & 0 \\ 0 & D_v \end{pmatrix} \begin{pmatrix} \partial_{xx} u \\ \partial_{xx} v \end{pmatrix} + \begin{pmatrix} F(u,v) \\ G(u,v) \end{pmatrix}$$

a plane wave perturbation:

$$\tilde{q}_k(x,t) = \begin{pmatrix} \tilde{u}(t) \\ \tilde{v}(t) \end{pmatrix} e^{ik\cdot x}$$

of the stationary homogeneous solution will satisfy:

$$\begin{pmatrix} \partial_t \tilde{u}_k(t) \\ \partial_t \tilde{v}_k(t) \end{pmatrix} = -k^2 \begin{pmatrix} D_u \tilde{u}_k(t) \\ D_v \tilde{v}_k(t) \end{pmatrix} + R' \begin{pmatrix} \tilde{u}_k(t) \\ \tilde{v}_k(t) \end{pmatrix}.$$

Turing's idea can only be realised in four equivalence classes of systems characterised by the signs of the Jacobian R' of the reaction function. In particular, if a finite wave vector k is supposed to be the most unstable one, the Jacobian must have the signs:

$$\begin{pmatrix} + & - \\ + & - \end{pmatrix}, \begin{pmatrix} + & + \\ - & - \end{pmatrix}, \begin{pmatrix} - & + \\ - & + \end{pmatrix}, \begin{pmatrix} - & - \\ + & + \end{pmatrix}.$$

This class of systems is named activator-inhibitor system after its first representative: close to the ground state, one component stimulates the production of both components while the other one inhibits their growth. Its most prominent representative is the FitzHugh–Nagumo equation:

$$\partial_t u = d_u^2 \nabla^2 u + f(u) - \sigma v,$$
$$\tau \partial_t v = d_v^2 \nabla^2 v + u - v$$

with $f(u) = \lambda u - u^3 - \kappa$ which describes how an action potential travels through a nerve. Here, d_u, d_v, τ, σ and λ are positive constants.

When an activator-inhibitor system undergoes a change of parameters, one may pass from conditions under which a homogeneous ground state is stable to conditions under which it is linearly unstable. The corresponding bifurcation may be either a Hopf bifurcation to a globally oscillating homogeneous state with a dominant wave number $k = 0$ or a Turing bifurcation to a globally patterned state with a dominant finite wave number. The latter in two spatial dimensions typically leads to stripe or hexagonal patterns.

For the Fitzhugh–Nagumo example, the neutral stability curves marking the boundary of the linearly stable region for the Turing and Hopf bifurcation are given by:

$$q_n^H(k): \frac{1}{\tau} + \left(d_u^2 + \frac{1}{\tau} d_v^2 \right) k^2 = f'(u_h),$$

$$q_n^T(k): \frac{\kappa_3}{1 + d_v^2 k^2} + d_u^2 k^2 = f'(u_h).$$

If the bifurcation is subcritical, often localised structures (dissipative solitons) can be observed in the hysteretic region where the pattern coexists with the ground state. Other frequently encountered structures comprise pulse trains, spiral waves and target patterns.

Three- and more-component reaction–diffusion equations

For a variety of systems, reaction–diffusion equations with more than two components have been proposed, e.g. as models for the Belousov–Zhabotinsky reaction, for blood clotting or planar gas discharge systems. While it is known that systems with more components allow for a variety of phenomena not possible in systems with one or two components (e.g. stable running pulses in more than one spatial dimension without global feedback), up to now a systematic overview of the possible phenomena in dependence on the properties of the underlying system is hardly present.

Applications and universality

In recent times, reaction–diffusion systems have attracted much interest as a prototype model for pattern formation. The above-mentioned patterns (fronts, spirals, targets, hexagons, stripes and dissipative solitons) can be found in various types of reaction–diffusion systems inspite of large discrepancies, e.g. in the local reaction terms. It has also been argued that reaction–diffusion processes are an essential basis for processes connected to morphogenesis in biology and may even be related to animal coats and skin pigmentation. Another reason for the interest in reaction–diffusion systems is that although they represent nonlinear partial differential equation, there are often possibilities for an analytical treatment.

Experiments

Well-controllable experiments in chemical reaction-diffusion systems have up to now been realised in three ways. First, gel reactors or filled capillary tubes may be used. Second, temperature pulses on catalytic surfaces have been investigated. Third, the propagation of running nerve pulses is modelled using reaction–diffusion systems.

Aside from these generic examples, it has turned out that under appropriate circumstances electric transport systems like plasmas or semiconductors can be described in a reaction-diffusion approach. For these systems various experiments on pattern formation have been carried out.

STEADY-STATE DIFFUSION EQUATION AND BOUNDARY CONDITIONS

In this section we discuss PDEs which are used to model certain steady-state diffusion phenomena. These model equations are typically derived by combining physical laws with constitutive equations.

Physical laws include conservation of energy, conservation of mass, conservation of momentum, and Newton's laws of motion. These are independent of the materials being modelled.

Constitutive equations are empirically derived and depend on the materials being modelled. Model for Motion of an Undamped Mass-Spring system. Newton's second law of motion gives the force associated with change in momentum:

$$F = \frac{d}{dt}\left(m \underbrace{\frac{dx}{dt}}_{\text{Velocity}} \right) = m\frac{d^2x}{dt^2}, \qquad \text{... (7.1)}$$

where, $x(t)$ is the displacement of the (constant) mass m. This is balanced by the restoring force of the spring:

$$F = k\,x(t), \qquad \text{...(7.2)}$$

where, k is the positive spring constant. Combining the physical law (Eq. 7.1) with the constitutive Eq. 7.2 gives the equation of motion:

$$m\frac{d^2x}{dt^2} + kx(t) = 0. \qquad \dots (7.3)$$

The undamped mass-spring (Eq. 7.3) is an ordinary differential equation. Of greater interest are diffusion models like the following.

Model for 'steady' heat flow: Molecular motion of the molecules of a material gives rise to heat flow. By 'steady' flow we mean that while the molecules are in motion, the steady-state temperature u of the material is independent of time t (u may vary with position x). The relevant physical law is conservation of thermal energy:

$$\text{div } q = f(x), \qquad \dots (7.4)$$

where $q(x)$ is the heat flow (vector) field and $f(x)$ represents external sources or sinks (loss mechanisms) for thermal energy. The relevant constitutive equation is the Newton–Fourier law of cooling,

$$q = -\kappa \nabla u. \qquad \dots (7.5)$$

Here κ is a positive scalar-valued function known as the thermal conductivity. Combining Eq. 7.4 with Eq. 7.5 gives the steady-state diffusion equation:

$$-\text{div } (\kappa \nabla u) = f(x). \qquad \dots (7.6)$$

Other Applications of the steady-state diffusion equation include:

1. Molecular diffusion: In this case $u(x)$ represents chemical concentration at position x; κ is called the diffusion coefficient, or chemical diffusivity; $q(x)$ represents the chemical flow field; and Eq. 7.5 is called Fick's Law in this context.
2. Fluid flow through a porous medium: Here $u(x)$ represents fluid pressure; κ is known as the hydraulic conductivity; $q(x)$ represents the fluid flow field; and Eq. 7.5 is called Darcy's Law.
3. Electron flow in a conductor: $u(x)$ is the voltage, or electrostatic potential; κ is called the electrical conductivity; $q(x)$ represents electrical current flow and Eq. 7.5 is called Ohm's Law.

Important Equations and Operators

Poisson's equation: This equation arises from the steady-state diffusion (Eq. 7.6) when $\kappa = 1$:

$$-\text{div } (\nabla u) = f(x). \qquad \dots (7.7)$$

Laplace's equation: This arises from Poisson's Eq. 7.7 when the external source/sink term $f(x)$ is zero,

$$\text{div } (\nabla u) = 0. \qquad \dots (7.8)$$

Laplacian operator: This appears on the left-hand-side of Laplace's Eq. 7.8:

$$\text{div}(\nabla(\cdot)) = \sum_{i=1}^{d} \frac{\partial^2}{\partial x_i^2}. \qquad \dots (7.9)$$

The Laplacian maps smooth scalar-valued functions to scalar-valued functions. In one space dimension ($d = 1$), the Laplacian reduces to the second derivative.

Linear operators: A differential operator L is called linear if:

$$L(\alpha u + \beta v) = \alpha L(u) + \beta L(v)$$

whenever u, v are smooth functions and α, $\beta \varepsilon$ IR.

Boundary Conditions

The steady-state diffusion (Eq. 7.6), or a simplified version like Poisson's equation or Laplace's equation, holds in a region Ω which corresponds to the extent of the material in which diffusive flow is being modelled. To completely determine the solution $u(x)$ we also need to describe mathematically what happens on the boundary $\partial\Omega$. Three important types of boundary conditions are:

Dirichlet boundary conditions: The solution u is specified on the boundary:

$$u(x) = g(x), \quad x \, \varepsilon \, \partial\Omega, \qquad \qquad \text{... (7.10)}$$

where, g is a known function. If $g = 0$, then we refer to Eq. 7.10 as homogeneous Dirichlet boundary conditions.

Neumann boundary conditions: The boundary flux, or rate of flow $q(x)$ in Eq. 7.5 which is normal to the boundary, is specified,

$$-\kappa\nabla u \cdot n(x) = g(x), \quad x \, \varepsilon \, \partial\Omega. \qquad \qquad \text{... (7.11)}$$

If $g = 0$, then Eq. 7.11 is referred to as a no-flux boundary condition, or homogeneous Neumann boundary condition.

Robin or 3rd kind, boundary conditions: The boundary flux is equal to the sum of term which is proportional to the solution u and a another term which is independent of u:

$$-\kappa\nabla u \cdot n(x) = \gamma u(x) + g(x); \quad x \, \varepsilon \, \partial\Omega. \qquad \qquad \text{... (7.12)}$$

MASS TRANSFER COEFFICIENT

In engineering, the mass transfer coefficient is a diffusion rate constant that relates the mass transfer rate, mass transfer area, and concentration gradient as driving force:

$$k_c = \frac{\dot{n}_A}{A\Delta c_A}$$

where,

k_c = is the mass transfer coefficient [mol/(s · m²)/(mol/m³) or m/s].

$\dot{n}_A$ = is the mass transfer rate [mol/s].

A = is the effective mass transfer area [m²].

ΔC_A = is the driving force concentration difference [mol/m³].

This can be used to quantify the mass transfer between phases, immiscible and partially miscible fluid mixtures (or between a fluid and a porous solid). Quantifying mass transfer allows for design and manufacture of separation process equipment that can meet specified requirements, estimate what will happen in real life situations (chemical spill), etc.

Mass transfer coefficients can be estimated from many different theoretical equations, correlations, and analogies that are functions of material properties, intensive properties and flow regime (laminar or turbulent flow). Selection of the most applicable model is dependent on the materials and the system or environment, being studied. Mass transfer coefficient units are: (i) mol/(s·m²)/(mol/m³), and (ii) m/s.

Packed Bed

In chemical processing, a packed bed is a hollow tube, pipe or other vessel that is filled with a packing material. The packing can be randomly filled with small objects like Raschig rings or else it can be a specifically designed structured packing.

The purpose of a packed bed is typically to improve contact between two phases in a chemical or similar process. Packed beds can be used in a chemical reactor, a distillation process, or a scrubber, but packed beds have also been used to store heat in chemical plants. In this case, hot gases are allowed to escape through a vessel that is packed with a refractory material until the packing is hot. Air or other cool gas is then fed back to the plant through the hot bed, thereby pre-heating the air or gas feed.

In industry, a packed column is a type of packed bed used to perform separation processes, such as absorption, stripping, and distillation. A packed column is a pressure vessel that has a packed section. The column can be filled with random dumped packing or structured packing sections, which are arranged or stacked. In the column, liquids tend to wet the surface of the packing and the vapours pass across this wetted surface, where mass transfer takes place. Packing material can be used instead of trays to improve separation in distillation columns. Packing offers the advantage of a lower pressure drop across the column (when compared to plates or trays), which is beneficial while operating under vacuum. Differently shaped packing materials have different surface areas and void space between the packing. Both of these factors affect packing performance.

Another factor in performance, in addition to the packing shape and surface area, is the liquid and vapour distribution that enters the packed bed. The number of theoretical stages required to make a given separation is calculated using a specific vapour to liquid ratio. If the liquid and vapour are not evenly distributed across the superficial tower area as it enters the packed bed, the liquid to vapour ratio will not be correct and the required separation will not be achieved. The packing will appear to not be working properly. The height equivalent to a theoretical plate (HETP) will be greater than expected. The problem is not the packing itself but the mal-distribution of the fluids entering the packed bed. These columns can contain liquid distributors and redistributors which help to distribute the liquid evenly over a section of packing, increasing the efficiency of the mass transfer. The design of the liquid distributors used to introduce the feed and reflux to a packed bed is critical to making the packing perform at maximum efficiency.

Packed columns have a continuous vapour-equilibrium curve, unlike conventional tray distillation in which every tray represents a separate point of vapour-liquid equilibrium. However, when modelling packed columns it is useful to compute a number of theoretical plates to denote the separation efficiency of the packed column with respect to more traditional trays. In design, the number of necessary theoretical equilibrium stages is first determined and then the packing height equivalent to a theoretical equilibrium stage, known as the height equivalent to a theoretical plate (HETP), is also determined. The total packing height required is the number theoretical stages multiplied by the HETP.

Columns used in certain types of chromatography consisting of a tube filled with packing material can also be called packed columns and their structure has similarities to packed beds.

Packed bed reactors can be used in chemical reaction. These reactors are tubular and are filled with solid catalyst particles, most often used to catalyse gas reactions. The chemical reaction takes place on the surface of the catalyst. The advantage of using a packed bed reactor is the higher conversion per weight of catalyst than other catalytic reactors. The reaction rate is based on the amount of the solid catalyst rather than the volume of the reactor.

The Ergun equation can be used to predict the pressure drop along the length of a packed bed given the fluid velocity, the packing size, and the viscosity and density of the fluid.

Chilton and Colburn J-factor Analogy

Chilton and Colburn J-factor analogy is probably the most successful and widely used analogy from heat, momentum, and mass transfer analogies. The basic mechanisms and mathematics of heat, mass, and momentum transport are essentially the same. Among many analogies (like Reynolds analogy, Prandtl–Taylor analogy) developed to directly relate heat transfer coefficients, mass transfer coefficients, and friction factors Chilton and Colburn J-factor analogy proved to be the most accurate.

It is written as follows:

$$\frac{f}{2} = J_H = \frac{h}{cp \cdot G} \cdot \Pr^{\frac{2}{3}} = J_D = \frac{k_c'}{\overline{\upsilon}} \cdot Sc^{\frac{2}{3}}$$

This equation permits the prediction of an unknown transfer coefficient when one of the other coefficients is unknown. The analogy is valid for fully developed turbulent flow in conduits with Re > 10000, 0.7 < Pr < 160, and tubes where L/d > 60 (the same constraints as the Sieder-Tate correlation). The wider range of data can be correlated by Friend-Metzner analogy.

Catalyst Regeneration

Catalytic crackers have long been utilised to extract additional gasoline from heavier components resulting from the distillation process. The distillation process is the physical separation of a mixture of different molecules, based upon the different boiling points of these molecules. The catalytic cracking process splits larger hydrocarbon molecules into lighter components using a catalyst, which aids the reaction or 'cracking' process. The cracking process produces carbon, or coke, which remains on the catalyst particle, reducing its effectiveness over time.

Spent catalyst is continuously routed from the reactor to a 'regenerator'. Oil remaining on the surface of the catalyst is stripped off with steam or solvent. The catalyst is then sent to the regenerator, where air is introduced to burn the coke off the catalyst.

There are many different variations in the regeneration process, and there are several applications where the *in situ* zirconium oxide probe works well.

Oxygen is measured in the flue gas resulting from the coke burn-off to assist in maintaining the most efficient fuel/air ratios. Some catalytic regenerators generate CO gas, and route this to a separate CO boiler (Fig. 7.8), while other units complete the combustion process within the regenerator (Fig. 7.9). In either case, the O_2 analyser is placed in the flue gas ductwork exiting the unit, or in the stack.

Enriching the air used in the regeneration process can increase the coke burn-off rate. Many refineries will mix pure oxygen with the air used for regeneration, resulting in a mixture of 21 to 25 per cent O_2, which increases the efficiency of the regeneration process. This oxygen measurement can be used for operator information, alarming, or automatic control of the oxygen injection valve (Fig. 7.8).

Pressures are normally 35PSI, so the probe must be pressure-balanced with reference air. Utilised with a 'moving bed' process, the CCR utilises a regeneration gas loop to control temperature and O_2 levels in the regenerator 'burn zone'. This recirculation loop heats or cools the combustion flue gas resulting from the regeneration process to control temperature, and air is also controlled to an optimum point (Fig. 7.10). The O_2 measurement is critical to maintaining optimum rate of regeneration, and for preventing thermal damage inside the burn zone.

Process pressures may be close to atmospheric or pressurised to approximately 35 PSI. Pressure balancing may be required, and an isolation valving system may be needed if probe insertion or withdrawal is required with the process on-line. Pressure balancing will adjust to pressure variations, while traditional O_2 systems place a fixed compensation into the electronics. Chlorine is injected into this unit, and a special HCL-resistant cell is required.

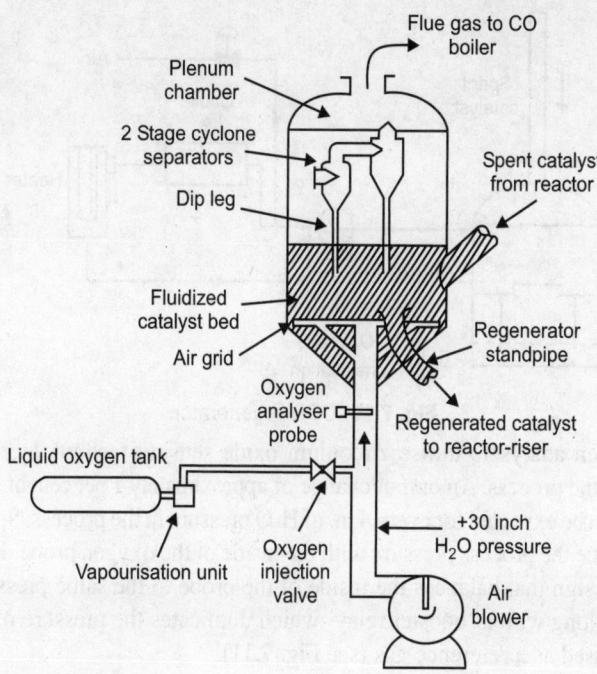

Fig. 7.8. Typical regenerator with separate CO boiler (not shown).

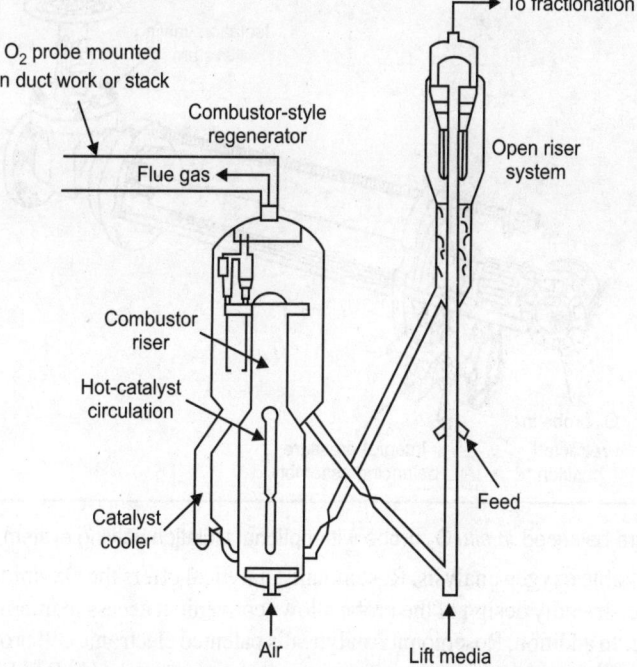

Fig. 7.9. Typical reactor with integral combustor section.

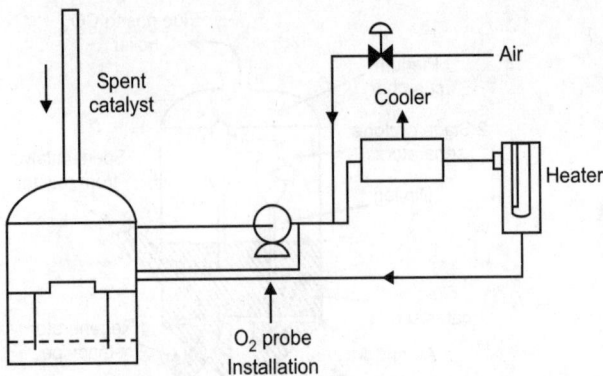

Fig. 7.10. CCR regenerator.

Most *in situ* oxygen analysers utilise zirconium oxide sensing technology, which is sensitive to pressure variations in the process. An output change of approximately 1 per cent of reading (not 1 per cent FS or 1 per cent O_2) can be expected for every 4 in. of H_2O pressure in the process. Special accommodation must be made to balance the process pressure with the inside of the oxygen probe. Rosemount Analytical has a special probe design that balances the inside of the probe to the same pressure as the process. A sealed probe is used along with a 'booster relay' which duplicates the pressure of the process with the instrument air being used as a reference gas (see Fig. 7.11).

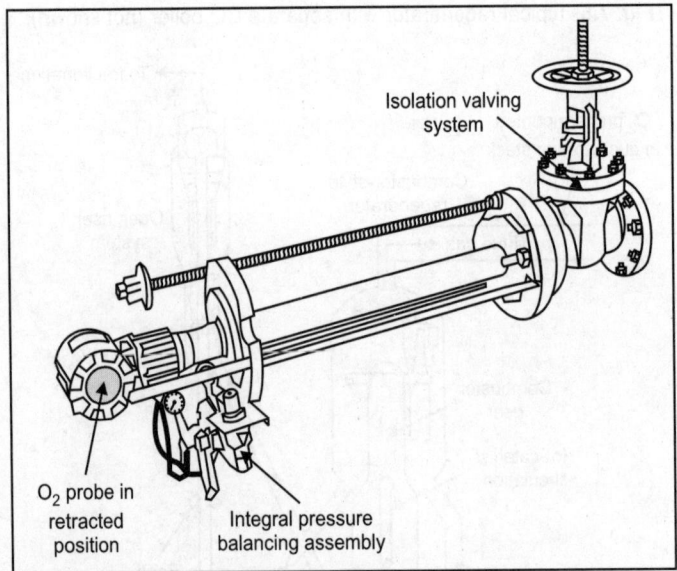

Fig. 7.11. Pressure balanced *in situ* O_2 probe with optional isolation valving system (probe withdrawn).

For accurate, reliable oxygen analysis, Rosemount Analytical offers the Oxymitter 4000 with integral electronics. The user-friendly design of the probe allows convenient access to internal probe components for in-house service. In addition, Rosemount Analytical's patented electronic cell protection automatically protects the sensor cell's electrodes from harmful corrosive gases. And, the HART® Field Communications Protocol permits all operator functions to be performed from the control room.

MASS DIFFUSIVITY

Diffusivity or diffusion coefficient is a proportionality constant between the molar flux due to molecular diffusion and the gradient in the concentration of the species (or the driving force for diffusion). Diffusivity is encountered in Fick's law and numerous other equations of physical chemistry.

It is generally prescribed for a given pair of species. For a multi-component system, it is prescribed for each pair of species in the system. The higher the diffusivity (of one substance with respect to another), the faster they diffuse into each other. This coefficient has an SI unit of m²/s (length²/time).

Temperature Dependence of the Diffusion Coefficient

Typically, a compound's diffusion coefficient is ~10,000x greater in air than in water. Carbon dioxide in air has a diffusion coefficient of 16 mm²/s, and in water its coefficient is 0.0016 mm²/s.

The diffusion coefficient in solids at different temperatures is often found to be well predicted by:

$$D = D_0 e^{-\frac{E_A}{RT}},$$

where,

D = is the diffusion coefficient.
D_0 = is the maximum diffusion coefficient (at infinite temperature).
E_A = is the activation energy for diffusion in dimensions of [energy (amount of substance)$^{-1}$].
T = is the temperature in units of [absolute temperature] (kelvins or degrees Rankine).
R = is the gas constant in dimensions of [energy temperature^{-1} (amount of substance)$^{-1}$].

An equation of this form is known as the Arrhenius equation.

An approximate dependence of the diffusion coefficient on temperature in liquids can often be found using Stokes-Einstein equation, which predicts that:

$$\frac{D_{T1}}{D_{T2}} = \frac{T_1}{T_2}\frac{\mu_{T2}}{\mu_{T1}}$$

where,

T_1 and T_2 = denote temperatures 1 and 2, respectively.
D = is the diffusion coefficient (cm²/s).
T = is the absolute temperature (K).
μ = is the dynamic viscosity of the solvent (Pa·s).

The dependence of the diffusion coefficient on temperature for gases can be expressed using the Chapman-Enskog theory (predictions accurate on average to about 8 per cent):

$$D = \frac{1.86 \cdot 10^{-3} T^{3/2}\sqrt{1/M_1 + 1/M_2}}{p\sigma_{12}^2 \Omega}$$

where,

1 and 2 = index the two kinds of molecules present in the gaseous mixture.
T = temperature (K).
M = molar mass (g/mol).
p = pressure (atm).

$\sigma_{12} = \dfrac{1}{2}(\sigma_1 + \sigma_2)$ = the average collision diameter (the values are tabulated) (Å).

Ω = a temperature-dependent collision integral (the values are tabulated but usually of order of 1) (dimensionless).

D = diffusion coefficient (which is expressed in cm²/s when the other magnitudes are expressed in the units as given above).

Pressure Dependence of the Diffusion Coefficient

For self-diffusion in gases at two different pressures (but the same temperature), the following empirical equation has been suggested:

$$\frac{D_{P1}}{D_{P2}} = \frac{\rho_{P2}}{\rho_{P1}}$$

where,

P_1 and P_2 = denote pressures 1 and 2, respectively.

D = is the diffusion coefficient (m²/s).

ρ = is the gas mass density (kg/m³).

Effective Diffusivity in Porous Media

The effective diffusion coefficient describes diffusion through the pore space of porous media. It is macroscopic in nature, because it is not individual pores but the entire pore space that needs to be considered. The effective diffusion coefficient for transport through the pores, D_e, is estimated as follows:

$$D_e = \frac{D \varepsilon_t \delta}{\tau}$$

where,

D = diffusion coefficient in gas or liquid filling the pores (m²s⁻¹).

ε_t = porosity available for the transport (dimensionless).

δ = constrictivity (dimensionless).

τ = tortuosity (dimensionless).

The transport-available porosity equals the total porosity less the pores which, due to their size, are not accessible to the diffusing particles, and less dead-end and blind pores (i.e. pores without being connected to the rest of the pore system). The constrictivity describes the slowing down of diffusion by increasing the viscosity in narrow pores as a result of greater proximity to the average pore wall. It is a function of pore diameter and the size of the diffusing particles.

Effective Diffusion Coefficient

The effective diffusion coefficient (also referred to as the apparent diffusion coefficient) of a diffusant in atomic diffusion of solid polycrystalline materials like metal alloys is often represented as a weighted average of the grain boundary diffusion coefficient and the lattice diffusion coefficient. Diffusion along both the grain boundary and in the lattice may be modelled with an Arrhenius equation.

The ratio of the grain boundary diffusion activation energy over the lattice diffusion activation energy is usually 0.4– 0.6, so as temperature is lowered, the grain boundary diffusion component increases. Increasing temperature often allows for increased grain size, and the lattice diffusion component increases with increasing temperature, so often at $0.8\ T_{melt}$ (of an alloy), the grain boundary component can be neglected.

Modelling

The effective diffusion coefficient can be modelled using Hart's equation when only grain boundary and lattice diffusion are dominant:

$$D^{eff} = f D_{gb} + (1 - f) D_1$$

where,

D^{eff} = effective diffusion coefficient.

D_{gb} = grain boundary diffusion coefficient.

D_1 = lattice diffusion coefficient.

f = $\dfrac{q}{d} \delta$.

q = value based on grain shape, 1 for parallel grains, 3 for square grains.

d = average grain size.

δ = grain boundary width, often assumed to be 0.5 nm.

Grain boundary diffusion is significant in face centered cubic metals below about $0.8\,T_{melt}$ (absolute). Line dislocations and other crystalline defects can become significant below ~$0.4\,T_{melt}$ in FCC metals.

Diffusion-Controlled Reaction

Diffusion-controlled (or diffusion-limited) reactions are reactions that occur so quickly that the reaction rate is the rate of transport of the reactants through the reaction medium (usually a solution). As quickly as the reactants encounter each other, they react. The process of chemical reaction can be considered as involving the diffusion of reactants until they encounter each other in the right stoichiometry and form an activated complex which can form the product species. The observed rate of chemical reactions is, generally speaking, the rate of the slowest or 'rate determining' step. In diffusion controlled reactions the formation of products from the activated complex is much faster than the diffusion of reactants and thus the rate is governed by diffusion. Diffusion control is rare in the gas phase, where rates of diffusion of molecules are generally very high. Diffusion control is more likely in solution where diffusion of reactants is slower due to the greater number of collisions with solvent molecules. Reactions where the activated complex forms easily and the products form rapidly are most likely to be limited by diffusion control. Examples are those involving catalysis and enzymatic reactions. Heterogeneous reactions where reactants are in different phases are also candidates for diffusion control.

One classical test for diffusion control is to observe whether the rate of reaction is affected by stirring or agitation; if so then the reaction is almost certainly diffusion controlled under those conditions.

Rates of diffusion-controlled reactions

The metabolism of the biological cell, the control of its development and its communication with other cells in the organism or with its environment involves a complex web of biochemical reactions. The efficient functioning of this web relies on the availability of suitable reaction rates. Biological functions are often controlled through inhibition of these reaction rates, so the base rates must be as fast as possible to allow for a wide range of control. The maximal rates have been increased throughout the long evolution of life, often surpassing by a wide margin rates of comparable test tube reactions. In this respect it is important to realise that the rates of biochemical reactions involving two molecular partners, e.g. an enzyme and its substrate, at their optimal values are actually determined by the diffusive process which leads to the necessary encounter of the reactants. Since many biochemical reactions are proceeding close to their optimal speed, i.e. each encounter of the two reactants leads to a chemical transformation, it is essential for an understanding of biochemical reactions to characterise the diffusive encounters of biomolecules. In this section we want to describe first the relative motion of two diffusing biomolecules subject to an interaction between the partners. We then determine the rates of reactions as determined by the diffusion process. We finally discuss examples of reactions for various interactions.

Relative diffusion of two free particles

We consider first the relative motion in the case that two particles are diffusing freely. One can assume that the motion of one particle is independent of that of the other particle. In this case the diffusion is described by a distribution function $p(r_1, r_2, t|r_{10}, r_{20}, t_0)$ which is governed by the diffusion equation:

$$\partial_t\, p(r_1, r_2, t|r_{10}, r_{20}, t_0) = (D_1\nabla_1^2 + D_1\nabla_2^2)\, p(r_1, r_2, t|r_{10}, r_{20}, t_0) \qquad \ldots (7.13)$$

where, $\nabla_j = \partial/\partial r_j$, $j = 1, 2$. The additive diffusion operators $D_j\nabla_j^2$ in Eq. 7.13 are a signature of the statistical independence of the Brownian motions of each of the particles.

Our goal is to obtain from Eq. 7.13 an equation which governs the distribution $p(r, t|r_0; t_0)$ for the relative position:

$$r = r_2 - r_1 \qquad \ldots (7.14)$$

of the particles. For this purpose we express Eq. 7.13 in terms of the coordinates r and

$$R = a\, r_1 + b\, r_2 \qquad \ldots (7.15)$$

which, for suitable constants a, b, are linearly independent. One can express

$$\nabla_1 = a\nabla_R - \nabla,\ \nabla_2 = b\nabla_R + \nabla \qquad \ldots (7.16)$$

where, $\nabla = \partial/\partial r$. One obtains, furthermore,

$$\nabla_1^2 = a^2\nabla_R^2 + \nabla^2 - 2a\nabla_R\nabla \qquad \ldots (7.17)$$

$$\nabla_2^2 = b^2\nabla_R^2 + \nabla^2 + 2b\nabla_R\nabla \qquad \ldots (7.18)$$

The diffusion operator:

$$\hat{D} = D_1\nabla_1^2 + D_2\nabla_2^2 \qquad \ldots(7.19)$$

can then be written:

$$\hat{D} = \left(D_1 a^2 + D_2 b^2\right)\nabla_R^2 + \left(D_1 + D_2\right)\nabla^2 + 2\left(D_2 b - D_1 a\right)\nabla_R\nabla \qquad \ldots (7.20)$$

If one defines:

$$a = \sqrt{D_2/D_1}, \qquad b = \sqrt{D_1/D_2} \qquad \ldots (7.21)$$

one obtains:

$$\hat{D} = \left(D_1 + D_2\right)\nabla_R^2 + \left(D_1 + D_2\right)\nabla^2. \qquad \ldots (7.22)$$

The operator (Eq. 7.22) can be considered as describing two independent diffusion processes, one in the coordinate R and one in the coordinate r. Thus, the distribution function may be written $p(R, t|R_0, t_0)$ $p(r, t|r_0, t_0)$. If one disregards the diffusion along the coordinate R the relevant remaining relative motion is governed by:

$$\partial_t\, p(r, t|r_0, t_0) = (D_1 + D_2)\, \nabla^2 p(r, t|r_0, t_0) \qquad \ldots (7.23)$$

This equation implies that the relative motion of the two particles is also governed by a diffusion equation, albeit for a diffusion coeffcient:

$$D = D_1 + D_2 \qquad \ldots (7.24)$$

Relative motion of two diffusing particles with interaction

We seek to describe now the relative motion of two molecules which diffuse while interacting according to a potential $U(r)$ where $r = r_2 - r_1$. The force acting on particle 2 is $-\nabla_2 U(r) = F$; the force acting on particle 1 is $-F$. The distribution function $p(r_1; r_2; t|r_{10}, r_{20}; t_0)$ obeys the Smoluchowski equation:

$$\partial_t\, p = [(D_1\nabla_1^2 + D_2\nabla_2^2) - D_2\beta\nabla_2 \cdot F(r) + D_1\beta\nabla_1 \cdot F]\, p. \qquad \text{... (7.25)}$$

The first two terms on the r.h.s. can be expressed in terms of the coordinates R and r according to Eq. 7.22. For the remaining terms holds, using Eqs 7.16 and 7.21,

$$D_2\nabla_2 - D_1\nabla_1 = (D_1 + D_2)\nabla \qquad \text{... (7.26)}$$

Hence, one can write the Smoluchowski equation (7.25):

$$\partial_t\, p = [(D_1 + D_2)\, \nabla_R^2 + (D_1 + D_2)\, \nabla \cdot (\nabla - \beta F)]\, p. \qquad \text{... (7.27)}$$

This equation describes two independent random processes, free diffusion in the R coordinate and a diffusion with drift in the r coordinate. Since we are only interested in the relative motion of the two molecules, i.e. the motion which governs their reactive encounters, we describe the relative motion by the Smoluchowski equation:

$$\partial_t p\, (r, t|r_0, t_0) = D\nabla \cdot (\nabla - \beta F)\, p(r, t|r_0, t_0) \qquad \text{... (7.28)}$$

Diffusion-controlled reactions under stationary conditions

We want to consider now a reaction vessel which contains a solvent with two types of particles, particle 1 and particle 2, which engage in a reaction:

$$\text{Particle 1 + Particle 2} \rightarrow \text{Products} \qquad \text{... (7.29)}$$

We assume that particle 1 and particle 2 are maintained at concentrations c_1 and c_2, respectively, i.e. the particles are replenished as soon as they are consumed by Eq. 7.29. We also consider that the reaction products are removed from the system as soon as they are formed.

One can view the reaction vessel as containing pairs of particles 1 and 2 at various stages of the relative diffusion and reaction. This view maintains that the concentration of particles is so small that only rarely triple encounters, e.g. of two particles 1 and one particle 2, occur, so that these occurrences can be neglected. The system considered contains then many particle 1 and particle 2 pairs described by the Smoluchowski Eq. 7.28. Since the concentration of the particles is maintained at a steady level one can expect that the system adopts a stationary distribution of inter-pair distances $p(r)$ which obeys (Eq. 7.28), i.e.

$$\nabla D\,(r) \cdot (\nabla - \beta F)\, p(r) = 0, \qquad \text{... (7.30)}$$

subject to the condition:

$$p(r) \times |r| \rightarrow \infty\, c_1\, c_2 \qquad \text{... (7.31)}$$

Equation 7.29 is described by the boundary condition:

$$\hat{n} \cdot D(r)(\nabla - \beta F)\, p(r) = wp(r) \qquad \text{at } |r| = R_o \qquad \text{... (7.32)}$$

for some constant w.

The occurrence of Eq. 7.29 implies that a stationary current develops which describes the continuous diffusive approach and reaction of the particles. We consider in the following case that the particles are governed by an interaction potential which depends solely on the distance $|r|$ of the particles and that the diffusion coeffcient D also depends solely on $|r|$. The stationary Smoluchowski Eq.7.30 reads then:

$$(\partial_r D(r))\, ((\partial_r - \beta F(r))\, p(r)) = 0 \qquad \text{... (7.33)}$$

to which is associated the radial current:

$$J_{\text{tot}}(r) = 4\pi r^2 D(r)(\partial_r - \beta F(r)) p(r) \qquad \text{... (7.34)}$$

where we have summed over all angles θ, ϕ obtaining the total current at radius r. For $F(r) = -\partial_r U(r)$ one can express this:

$$J_{\text{tot}}(r) = 4\pi r^2 D(r) \exp[-\beta U(r)] (\partial_r \exp[\beta U(r)] p) \qquad \text{... (7.35)}$$

However, $J_{\text{tot}}(r)$ must be the same at all r since otherwise $p(r)$ would change in time, in contrast to the assumption that the distribution is stationary. It must hold, in particular,

$$J_{\text{tot}}(R_o) = J_{\text{tot}}(r) \qquad \text{... (7.36)}$$

The boundary condition (Eq. 7.32), together with (Eq. 7.35), yields:

$$4\pi R_o^2 w\, p(R_o) = 4\pi r^2 D(r) \exp[-\beta U(r)] (\partial_r \exp[\beta U(r)] p(r)) \qquad \text{... (7.37)}$$

This relationship, a first order differential equation, allows one to determine $p(r)$.
For the evaluation of $p(r)$ we write (7.37):

$$\partial_r \left(e^{\beta U(r)} p(r) \right) = \frac{R_0^2 w}{r^2 D(r)} p(R_o) e^{\beta U(r)}. \qquad \text{... (7.38)}$$

Integration $\int_r^{\infty} dr \cdots$ yields:

$$p(\infty) e^{\beta U(\infty)} - p(r) e^{\beta U(r)} = R_o^2 w\, p(R_o) \int_r^{\infty} dr' \frac{e^{\beta U(r')}}{r'^2 D(r')} \qquad \text{... (7.39)}$$

or, using Eq. 7.31 and $U(\infty) = 0$:

$$p(r) e^{\beta U(r)} = c_1 c_2 - R_o^2 w\, p(R_o) \int_r^{\infty} dr' \frac{e^{\beta U(r')}}{r'^2 D(r')} \qquad \text{... (7.40)}$$

Evaluating this at $r = R_o$ and solving for $p(R_o)$ yields:

$$p(R_o) = \frac{c_1 c_2 e^{-\beta U(R_o)}}{1 + R_o^2 w\, e^{-\beta U(R_o)} \int_{R_o}^{\infty} dr\, e^{-\beta U(r)}/r^2 D(r)}. \qquad \text{... (7.41)}$$

Using this in Eq. 7.40 leads to an expression of $p(r)$.
We are presently interested in the rate at which Eq. 7.29 proceeds. This rate is given by $J_{\text{tot}}(R_o) = 4\pi R_o^2 w\, p(R_o)$. Hence, we can state:

$$\text{Rate} = \frac{4\pi R_o^2 w\, c_1 c_2\, e^{-\beta U(R_o)}}{1 + R_o^2 w\, e^{-\beta U(R_o)} \int_{R_o}^{\infty} dr\, e^{\beta U(r)}/r^2 D(r)}. \qquad \text{... (10.42)}$$

This expression is proportional to $c_1 c_2$, a dependence expected for a bimolecular reaction of the type Eq. 7.29. Conventionally, one defines a bimolecular rate constant k as follows:

$$\text{Rate} = k\, c_1 c_2. \qquad \text{... (7.43)}$$

This constant is then, in the present case,

$$k = \frac{4\pi}{e^{\beta U(R_o)}/R_o^2 w + \int_{R_o}^{\infty} dr\, R(r)}. \qquad \text{... (7.44)}$$

Here, we defined:

$$R(r) = e^{\beta U(r)}/r^2 D(r) \qquad \qquad ... (7.45)$$

a property which is called the resistance of the diffusing particle, a name suggested by the fact that $R(r)$ describes the ohmic resistance of the system as shown further below.

Examples: We consider first the case of very ineffective reactions described by small w values. In this case the time required for the diffusive encounter of the reaction partners can become significantly shorter than the time for the local reaction to proceed, if it proceeds at all. In this case it may hold:

$$\frac{e^{\beta U(R_o)}}{R_o^2 w} \gg \int_{R_o}^{\infty} dr\, R(r) \qquad \qquad ... (7.46)$$

and the reaction rate (Eq. 7.44) becomes:

$$k = 4\pi R_o^2 w\, e^{-\beta U(R_o)} \qquad \qquad ... (7.47)$$

This expression conforms to the well-known Arrhenius law.

We want to apply Eqs 7.44, 7.45 to two cases, free diffusion ($U(r) \equiv$) and diffusion in a Coulomb potential ($U(r) = q_1 q_2 = /\epsilon\, r$, ϵ = dielectric constant). We assume in both cases a distance independent diffusion constant. In case of free diffusion holds $R(r) = D^{-1} r^{-2}$ and, hence,

$$\int_{R_o}^{\infty} dr\, R(r) = 1/DR_o. \qquad \qquad ... (7.48)$$

From this results:

$$k = \frac{4\pi DR_o}{1 + D/R_o w}. \qquad \qquad ... (7.49)$$

In case of very effective reactions, i.e. for very large w, this becomes

$$k = 4\pi DR_o \qquad \qquad ... (7.50)$$

which is the well-known rate for diffusion-controlled reaction processes. No bi-molecular rate constant involving a diffusive encounter in a three-dimensional space without attracting forces between the reactants can exceed Eq. 7.50. For instance, in a diffusion-controlled reaction in a solvent with relative diffusion constant $D = 10^{-5}$ cm^2 s^{-1} and with reactants such that $R_o = 1$ nm, the maximum possible reaction rate is 7.56×10^9 L mol^{-1} s^{-1}.

In case of a Coulomb interaction between the reactants one obtains:

$$\int_{R_o}^{\infty} dr\, R(r) = \frac{1}{D} \int_{R_o}^{\infty} dr\, \frac{1}{r^2} \exp\left[\frac{\beta q_1 q_2}{\epsilon\, r} \right]$$

$$= \frac{1}{D} \int_{0}^{1/R_o} dy \exp\left[\frac{\beta q_1 q_2 y}{\epsilon} \right]$$

$$= \frac{1}{R_L D} \left(e^{R_L/R_o} - 1 \right) \qquad \qquad ... (7.51)$$

where,

$$R_L = \beta q_1 q_2/\epsilon \qquad \qquad ... (7.52)$$

defines the so-called Onsager radius. Note that R_L can be positive or negative, depending on the sign of $q_1 q_2$, but that the integral over the resistance Eq. 7.51 is always positive. The rate constant Eq. 7.44 can then be written:

$$k = \frac{4\pi D R_L}{\dfrac{R_L D}{R_o^2 w} e^{R L/R_o} + e^{R L/R_o} - 1} \qquad \text{... (7.53)}$$

For instance, suppose we wish to find the maximum reaction rate for a reaction between pyrene-N and N-dimethylaniline in acetoneitrile. The reaction consists of an electron exchange from pyrene to dimethylaniline, and the reactants have charges of $\pm e$. The relative diffusion constant of both reactants in acetonitrile at 25°C is 4.53×10^{-5} cm^2 s^{-1}, the dielectric constant of acetonitrile at that temperature is 37.5, and the effective reaction radius R_o of the reactants is 0.7 nm. Using these values, and assuming $w \to \infty$ in (Eq. 7.53) we obtain an Onsager radius of -10.8 nm, and a maximum reaction rate of $k = 6.44 \times 10^{11}$ L mol^{-1} s^{-1}.

In a different solvent, C_3H_7OH, with relative diffusion constant $D = 0.77 \times 10^{-5}$ cm^2 s^{-1} at 25°C and a dielectric constant of 19.7, the Onsager radius is -35.7 nm and the maximum reaction rate is $k = 2.08 \times 10^{11}$ L mol^{-1} s^{-1}.

Tortuosity

Tortuosity is a property of curve being tortuous (twisted; having many turns). There have been several attempts to quantify this property. Tortuosity is commonly used to describe diffusion in porous media. This concept is also used for porous media as soils and snow.

Tortuosity in 2-D

Subjective estimation (sometimes aided by optometric grading scales) is often used. The most simple mathematic method to estimate tortuosity is arc-chord ratio: ratio of the length of the curve (L) to the distance between the ends of it (C):

$$\tau = \frac{L}{C}$$

Arc-chord ratio equals 1 for a straight line and is infinite for a circle.

Another method, proposed in 1999, is to estimate the tortuosity as integral of square (or module) of curvature. Dividing the result by length of curve or chord has also been tried. In 2006 several Italian scientists proposed one more method. At first, the curve is divided into several (N) parts with constant sign of curvature (using hysteresis to decrease sensitivity to noise).

Then the ar–chord ratio for each part is found and the tortuosity is estimated by:

$$\tau = \frac{N-1}{L} \cdot \sum_{i=1}^{N} \left(\frac{L_i}{S_i} - 1 \right)$$

In this case tortuosity of both straight line and circle is estimated to be 0.

In 1993 Swiss mathematician Martin Mächler proposed an analogy: it is relatively easy to drive a bicycle or a car in a trajectory with a constant curvature (an arc of a circle), but it is much harder to drive where curvature changes. This would imply that roughness (or tortuosity) could be measured by relative change of curvature. In this case the proposed 'local' measure was derivative of logarithm of curvature:

$$\frac{d}{dx}\log(\kappa) = \frac{\kappa'}{\kappa}$$

However, in this case tortuosity of a straight line is left undefined.

In 2007 it was proposed to measure tortuosity by an integral of square of derivative of curvature, divided by the length of a curve:

$$\tau = \frac{\int_{t_1}^{t_2} \left(\kappa'(t)\right)^2 dt}{L}$$

In this case tortuosity of both straight line and circle is estimated to be 0. In most of these methods digital filters and approximation by splines can be used to decrease sensitivity to noise.

Tortuosity in 3-D

Usually subjective estimation is used. However, several ways to adapt methods estimating tortuosity in 2-D have also been tried. The methods include arc-chord ratio, arc-chord ratio divided by number of inflection points and integral of square of curvature, divided by length of the curve (curvature is estimated assuming that small segments of curve are planar). Another method used for quantifying tortuosity in 3D has been applied in 3D reconstructions of solid oxide fuel cell cathodes where the Euclidean distance sums of the centroids of a pore were divided by the length of the pore.

Applications of tortuosity

Tortuosity of blood vessels (for example, retinal and cerebral blood vessels) is known to be used as a medical sign. In mathematics, cubic splines minimise the functional, equivalent to integral of square of curvature (approximating the curvature as the second derivative).

In many engineering domains dealing with mass transfer in porous materials, such as hydrogeology or heterogeneous catalysis, the tortuosity refers to the ratio of the diffusivity in the free space to the diffusivity in the porous medium (analogous to arc-chord ratio of path). Strictly speaking, however, the effective diffusivity is proportional to the reciprocal of the square of the geometrical tortuosity.

In acoustics and following initial works by Maurice Anthony Biot in 1956, the tortuosity is used to describe sound propagation in fluid-saturated porous media. In such media, when frequency of the sound wave is high enough, the effect of viscous drag force between the solid and the fluid can be ignored. In this case, velocity of sound propagation in the fluid in the pores is non-dispersive and compared with the value of the velocity of sound in the free fluid is reduced by a ratio equal to the square root of the tortuosity. This has been used for a number of applications including the study of materials for acoustic isolation, and for oil prospection using acoustics means.

In analytical chemistry applied to polymers and sometimes small molecules tortuosity is applied in Gel permeation chromatography (GPC) also known as Size exclusion chromatography (SEC). As with any chromatography it is used to separate mixtures. In the case of GPC the separation is based on molecular size and it works by the use of stationary media with appropriately dimensioned pores. The separation occurs because larger molecules take a shorter, less tortuous path and elute more quickly and smaller molecules can pass into the pores and take a longer, more tortuous path and elute later.

Diffusion Equation

The diffusion equation is a partial differential equation which describes density fluctuations in a material undergoing diffusion. It is also used to describe processes exhibiting diffusive-like behaviour, for instance the 'diffusion' of alleles in a population in population genetics. The equation is usually written as:

$$\frac{\partial \phi(r,\ t)}{\partial t} = \nabla \cdot \left[D(\phi, r) \nabla \phi(r, t) \right],$$

where, $\phi(r, t)$ is the density of the diffusing material at location r and time t and $D(\phi, r)$ is the collective diffusion coefficient for density ϕ at location r; the nabla symbol ∇ represents the vector differential operator del acting on the space coordinates. If the diffusion coefficient depends on the density then the equation is nonlinear, otherwise it is linear. If D is constant, then the equation reduces to the following linear equation:

$$\frac{\partial \phi(r,\ t)}{\partial t} = D \nabla^2 \phi(r,\ t),$$

also called the heat equation. More generally, when D is a symmetric positive definite matrix, the equation describes anisotropic diffusion, which is written (for three-dimensional diffusion) as:

$$\frac{\partial \phi(r,\ t)}{\partial t} = \sum_{i=1}^{3} \sum_{j=1}^{3} \frac{\partial}{\partial x_i} \left[D_{ij}(\phi, r) \frac{\partial \phi(r,\ t)}{\partial x_j} \right]$$

Derivation

The diffusion equation can be derived in a straightforward way from the continuity equation, which states that a change in density in any part of the system is due to inflow and outflow of material into and out of that part of the system. Effectively, no material is created or destroyed:

$$\frac{\partial \phi}{\partial t} + \nabla \bullet j = 0,$$

where, j is the flux of the diffusing material. The diffusion equation can be obtained easily from this when combined with the phenomenological Fick's first law, which assumes that the flux of the diffusing material in any part of the system is proportional to the local density gradient:

$$j = -D\ (\phi)\ \nabla \phi\ (r, t).$$

Discretisation (Image)

The product rule is used to rewrite the anisotropic tensor diffusion equation, in standard discretisation schemes. Because direct discretisation of the diffusion equation with only first order spatial central differences leads to checkerboard artifacts. The rewritten diffusion equation used in image filtering:

$$\frac{\partial \phi(r,\ t)}{\partial t} = \mathrm{div} \left[D(\phi,\ r) \right] \nabla \phi(r,\ t) + \mathrm{trace} \left[D(\phi,\ r) \left(\nabla \nabla^T \phi(r,\ t) \right) \right]$$

In which in image filtering $D(\phi, r)$ are symmetric matrices constructed from the eigenvectors of the image structure tensors. The spatial derivatives can then be approximated by two first order and a second order central finite differences. The resulting diffusion algorithm can be written as an image convolution with an varying kernel (stencil) of size 3×3 in 2D and $3 \times 3 \times 3$ in 3D.

Reaction-Diffusion Processes in Tissue Engineering

Microscopy

Complex networks of reactions occurring at various scales form the fundamental machinery by which any living organism can perform essential functions to maintain and propagate itself. For example, the process of metabolism is but a series of chemical reactions which result in breakdown of large complex molecules resulting in the generation of energy or making building blocks for cellular function. Similarly, the biological signalling transduction occurs via series of reactions occurring within/among cells by which the cells can respond to the changes in the environment. The understanding of biological reaction networks, therefore, occupies a significant position in the area of biological engineering, and many efforts have been made to get a better understanding of cellular metabolic and signalling networks both from an experimental as well as modelling standpoint. Further, the effect of spatial arrangement and diffusion in such systems is also a subject of active research, and the role of spatial dimensional and diffusion is shown to be important for signalling systems.

As in usual chemical systems, the characterisation of the equilibrium and dynamic reaction parameters remain crucial to the success of any modelling approach for biochemical reaction networks. Along with the usual methods of determination of reaction parameters, quantitative fluorescent microscopy imaging of live cells offers the promise to measure the reaction parameters in its native environment and hence is the center of attention of various research groups. Among them, visualisation of protein-protein binding by Fluorescence resonance energy transfer (FRET) is a effective way to gather data on cellular protein reactions. However, the process of inferring information about cellular reactions from the imaging data is far from understood.

Recently, there has been a focus on developing *in vitro* systems that can mimic the native cellular conditions. Such systems can be of vital importance, as they can offer the possibility of probing the cells to characterise their behaviour under controlled environmental conditions and also quantitatively measure the associated responses. From a medical perspective, they can be used to grow functional tissues that can be further used for transplantation. This research area, termed 'tissue engineering', aims at developing replacement organs in a laboratory (*in vitro* organs) starting from a small population of donor cells and providing them with appropriate microenvironment for development of target tissues. The development of such tissue engineering methods will involve extensive experimentation for screening, optimisation, and implementation of the final tissue. Considering that most of these systems have an underlying reaction-diffusion mechanism, mathematical analysis and predictive models can play a crucial role in reducing the expensive experimentation and successful application of tissue engineered therapy.

In particular, the development of *in vitro* culture systems to maintain and expand stem cells (both, adult as well as embryonic) has great therapeutic potential, considering that, in principle, the stem cells are multipotent and can be made to proliferate and differentiate into almost any tissue. Research efforts in engineering stem cells for developing functional tissues like bone and blood have been fairly successful and well understood. However, optimal strategies to maintain and expand stem cells still remain elusive. Microfabricated perfusion based bioreactors have emerged as strong candidates for developing *in vitro* cell culture systems for both tissue engineering as well as developing new platforms to be used as biosensors. Compared to traditional static cell culture methods, the media for such new bioreactors is continuously perfused, to provide a dynamically controlled microenvironment for the cells in the culture. Considering the basic mechanisms involved in the growth and culture of cells in such bioreactors would be based on diffusional and convective transport coupled with the reactive uptake/secretion, a mathematical analysis of these systems can help with improving the design without expensive experimentation.

Cartilage

Cartilage is a stiff and inflexible connective tissue found in many areas in the bodies of humans and other animals, including the joints between bones, the rib cage, the ear, the nose, the elbow, the knee, the ankle, the bronchial tubes and the intervertebral discs. It is not as hard and rigid as bone but is stiffer and less flexible than muscle. Cartilage is composed of specialised cells called chondrocytes that produce a large amount of extracellular matrix composed of Type II collagen (except Fibrocartilage which also contains type I collagen) fibres, abundant ground substance rich in proteoglycan, and elastin fibres. Cartilage is classified in three types, elastic cartilage, hyaline cartilage and fibrocartilage, which differ in the relative amounts of these three main components. Unlike other connective tissues, cartilage does not contain blood vessels. The chondrocytes are supplied by diffusion, helped by the pumping action generated by compression of the articular cartilage or flexion of the elastic cartilage. Thus, compared to other connective tissues, cartilage grows and repairs more slowly.

Growth and development

In embryogenesis, the skeletal system is derived from the mesoderm germ layer. Chondrification (also known as chondrogenesis) is the process by which cartilage is formed from condensed mesenchyme tissue, which differentiates into chondrocytes and begins secreting the molecules that form the extracellular matrix. Skeletal blast cells that express the Sox9 transcription factor, followed by continued expression of Sox5 and Sox6, develop into chondroblast precursors, while those that express Runx2, followed by osterix develop into chondroblast precursors, while those that express Runx2, followed by osterix develop into osteogenic precursors. The condroblastic differentiation is favoured in regions under compressive forces and low pO_2 because these down regulate BMP3, which normally inhibits cartilage differentiation. Osteogenic differentiation is favoured under neutral or mild, intermittent tensile forces and relatively high pO_2, which leads to upregulation of BMP4. High tensile strength favours the formation of tendinous connective tissue.

Weisz–Prater Criterion

The Weisz–Prater criterion is a method used to estimate the influence of pore diffusion on reaction rates in heterogeneous catalytic reactions. If the criterion is satisfied, pore diffusion limitations are negligible. The criterion is:

$$N_{w-p} = \frac{\Re R_p^2}{C_s D_{eff}} \leq 3\beta$$

where, $\Re$ is the reaction rate per volume of catalyst, R_p is the catalyst particle radius, C_s is the reactant concentration at the particle surface, and D_{eff} is the effective diffusivity. Diffusion is usually in the Knudsen regime when average pore radius is less than 100 nm.

For a given effectiveness factor, η, and reaction order, n, the quantity β is defined by the equation:

$$\eta = \frac{3}{R_p^3} \int_0^{R_p} \left[1 - \beta \left(1 - r/R_p \right)^n \right] r^2 \, dr$$

for small values of beta this can be approximated using the binomial theorem:

$$\eta = 1 - \frac{n\beta}{4}$$

Assuming $\eta \geq 0.95$ with a 1st or zero order reaction gives values of β, 0.6 and 6 respectively. Therefore for many conditions, if $N_{W-P} \leq 0.3$ then pore diffusion limitations can be excluded.

Multiphase Reactors

Multiphase reactors are very common in the chemical, petroleum, food, mining, pharmaceutical, and semiconductor industries. They are also prominent in many civilian processes. The production of drinking water can consist of several multiphase reactors and separators, removing soluble salts, organic matter, live organisms, and insoluble material. Our waste treatment systems usually have an activated sludge reactor in which air is forced into the solid-liquid waste stream. In the past, when untreated wastes fouled potential drinking water, people produced beer as a method of purification using a multiphase reactor in a procedure called fermentation.

Antibiotics are produced by a three-phase process: Air is sparged into a vessel or tank to supply oxygen, keep the producing cells suspended in a liquid broth, and mix the contents of the vessel. Coal liquefaction is another three-phase process with hydrogen as the gas phase, coal as the solid phase. Fuel cells, a key component of distributed energy systems, are three- or more-phase reactors depending on the type of fuel cell: solid oxide, molten carbonate, phosphoric acid or proton exchange membrane. They can be catalytic electrochemical reactors with membranes and multiple solid phases. Genetic engineering requires many multiphase reactor systems. Enzymatic peptide synthesis is a multiphase reaction. The reactors can be divided into just a few broad categories. These are column contactors with one phase fixed, fluidised or transported, stirred-tank reactors with at least one phase kept distributed by agitation, film reactors, membrane reactors, and large diameter tank reactors, including open top tanks and ponds. The goals for each reactor design are usually identical, maximise conversion of the reactants and yield of product.

Trickle Bed Reactors

Trickle beds (Fig. 7.12) with countercurrent flow of gas and liquid are used on a large scale for vinegar production, as biofilters for gas clean-up and deodourisation, for water purification and for ore leaching. Trickle beds can be operated with or without recycling, but recycling allows higher loading and gives better flow distribution, which is even more critical than in submerged packed bed operation. Cleaning is again possible by flooding the filter and proceeding as with submerged fixed bed reactors. In the case of gas purification, the gas is cleaned in a single passage and the liquid is there both as an absorption fluid and as nutrient supply to the biomass on the packing (usually wood shavings or bark). Excess biomass is settled out of this stream and no other cleaning is needed.

Fluidised Bed Reactor

A fluidised bed reactor (FBR) is a type of reactor device that can be used to carry out a variety of multiphase chemical reactions. In this type of reactor, a fluid (gas or liquid) is passed through a granular solid material (usually a catalyst possibly shaped as tiny spheres) at high enough velocities to suspend the solid and cause it to behave as though it were a fluid. This process, known as fluidisation, imparts many important advantages to the FBR. As a result, the fluidised bed reactor is now used in many industrial applications (Fig. 7.13).

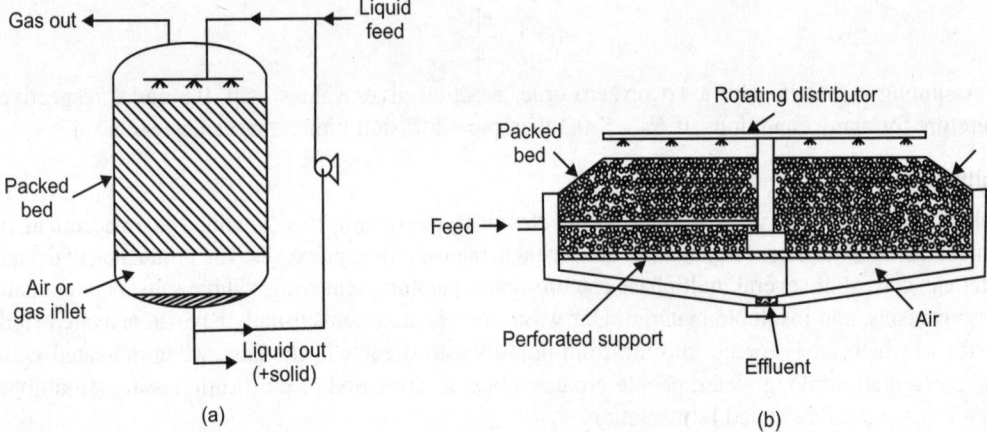

Fig. 7.12. Trickle bed reactors: (a) trickle bed or biofilter; (b) waste water trickling filter.

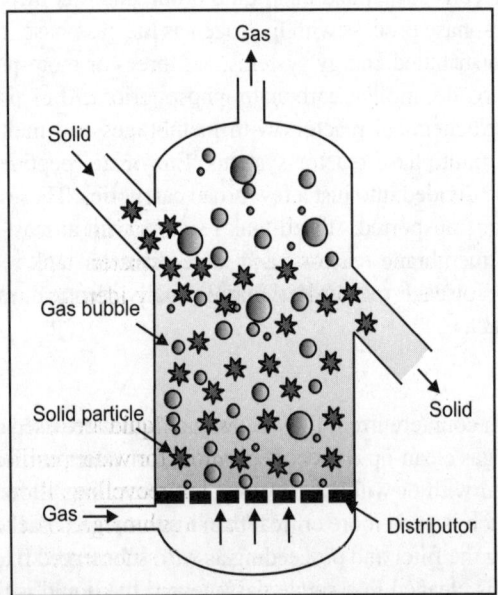

Fig. 7.13. Basic diagram of a fluidised bed reactor.

The solid substrate (the catalytic material upon which chemical species react) material in the fluidised bed reactor is typically supported by a porous plate, known as a distributor. The fluid is then forced through the distributor up through the solid material. At lower fluid velocities, the solids remain in place as the fluid passes through the voids in the material. This is known as a packed bed reactor. As the fluid velocity is increased, the reactor will reach a stage where the force of the fluid on the solids is enough to balance the weight of the solid material. This stage is known as incipient fluidisation and occurs at this minimum fluidisation velocity. Once this minimum velocity is surpassed, the contents of the reactor bed begin to expand and swirl around much like an agitated tank or boiling pot of water. The reactor is now a fluidised bed. Depending on the operating conditions and properties of solid phase various flow regimes can be observed in this reactor.

Today fluidised bed reactors are still used to produce gasoline and other fuels, along with many other chemicals. Many industrially produced polymers are made using FBR technology, such as rubber, vinyl chloride, polyethylene, and styrenes. Various utilities also use FBR's for coal gasification, nuclear power plants, and water and waste treatment settings. Used in these applications, fluidised bed reactors allow for a cleaner, more efficient process than previous standard reactor technologies.

Advantages

The increase in fluidised bed reactor use in today's industrial world is largely due to the inherent advantages of the technology.

1. Uniform particle mixing: Due to the intrinsic fluid-like behaviour of the solid material, fluidised beds do not experience poor mixing as in packed beds. This complete mixing allows for a uniform product that can often be hard to achieve in other reactor designs. The elimination of radial and axial concentration gradients also allows for better fluid-solid contact, which is essential for reaction efficiency and quality.

2. Uniform temperature gradients: Many chemical reactions require the addition or removal of heat. Local hot or cold spots within the reaction bed, often a problem in packed beds, are avoided in a fluidised situation such as an FBR. In other reactor types, these local temperature differences, especially hotspots, can result in product degradation. Thus FBRs are well suited to exothermic reactions. Researchers have also learned that the bed-to-surface heat transfer coefficients for FBRs are high.

3. Ability to operate reactor in continuous state: The fluidised bed nature of these reactors allows for the ability to continuously withdraw product and introduce new reactants into the reaction vessel. Operating at a continuous process state allows manufacturers to produce their various products more efficiently due to the removal of startup conditions in batch processes.

Disadvantages

As in any design, the fluidised bed reactor does have it drawbacks, which any reactor designer must take into consideration.

1. Increased reactor vessel size: Because of the expansion of the bed materials in the reactor, a larger vessel is often required than that for a packed bed reactor. This larger vessel means that more must be spent on initial capital costs.

2. Pumping requirements and pressure drop: The requirement for the fluid to suspend the solid material necessitates that a higher fluid velocity is attained in the reactor. In order to achieve this, more pumping power and thus higher energy costs are needed. In addition, the pressure drop associated with deep beds also requires additional pumping power.

3. Particle entrainment: The high gas velocities present in this style of reactor often result in fine particles becoming entrained in the fluid. These captured particles are then carried out of the reactor with the fluid, where they must be separated. This can be a very difficult and expensive problem to address depending on the design and function of the reactor. This may often continue to be a problem even with other entrainment reducing technologies.

4. Lack of current understanding: Current understanding of the actual behaviour of the materials in a fluidised bed is rather limited. It is very difficult to predict and calculate the complex mass and heat flows within the bed. Due to this lack of understanding, a pilot plant for new processes is required. Even with pilot plants, the scale-up can be very difficult and may not reflect what was experienced in the pilot trial.

5. Erosion of internal components: The fluid-like behaviour of the fine solid particles within the bed eventually results in the wear of the reactor vessel. This can require expensive maintenance and upkeep for the reaction vessel and pipes.

Chemical Vapour Deposition

Chemical vapour deposition (CVD) is a chemical process used to produce high-purity, high-performance solid materials. The process is often used in the semiconductor industry to produce thin films. In a typical CVD process, the wafer (substrate) is exposed to one or more volatile precursors, which react and/or decompose on the substrate surface to produce the desired deposit. Frequently, volatile by-products are also produced, which are removed by gas flow through the reaction chamber.

Microfabrication processes widely use CVD to deposit materials in various forms, including: monocrystalline, polycrystalline, amorphous, and epitaxial. These materials include: silicon, carbon fibre, carbon nanofibres, filaments, carbon nanotubes, SiO_2, silicon-germanium, tungsten, silicon carbide, silicon nitride, silicon oxynitride, titanium nitride, and various high-k dielectrics. The CVD process is also used to produce synthetic diamonds.

Types of Chemical Vapour Deposition

A number of forms of CVD are in wide use and are frequently referenced in the literature. These processes differ in the means by which chemical reactions are initiated (e.g. activation process) and process conditions.

1. Classified by operating pressure.
   (a) Atmospheric pressure CVD (APCVD): CVD processes at atmospheric pressure.
   (b) Low-pressure CVD (LPCVD): CVD processes at subatmospheric pressures. Reduced pressures tend to reduce unwanted gas-phase reactions and improve film uniformity across the wafer. Most modern CVD processes are either LPCVD or UHVCVD.
   (c) Ultrahigh vacuum CVD (UHVCVD): CVD processes at a very low pressure, typically below 10^{-6} Pa ($\sim 10^{-8}$ torr). Note that in other fields, a lower division between high and ultra-high vacuum is common, often 10^{-7} Pa.

2. Classified by physical characteristics of vapour:
   (a) Aerosol assisted CVD (AACVD): A CVD process in which the precursors are transported to the substrate by means of a liquid/gas aerosol, which can be generated ultrasonically. This technique is suitable for use with non-volatile precursors.
   (b) Direct liquid injection CVD (DLICVD): A CVD process in which the precursors are in liquid form (liquid or solid dissolved in a convenient solvent). Liquid solutions are injected in a vapourisation chamber towards injectors (typically car injectors). Then the precursor vapours are transported to the substrate as in classical CVD process. This technique is suitable for use on liquid or solid precursors. High growth rates can be reached using this technique.

3. Plasma methods.
   (a) Microwave plasma-assisted CVD (MPCVD).
   (b) Plasma-Enhanced CVD (PECVD): CVD processes that utilise plasma to enhance chemical reaction rates of the precursors. PECVD processing allows deposition at lower temperatures, which is often critical in the manufacture of semiconductors.

(d) Remote plasma-enhanced CVD (RPECVD): Similar to PECVD except that the wafer substrate is not directly in the plasma discharge region. Removing the wafer from the plasma region allows processing temperatures down to room temperature.

4. Atomic layer CVD (ALCVD): Deposits successive layers of different substances to produce layered, crystalline films.

5. Combustion chemical vapour deposition (CCVD): nGimat's proprietary Combustion chemical vapour deposition process is an open-atmosphere, flame-based technique for depositing high-quality thin films and nanomaterials.

6. Hot wire CVD (HWCVD): Also known as catalytic CVD (Cat-CVD) or hot filament CVD (HFCVD). Uses a hot filament to chemically decompose the source gases.

7. Metalorganic chemical vapour deposition (MOCVD): CVD processes based on metalorganic precursors.

8. Hybrid physical-chemical vapour deposition (HPCVD): Vapour deposition processes that involve both chemical decomposition of precursor gas and vapourisation of a solid source.

9. Rapid thermal CVD (RTCVD): CVD processes that use heating lamps or other methods to rapidly heat the wafer substrate. Heating only the substrate rather than the gas or chamber walls helps reduce unwanted gas phase reactions that can lead to particle formation.

10. Vapour phase epitaxy (VPE).

Substances commonly deposited for ICs

This section discusses the CVD processes often used for integrated circuits (ICs). Particular materials are deposited best under particular conditions.

Polysilicon

Polycrystalline silicon is deposited from silane (SiH_4), using the following reaction:

$$SiH_4 \rightarrow Si + 2H_2$$

This reaction is usually performed in LPCVD systems, with either pure silane feedstock or a solution of silane with 70–80 per cent nitrogen. Temperatures between 600° and 650°C and pressures between 25 and 150 Pa yield a growth rate between 10 and 20 nm per minute. An alternative process uses a hydrogen-based solution. The hydrogen reduces the growth rate, but the temperature is raised to 850° or even 1050°C to compensate.

Polysilicon may be grown directly with doping, if gases such as phosphine, arsine or diborane are added to the CVD chamber. Diborane increases the growth rate, but arsine and phosphine decrease it.

Silicon dioxide

Silicon dioxide (usually called simply 'oxide' in the semiconductor industry) may be deposited by several different processes. Common source gases include silane and oxygen, dichlorosilane ($SiCl_2H_2$) and nitrous oxide (N_2O), or tetraethylorthosilicate (TEOS; $Si(OC_2H_5)_4$). The reactions are as follows:

$$SiH_4 + O_2 \rightarrow SiO_2 + 2H_2$$

$$SiCl_2H_2 + 2N_2O \rightarrow SiO_2 + 2N_2 + 2HCl$$

$$Si(OC_2H_5)_4 \rightarrow SiO_2 + by\text{-}products$$

The choice of source gas depends on the thermal stability of the substrate; for instance, aluminium is sensitive to high temperature. Silane deposits between 300° and 500°C, dichlorosilane at around 900°C, and TEOS between 650° and 750°C, resulting in a layer of low-temperature oxide (LTO). However, silane produces a lower-quality oxide than the other methods (lower dielectric strength, for instance), and it deposits nonconformally. Any of these reactions may be used in LPCVD, but the silane reaction is also done in APCVD. CVD oxide invariably has lower quality than thermal oxide, but thermal oxidation can only be used in the earliest stages of IC manufacturing.

Oxide may also be grown with impurities (alloying or 'doping'). This may have two purposes. During further process steps that occur at high temperature, the impurities may diffuse from the oxide into adjacent layers (most notably silicon) and dope them. Oxides containing 5–15 per cent impurities by mass are often used for this purpose.

In addition, silicon dioxide alloyed with phosphorus pentoxide (P-glass) can be used to smooth out uneven surfaces. P-glass softens and reflows at temperatures above 1000°C. This process requires a phosphorus concentration of at least 6 per cent, but concentrations above 8 per cent can corrode aluminium. Phosphorus is deposited from phosphine gas and oxygen:

$$4PH_3 + 5O_2 \rightarrow 2P_2O_5 + 6H_2$$

Glasses containing both boron and phosphorus (borophosphosilicate glass, BPSG) undergo viscous flow at lower temperatures; around 850°C is achievable with glasses containing around 5 weight per cent of both constituents, but stability in air can be difficult to achieve. Phosphorus oxide in high concentrations interacts with ambient moisture to produce phosphoric acid. Crystals of BPO_4 can also precipitate from the flowing glass on cooling; these crystals are not readily etched in the standard reactive plasmas used to pattern oxides, and will result in circuit defects in integrated circuit manufacturing.

Besides these intentional impurities, CVD oxide may contain by-products of the deposition process. TEOS produces a relatively pure oxide, whereas silane introduces hydrogen impurities, and dichlorosilane introduces chlorine.

Lower temperature deposition of silicon dioxide and doped glasses from TEOS using ozone rather than oxygen has also been explored (350° to 500°C). Ozone glasses have excellent conformality but tend to be hygroscopic—that is, they absorb water from the air due to the incorporation of silanol (Si-OH) in the glass. Infrared spectroscopy and mechanical strain as a function of temperature are valuable diagnostic tools for diagnosing such problems.

Silicon nitride

Silicon nitride is often used as an insulator and chemical barrier in manufacturing ICs. The following two reactions deposit nitride from the gas phase:

$$3SiH_4 + 4NH_3 \rightarrow Si_3N_4 + 12H_2$$

$$3SiCl_2H_2 + 4NH_3 \rightarrow Si_3N_4 + 6HCl + 6H_2$$

Silicon nitride deposited by LPCVD contains up to 8 per cent hydrogen. It also experiences strong tensile stress, which may crack films thicker than 200 nm. However, it has higher resistivity and dielectric strength than most insulators commonly available in microfabrication (10^{16} $\Omega \cdot$cm and 10 MV/cm, respectively).

Another two reactions may be used in plasma to deposit SiNH:

$$2SiH_4 + N_2 \rightarrow 2SiNH + 3H_2$$

$$SiH_4 + NH_3 \rightarrow SiNH + 3H_2$$

These films have much less tensile stress, but worse electrical properties (resistivity 10^6 to 10^{15} $\Omega \cdot$cm, and dielectric strength 1 to 5 MV/cm).

Metals

Some metals (notably aluminium and copper) are seldom or never deposited by CVD. As of 2006, a commercially, cost effective, viable CVD process for copper did not exist, though copper formate, copper(hfac)2, Cu(II) ethyl acetoacetate, and other precursors have been used. Copper deposition of the metal has been done mostly by electroplating, in order to reduce the cost. Aluminium can be deposited from tri-isobutyl aluminium (TIBAL), triethyl/methyl aluminium (TEA,TMA), or dimethylaluminium hydride (DMAH), but physical vapour deposition methods are usually preferred.

However, CVD processes for molybdenum, tantalum, titanium, nickel, and tungsten are widely used. These metals can form useful silicides when deposited onto silicon. Mo, Ta and Ti are deposited by LPCVD, from their pentachlorides. Nickel, molybdenum, and tungsten can be deposited at low temperatures from their carbonyl precursors. In general, for an arbitrary metal M, the reaction is as follows:

$$2MCl_5 + 5H_2 \rightarrow 2M + 10HCl$$

The usual source for tungsten is tungsten hexafluoride, which may be deposited in two ways:

$$WF_6 \rightarrow W + 3F_2$$

$$WF_6 + 3H_2 \rightarrow W + 6HF$$

Thus, the matter discussed in this chapter should be applied to catalyst particles as well as biomaterial tissue engineering.

Chapter 8

Residence Time Distribution in Chemical Reactors

INTRODUCTION

Residence time is the average amount of time that a particle spends in a particular system. Residence time is a widely used term that is mostly seen in science, technological and medical disciplines. Every discipline that uses residence time in some way adapts the definition in order to make it more specific to the application to which it is referring. The base definition for residence time also has a universal mathematical equation that can be added to and adapted for different disciplines. This is as follows:

$$\frac{\text{The capacity of a system to hold a substance}}{\text{The rate of flow of the substance into the system}}$$

The generic variable form of this equation is as follows:

$$\tau = \frac{V}{q}$$

where, τ is used as the variable for residence time, V is the capacity of the system, and q is the flow for the system.

Residence time begins from the moment that a particle of a particular substance enters the system and ends the moment that the same particle of that substance leaves the system. The system in question is arbitrary and can be defined as needed according to the application. If a large amount of a substance enters a system, the longer it will take for the substance to leave the system, resulting in a longer residence time. This is assuming that the inflow and outflow for the system are kept constant. By this same logic, the smaller the amount of a substance in a particular system, the shorter the residence time will be. Inflow and outflow will also have an effect on the residence time of a system. If the inflow and outflow are increased, the residence time of the system will be shorter. However, if the inflow and the outflow of a system are decreased, the residence time will be longer. This is assuming that the concentration of the substance in the system and the size of the system remain constant, and assuming steady-state conditions.

If the size of the system is changed, the residence time of the system will be changed as well. The larger the system, then larger the residence time, assuming the inflow and outflow rates are held constant. The smaller the system, the shorter the residence time will be, again assuming steady-state conditions.

The residence time of a reservoir within the hydrologic cycle is the average time a water molecule will spend in that reservoir (Table 8.1). It is a measure of the average age of the water in that reservoir.

Table 8.1. Average reservoir residence times.

Reservoir	Average residence time
Antarctica	20,000 years
Oceans	3200 years
Glaciers	20 to 100 years
Seasonal snow cover	2 to 6 months
Soil moisture	1 to 2 months
Groundwater: shallow	100 to 200 years
Groundwater: deep	10,000 years
Lakes (see lake retention time)	50 to 100 years
Rivers	2 to 6 months
Atmosphere	9 days

Groundwater can spend over 10,000 years beneath earth's surface before leaving. Particularly old groundwater is called fossil water. Water stored in the soil remains there very briefly, because it is spread thinly across the earth, and is readily lost by evaporation, transpiration, stream flow or groundwater recharge. After evaporating, the residence time in the atmosphere is about 9 days before condensing and falling to the earth as precipitation.

The major ice sheets—Antarctica and Greenland store ice for very long periods. Ice from Antarctica has been reliably dated to 6,50,000 years before present, though the average residence time is shorter.

In hydrology, residence times can be estimated in two ways. The more common method relies on the principle of conservation of mass and assumes the amount of water in a given reservoir is roughly constant. With this method, residence times are estimated by dividing the volume of the reservoir by the rate by which water either enters or exits the reservoir. Conceptually, this is equivalent to timing how long it would take the reservoir to become filled from empty if no water were to leave (or how long it would take the reservoir to empty from full if no water were to enter).

An alternative method to estimate residence times, which is gaining in popularity for dating groundwater, is the use of isotopic techniques. This is done in the subfield of isotope hydrology.

In chemical oceanography, residence time (t) of each element expresses how long it takes to add an amount of the element to the ocean that is equal to the amount of the element in the ocean at steady-state:

t = Mean Concentration in Ocean × Ocean Volume/Input per year where the ocean volume is $(1.37 \times 1021$ L). The input sum all inputs to the ocean. For many elements, the major input is from rivers and the input per year is the Mean River Concentration × Continental Runoff Rate $(3.6 \times 1016$ L/yr). If the concentration of an element is not changing, then the Input and Output of an element must be equal. The residence time can then be calculated using the estimated output, if that is known.

In the designing of chemical reactors, the residence time is regarded as the average time for the processing of the feed in one reactor volume measured at specified conditions. It is also known as space time and is denoted by τ.

It is related to the volume and the volumetric flow rate v in the mathematical relation. The τ is also related to space velocity s, which is the number of reactor volumes of feed treated per unit time at specified conditions.

Residence time not only relates to hydraulic residence time but bacterial residence time as well. It has a symbol Γ (tow). It is the inverse of the eigen value derived from the mass balance method.

Both space time and space velocity are adequate performance measures for the mixed flow reactor and the plug flow reactor. When using the residence time equation, a variety of assumptions are made to reduce the complexity of the system being modelled. These assumptions include, but are not limited to: steady-state inflow and outflow, constant volume, constant temperature, and uniform distribution of the substance throughout the volume of the system. It is also assumed that chemical degradation does not occur in the system in question and that particles do not attach to surfaces that would hinder their flow. If chemical degradation (chemical decomposition) were to occur in a system, the substance that originally entered the system may react with other existing compounds in the system, causing the residence time to be significantly shorter since the substance would be chemically broken down and effectively be removed from the system before it was able to naturally flow out of the system.

APPLICATIONS OF RESIDENCE TIME

Depending on the complexity of the system being modelled and the application for which it is being used, the residence time equation can be altered significantly or even used as a factor.

Engineering

Residence time is widely used across all engineering disciplines, including chemical engineering, biological systems engineering, biomedical engineering, environmental engineering and geological engineering. The residence time formula is adapted for each of these disciplines depending on the system, the complexity, and the substance involved. In environmental engineering, residence time applies to water treatment and waste-water treatment. It refers to the amount of time that water spends in a batch reactor, plug flow reactor, completely mixed flow reactor (CMFR), and/or flocculation tanks. Batch reactors, plug flow reactors, and CMFR's are used in waste-water treatment plants as a means of treating waste-water. Flocculation tanks are part of drinking water treatment facilities where the chemically treated water needs enough time to form flocs before reaching the sedimentation basin. These processes are dependent on an adapted version of residence time. In this situation, the important parameter is how long a concentration of fluid needs to remain in the system for in order to be adequately treated.

$$C = C_0 \times \exp^{(-k \times \tau)}$$

where,

C = concentration.
C_0 = initial concentration.
k = reaction rate constant.
τ = batch reactor residence time.

Here the residence time is being used to determine the changing concentration of a contaminant in a system. This residence time is based on the inflow, outflow, volume, initial concentration of contaminant, the added chemical for treatment, and the rate at which the reactions take place. This is particularly useful for a flash mixer in a water treatment facility to determine if too little or too much of a chemical is initially being introduced into the system.

Environmental

In environmental terms, the residence time definition is adapted to fit with groundwater, the atmosphere, glaciers, lakes, streams, and oceans. Groundwater residence time applications are useful for determining the amount of time it will take for a pollutant to reach and contaminate a groundwater drinking water source and at what concentration it will arrive. This can also work to the opposite effect to determine how long until a groundwater source becomes uncontaminated via inflow, outflow, and volume. The residence time of lakes and streams is important as well to determine the concentration of pollutants in a lake and how this may affect the local population and marine life. Hydro-science, the study of water, discusses the water budget in terms of residence time. The amount of time that water spends in each different stage of life (glacier, atmosphere, ocean, lake, stream, river), is used to show the relation of all of the water on the earth and how it relates in its different forms.

Pharmaceutical

For the medical field, residence time often refers to the amount of time that a pharmaceutical spends in the body. This is dependent on an individual's body size, the rate at which the pharmaceutical will move through and react within the person's body, and the amount of the pharmaceutical administered. The Mean Residence Time (MRT) in Pharmaceuticals deviates from the previous equations as it is based on a statistical derivation. This still runs off a steady-state volume assumption but then uses the area under a distribution curve to find the average drug dose clearance time. The distribution is determined by numerical data derived from either urinary or plasma data collected. Each drug will have a different residence time based on its chemical composition and technique of administration. Some of these drug molecules will remain in the system for a very short time while others may remain for a lifetime. Since individual molecules are hard to trace, groups of molecules are tracked and the distribution of these is plotted to find a mean residence time. The equation for this distribution comes from the following equation:

$$MRT = \frac{\sum_{i=1}^{m} t_i \times n_i}{N}$$

where,
- m = total number of groups.
- t_i = the average time in the body.
- n_i = number of molecules in the ith group.
- N = total number of molecules introduced into the system.

RESIDENCE TIME DISTRIBUTION

The residence time distribution (RTD) of a chemical reactor is a probability distribution function that describes the amount of time a fluid element could spend inside the reactor. Chemical engineers use the RTD to characterise the mixing and flow within reactors and to compare the behaviour of real reactors to their ideal models. This is useful, not only for troubleshooting existing reactors, but in estimating the yield of a given reaction and designing future reactors.

Theory

The theory of residence time distributions generally begins with three assumptions:
1. The reactor is at steady-state.
2. Transports at the inlet and the outlet takes place only by advection.
3. The fluid is incompressible.

The incompressibility assumption is not required, but compressible flows are more difficult to work with and less common in chemical processes. A further level of complexity is required for multi-phase reactors, where a separate RTD will describe the flow of each phase, for example bubbling air through a liquid.

The distribution of residence times is represented by an exit age distribution, $E(t)$. The function $E(t)$ has the units of time^{-1} and is defined such that:

$$\int_0^\infty E(t)dt = 1.$$

The fraction of the fluid that spends a given duration, t inside the reactor is given by the value of $E(t)dt$.

The fraction of the fluid that leaves the reactor with an age less than t_1 is:

$$\int_0^{t_1} E(t)dt.$$

The fraction of the fluid that leaves the reactor with an age greater than t_1 is:

$$\int_{t_1}^\infty E(t)dt = 1 - \int_0^{t_1} E(t)dt.$$

The average residence time is given by the first moment of the age distribution:

$$\bar{t} = \int_0^\infty t \cdot E(t)dt.$$

If there are no dead or stagnant, zones within the reactor then $\bar{t}$ will be equal to τ, the residence time calculated from the total reactor volume and the volumetric flow rate of the fluid:

$$\tau = \frac{V}{\upsilon}.$$

The higher order central moments can provide significant information about the behaviour of the function $E(t)$. For example, the second central moment indicates the variance (σ^2), the degree of dispersion around the mean:

$$\sigma^2 = \int_0^\infty (t - \bar{t})^2 \cdot E(t)dt$$

The third central moment indicates the skewness of the RTD and the fourth central moment indicates the kurtosis (the peakedness).

One can also define an internal age distribution $I(t)$ that describes the reactor contents. This function has a similar definition as $E(t)$: the fraction of fluid within the reactor with an age of t is $I(t)dt$. As shown by Danckwerts, the relation between $E(t)$ and $I(t)$ can be found from the mass balance:

$$I(t) = \frac{1}{\tau}\left(1 - \int_0^t E(t)dt\right) \qquad E(t) = -\tau \frac{dI(t)}{dt}$$

Determining the RTD Experimentally

Residence time distributions are measured by introducing a non-reactive tracer into the system at the inlet. The concentration of the tracer is changed according to a known function and the response is found by measuring the concentration of the tracer at the outlet. The selected tracer should not modify the physical characteristics of the fluid (equal density, equal viscosity) and the introduction of the tracer should not modify the hydrodynamic conditions.

In general, the change in tracer concentration will either be a pulse or a step. Other functions are possible, but they require more calculations to deconvolute the RTD curve, $E(t)$.

Pulse experiments

This method required the introduction of a very small volume of concentrated tracer at the inlet of the reactor, such that it approaches the dirac delta function. Although an infinitely short injection cannot be produced, it can be made much smaller than the mean residence time of the vessel. If a mass of tracer, M, is introduced into a vessel of volume V and an expected residence time of τ, the resulting curve of $C(t)$ can be transformed into a dimensionless residence time distribution curve by the following relation:

$$E(t) = \frac{C(t)}{\int_0^\infty C(t)\,dt}$$

Step experiments

In a step experiment the concentration of tracer at the reactor inlet changes abruptly from 0 to C_0. The concentration of tracer at the outlet is measured and normalised to the concentration C_0 to obtain the non-dimensional curve $F(t)$ which goes from 0 to 1:

$$F(t) = \frac{C(t)}{C_0}.$$

The step- and pulse-responses of a reactor are related by the following:

$$F(t) = \left(\int_0^t E(t)\,dt \right) \qquad E(t) = \frac{dF(t)}{dt}$$

The value of the mean residence time and the variance can also be deduced from the function $F(t)$:

$$\bar{t} = \int_0^\infty t \cdot E(t)\,dt = \int_0^1 t\,dF(t) = -\int_0^1 t\,d(1-F(t)) = \int_0^\infty (1-F(t))\,dt$$

$$\sigma^2 = \int_0^\infty (t-\bar{t})^2 \cdot E(t)\,dt = \int_0^1 (t-\bar{t})^2\,dF(t) = \int_0^1 t^2\,dF(t) - \bar{t}^2 = 2\int_0^\infty t(1-F(t))\,dt - \bar{t}^2$$

A step experiment is often easier to perform than a pulse experiment, but it tends to smooth over some of the details that a pulse response could show. It is easy to numerically integrate an experimental pulse response to obtain a very high-quality estimate of the step response, but the reverse is not the case because any noise in the concentration measurement will be amplified by numeric differentiation.

RTDs of Ideal and Real Reactors

The residence time distribution of a reactor can be used to compare its behaviour to that of two ideal reactor models: the plug-flow reactor and the continuous stirred-tank reactor (CSTR) or mixed-flow reactor. This characteristic is important in order to calculate the performance of a reaction with known kinetics.

Plug flow reactors

In an ideal plug-flow reactor there is no mixing and the fluid elements leave in the same order they arrived. Therefore, fluid entering the reactor at time t will exit the reactor at time $t + \tau$, where, τ is the residence time of the reactor. The residence time distribution function is, therefore, a dirac delta function at τ.

$$E(t) = \delta(t - \tau)$$

The variance of an ideal plug-flow reactor is zero. The RTD of a real reactor deviate from that of an ideal reactor, depending on the hydrodynamics within the vessel. A non-zero variance indicates that there is some dispersion along the path of the fluid, which may be attributed to turbulence, a non-uniform velocity profile or diffusion. If the mean of the $E(t)$ curve arrives earlier than the expected time τ it indicates that there is stagnant fluid within the vessel. If the RTD curve shows more than one main peak it may indicate channelling, parallel paths to the exit or strong internal circulation.

Mixed flow reactors

An ideal continuous stirred-tank reactor is based on the assumption that the flow at the inlet is completely and instantly mixed into the bulk of the reactor. The reactor and the outlet fluid have identical, homogeneous compositions at all times. An ideal CSTR has an exponential residence time distribution:

$$E(t) = \frac{1}{\tau} e^{-t/\tau}$$

In reality, it is impossible to obtain such rapid mixing, especially on industrial scales where reactor vessels may range between 1 and several tens of cubic meters, and hence the RTD of a real reactor will deviate from the ideal exponential decay. For example, there will be some finite delay before $E(t)$ reaches its maximum value and the length of the delay will reflect the rate of mass transfer within the reactor. Just as was noted for a plug-flow reactor, an early mean will indicate some stagnant fluid within the vessel, while the presence of multiple peaks could indicate channelling, parallel paths to the exit, or strong internal circulation. Short-circuiting fluid within the reactor would appear in an RTD curve as a small pulse of concentrated tracer that reaches the outlet shortly after injection.

RESIDENCE TIME DISTRIBUTION MEASUREMENTS

Residence time distribution (RTD) measurements provide an effective technique to diagnose flow behaviour within a wide range of flow systems. RTD analysis of a controlled tracer addition into a system can reveal flow distribution characteristics such as transit times, short-circuiting, re-circulation zones, and dead zones. The system under study can range from reactants flowing through a process vessel to bioagents dispersing through a building. RTD methods may also be used to develop and validate flow models, and can be applied to field and lab-scale systems from the size of millimetres to several metres representing residence times of milliseconds to hours.

RTD Technique

For a simple RTD experiment, a known tracer distribution is introduced into the inlet of a system, and the tracer concentration is recorded at the outlet after it has been modified by the system processes. Analysis of this convoluted output tracer distribution allows insight into the processes that brought those changes about. In addition to this simple case, tracer addition and detection may be performed at locations other than the system inlet and outlet, allowing for the isolation of particular flow phenomena of interest. Several examples of system behaviours are illustrated in Fig. 8.1. In the plug flow reactor shown in Figure 8.1(a) the inlet tracer distribution is not modified at all by the system process, and is simply convected through while maintaining its original shape. Experimental data demonstrating this effect is shown in Fig. 8.2 where the configuration 1 outlet RTD is identical to the inlet step RTD. If short circuiting occurs, as in the baffled reactor shown in Fig. 8.1(b), the step change moves through the system much faster than anticipated due to the nonparticipation of a large portion of the vessel volume.

In the well-mixed system shown in Fig. 8.1(c) the tracer concentration is always uniformly distributed throughout the flow volume as tracer is added and exhausted. In this case the response to a step input is an exponential rise beginning at the moment of tracer introduction.

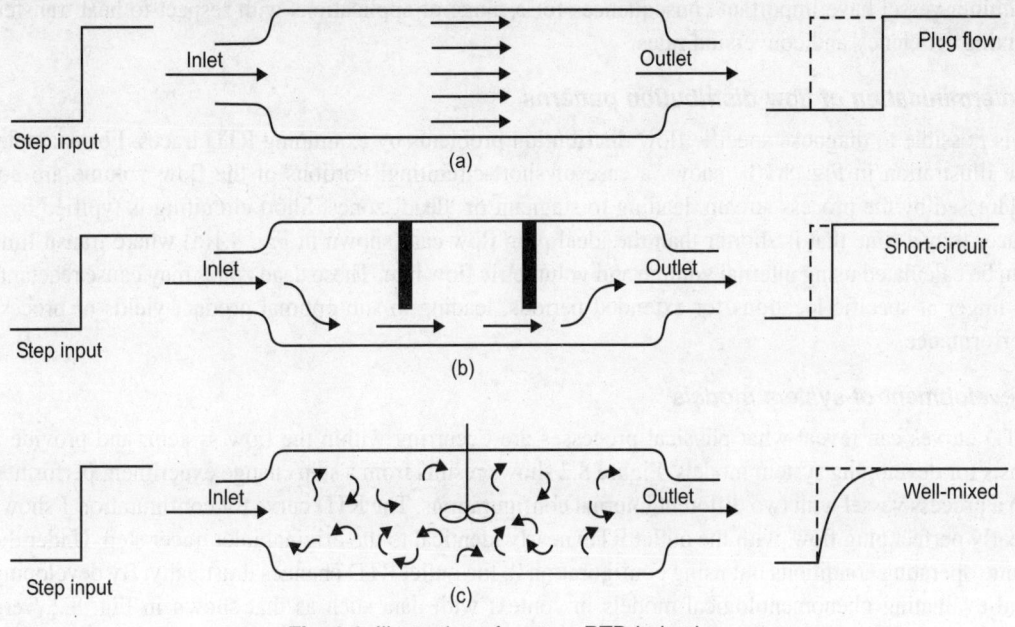

Fig. 8.1. Illustration of system RTD behaviour.

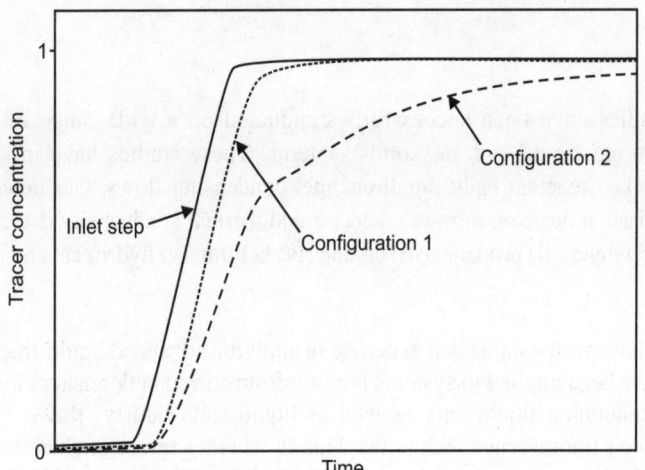

Fig. 8.2. Experimental step-change RTD obtained from process vessel.

Application of RTD Methods

RTD applications range from the characterisation of flow systems, to determining flow distribution patterns, to developing system models. These elements are combined in an integrated approach designed to address the specific needs of a particular client.

Characterisation of flow systems

Knowledge of the type of flow behaviour occurring within a process vessel allows for verification of a system design or indication of possible design improvements. For example, the mixing characteristics within a vessel have important consequences for a range of applications with respect to heat transfer, mixing efficiency and conversion rates.

Determinination of flow distribution patterns

It is possible to diagnose specific flow distribution problems by examining RTD traces. For example, the illustration in Fig. 8.1(b) shows a case of short-circuiting. Portions of the flow volume are not addressed by the process stream, leading to stagnant or 'dead' zones. Short circuiting is typified by a tracer transit time that is shorter than the ideal plug flow case shown in Fig. 8.1(a) where transit time can be calculated using internal volume and volumetric flow rate. These dead zones may cause reactants to linger at specific locations for extended periods, leading to sub-optimal product yields or process performance.

Development of system models

RTD curves can reveal what physical processes are occurring within the flow system, and provide a basis for developing system models. Figure 8.2 shows results from a step change experiment performed on a process vessel with two different internal configurations. The RTD curve for configuration 1 shows nearly perfect plug flow, with the outlet RTD nearly identical to the original inlet tracer step. Under the same operating conditions but using configuration 2, the outlet RTD changes drastically. By developing and evaluating phenomenological models in context with data such as that shown in Fig. 8.2, very subtle and complex behaviours within the flow system can be captured and predicted.

Tracer Methods

Gas tracing

Gas-tracer RTD studies have been successfully conducted on a wide range of systems including gas-only, multiphase gas-liquid and gas-solids systems. These studies have included field and lab measurements in packed reactant beds, small-channel condensing flows. Gas detection systems have included a photo-ionisation detector, infra-red detector and thermal conductivity detector. Available tracer gases include: (i) propylene, (ii) propane, (iii) ethane (iv) helium, (v) hydrogen, and (vi) carbon dioxide.

Liquid tracing

Now the companies have well-established expertise in applying advanced liquid-tracer RTD techniques. These techniques have been applied to systems ranging from stirred tank reactors to inertial mixing and separation vessels containing liquid-only as well as liquid-solid (slurry) flows. Primary diagnostics include a laser induced fluorescence technique. This is where a tracer of fluorescent dye solution is introduced into the liquid system and then traced using laser illumination. In addition, conductivity probes are available that use saline solution as a tracer, and are effective in obscured flows or opaque liquids.

Solids tracing

A novel method of solids tracer addition and sampling has been developed for performing solid-phase RTD studies in gas–solid (fluidised bed) reactors. This technique was successfully applied to the fluidised bed. Other available techniques for solids tracing include heat tracing as well as fluorescent or visible dye tracing.

RTD MEASUREMENT ON A SUBMERGED BIO-REACTOR USING A RADIOISOTOPE TRACER AND THE RTD ANALYSIS

The residence time distribution (RTD) function was defined by Danckwerts using the application of the population balance principle. The knowledge of the RTD function is very important because it provides information about the fraction of the fluid that spends a certain time in a reactor. A considerable bypassing flow is an indication of poor design, and residual stagnant regions may serve to cut down the effective or useful volume of the reactor. Therefore, a lot of RTD models have been investigated for the last decade as a useful means of understanding mixing characteristics, such as a bypassing flow or the amount of stagnant regions in a reactor. The radioisotope tracer is one of the efficient methods for RTD function determination especially in waste-water treatment facilities where the waste-water is opacified and its colour is very dark.

This section describes a method of obtaining the parameters of RTD model that represents the flow behaviour and the mixing characteristics of a reactor. The numerical approach allows the implementation of time domain based parameter estimation for the evaluation of RTD model parameters. Newton's method, one of the least squares estimation methods, has several drawbacks that can cause numerical difficulties.

This section suggests that drawbacks of Newton's method have been improved by introducing Marquardt's constant. In addition, by fitting a RTD model response to the RTD response obtained from the radioisotope tracer experiment, the submerged bio-reactor is analysed and diagnosed.

RTD Model

A lot of mathematical RTD models have been developed to represent RTD characteristics of a reactor. The RTD model with a stagnant region will be used in this section to know the stagnant region that may exist the submerged bio-reactor.

Model with a stagnant region

This model was suggested by Adler and Hovorka and is sketched in Fig. 8.3.

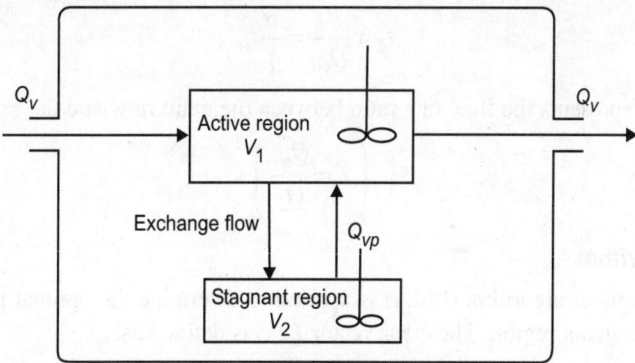

Fig. 8.3. Model with a stagnant region.

This model has a slow exchange or cross flow between the fluid in the stagnant region and the active fluid passing through the reactor. The stagnant region is considered as a perfect mixer and slowly interchanges fluid with the active flow region. This model is expressed as the following differential equation form.

$$V_1 \frac{dc_1(t)}{dt} = Q_v \delta(t) + c_2(t) Q_{vp} - c_{1(t)} \left[Q_v + Q_{vp} \right], \qquad \dots (8.1)$$

$$V_2 \frac{dc_2(t)}{dt} = Q_{vp} \left[c_1(t) - c_2(t) \right] \qquad \dots (8.2)$$

where,

V_1 = is the volume of the active region (m³) ,
Q_v = is the flow rate of the main flow (m³/s) ,
V_2 = is the volume of the stagnant region (m³),
Q_{vp} = is the flow rate of the exchange flow(m³/s).

The above equations can be rewritten by dividing Eq. 8.1 by Q_v and Eq. 8.2 by Q_{vp}.

$$t_1 \frac{dc_1(t)}{dt} = \delta(t) + f c_2(t) - (1 + f) c_1(t), \qquad \dots (8.3)$$

$$t_2 \frac{dc_2(t)}{dt} = c_1(t) - c_2(t). \qquad \dots (8.4)$$

The t_1 in Eq. 8.3 represents the mean residence time of the active region:

$$t_1 = \frac{V_1}{Q_v} = \frac{\bar{t}}{(1+\alpha)} \qquad \dots (8.5)$$

The $\bar{t}$ in Eq. 8.5 is the mean residence time of this model and the α represents the volume ratio between the active region and the stagnant region:

$$\left(\alpha = \frac{V_2}{V_1} \right)$$

The t_2 in Eq. 8.4 represents the mean residence time of the stagnant region:

$$t_2 = \frac{V_2}{Q_{vp}} = \frac{\alpha}{f} t_1. \qquad \dots (8.6)$$

The f in Eq. 8.6 represents the flow rate ratio between the main flow and the exchange flow:

$$\left(f = \frac{Q_{vp}}{Q_v} \right)$$

Optimisation algorithm

The Levenberg-Marquardt algorithm (LMA) is applied to determine the optimal parameter set of the RTD model with a stagnant region. The error vector $F(k)$ is defined as:

$$F_{(k)} = (Y^s - Y^c(k)) \qquad \dots (8.7)$$

where, $Y^c(k)$ is the RTD model response obtained from the k-th parameter vector $p(k)$ and Y^s is the output response obtained from an experiment. $F(k)$ can be linearly approximated by Taylor's expansion to compute the $(k+1)$-th parameter vector $p(k+1)$ from the k-th parameter vector $p(k)$.

$$F(k+1) = F(k) + J(k)[p(k+1) - p(k)] \qquad \text{... (8.8)}$$

where, $J(k)$ is a gradient vector of $F(k)$.

The cost function $f(k+1)$ is defined by using linear approximation of $F(k+1)$:

$$f(k+1) = \frac{1}{2} F(k+1)^T F(k+1). \qquad \text{... (8.9)}$$

$F(k+1)$ in Eq. 8.9 is substituted for Eq. 8.8 and then the quadratic form of Eq. 8.9 becomes:

$$f(k+1) = \frac{1}{2}\left[F(k) + J(k)\Delta p(k)\right]^T \left[F(k) + J(k)\Delta p(k)\right] \qquad \text{... (8.10)}$$

where, $\Delta p(k) = p(k+1) - p(k)$.

To calculate the gradient vector of $f(k+1)$, it is differentiated with respect to $\Delta p(k)$:

$$\nabla f(k+1) = J(k)^T F(k) + J(k)^T J(k)\Delta p(k). \qquad \text{... (8.11)}$$

We set $\nabla f(k+1) = 0$ in Eq. 8.11 to determine $p(k+1)$.

$$\Delta p(k) = -\left[J(k)^T J(k)\right]^{-1} J(k)^T F(k). \qquad \text{... (8.12)}$$

Using the relation of:

$$\Delta p(k) = p(k+1) - p(k),$$

$p(k+1)$ becomes:

$$p(k+1) = p(k) - \left[J(k)^T J(k)\right]^{-1}\left[J(k)^T F(k)\right]. \qquad \text{... (8.13)}$$

The Levenberg-Marquardt algorithm (LMA) adds λ to Eq. 8.13 to improve stability and convergence (Fig. 8.4):

$$p(k+1) = p(k) - \left[J(k)^T J(k) + \lambda(k)I\right]^{-1}\left[J(k)^T F(k)\right] \qquad \text{... (8.14)}$$

where, $\lambda(k)$ is Marquardt's constant and I is an identity matrix. This LMA is executed by the following steps:

Step 1: Set the iteration index as $k = 0$ and select the parameter vector $p(0)$ with an engineering point view. Also, select a tolerance ε as the stoping criterion and $\lambda(0)$ as a constant.

Step 2: Calculate the RTD model response $Y^c(k)$ using the k-th parameter vector $p(k)$, and then calculate the error function $F(k)$ and the cost function $f(k)$.

Step 3: Calculate the gradient vector $J(k)$ of the error function and calculate the $(k+1)$-th parameter vector by Eq. 8.14.

Step 4: If $\|\Delta p(k)\| < \varepsilon$, stop the iterative process. Otherwise, continue.

Step 5: If $f(k+1) < f(k)$, go to Step 6. Otherwise, let $\lambda(k) = 10\lambda(k)$ and go to Step 3.

Step 6: Set $\lambda(k+1) = 0.1\lambda(k)$, let the iteration index be $k = k+1$, and go to Step 2.

Radioisotope Tracer Experiment

The waste-water treatment facility consists of six compartments as shown in Fig. 8.5. A radioisotope tracer experiment was carried out in the submerged bio-reactor processing a dye waste-water. This section investigated the first of six compartments. The volume of the submerged bio-reactor is 273.3 (m³) and the dye-waste water is flowing into the reactor.

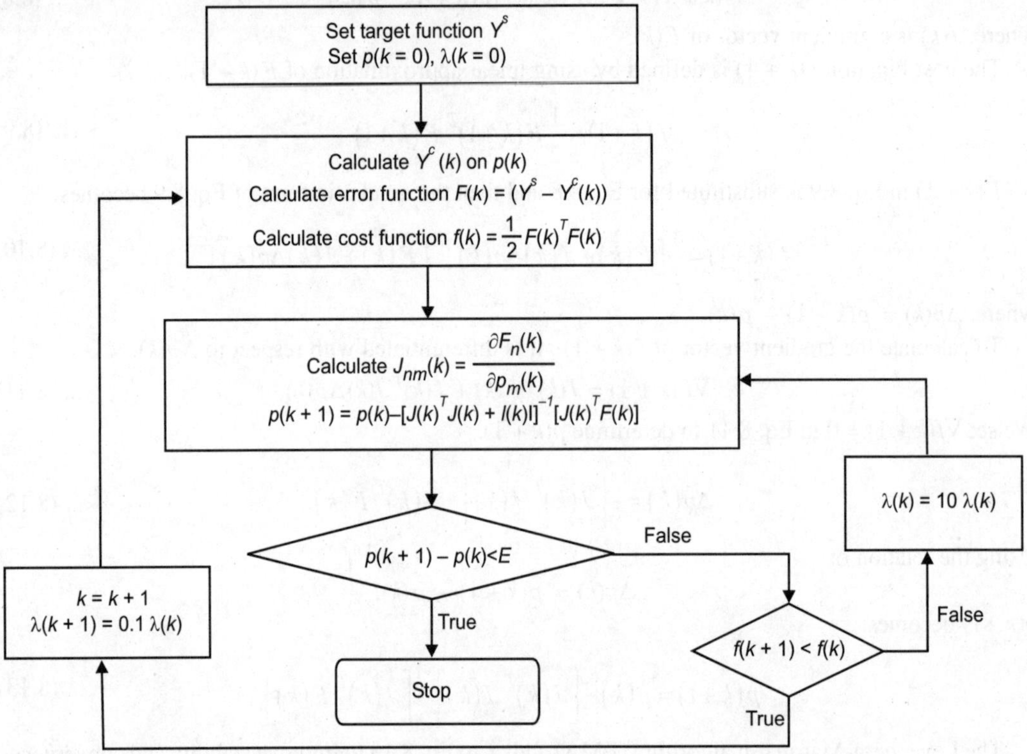

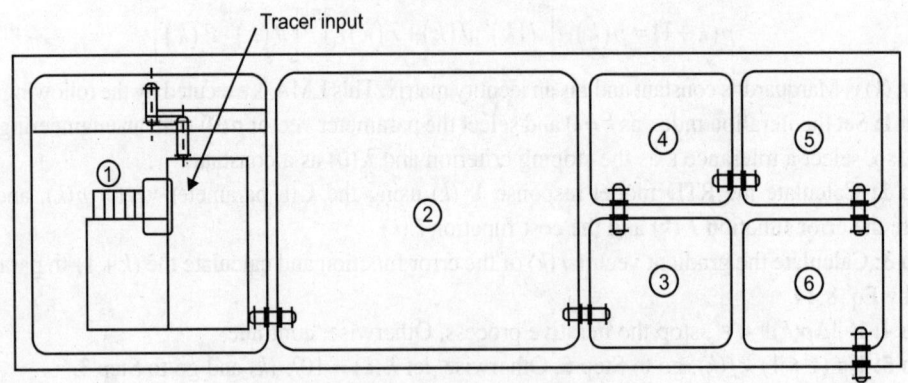

Set target function Y^s
Set $p(k = 0)$, $\lambda(k = 0)$

Calculate $Y^c(k)$ on $p(k)$
Calculate error function $F(k) = (Y^s - Y^c(k))$
Calculate cost function $f(k) = \dfrac{1}{2} F(k)^T F(k)$

Calculate $J_{nm}(k) = \dfrac{\partial F_n(k)}{\partial p_m(k)}$
$p(k + 1) = p(k) - [J(k)^T J(k) + I(k)I]^{-1} [J(k)^T F(k)]$

$\lambda(k) = 10\,\lambda(k)$

$p(k + 1) - p(k) < E$

False

True

$k = k + 1$
$\lambda(k + 1) = 0.1\,\lambda(k)$

False

Stop

$f(k + 1) < f(k)$

True

Fig. 8.4. Represents the flow chart of the optimisation algorithm using the LMA.

Tracer input

① ② ③ ④ ⑤ ⑥

Fig. 8.5. The waste-water treatment facility.

Tracer experiment

Approximately 20 (mCi) of ^{131}I as a tracer was instantaneously injected at the inlet of the submerged bio-reactor. To measure the change of the radioisotope tracer concentration by detecting gamma radiation emitted from the tracer, NaI (Tl) scintillation detectors (SPA-3, Eberline) were installed at the outlets of each compartment.

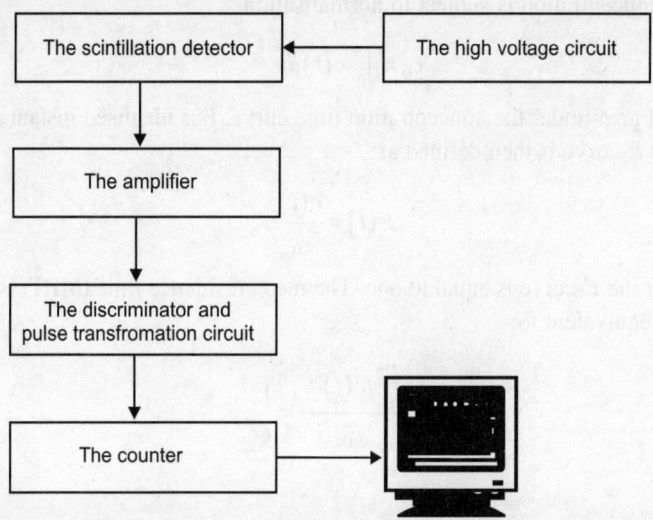

Fig. 8.6. Representation for the measurement device for λ-radiation.

Experimental equipments

The NaI(Tl) scintillation detector (SPA-3, Eberline) was used to measure gamma radiation. The gamma radiation is converted into visible light by the scintillation, and the light is converted into electric signals by the photo-multiplier. The signals from the photo-multiplier are amplified and transformed into TTL-level pulses by the discriminator and pulse transformation circuit. Then these pulses are counted by the rate meter for a preset counting time and transferred into a notebook computer via serial communication (RS-232C).

Experiment Result and RTD Simulation

The experimental data was recorded with a sampling time of 60 [s] and the number of recorded data was 3184. To use the experimental data for the RTD analysis of the bio-reactor, the background radiation and the spontaneous decay should be corrected.

Data correction

The radioisotope tracer concentration obtained from the experiment was corrected to decrease the error arising from the background radiation and the spontaneous decay of the radioisotope. Let $c_m(t)$ be the radioisotope tracer concentration obtained from the experiment and $c_{bc}(t)$ be the radioisotope tracer concentration after subtracting the influence of the background radiation from $c_m(t)$. Then the concentration $c(t)$, which is corrected the effect of spontaneous decay, is given by:

$$c(t) = c_{bc}(t)e^{\frac{ln2}{T_{half}t}} \qquad \qquad \text{... (8.15)}$$

where, $c_{bc}(t) = c_m(t) - c_{background}$ and T_{half} is the half-life of the radioisotope.

Normalisation

To compare the RTD response obtained from the experiment with the RTD model response, the radioisotope tracer concentration is subject to normalisation.

$$c_0 = \int_0^\infty c(t)dt \qquad \text{... (8.16)}$$

where, c_0 is the total area under the concentration time curve. For idealised instantaneous pulse input, the RTD function or E-curve is then defined as:

$$E(t) = \frac{c(t)}{c_0} \qquad \text{... (8.17)}$$

where the area under the E-curve is equal to one. The mean residence time (MRT) is determined as the first moment and is equivalent to:

$$\bar{t} = \frac{\int_0^\infty tc(t)dt}{c_0} = \frac{V}{Q}. \qquad \text{... (8.18)}$$

Simulation

The radioisotope tracer concentration obtained from the experiment was corrected by Eq. 8.15. Figure 8.7 represents the experimental data and the corrected data. In addition, the corrected data was normalised by Eq. 8.17 to compare it with the RTD model response. The RTD model suggested by Adler-Hovorka was used and the differential equation was solved by the four-stage Runge-Kutta method. The parameters of the model is determined by fitting the model response to the response obtained from the experiment. The LMA method was applied to determine optimal parameters.

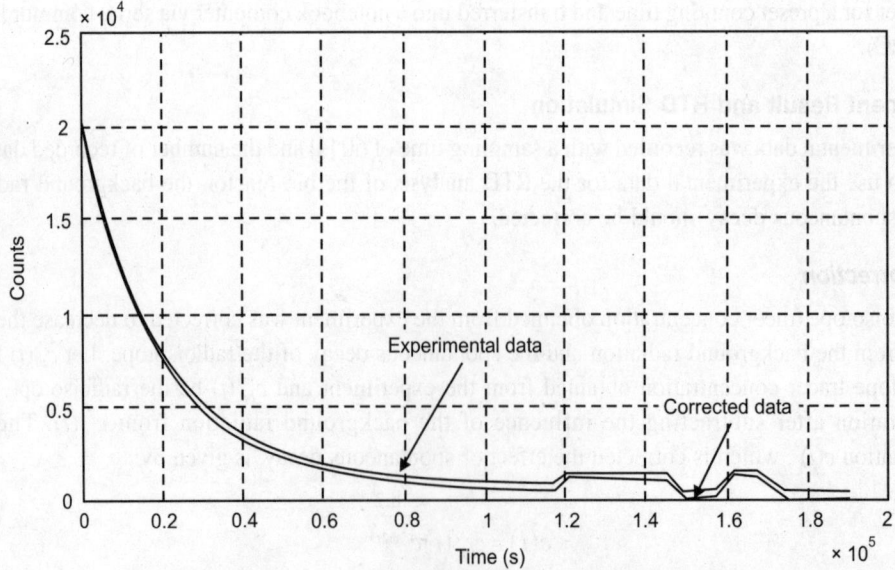

Fig. 8.7. Comparison between experimental data and corrected data.

The initial values were $t_1 = 10000(s)$, $t_2 = 23000$ (s), $f = 0.4$, and the specified tolerance $\varepsilon = 0.001$ for the stop criterion of RTD simulation. The optimal parameter values are calculated by adjusting the initial parameter values in an iterative procedure. The process continues until the RTD model response matches the output response obtained from the experiment within a specified tolerance. The simulation was iterated forty times to reach convergence. Figure 8.8 represents the change in parameters according to the iteration number during the RTD simulation using the LMA. The simulation result is summarised in Table 8.2.

Table 8.2. The parameters of the RTD model.

MRT t_1	MRT t_2	$f = \dfrac{Q_{vp}}{Q_v}$
25165 [s]	66347 [s]	0.2943

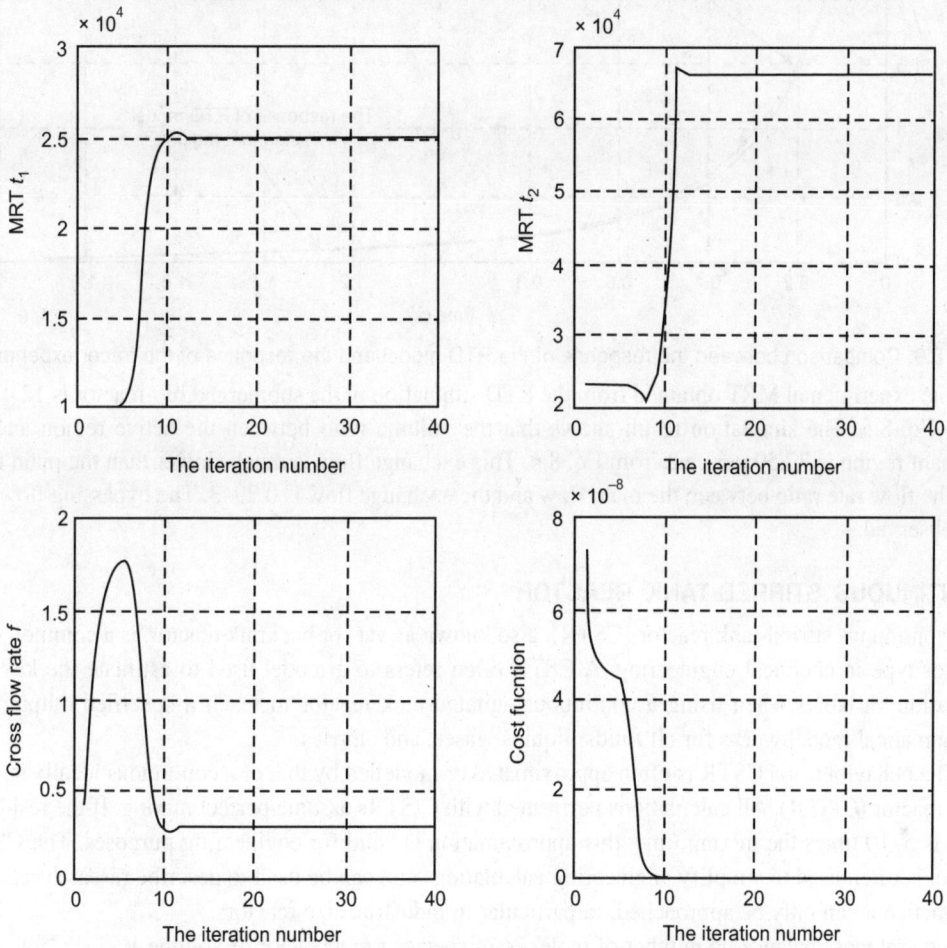

Fig. 8.8. The parameters change according to the iteration number.

The RTD simulation result is shown in Fig. 8.9. The root mean square error between the response of the radioisotope tracer experiment and the response of the RTD model is 6.6318×10^{-7}.

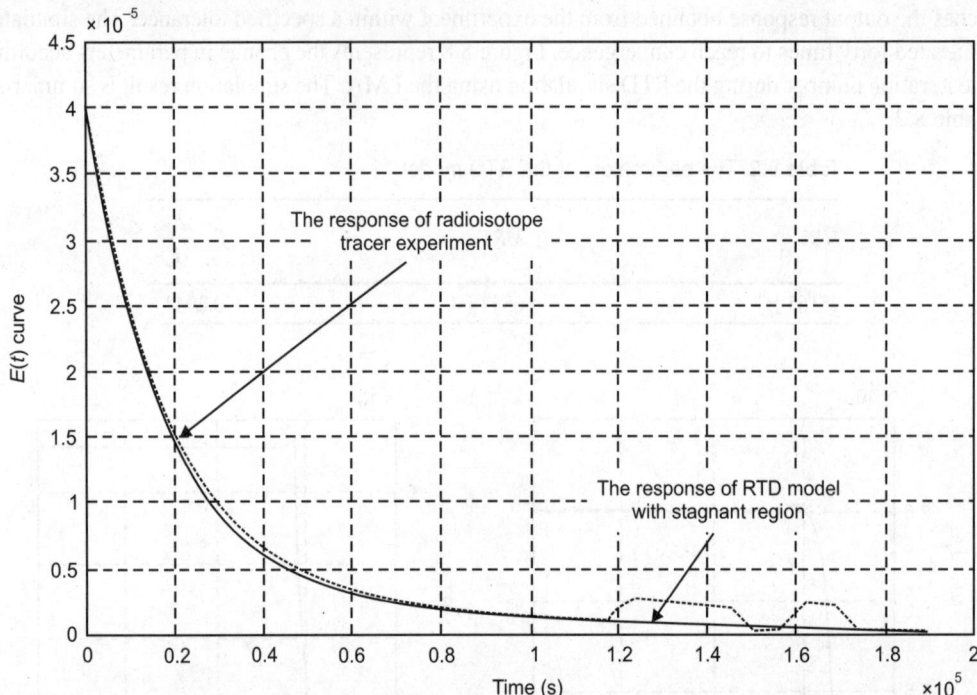

Fig. 8.9. Comparison between the response of the RTD model and the response of the tracer experiment.

The experimental MRT obtained from the RTD simulation in the submerged bio-reactor is 12.41 (h) from Eq. 8.5. The simulation result shows that the volume ratio between the active region and the stagnant region is 77.59 per cent from Eq. 8.6. This exchange flow is much slower than the main flow, and the flow rate ratio between the main flow and the exchange flow is 0.2943. The bypassing flow was not observed.

CONTINUOUS STIRRED-TANK REACTOR

The continuous stirred-tank reactor (CSTR), also known as vat- or backmix reactor, is a common ideal reactor type in chemical engineering. A CSTR often refers to a model used to estimate the key unit operation variables when using a continuous agitated-tank reactor to reach a specified output. The mathematical model works for all fluids: liquids, gases, and slurries.

The behaviour of a CSTR is often approximated or modelled by that of a continuous ideally stirred-tank reactor (CISTR). All calculations performed with CISTRs assume perfect mixing. If the residence time is 5–10 times the mixing time, this approximation is valid for engineering purposes. The CISTR model is often used to simplify engineering calculations and can be used to describe research reactors. In practice it can only be approached, in particular in industrial size reactors.

Integral mass balance on number of moles N_i of species i in a reactor of volume V.

$$[\text{Accumulation}] = [\text{in}] - [\text{out}] + [\text{generation}]$$

$$\frac{dN_i}{dt} = F_{io} - F_i + V_{vi}r_i \qquad \qquad \text{... (8.19)}$$

where, F_{io} is the molar flow rate inlet of species i, F_i the molar flow rate outlet, and v_i stoichiometric coefficient. The reaction rate, r, is generally dependent on the reactant concentration and the rate constant (k). The rate constant can be figured by using the Arrhenius temperature dependence. Generally, as the temperature increases so does the rate at which the reaction occurs. Residence time, τ, is the average amount of time a discrete quantity of reagent spends inside the tank.

Assume:

1. Constant density (valid for most liquids; valid for gases only if there is no net change in the number of moles or drastic temperature change).
2. Isothermal conditions, or constant temperature (k is constant).
3. Steady-state.
4. Single, irreversible reaction ($v_A = -1$).
5. First-order reaction ($r = kC_A$).

$$A \rightarrow \text{products}$$

$N_A = C_A V$ (where C_A is the concentration of species A, V is the volume of the reactor, N_A is the number of moles of species A):

$$C_A = \frac{C_{Ao}}{1 + k\tau} \qquad \qquad \text{... (8.20)}$$

The values of the variables, outlet concentration and residence time, in Eq. 8.20 are major design criteria. To model systems that do not obey the assumptions of constant temperature and a single reaction, additional dependent variables must be considered. If the system is considered to be in unsteady-state, a differential equation or a system of coupled differential equations must be solved.

CSTR's are known to be one of the systems which exhibit complex behaviour such as steady-state multiplicity, limit cycles and chaos.

In environmental engineering, a continuous or continuously stirred tank reactor (CSTR) is a system that has the following properties:

1. There is inflow and outflow or matter.
2. Chemical reactions occur within the system's boundary.
3. The accumulation rate of any substance.
4. The system is in a steady-state, i.e. the concentration of any substance remains constant or equivalently, the accumulation rate is zero.
5. Any substance on the system is assumed to be homogeneously distributed.

With the above assumptions the law of conservation of mass can be written in the generic form:

Accumulation rate = 0 = input rate – output rate + reaction rate.

RTD IN IDEAL REACTORS

Batch and Continuous Flow Reactors

Batch and continuous flow reactors:

1. The productivity of continuous flow reactors, as defined in this problem, may be evaluated by solving the conversion expressions for Q.

In plug flow, this gives:

$$\frac{C_{A,out}}{C_{A,in}} = \exp\left(-\frac{kV}{Q}\right)$$

$$Q = \frac{kV}{\ln\dfrac{C_{A,\,out}}{C_{A,\,in}}}$$

And for perfect mixing/CSTR:

$$\frac{C_{A,\,out}}{C_{A,\,in}} = \frac{1}{1+\dfrac{kV}{Q}}$$

$$1+\frac{kV}{Q} = \frac{C_{A,\,in}}{C_{A,\,out}}$$

$$Q = \frac{kV}{\dfrac{C_{A,\,in}}{C_{A,\,out}}-1}$$

2. For the batch reactor with downtime t_d and residence time t_R, the productivity is the volume divided by the total time $V/(t_R + t_d)$. The residence time can be determined from the batch conversion equation:

$$\frac{C_{A,\,out}}{C_{A,\,in}} = e^{-kt_R}$$

$$t_R = -\frac{1}{k}\ln\frac{C_{A,\,out}}{C_{A,\,in}}$$

The productivity is then:

$$Q = \frac{V}{t_R+t_d} = \frac{V}{-\dfrac{1}{k}\ln\dfrac{C_{A,\,out}}{C_{A,\,in}}+t_d} = \frac{kV}{-\ln\dfrac{C_{A,\,out}}{C_{A,\,in}}+kt_d}$$

In the limit where t_d/t_R goes to zero, the second term in the denominator vanishes and this approaches the productivity of the plug flow reactor.

Another way to express this is:

$$Q = \frac{V}{t_R+t_d} = \frac{V}{t_R}\frac{t_R}{t_R+t_d} = \frac{kV}{\ln\dfrac{C_{A,\,out}}{C_{A,\,in}}}\frac{t_R}{t_R+t_d}$$

This is fine, but since t_R is a dependent variable, putting it on the right side does not allow direct calculation of the productivity from reactor characteristics as can be done with the expression above.

3. In this case with a target $C_{A, \text{out}}/C_{A, \text{in}} = 0.05$, the plug flow reactor's productivity is:

$$Q = \frac{kV}{\ln 0.05} = \frac{kV}{3.0}$$

And for perfect mixing/CSTR:

$$Q = \frac{kV}{\dfrac{1}{0.05} - 1} = \frac{kV}{19}$$

For a batch reactor:

$$Q = \frac{kV}{-\ln \dfrac{C_{A, \text{out}}}{C_{A, \text{in}}} + kt_d} = \frac{kV}{3.0 + kt_d}$$

The plug flow reactor is clearly $\dfrac{1/3}{1/19} = 6.34$ times more productive than the perfect mixing reactor. And since h_D, A, V and t_d are all positive, the batch reactor will have slightly lower productivity than the plug flow reactor as well, with the ratio depending on the downtime.

4. The perfect mixing reactor is likely to give the highest quality output for a couple of reasons:
    (a) In the batch reactor, significant concentration in homogeneities can develop, particularly if k is temperature-dependent and the reaction gives off or consumes a significant amount of heat (much more so in a heterogeneous/surface reactor). Reactions with the containment vessel can also introduce location-dependent variations in composition. For these reasons, it is often desirable to stir a batch reactor.
    (b) A plug flow reactor will always exhibit significant concentration inhomogeneities logitudinally, and sometimes across the reactor for the same reason as the batch reactor. If flow conditions deviate at all from steady-state, changes in flow patterns can result in inhomogeneity in the output.
    (c) If the raw material in a plug flow reactor enters at varying concentration, that variation will carry through to the end of the reactor, and show up as inhomogeneous output. A perfect mixing reactor, on the other hand, will rapidly mix in any concentration departures in the raw material, smoothing out their influence over time.

5. Batch reactors have several advantages over continuous flow reactors:
    (a) They can be stopped between batches, so the production rate is flexible and can be varied if economically desirable. If a continuous reactor is stopped, it takes time and wastes material to return it to consistent steady-state production.
    (b) Batch reactors are also more flexible, in that one can easily use different compositions in different batches to produce products with different specifications.
    (c) If the process degrades the reactor in some way, a batch reactor can be cleaned, relined, etc. between batches, where continuous reactors must run for a long time before that can be done.
    (d) Because flaws tend to be isolated to a given batch, one can track the quality more easily in a batch process. For example, when the cause of a plane is traced to a flaw in a titanium part, it

is common practice to ground all planes with parts made from the same vacuum arc remelt (VAR) titanium ingot, and other defective parts are often found in the subsequent inspections.

(e) If the reactants are stirred, a batch reactor can often achieve better quality than a plug flow reactor (which cannot be well-mixed in order to maintain plug flow), and better productivity than a continuous flow reactor.

Ideal Plug Flow Reactor

Characteristics of ideal plug flow are:
1. Perfect mixing in the radial dimension (uniform cross section concentration).
2. No mixing in the axial direction, or no axial dispersion (segregated flow)

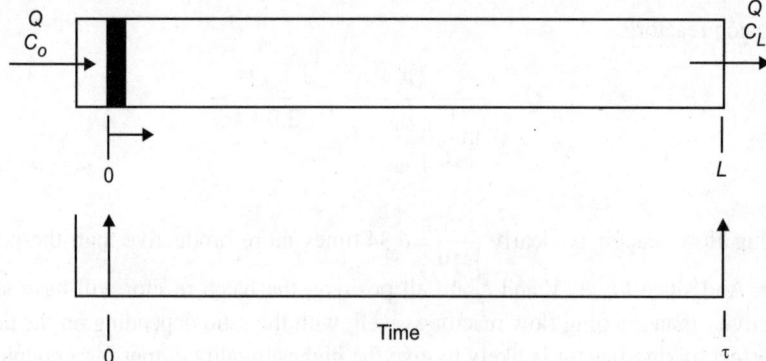

Tracer pulse input at $t = 0$ translated to equal pulse output at $t = \tau$, $\tau = L/v$ (L = PFR length, v = average velocity).

Compare with CSTR response to tracer pulse dispersion:

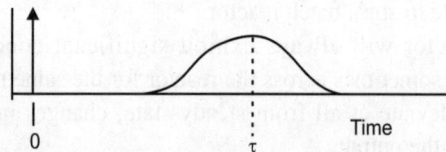

In an ideal PFR, concentration is a function of both distance along the flow path, x, and time, t:

$$C = C(x,t)$$

For a mass balance on a reacting compound, take mass balance on differential axial element with uniform reaction potential (concentration),

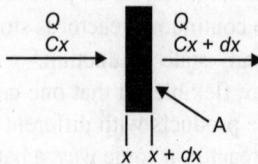

and

$$dV = Adx$$

where,

dV = differential volume.

A = cross sectional area.

dx = differential distance.

Mass balance over differential element on a reactant, C

In $= QCx$

Out $= QCx + dx$

Generation $= dVr_C = Adxr_C$

$$\text{Accumulation} = dV\frac{\delta Cx}{\delta t} = Adx\frac{\delta Cx}{\delta t}$$

$$QCx - QCx + dx + dVr_C = dV\frac{\delta Cx}{\delta t}$$

$$Cx + dx = Cx + dCx$$

$$Q(Cx - Cx - dCx) + dVr_C = dV\frac{\delta Cx}{\delta t}$$

$$-Q\frac{\delta Cx}{\delta t} + r_C = \frac{\delta Cx}{\delta t} = -\frac{\delta Cx}{\delta\left(V\!\!\big/\!Q\right)} + r_C \text{ since } Q \text{ is constant}$$

$$\delta(V/Q) = \delta\tau$$

$$-\frac{\delta Cx}{\delta\tau} + r_C = \frac{\delta Cx}{\delta t}$$

is the non-steady-state ideal PFR mass balance for a reactant.

At steady-state,

$$\frac{\delta Cx}{\delta t} = 0$$

And the ordinary differential can be substituted for the partial differential:

$$\frac{dCx}{d\tau} = r_C$$

Comments:

1. At steady-state, the concentration of a reactant at any single point along the PFR is constant at Cx. Overall a stable concentration profile is obtained at steady-state, with the concentration varying in space as the reaction occurs along the flow path.

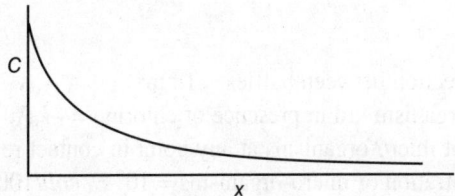

2. In an ideal PFR, τ is the absolute residence time for mass flowing through the reactor, not the average residence time as in a CSTR.

3. Compare ideal batch and ideal PFR mass balances:

$$\text{Ideal PFR} : \frac{dC}{d\tau} r_C$$

$$\text{Ideal batch} : \frac{dC}{dt} r_C$$

Position in a PFR is equivalent to time in a batch reactor.

For a 1st order reaction, $r = -kC$, in a PFR at steady-state:

$$\frac{dC}{d\tau} = -kC$$

$$\int_{Co}^{C_L} \frac{dC}{C} = \int_0^\tau -kd\tau$$

$$C_L = C_0 \exp(-k\tau)$$

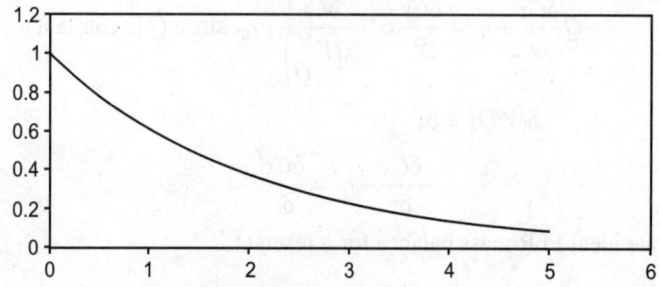

Ideal PFR, steady-state 1st order reaction profile

Example: Chlorine contact basin for disinfection.

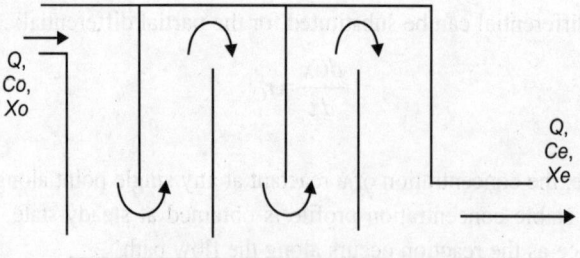

where,

Q = 0.25 m³/s.

A = channel cross section between baffles = 18 m².

r_d = rate of micro-organism kill in presence of chlorine = $-k_d X$.

X = concentration of micro-organisms at any point in contact reactor.

X_o = influent concentration of micro-organisms = 10^6 E. coli/100 ml.

$k_d = 5 \text{ hr}^{-1}$.

r_c = rate of chlorine decay (from micro-organism Cl-demand) $= -k_c X$.

$k_c = 10^{-5} \text{ (mg-chlorine/L)}(\#/100 \text{ ml})^{-1}\text{hr}^{-1}$.

2 rate expressions, 2 constituents, 2 coupled mass balances.

Find

1. Reactor volume and flow path length, L, such that $X_L < 10^3$ cells/100 ml.
2. Chlorine concentration which must be added to insure that there is detectable chlorine at PFR exit (detection level $= C_L = 0.05$ mg/l).

Steady-state mass balance on cells:

$$X_L = X_o \exp(-k_d \tau)$$

$$\tau = (1/k_d) \ln (X_o/X_L) = (1/5)(\text{hr}) \ln (10^6/10^3) = 1.4 \text{ hr}$$

$$V = Q\tau = 0.25 \text{ m}^3/\text{s} \times 3600 \text{ s/hr} \times 1.4 \text{ hr} = 1{,}260 \text{ m}^3$$

$$L = V/A = 1{,}260 \text{ m}^3/18 \text{ m}^2 = 70 \text{ m}$$

3. Steady-state mass balance on chlorine.

$$\frac{dC_c}{d\tau} = -k_c X = -k_c X_o \exp(-k_d d\tau)$$

$$\int_{C_{co}}^{C_L} dC_c = -k_c X_o \int_0^{\tau} \exp(-k_d \tau) d\tau$$

$$C_L = C_{co} - \frac{(k_c X_o)}{k_d} + \frac{k_c X_o \exp(-k_d \tau)}{k_d}$$

$$C_L = C_{co} - \frac{(k_c X_o)}{k_d}(1 - \exp(-k_d \tau))$$

$$C_{CO} = 0.05 + (10^{-5}(10^6)/5)(1 - \exp(-5(1.4))) = 2.05 \text{ mg/l}$$

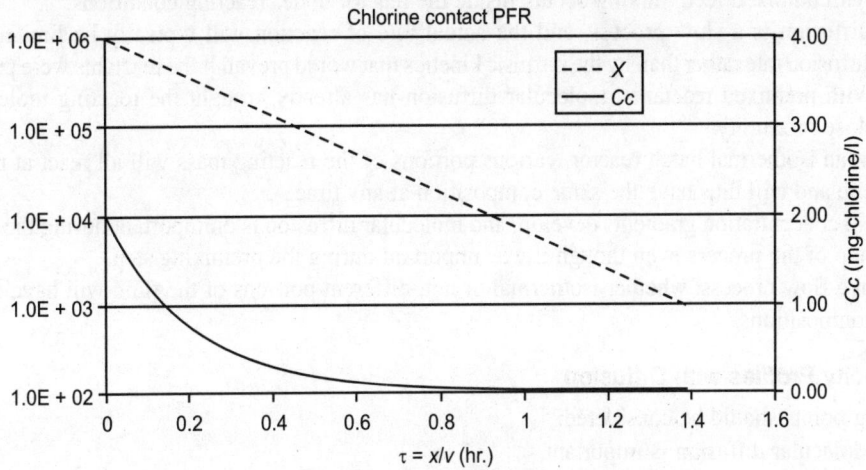

Chlorine contact PFR

LAMINAR FLOW REACTOR

Isothermal Laminar Flows with Negligible Diffusion

Isothermal laminar flows with negligible diffusion:
1. In laminar flow reactors, there will be a pronounced velocity gradient across the tube, with zero velocity at the wall and high velocities near the centerline.
2. Molecules near the center will follow high-velocity streamlines and will undergo relatively little reaction.
3. Those near the tube wall will be on low-velocity streamlines, will remain in the system for long times, and will react to near completion.
4. In the absence of diffusion, each streamline through a laminar flow reactor can be treated as if it were a piston flow reactor.
5. The system as a whole can be regarded as a large number of piston flow reactors in parallel.
6. Concentrations along the streamlines are:

$$u_z(r)\frac{\partial C}{\partial z} = R_A$$

Convective Diffusion of Mass

Following points should be considered:
1. For chemical reactions to occur, molecules must come into contact; and the mechanism for this contact is molecular motion. (This is also the mechanism for diffusion.)
2. With unmixed feed, mixing occurs inside the reactor under reacting conditions.
3. Diffusion is a slow process, and the actual rate of reaction will typically be limited by the diffusion rate rather than by the intrinsic kinetics that would prevail if the reactants were premixed.
4. Thus diffusion is important in unmixed feed reactors unless the reaction is very low.
5. For chemical reactions to occur, molecules must come into contact.
6. The mechanism for this contact is molecular motion.
7. This is also the mechanism for diffusion.
8. There are some reactor design problems where diffusion need not be explicitly considered.
9. With unmixed feed, mixing occurs inside the reactor under reacting conditions.
10. Diffusion is a slow process, and the actual rate of reaction will typically be limited by the diffusion rate rather than by the intrinsic kinetics that would prevail if the reactants were premixed.
11. With premixed reactants, molecular diffusion has already brought the reacting molecules in close proximity.
12. In an isothermal batch reactor, various portions of the reacting mass will all react at the same rate and will thus have the same composition at any time.
13. No concentration gradients develop, and molecular diffusion is unimportant during the reaction step of the process even though it was important during the premixing step.
14. In a flow process, whether isothermal or not, different portions of the fluid will have different compositions.

1-D Velocity Profiles with Diffusion

Following points should be considered:
1. Molecular diffusion is important.

2.   For reactions in circular tubes with 1-D velocity profiles (convective diffusion equation):

$$u_z(r)\frac{\partial C}{\partial z} = D_A\left[\frac{1}{r}\frac{\partial C}{\partial r} + \frac{\partial^2 C}{\partial r^2} + \frac{\partial^2 C}{\partial z^2}\right] + R_A$$

Normally $(R/L)^2 < 10^{-2}$
Typically $(R/L)^2 < 10^{-4}$

$$\frac{u_z(r^*)}{\bar{u}}\frac{\partial C}{\partial z^*} = \left(\frac{D_A\bar{t}}{R^2}\right)\left[\frac{1}{r^*}\frac{\partial C}{\partial r^*} + \frac{\partial^2 C}{\partial r^{*2}} + \left(\frac{R}{L}\right)^2\frac{\partial^2 C}{\partial z^{*2}}\right] + \bar{t}R_A$$

$z^* = z/L$
$r^* = r/R$

Determine the significant of molecular diffusion

Determine the importance of diffusion in the radial direction compared to that in the axial direction

Negligible compared to radial direction

Boundary Conditions

$$C = C_{in}$$

$$\frac{\partial C}{\partial r^*} = 0 \quad \text{at} \quad r^* = 0$$

$$\frac{\partial C}{\partial r^*} = 0 \quad \text{at} \quad r^* = 1$$

$$T = T_{in} \quad \text{at} \quad z^* = 0 \quad \text{Initial condition}$$

$$\frac{\partial T}{\partial r^*} = 0 \quad \text{at} \quad r^* = 0 \quad \text{Radial symmetry}$$

$$\frac{\partial T}{\partial r^*} = 0 \quad \text{at} \quad r^* = 1 \quad \text{Adiabatic operation}$$

$$T = T_{wall} \quad \text{at} \quad r^* = 1 \quad \text{Constant wall temperature}$$

Temperature Profiles in Laminar Flow

Thermal diffusivity

$$\frac{u_z(r^*)}{\bar{u}}\frac{\partial T}{\partial z^*} = \left(\frac{\alpha_T\bar{t}}{R^2}\right)\left[\frac{1}{r^*}\frac{\partial T}{\partial r^*} + \frac{\partial^2 T}{\partial r^{*2}} + \left(\frac{R}{L}\right)^2\frac{\partial^2 T}{\partial z^{*2}}\right] - \frac{\Delta H_R R_A\bar{t}}{\rho C_p}$$

$z^* = z/L$
$r^* = r/R$

R/L is small. Therefore this term is dropped out

If the overall reaction is exothermic, $-\Delta H_R R$ will be positive so that T will be an increasing function of z^*.

In order to determine the RTD, $f(t)$, for the CSTR we consider a simpler situation where a concentration C_0 of a component in a flow stream Q flows into a tank of volume V. At an instant of time all of the concentration C_0 is tagged red so that it can be distinguished from the other reactant in the stirred tank. We then look for the red tagged reactant in the outflow stream to determine the residence time of the reactant in the tank. The amount of tagged material which has left the tank at time 't' is given by the cumulative residence time distribution function, $F(t)$, $F(t)C_0Q$. This is related to the concentration of tagged material in the effluent stream, $QC(t)$,

$$QC(t) = F(t)C_0Q$$

so

$$F(t) = C(t)/C_0$$

Then $F(t)$ is the response, efflux, of the system to a pulse of concentration in the influx.

A material balance for the CSTR under the assumption of perfect mixing yields,

$$V\frac{dC(t)}{dt} = QC_0 - QC(t)$$

with the starting condition that the concentration of the tagged component in the effluent is 0 at $t = 0$, $C(t = 0) = 0$. The solution to this differential equation is,

$$F(t)\frac{C(t)}{C_0} = 1 - e^{-tQ/V}$$

where, $V/Q = \tau$ is a kind of time constant for the system. Under the assumption of perfect mixing, this time constant is the mean residence time for the CSTR, $\bar{t} = \dfrac{V}{Q}$. The residence time distribution function is the derivative of the cumulative residence time distribution function,

$$f(t)dt = \frac{1}{\bar{t}}e^{-t/\bar{t}}dt$$

The function has a value at $t = 0$ of $\dfrac{1}{\bar{t}} = \dfrac{Q}{V}$ which decays exponentially with time. The function has a value at $t = 0$ because mixing is perfect, that is some material is instantaneously in the effluent at the instant material is introduced to the tank. Obviously this is not realistic. Nonetheless, the exponential approximation for a CSTR is a common assumption both in polymer processing and in the chemical process industry as a whole. It is widely used in a wide range of scientific fields as for a first approximation for quantities such as residence time in a lake or ocean or for an approximation of a drug or toxins residence in the human body since it depends only on the system volume and rate of dilution. The function can be modified for dead space using an effective volume rather than the actual volume of the system. Alternatively, tracer studies can be used to measure the mean residence time.

LAMINAR-FLOW REACTORS: FROM EXPERIMENTATION TO CFD SIMULATION

Reaction between crystal violet and sodium hydroxide has been widely used for pedagogical purposes. Some important reasons are the change in colour as the reaction progresses, which is a visual reinforcement of the measured results, and also the fact of being operable at room temperature, thus reducing the complexity and cost of the experimental apparatus. It is well known that this hydrolysis reaction is second-order (first-order towards each reactant), but, if sodium hydroxide is present in great excess, it can be assumed to be pseudo first-order:

$$(-r)_{CV} = kC_{NaOH}C_{CV} \approx k'C_{CV} \qquad \ldots (8.21)$$

where, k' is the pseudo rate constant.

Crystal violet is a highly coloured dye with a maximum absorbance at a wavelength of 588 nm. On the other hand, both sodium hydroxide and the hydrolysis product (carbinol derivative) and colourless, the corresponding aqueous solution having an absorbance similar to that of water. So, it is possible to follow this reaction by spectrophotometry. Direct visualisation is also of interest, as it provides an intuitive qualitative description of the hydrolysis progression.

This reaction was conducted by Hudgins and Cayrol in a tubular reactor. The authors reported that students were very interested in the experiment, which led us to adopt it. However, authors have incorporated tracer experiments for flow pattern characterisation in the laminar-flow tubular reactor (LFTR). The idea is to obtain an integrated chemical reaction engineering lab experiment. The residence time distribution (RTD) then determined is used in a subsequent lab class, together with reaction kinetics data collected in a batch reactor, to predict the theoretical conversion in the same reactor, which is compared to experiment results. This experiment work is performed in laboratory (one for the kinetic analysis, the second for RTD determination, and the final one for measurement of steady-state conversion).

Laminar-Flow Tubular Reactors: Theoretical Background

Operation of tubular reactors at high flow rates may necessitate the use of long vessels to ensure that sufficient retention time is provided for the reaction mixture. Otherwise, low conversion values are obtained. To avoid this, operation is frequently done at low flow rates (in some cases under a laminar regime), as such rates allow a greater reaction time for a given reactor, or allow the design of a shorter reactor to perform a desired conversion. Apart for the recognised industrial interest of LFTRs, particularly in processing polymer, food or other types of viscous fluids, or in emerging areas such as monolith reactors or micro-process engineering (because flow in the channels is laminar), this type of reactor has been widely used for instructional purposes. In this section we present the main equations that students need to use in order to: (i) verify if the flow pattern in the tubular reactor can be modelled as a laminar-flow tubular reactor (LFTR), and (ii) predict the steady-state conversion in the real reactor.

In an LFTR with radius R, the velocity profile is parabolic and can be mathematically described by the well-known Hagen-Poiseuille equation:

$$u(r) = u_{max}\left[1 - \left(\frac{r}{R}\right)^2\right] \qquad \ldots (8.22)$$

where, u_{max} is the fluid velocity at the centre of the pipe ($u_{max} = 2u_{mean}$). From this profile, and assuming negligible molecular diffusion of the species, the residence time distribution in a LFTR can be easily deduced:

$$E(t) = \frac{\tau^2}{2t^3}H\left(t - \frac{\tau}{2}\right) \qquad \ldots (8.23)$$

where, τ is the space-time and $H(t)$ the Heaviside function. Integration of this equation provides the well-known Danckwerts' F curve, which is the normalised reactor response to a step change at the entrance:

$$F(t) = \frac{C_{out}(t)}{C_{in}} = \int_0^t E(t)dt$$

$$= \left[1 - \left(\frac{\tau}{2t}\right)^2\right]H\left(t - \frac{\tau}{2}\right)$$

... (8.24)

$$= \begin{cases} 0 & t < \tau/2 \\ \left[1 - \left(\frac{\tau}{2t}\right)^2\right] & t \geq \tau/2 \end{cases}$$

In the pure convection regime (negligible molecular diffusion), each element of fluid follows its streamline with no intermixing with neighbouring elements. In essence this gives macrofluid behaviour, for which the average conversion $(\bar{X})$ can be predicted by the total segregation model. This model assumes that fluid elements having the same age (residence time) 'travel together' in the reactor and do not mix with elements of different ages until they exit the reactor. Because no mass interchange occurs between fluid elements, each one acts as a batch reactor and the mean steady-state conversion in the real vector is given by:

$$\bar{X} = \int_0^\infty X_{element\ of\ fluid}E(t)dt = \int_0^\infty X_{batch}E(t)dt$$ ... (8.25)

For first-order kinetics, conversion in a batch reactor is given by:

$$X_{batch} = 1 - e^{-k't}$$ ... (8.26)

where, k' is the appropriate first-order rate constant (in our case $k' = kC_{NaOH_m}$, which must be determined at the reaction temperature and can be assumed to be constant because NaOH is considerably in excess compared with the dye). If the flow pattern in the real reactor corresponds to an LFTR, Eq. 8.23 can be used in the integration of Eq. 8.25. Otherwise, the experimental $E(t)$ data must be used.

In addition to the total segregation model, other approaches can be used for prediction of the reactor's conversion which include the mass balance or the maximum mixedness model (Zwietering equation). Since the reaction is first-order (linear system), all models provide the same result, so students may freely choose which one to use.

Experimental Set-up

The experiment set-up used for the tracer and conversion experiments is basically composed of a transparent acrylic jacketed tubular reactor (length = 1.01 m; internal diameter (ID) = 2.2×10^{-2} m), which is connected to a thermostatic bath with heating and cooling system. Solutions are fed to the reactor using a peristaltic pump with eight rolls and two channels, from Ismatec (model Reglo-Analog MS-2/8–160), and the exit stream is directed to a flow-through cell of a Jenway 6300 spectrophotometer, operating at 588 nm. Connection of the RS232 port of the spectrophotometer to a computer allows for absorbance data to be collected, saved and displayed in the monitor, at a frequency of 0.05 Hz. A simple data acquisition program was developed for this purpose in LabVIEW (National Instruments). It is noteworthy that a static mixer was introduced at the reactor's inlet (bottom) for homogenisation of the reactant streams and to ensure a good distribution of the feed through the entire reactor cross-section (Fig 8.10).

This mixer consists of a small cylinder (20 mm length and 8 mm ID) filled with small glass beads (with diameters in the range of 850–1230 μm), followed by a conical tube (with internal diameters of 3 mm and 22 mm and a length of 10 cm), containing glass beads with $d_p \approx 3$ mm. At the reactor outlet, an identical conical tube containing glass beads ensures good mixing of the exiting streamlines.

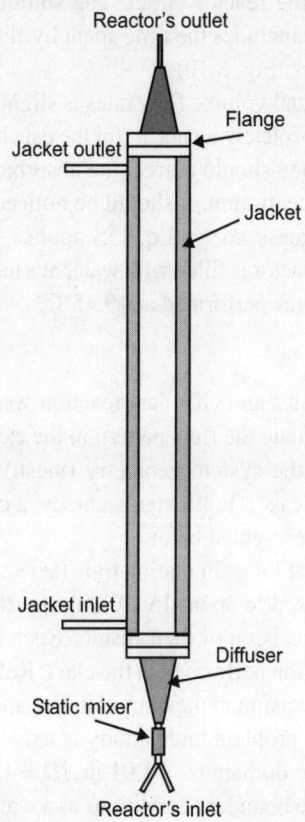

Reactor's outlet

Flange

Jacket outlet

Jacket

Jacket inlet

Diffuser

Static mixer

Reactor's inlet

Fig. 8.10. Sketch of the tubular reactor.

For the flow pattern experiments, a tracer step input is made at the reactor inlet. The tubular reactor is first filled with distilled water and then one switches to the tracer solution (crystal violet at a concentration of about 1×10^{-5} M), which is fed until absorbance at the reactor outlet approaches the value previously recorded for that solution. Students' attention is drawn to the fact that a much diluted tracer solution must be used, so that its density is close to that of water and therefore does not affect the reactor hydrodynamics.

Otherwise, the expected laminar profile will not be obtained, due to the action of gravity. Another crucial point is that the temperature of both solutions and that of the reactor's wall must not differ by more than about 0.5°C. Higher differences will lead to natural convection streams induced by the temperature gradients, which would also significantly affect the parabolic velocity profile.

During the tracer experiment the outlet flow rate must be determined at least three times. For the data herein provided, $Q = 0.879 \times 10^{-6}$ m³/s. The time required for the tracer solution to reach the reactor inlet (i.e. the time spent along the path vessel → reactor), as well as the time spent between the reactor outlet and the detector, must be known to correct the $F(t)$ curve. These values are about 48.8s and 18.4s, respectively. Moreover, this last value is also crucial to rectify the steady-state conversion, once reaction progresses in the connection tubes. This issue will be discussed further below.

For determination of the stedy-state coversion in the continuous-flow tubular reactor, the two reactant solutions (sodium hydroxide 0.05 M and crystal violet 2.0×10^{-3} M) are fed to the bottom until a constant absorbance is recorded at the reactor outlet. The solutions were fed at a total flow rate of 0.849×10^{-6} m^3/s ($\tau = 452.2$ s, which includes the time spent by fluid elements in the reactor and in the inlet and outlet static mixers/diffusers; Fig. 8.10).

The ratio between the two individual volume flow rates is slightly different from unity, since the two heads of the peristaltic pump are not absolutely identical (for the data herein presented, $Q_{NaOH}/Q_{CV} = 1.124$). Students are warned about this, so they should correct the absorbance of the fed crystal violet solution and the inlet soldium hydroxide concentration. It should be noticed that the basic solution is present in a concentration about 2500-fold in excess, so that Eq. 8.21 applies, i.e. the reaction can be considered as pseudo first-order. The jacket of the reactor is filled with water at a temperature close to room temperature. In the present case, the experiment was performed at 19.45°C.

CFD Simulations

Simulations of both the hydrodynamics and chemical reaction were run with the commercial package Fluent 6.0, from Fluent INC. To simulate the flow pattern in the experimental tubular reactor, a laminar 2D model was adopted, this being the system geometry (mesh) previoulsy constructed with a pre-processing software (Gambit, in our case). For better accuracy, a computational grid containing 10000 elements was adopted in the reasult presented below.

Although this involves a somewhat long simulation time (several hours with a 2.00 GHz Pentium IV processor), this can easily be decreased to about 15–20 minutes if students use less elaborate meshes, without significantly affecting the precision of their results. Such less refined meshes are provided for students, who use them in the simulation performed in the class. Refining of the meshes is only advisable if students wish to obtain a higher precision in their simulations, and this can be performed outside class time. It is noteworthy that, since the problem under study is axis-symmetric, only one half of the tube has to be considered in the geometric domain ($L = 1.01$ m, $ID = 1.1 \times 10^{-2}$ m). One face (upper, in the figures shown below) is defined in the boundary conditions as a wall (no-slip conditions and null fluxes), while the other is an axis (the symmetry axis, which corresponds to the centre line of the cylindrical reactor). At the inlet, a perfectly developed parabolic profile was imposed, computed using the Hagen-Poiseuille equation (Eq. 8.22).

The components considered were water and a tracer solution, both having identical properties, so that there is no interference with the reactor hydrodynamics. Mixture properties were computer based on fluid characteristics, using mixing-law (mass-weighted or volume-weighted) formulations.

To simulate the tracer experiments, the system is initialised imposing a null tracer concentration throughout the domain. At time $t = 0$ a step change is introduced at the reactor inlet, with uniform tracer concentration. The tracer concentration history at the reactor outlet (obtained from a mass-weighted average formulation) is then computed and saved in an ASCII file, allowing subsequent manipulation with any spreadsheet software, such as Microsoft Excel^{TM}.

For simulation of the reaction process, fluent's steady scheme was adopted. In addition, the volumetric reaction menu was activated, requiring introduction of the reaction rate parameters (reaction order, stoichiometric coefficients, frequency factor and activation energy). The new compounds (reactants and products) and their properties were also included in the materials database, as well as the operating temperature (but an isothermal process was considered). Obviously, the mass fraction of each compound was introduced in the inlet boundary conditions.

Finally, it should be noted that all the simulation results presented here required a preliminary analysis in order to choose the best conditions and numerical algorithms to adopt, so that higher accuracy could be achieved. Students are warned about this, and to obtain smaller computational times the numerical algorithms that should be adopted are explained in their CFD tutorial. The simulations discussed below were run using the QUICK scheme, convergence criterion of 1×10^{-4} (although residuals were frequently much smaller). In the specific case of unsteady-state simulations the second order-implicit formulation was adopted with a time step of 1s. In this manner, the results are independent of the time-step size. The same criterion was used for the selection of the mesh size (grid with 10,000 elements).

Results and Discussion

Flow pattern characterisation in the tubular reactor

The data shown in Fig. 8.11 illustrate the history of the normalised absorbance data at the reactor outlet after performing a step change in the tracer concentration at the feed. That is, it represents the experimentally recorded Danckwerts' F curve ($F(t) = C_{out}(t)/C_{in}$), which shows the expected shape, only reaching the asymptotic value of 1 for very long times. In addition, the tracer reaches the reactor outlet at about $t = 216$s, which is practically half the space-time, as is to be expected for a laminar velocity profile (fluid elements in the centre of the pipe have the highest velocity, $u_{max} = 2u_{mean}$, thus reaching the exit at $t = \tau/2$). While recording the data, students should pay particular attention to the parabolic velocity profile that can be seen travelling along the reactor, since its wall is transparent and the tracer solution coloured.

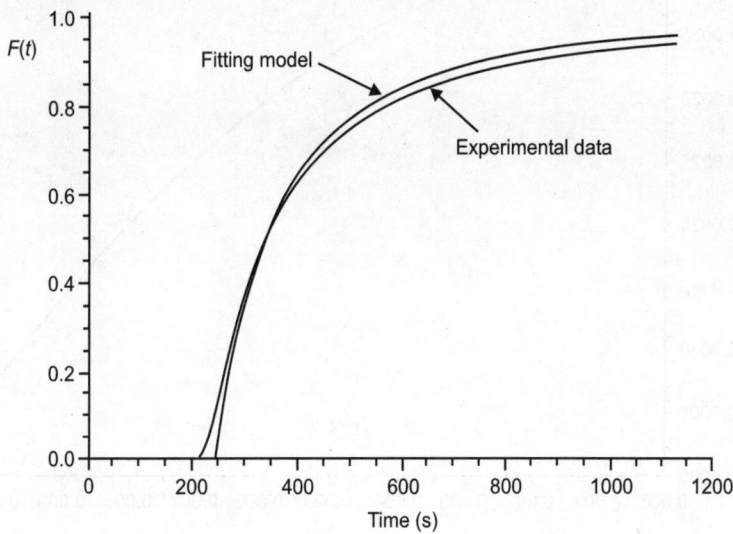

Fig. 8.11. Experimental Danckwerts' F curve ($\tau = V/Q = 436.8$ s) and theoretical fitting using Eq. 8.24 ($\tau = 483.3$ s).

Because one is dealing with laminar flow in a tube, an analytical solution exists for both the residence time distribution and the Danckwerts' F curve (Eqs 8.23, 8.24). In order to fit the theoretical model to their experiment results and therefore determine the value of the single model parameter (τ), students can use the 'Solver' add-in available in Microsoft Excel™ and perform a nonlinear fitting. In this case, optimisation of the least-squares provided the following value: $\tau = 483.3$. The corresponding fitting curve is also shown in Fig. 8.11. Although some discrepancy exists between the experimental space-time value ($\tau = V/Q = 436.8$s, where V is the total volume of the reactor and Q the volumetric flow rate) and that obtained from the fitting process, this is certainly not due to irregularities in the reactor's operation, like stagnant regions or by-passes. This difference is most probably a consequence of experimental errors, particularly those associated with V and Q determination.

When using the CFD software to characterise the hydrodynamics in the reactor, one can start by analysing the steady-state contours of the stream function, which illustrate the trajectories of the fluid elements. The streamlines obtained must be absolutely parallel, as expected for laminar flow, so that no intermixing occurs with neighbouring elements. The simulated velocity profiles at any axial position are also interesting to observe.

Figure 8.12 illustrates a typical profile, after the convergence criteria are satisfied. In this case the tube's outlet boundary was chosen to build the velocity plot, although any other position may be created for that purpose. Data obtained show the expected parabolic profile, with the higher velocities at the center of the tube and with null velocities near the wall.

The simulated data are very well described by wall. The simulated data are very well described by the Hagen-Poiseuille law (Eq. 8.22).

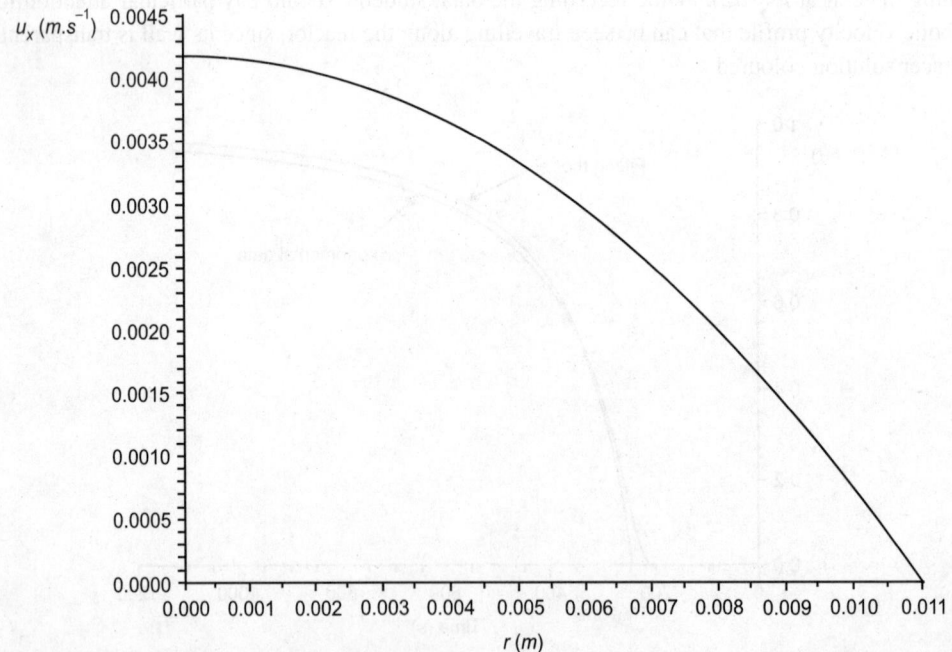

Fig. 8.12. Axial velocities at the reactor outlet as a function of the radial position, obtained from Fluent simulation (points) and from the Hagen–Poiseuille equation (solid line) ($\tau = 483.3$s).

After performing steady-state simulations, one may proceed to transient runs. After defining the tracer step input at the inlet boundary, the software code solves the convection-diffusion equation that describes the tracer transport and the student obtains the concentration field of tracer under transient regime. The contours of tracer concentration throughout the reactor along time are quite interesting to observe because they provide a good perspective of how the concentration front evolves. The frames recorded and shown in Fig. 8.13 show that it is only for $t = \tau/2$ that one starts 'seeing' tracer at the reactor outlet, as expected for laminar flow in pipes (see Eq. 8.24). In addition, even for very long times (about five times the space-time) the tube is not completely full of tracer, due to the very low velocities near the walls.

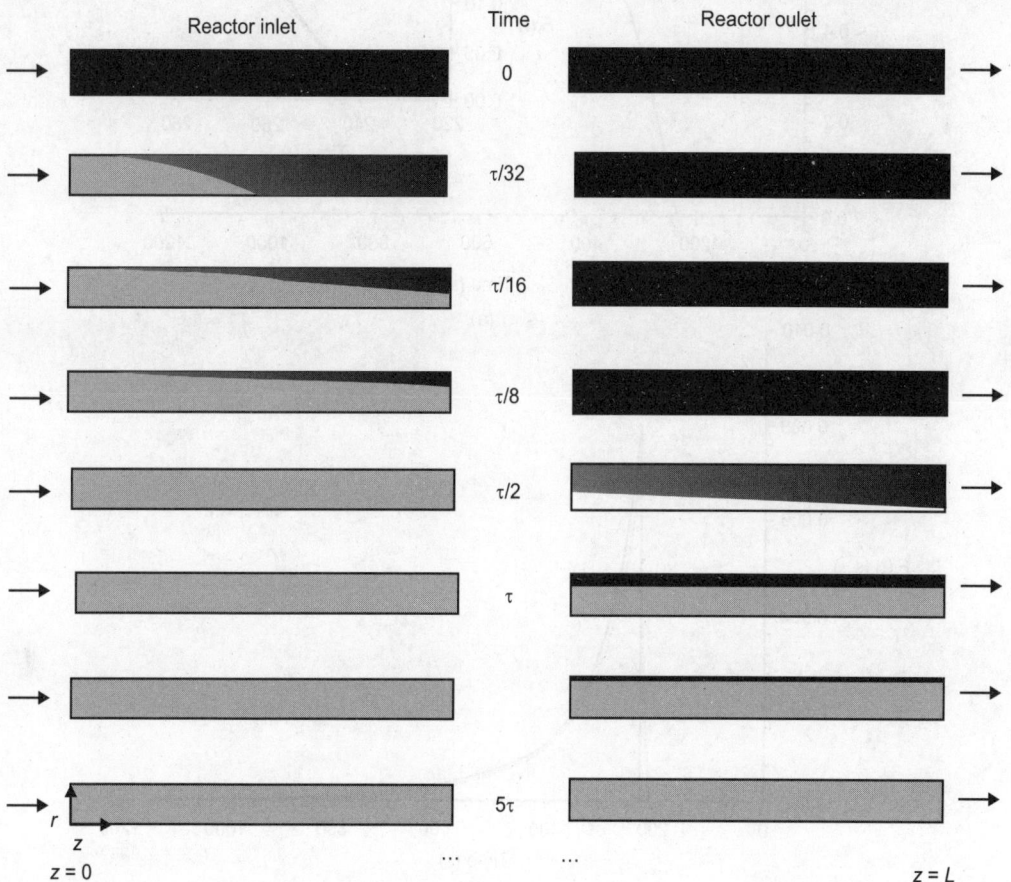

Fig. 8.13. Transient tracer concentration contours at the inlet and outlet sections of the tubular reactor (τ = 483.3 s). The entire reactor length cannot be shown because of the high L/D ratio.

As mentioned above, during unsteady simulations the student may save the tracer concentration at the outflow boundary in an ASCII file, so that the Danckwerts' F curve can be subsequently computed. Figure 8.14(a) shows the $F(t)$ curve obtained from Fluent simulation. It is noteworthy that the simulated data practically coincide with the analytical solution (Eq. 8.24), the difference being more evident at times around 240 s [see detail of Fig. 8.14(a)]; i.e. when the tracer concentration front reaches the reactor outlet. Figure 8.14(b) shows the $E(t)$ curve, obtained by derivation of the $F(t)$ data, which exhibits the typical very long tail.

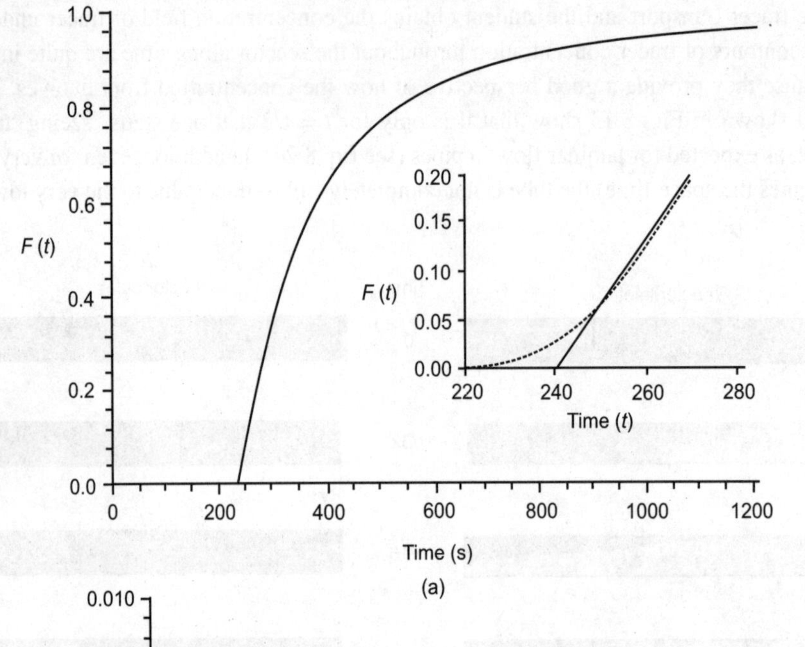

Fig. 8.14. Fluent simulation data (dotted lines) versus analytical solutions (black lines) for: (a) Danckwerts' F curve and (b) residence time distribution function ($\tau = 483.3$s).

It should be pointed out that, although more accurate data could be obtained with CFD simulation (for instance with a more refined mesh), this is completely unnecessary. Indeed, with the numerical algorithms used and the time-step and grid adopted, the steady-state conversion prediction leads to a relative error below 1 per cent, when comparing the fluent simulation with the analytical solution.

Students are also warned that the $E(t)$ curve does not need to be obtained from their experiment data. First, one introduces too much numerical noise when computing the derivative of the $F(t)$ curve. Second, to accurately predict the steady-state conversion directly from the RTD data, tracer experiments should be performed for very long times. Otherwise, the area under the $E(t)$ curve is not duly taken into account in the total segregation model (compare Eq. 8.25). Indeed, due to the very long tail of the $E(t)$ curve, the normalisation condition $\left(\int_0^\infty E(t)dt = 1 \right)$ is only satisfied if experiments/simulations are performed for very protracted times (typically about five times the space-time, such that relative errors are below 1–2 per cent). Therefore, conversion is predicted using the analytical RTD equation, since the flow pattern corresponds to that of an LFTR.

Determination of steady-state conversion in the laminar-flow tubular reactor

This is done in a batch reactor, at three different temperatures, so that the activation energy can also be obtained. Experiments performed in the range $15°$–$25°C$ provided the following results for the frequency factor and activation energy: $k_o = 1.193 \times 10^7$ m^3 mol^{-1} s^{-1} and $E_a = 6.195 \times 10^4$ J mol^{-1}. These values will subsequently be used for prediction of the theoretical conversion.

In the lab, students operate the continuous flow reactor until a constant absorbance is recorded at the reactor outlet. At this stage, a strong change in colour may be observed along the axial position, due to the crystal violet concentration gradient. From such steady-state value and the absorbance at the reactor inlet (which must be corrected taking into account the flow rate ratio of both reactant streams), a first estimate for the crystal violet conversion ($X_{overall}$) is obtained. However, since the reaction continues in the exit line before reaching the spectrophotometer cell, that value must be corrected. For reasons of simplicity, plug flow behaviour can be assumed in the tube ($X_{tube} = 1 - e^{-k'\tau_{tube}}$), and so the experiment conversion is obtained from the following equation:

$$X_{CV} = \frac{X_{overall} - X_{tube}}{1 - X_{tube}} \quad \text{... (8.27)}$$

which can easily be deduced by students, assuming a series-association of two reactors with space-times τ and τ_{tube}. The value obtained in the experiment was $X_{CV} = 0.609$ (Table 8.3).

Table 8.3. Crystal violet conversion obtained experimentally by CFD simulation and theoretical values*.

Calculation procedure	X_{CV} (%)
Experimental value	60.9
Fluent simulation	62.6
Segregation model—Eq. 8.25	63.0
Mass balance—Eq. 8.28	63.0

*$\tau = 452.2$ s, $T = 19.45°C$.

As mentioned above, fluent also simulates the flow system with a chemical reaction present, which simply requires the introduction of a reactive feed stream (with specification of reaction parameters, as explained). Simulation of the continuous-flow reactor via CFD can also be used to evidence the contours of species concentration along the reactor, which are particularly interesting to observe. To obtain the steady-state conversion, the surface integral for the limiting reactant (crystal violet) must be evaluated at the outer interface, using a mass-weighted average formulation. The result obtained is shown in Table 8.3, being less than 3 per cent higher than the experimentally recorded value.

Once again, it is very important that students may compare their experiment and/or simulated results with those predicted from theory, because in this way they will feel more confident about their own work. First, they can use the above-described total segregation model. Once the tracer experiments have shown that the tubular reactor can be modelled as an LFTR, the $E(t)$ expression given in Eq. 8.23 can be used and Eq. 8.25 can be easily integrated. This was done with the Maple 7.00 software, and the result obtained is shown in Table 8.3 (X_{CV} = 63.0 per cent). It is worth noting that there is very good agreement between this value and both the experiment one and, particularly, that achieved by CFD simulation.

Finally, students can still use another approach to predict the theoretical conversion. Indeed, the mass balance for first-order irreversible kinetics in a laminar-flow reactor has an analytical solution given by:

$$\frac{\overline{C_{out}}}{C_{in}} = 1 - X = e^{-\lambda}(1-\lambda) + \lambda^2 \int_{\lambda}^{\infty} \frac{e^{-z}}{z} dz \qquad \ldots (8.28)$$

$$= e^{-\lambda}(1-\lambda) + \lambda^2 ei(\lambda)$$

where, $\overline{C_{out}}$ is the reactant average exit concentration, $\lambda = k' L/u_{max} = k'\tau/2$ and $ei(\lambda)$ the exponential integral (which is tabulated in most mathematical handbooks or can be obtained from Maple software, for instance). As expected, the value obtained for the crystal violet conversion is identical to that predicted by the segregation model (Table 8.3). Although application of Eq. 8.28 is also simple, the use of the segregation model has the advantage of also allowing students to predict the transient behaviour of the LFTR. For that, they should simply replace the superior limit of integration in Eq. 8.25 by t. Such a dynamic response can also be easily compared with either experiment or CFD-simulated data, although they have not been shown here.

SOLVED EXAMPLES

Example 8.1. A sample of tracer was injected as a pulse to a reactor and effluent concentration (C) measured as a function of time, resulting in the following data:

Time (min)	0	1	2	3	4	5	6	7	8	9	10	12	14
C(g/m^3)	0	1	5	8	10	8	6	4	3	2.2	1.5	0.6	0

Plot the exit age distribution, E (the RTD) for the system and determine the fraction of material leaving the reactor that has spent time between 3 and 6 minutes in the reactor.

Solution:

Time (t) (minute)	C(g/m^3)	$C\Delta t$	$E = \dfrac{C}{Q} = \dfrac{C}{\Sigma C\Delta t}$
0	0	0	0
1	1	1	0.019
2	5	5	0.097
3	8	8	0.155
4	10	10	0.194
5	8	8	0.155

(Contd ...)

Time (t) (minute)	$C(g/m^3)$	$C\Delta t$	$E = \dfrac{C}{Q} = \dfrac{C}{\Sigma C\Delta t}$
6	6	6	0.116
7	4	4	0.078
8	3	3	0.058
9	2.2	2.2	0.029
10	1.5	3	0.029
12	0.6	1.2	0.011
14	0	0	0

$$\Sigma C\Delta t = 51.4$$

Plot t vs E is known as Exit age distribution. Fraction of material leaving the reactor that has spent time between 3 and 6 minutes in the reactor:

$$= \sum_{t=3}^{6} E\Delta t = 0.0194(1) + 0.0155(1) + 0.116(1) = 0.465.$$

Example 8.2. Response measurements to a pulse input are made for a reactor and following data are obtained.

Time (sec)	0	10	20	30	40	50	60	70	80	90	100	110
Tracer conc. (gm/cm³)	0	0.2	0.4	0.7	1.5	1.8	1.2	0.8	0.4	0.3	0.1	0

1. Evaluate the mean residence time.
2. Determine the integral value of N if the stirred tank-in-series model is used to approximate the data.

t (sec)	$C(g/cm^3)$	$C\Delta t$	$tC\Delta t$	$t^2 C\Delta t$
0	0	0	0	0
10	0.2	2	20	200
20	0.4	4	80	1600
30	0.7	7	210	6300
40	1.5	15	600	24000
50	1.8	18	900	45000
60	1.2	12	720	43200
70	0.8	8	560	39200
80	0.4	4	320	25600
90	0.3	3	270	24300
100	0.1	1	100	10000
110	0	0	0	0
		$\Sigma C\Delta t = 74$	$\Sigma tC\Delta t = 3780$	$\Sigma t^2 C\Delta t = 219400$

1. Mean residence time,

$$\bar{t} = \frac{\Sigma tC\Delta t}{\Sigma C\Delta t} = \frac{3780}{74} = 51.08 \text{ sec}$$

2.
$$N = \frac{\left(\bar{t}\right)^2}{\sigma^2}, \quad \sigma^2 = \frac{\Sigma t^2 C \Delta t}{\Sigma C \Delta t} \cdot \left(\bar{t}\right)^2$$

$$= \frac{219400}{74} - \left(51.08\right)^2 = 355.7$$

∴
$$N = \frac{\left(51.08\right)^2}{355.7} = 7.33 \approx 8$$

Nomenclature

E_a	:	Activation energy ($J.\ mol^{-1}$).
$ei(\lambda)$	:	Exponential integral function.
$E(t)$	:	Residence time distribution function (s^{-1}).
$F(t)$	:	Danckwerts' F curve.
$H(t)$	:	Heaviside function.
k'	:	Pseudo first-order rate constant (s^{-1}).
k	:	Second-order rate constant ($m^3\ mol^{-1}\ s^{-1}$).
k_o	:	Pre-exponential factor ($m^3\ mol^{-1}\ s^{-1}$).
L	:	Length of the reactor (m).
Q	:	Volumetric flow rate ($m^3 s^{-1}$).
r	:	Radial position in the tubular reactor (m).
$(-r)$	:	Reaction rate ($mol.\ m^{-3} s^{-1}$).
R	:	Internal radius of the reactor (m).
t	:	Time (s).
$u(r)$	:	Fluid velocity at the radial position r ($m.s^{-1}$).
u_{max}	:	Maximum fluid velocity ($m.s^{-1}$).
u_{mean}	:	Mean fluid velocity ($m.s^{-1}$).
V	:	Reactor volume (m^3).
X	:	Conversion.
$\bar{X}$	:	Average conversion.
$X_{overall}$	:	Overall experimental conversion (considering both reactor and tube).
X_{tube}	:	Experimental conversion reached in the connection tube.
z	:	Axial position in the tubular reactor (m).

Subscripts

CV	:	Crystal violet.
in	:	Inlet conditions.
out	:	outflow conditions.
batch	:	Batch reactor.

Greek Symbols

τ	:	Space-time (s).
λ	:	$\dfrac{kL}{u_{max}} = \dfrac{k\tau}{2}$.

Chapter 9

Models for Nonideal Reactors

INTRODUCTION

Not all tank reactors are perfectly mixed nor do all tubular reactors exhibit plug-flow behaviour. In these situations, some means must be used to allow for deviations from ideal behaviour. We use the segregation and maximum mixedness models to bound the conversion when no adjustable parameters are used. For non-first-order reactions in a fluid with good micromixing, more than just the RTD is needed. These situations compose a great majority of reactor analysis problems and cannot be ignored. For example, we may have an existing reactor and want to carry out a new reaction in that reactor. To predict conversions and product distributions for such systems, a model of reactor flow patterns is necessary. To model these patterns, we use combinations and/or modifications of ideal reactors to represent real reactors. With this technique, we classify a model as being either a one-parameter model (e.g. tanks-in series model or dispersion model) or a two-parameter model (e.g. reactor with bypassing and dead volume). The RTD is then used to evaluate the parameter(s) in the model.

The choice of the particular model to be used depends largely on the engineering judgement of the person carrying out the analysis. It is this person's job to choose the model that best combines the conflicting goals of mathematical simplicity and physical realism. There is a certain amount of art in the development of a model for a particular reactor, and the examples presented here can only point toward a direction that an engineer's thinking might follow.

For a given real reactor, it is not uncommon to use all the models discussed previously to predict conversion and then make a comparison. Usually, the real conversion will be bounded by the model calculation.

DEVIATIONS FROM IDEAL REACTOR BEHAVIOUR

Tank reactors: Inadequate mixing, stagnant regions, bypassing or short-circuiting.
Tubular reactors: Mixing in longitudinal direction, incomplete mixing in radial direction, bypassing (especially in fixed bed reactors).

Definitions

1. Segregated flow: Fluid elements do not mix, have different residence times.
   (a) Need 'residence time distribution'.
2. Micromixing: Adjacent elements mix partially.
   (a) Need extent of micromixing.

3. Effects of non-ideality are higher for viscous reaction mixtures.
4. Study limited to non-idealities for single reactions in homogeneous reactors here.

RESIDENCE TIME DISTRIBUTION

The residence time distribution (RTD) of a chemical reactor is a probability distribution function that describes the amount of time a fluid element could spend inside the reactor. Chemical engineers use the RTD to characterise the mixing and flow within reactors and to compare the behaviour of real reactors to their ideal models. This is useful, not only for troubleshooting existing reactors, but in estimating the yield of a given reaction and designing future reactors.

1. Residence time is the time it takes for a fluid element to pass through the reactor.
2. Age is the time since the element entered the reactor.
3. Residual lifetime — remaining time the element will spend in the reactor.
4. Age + residual lifetime = residence time.

$F(t)$ = Fraction of the effluent stream that has residence time less than t.

$F(0)$ = 0.

$F(\mathrm{inf})$ = 1.

$E(t)dt = dF$ is the fraction of the effluent stream that has residence time between t and $t + dt$.

$E(t)$ is typically called the residence time distribution function.

Based on the concentration of tracer species measured at effluent, $E(t)$ can be defined as:

$$E(t) = \frac{C(t)}{\int_0^\infty C(t)dt} \text{ if the volumetric flow rate can be assumed to be constant, for a pulse input.}$$

$$E(t) = \frac{d}{dt}\frac{C(t)}{C_0} \text{ for a step input.}$$

Properties of RTD

Cumulative RTD: $F(t) = \int_0^\infty E(t)dt$ (fraction of effluent that has been in reactor for time less than t).

Mean residence time: $\tau_m = \int_0^\infty tE(t)dt = V/v$ (for closed system).

Variance of the RTD: $\sigma^2 = \int_0^\infty (t - \tau_m)^2 E(t)dt$.

Internal age distribution function $I(t)$ (fraction of fluid in the reactor that has age between t and $t + dt$).

RTD for Ideal Reactors

Since the mixing conditions are well known for these ideal reactors, the RTD can be predicted without real experimentation:

PFR — All elements come out at same time.

$E(t) = \delta(t - \tau)$

$\delta(t)$ is the Dirac Delta function, has a value of ∞ at $t = 0$ and is zero otherwise.

Furthermore, $\int_0^\infty g(x)\delta(x-\tau)dx = g(\tau)$.

CSTR-Material balance on inert tracer can be performed: in-out = accumulation

$$0 - \upsilon C = V \frac{dC}{dt}$$

$$C = C_0 \exp(-t/\tau)$$

Thus,

$$E(t) = \frac{\exp(-t/\tau)}{\tau}$$

$$\tau_m = \tau$$

Laminar flow reactor: Similarly, the RTD for a laminar flow reactor can be derived as:

$$E(t) = 0; \; t < \tau/2$$

$$E(t) = \tau^2/2t^3; \; t >= \tau/2$$

Reactor modelling with RTD

Following steps are involved:
1. Chance of a molecule reacting depends on the other molecules it encounters during its stay in the reactor.
   (a) Need residence time and mixing pattern.
2. If velocity and local rate of mixing of every molecule in the reactor were known, differential mass balance can be written and integrated to get the final conversion.
3. In the absence of such information approximations are necessary for reactor modelling.
4. So, for reactors with known RTD, some assumptions regarding mixing have to be made.

Mixing conditions: extremes

The two extreme situations are:
1. Incoming fluid remains completely segregated, i.e. it forms small globules that are uniformly dispersed, and such that all molecules in the globule have the same residence time (Segregation Model).
2. Incoming fluid is completely mixed on a molecular scale (Maximum Mixedness Model)
   (a) It turns out that for first order reactions either assumption amounts to the same thing in terms of final conversion, since conversion is independent of local concentration.
   (b) In case of the CSTR, maximum mixedness also implies perfect mixing. If the RTD of the real reactor is the same as that for a CSTR, then the maximum mixedness assumption on top of it, leads to equations matching that of a CSTR.

Estimating Deviations from Ideal Behaviour

Zero and one-parameter models

1. Wholly segregated flow:
   (a) Measure RTD from experiment.

   (b) Calculate conversion by assuming wholly segregated flow (good for PFR; not good for nearly ideal CSTR).
2. Maximum mixedness model:
   (a) Measure RTD.
   (b) Assume maximum mixedness (good for nearly ideal CSTR).
3. Axial dispersion model:
   (a) Assume ideal PFR + axial dispersion occurs.
   (b) Actual RTD is required.
   (c) Good for PFR in turbulent flow.
4. Series of Ideal CSTRs of equal volume:
   (a) Actual reactor response to be used to determine number of tanks.
5. PFR with recycle (not covered in class):
   (a) Mixing introduced through recycle.
   (b) Extent of recycle determines whether close to PFR behaviour (zero recycle) or CSTR behaviour (infinite recycle).

Degree of Segregation

Danckwerts' two limiting mixing conditions can be further clarified as:
1. Incoming fluid is broken up into discrete fragmants which are small compared to the size of the reactor, and are uniformly dispersed in the reactor, the molecules of the fragments which have entered together remain together indefinitely (fully segregated fluid).
2. Incoming material is dispersed on a molecular scale in a time less than the average residence time; the mixture is chemically uniform.

Segregation' is a measure of mixing in the reactor. It is understood to be the degree of departure of reactor from the perfectly mixed behaviour.

Suppose we superimpose the F vs. time diagram of a perfectly mixed reactor over that for the real reactor. The difference between the two curves can be said to be a measure of segregation in the system.
1. A useful definition of perfect mixing is: If the age distribution of material in the reactor is the same as that in the outgoing stream, the reactor is perfectly mixed (CSTR). The other extreme is when the exit age is the same for all internal ages (PFR).
2. If the age of a molecule is: a = time elapsed since the molecule entered the reactor.

   $var\ a = \overline{(a - \bar{a})^2}$—average over all the molecules, $\bar{a}$ being the mean age of molecules currently in the reactor

3. Now let the mean age of molecules occupying some point in the reactor be a_p. Then,

   $var\ a_p = \overline{(a_p - \bar{a})^2}$—average over all points in the reactor.

Now, for mixed flow, there is no variation in ages of molecules at different points in the reactor, therefore $var\ a_p = 0$. (Note that this cannot be strictly true except for the CSTR.)

For 'Fully segregated flow' the variance in ages between points in the reactor equals the variance in ages, as molecules that enter at the same time stay together and pass through the reactor all together. This, $var\ a = var\ a_p$ for this case.

Based on these observations, 'degree of segregation' can be defined as:

$J = var\, a_P / var\, a$, such that $J = 1$ for fully segregated flow, and $J = 0$ for mixed flow.

The degree of segregation which can be estimated from experiments, serves as a useful means of estimating the mixing characteristics of the real reactor.

Qualitative analysis of the effect of segregation/mixing

Consider a second order-reaction in two reactor sequences:

1. A PFR followed by CSTR.
2. A CSTR followed by PFR.

Let the concentration of reactant at inlet be 1, and the reaction rate constant and residence time in each reactor also 1, all in self-consistent units.

Then it can be shown that in case: (1) the final conversion is 63.4 per cent while in case, and (2) the final conversion is 61.8 per cent. Note that it can be further shown the RTD for both (1) and (2) is the same.

This difference in conversion in the two set-ups is attributable to the fact that in (1) the mixing between molecules of different ages happens later than in case, and (2) this in turn implies that the degree of segregation is higher in (1) than in (2). For second and higher order reactions a higher degree of segregation is suitable, while the opposite is true for negative reaction orders.

Estimating conversion

1. In the general, the complex mixing pattern has to be known in addition to the residence time function for the evaluation of conversion.
2. In the extreme cases of mixing this estimation is quite simplified.
3. Typically, we can say that estimating conversion using the RTD of the real reactor, along with the completely segregated and maximum mixedness models will give the minimum and maximum bounds for the 'real' conversion.

Completely Segregated Model

1. In this case, each 'globule', i.e. bunch of molecules that enter the reactor at same time, can be thought of as a separate batch reactor.
2. Each globule is a lump of all the molecules that have the same residence time in the reactor.
3. Let the conversion in the batch reactor be $X(t)$, where t refers to the time spent in reactor, as relevant to the globules, this t is the corresponding residence time.

Mean conversion of globules with residence time between t and $t + dt$ = Conversion in batch reactor at time t x fraction of globules that have residence time between t and $t + dt$.

In other words, reactor conversion for this model can be written as:

$$\bar{X} = \int_0^\infty X(t)E(t)\,dt$$

Maximum Mixedness Model

1. This limiting case is not as easy to model.
2. If the residence time distribution function for the real reactor matches that for a CSTR, then in the limit of maximum mixedness, we can assume that the real reactor behaves like a CSTR, and just solve the CSTR equations.
3. It is not possible to apply this model if the RTD is that of a PFR.

4. If the residence time distribution is some other function, then the minimum value of the degree of segregation is not zero, we hypothesise a tubular reactor, with fluid entering along the length of the tube, through side entrances, in such a way that the RTD remains same as that of the real reactor. In addition, plug flow is assumed so that there is complete radial mixing as soon as the fluid enters the reactor.

5. Then, the differential equation of mass balance can be derived to be:

$$\frac{dC_A}{d\lambda} = -r_A + (C_A - C_{A0})\frac{E(\lambda)}{1 - F(\lambda)}$$

where, C_{A0} refers to the inlet concentration of the reactant, r_A is the reaction rate, E is the RTD, F the cumulative RTD, and λ is a time variable such that its value is 0 at the exit of the reactor and infinite at the entrance of the reactor.

6. Two initial conditions are common for this hypothetical reactor,

$$\lambda = \infty; \ C_A = C_{A0} \ \text{or} \ \lambda = \infty; \ dC_A/d\lambda = 0$$

Either can be used.

7. To solve this equation, we have to integrate backwards (typically numerically), starting at a large value of λ (usually 4–5 times the average residence time), and finally reaching $\lambda = 0$. The conversion can thus be calculated.

8. The process is little more involved than the completely segregated model, but turns out to be really simple if the RTD matches that of CSTR.

9. Note that this model cannot be used of the RTD matches that of a PFR, a quick substitution into the differential equation will show you the inconsistency.

10. This is because the micro- and macro-mixing aspects of a reactor are not exactly independent of each other, if the RTD is that of a PFR then maximum mixedness is impossible as the RTD itself implies virtually no mixing between adjacent elements.

11. Detailed mathematical analysis can be found in the journal articles by Danckwerts and Zwitterieg, and is closely related to the analysis of degree of segregation for real reactors.

Tanks-in-Series Model

1. This model has one adjustable parameter, and also requires the RTD function.
2. Typical application is for tubular reactors.
3. The assumption is that the real reactor can be approximated as a series of equal-volume CSTRs.
4. The job is the find the number of such CSTRs required.
5. If the average residence time and variance are evaluated from from the RTD function, the number of reactors can be obtained from a simple relation:

$$n = \frac{\tau^2}{\sigma^2}$$

6. Further, the conversion can be calculated easily as all the CSTRs are of same size (by repeated application of the CSTR mass balance equations).
7. For first-order reactions it is simply given below:

$$X = 1 - \frac{1}{(1 + k\tau_i)^n}$$

where, n is number of tanks from above, and t_i is the residence time in each tank ($= t/n$).

8. Note that for non-first-order reactions, numerical solution of the system of algebraic equations from mass balance for each tank, may be required.

Description of tanks in series model

This model can be used whenever the dispersion model is used; and for not too large a deviation from plug flow both models give identical results, for all practical purposes. The dispersion model has advantage in that all correlations for flow in real reactors invariably use this model. On the other hand the tanks in series model is simple, can be used with any kinetics, and it can be extended without too much difficulty to any arrangement of compartments, with or without recycle. The number of tanks in series is:

$$N = \frac{1}{\sigma^2}$$

This expression represents the number of tanks necessary to model the real reactor as N ideal tanks in series. As the number of tanks increases, the variance decreases.

Comments

1. For small deviation from plug flow, $N > 50$, the RTD becomes symmetrical and gaussian.
2. How to find N which fits an experimental curve.
   (a) Draw the RTD curves for various N and see which matches the experimental curve most closely.
   (b) Calculate σ^2 from experiment and compare with theory.
   (c) Evaluate the width of the curve at 61 per cent of maximum height.
   (d) Match the maximum height.
3. Independence: If M tanks are connected to N more tanks (all of same size) then the individual means and variances (in ordinary time units) are additive, or

$$\bar{t}_{M+N} = \bar{t}_M + \bar{t}_N \quad \text{and} \quad \sigma^2_{t,M+N} = \sigma^2_{t,M} + \sigma^2_{t,N}$$

4. From the above variance expression, we see that in adding tanks we get:

$$\sigma^2_t \propto N \quad \text{or} \quad \sigma^2_t \propto L \left(\frac{\text{Spread}}{\text{of curve}} \right) \propto \sqrt{L}$$

5. For any one shot tracer input:

$$\frac{\Delta \sigma^2_t}{\bar{t}^2} = \frac{\sigma^2_{t,\text{ out}} - \sigma^2_{t,\text{ in}}}{\bar{t}^2} = \frac{1}{N}$$

The increase in variance between input and output points is the same for any kind of input, whether pulse, double peaked, etc.

6. Relationship between dispersion and tanks in series model:

$$\left. \begin{array}{l} \text{Dispersion model}: \sigma^2 = 2 \left(\frac{D}{UL} \right) \\[2mm] \text{Tanks in series}: \quad \sigma^2 = \dfrac{1}{N} \end{array} \right\} \text{thus} \quad \frac{1}{N} = 2 \left(\frac{D}{UL} \right)$$

7. Value of N for flow of gases in fixed beds.

$$\frac{D\varepsilon}{U_0 d_p} = \frac{D}{U d_p} = \frac{D}{U\left(L/n_{particles}\right)} \cong \frac{1}{2} \text{ or } \frac{D}{UL} = \frac{1}{2n_{particles}}$$

8. Before deciding to use this model be sure to check the shape of the experimental curve to see if the model really applies. Do not use the model indiscriminately.

Dispersion Model

The dispersion model is used to describe nonideal tubular reactors. In this model, there is an axial dispersion of the material, which is governed by an analogy to Fick's law of diffusion, superimposed on the flow. So in addition to transport by bulk flow, $UA_C C$, every component in the mixture is transported through any cross section of the reactor at a rate equal to $[-D_a A_c (dC/dz)]$ resulting from molecular and convective diffusion. By convection diffusion we mean either Aris-Taylor dispersion in laminar flow reactors or turbulent diffusion resulting from turbulent eddies. This model applies to turbulent flow in pipes, laminar flow in long tubes, packed beds, shaft kilns, long channels, etc.

For laminar flow in short tubes or laminar flow of viscous materials these models may not apply, and it may be that the parabolic velocity profile is the main cause of deviation from plug flow.

To illustrate how dispersion affects the concentration profile in a tubular reactor, we consider the injection of a perfect tracer pulse. Figure 9.1 shows how dispersion causes the pulse to broaden as it moves down the reactor and becomes less concentrated.

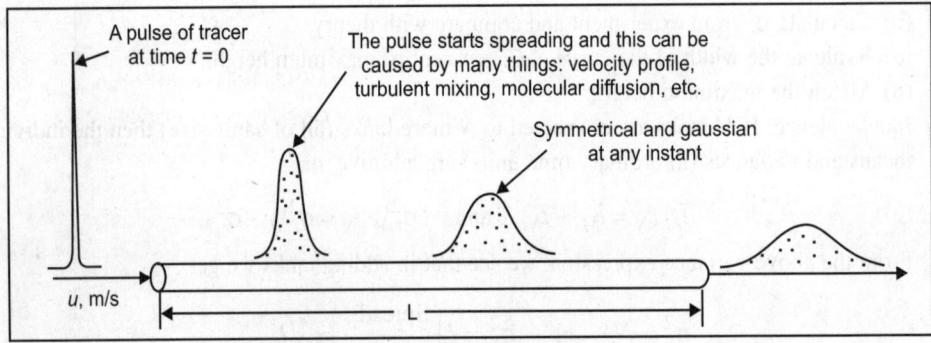

Fig. 9.1. Dispersion in a tubular reactor.

To characterise the spreading we assume a diffusion-like process superimposed on plug flow. We call this dispersion or longitudinal dispersion to distinguish it from molecular diffusion. The dispersion coefficient D (m²/s) represents this spreading process.

Thus:

1. Large D means rapid spreading of the tracer curve.
2. Small D means slow spreading.
3. $D = 0$ means no spreading hence plug flow.

Also

$$\frac{1}{Pe} = \left(\frac{D}{UL}\right)$$

is the dimensionales group characterising the spreading rate for the vessel.

We evaluate D or D/UL by recording the shape of the tracer curve as it passes the exit of the vessel. In particular, we measure:

$\bar{t}$ = mean time of passage, or when the curve passes by.

σ^2 = variance, or [how long]^2 it takes for the curve to pass by.

Dispersion number D/UL	Dispersion coefficient	
$D/UL < 0.01$ (Small deviation from plug flow)	$\sigma^2 = \dfrac{2}{Pe} = 2\left(\dfrac{D}{UL}\right)$	
$D/UL > 0.01$ (Large deviation from plug flow)	Closed vessel (for packed and open tube reactors)	Open vessel (for coiled tubular reactors)
	$\sigma^2 = 2\left(\dfrac{D}{UL}\right) - 2\left(\dfrac{D}{UL}\right)^2\left(1 - e^{-\frac{UL}{D}}\right)$	$\sigma^2 = 2\left(\dfrac{D}{UL}\right) + 8\left(\dfrac{D}{UL}\right)^2$

CASE STUDY FORMULATION AND SIMULATION WITH FLUENT

The case study described here can be implemented as homework for students taking a chemical reaction engineering (CRE) course dealing with nonideal reactors. Simulations can be performed using a commercial package such as Fluent, but it is advisable that a brief tutorial be provided so students can quickly familiarise themselves with the program. In this tutorial, the essential steps that must be followed for any simulation should be outlined. Different reservoir/reactor geometries can be provided for different groups of two-to-three students, but each group should perform its own parametric study (e.g. evaluate the effect of space-time, or Reynolds number, in flow pattern characterisation, or Damköhler number in reaction simulations). Finally, the students should compile the results obtained by other colleagues and discuss them in a final written report.

Using 2-D reservoirs/reactors with laminar flow (described in detail in the next section), students should perform the following tasks, which are also the objectives of this paper:

1. Characterise the hydrodynamics in the vessel(s).
2. Determine the RTD from tracer experiments, which includes diagnosis of reservoir/reactor operation.
3. Predict the conversion in a continuous-flow system.

It is well known that the mean residence time $(\bar{t}_r)$ has an important effect on the performance of some large natural conversion systems, such as biological lagoons, since it affects the biological conversion of biodegradable matter. Moreover, the geometry of the reservoirs seems to affect the RTD, and thus $(\bar{t}_r)$. It is, therefore, very important to perform tracer experiments in these systems (or try to estimate the RTD function) in order to better design such waste-water treatment plants. The mean residence times in these large reservoirs or lagoons are extremely high (ranging from one day up to several months, however, so tracer experiments are impractical.

Possible strategies to overcome this problem involve obtaining the RTD on pilot-scale setups with various geometries, where tracer experiments are easily conducted, and performing a scale-up analysis or deriving the RTD by solving the Navier-Stokes and diffusion-convection equations that students learned in the fluid mechanics curricula. Commercially available CFD packages (e.g. Fluent, CFX, Fidap, Phoenics, STAR-CD, FLOW3D, etc.) can readily solve the balance equations for reactor operation coupled with the Navier-Stoke equations. Thus, we adopted the second approach.

The case study considers a 2-D reservoir with dimensions L (length) and H (height) in laminar flow and isothermal conditions (Fig. 9.2). Reservoirs with different aspect ratios (L/H) were considered, varying between 0.5 and 20 (with H = 0.1 m). Both the inlet and the outlet boundaries of the reservoirs have a height of 0.01 m, with distances from the bottom of the reservoir of 0 and 0.02 m, respectively. A fully developed parabolic velocity profile is imposed at the inlet boundary:

$$u_x = U_{max}\left[1-\left(\frac{y-H/20}{H/20}\right)^2\right] \qquad \ldots (9.1)$$

where, U_{max} is the fluid velocity at the centre of the inlet boundary, $U_{max} = 1.5\,U_{mean}$. A constant species concentration profile is assumed at the inlet boundary. On the walls (Fig. 9.2), no-slip conditions are assumed ($u_x = u_y = 0$) and a null flux (zero gradient) of species concentration is imposed. At the outflow boundary condition, the CFD code extrapolates the required information from the interior cells (a zero diffusion flux is assumed for all flow variables in the direction normal to the exit plane).

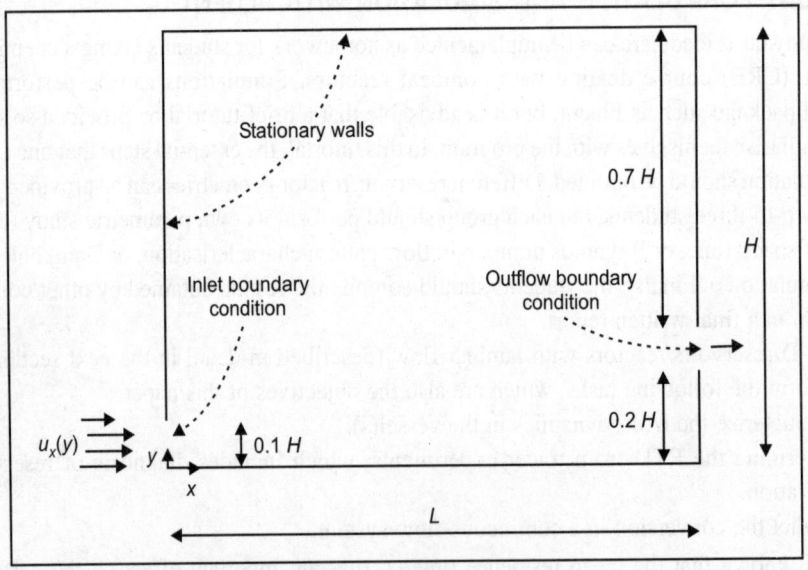

Fig. 9.2. Sketch of reservoir geometry.

In the simulations, the Reynolds number, here defined based on inlet conditions, ranged from 1 to 100. Changing Re for a given fluid and reservoir is equivalent to changing the fluid velocity (and thus the residence time). The governing equations to be solved are (for an incompressible fluid).

Continuity:

$$\frac{\partial u_x}{\partial x}+\frac{\partial u_y}{\partial y}=0 \qquad \ldots (9.2)$$

Momentum:

$$\rho\frac{\partial u_x}{\partial t}+\rho\left[\frac{\partial\left(u_x^2\right)}{\partial x}+\frac{\partial\left(u_x u_y\right)}{\partial y}\right]=-\frac{\partial P}{\partial x}+\mu\left(\frac{\partial^2 u_x}{\partial x^2}+\frac{\partial^2 u_x}{\partial y^2}\right) \qquad \text{... (9.3)}$$

$$\rho\frac{\partial u_y}{\partial t}+\rho\left[\frac{\partial\left(u_x u_y\right)}{\partial x}+\frac{\partial\left(u_y^2\right)}{\partial y}\right]=-\frac{\partial P}{\partial y}+\mu\left(\frac{\partial^2 u_y}{\partial x^2}+\frac{\partial^2 u_y}{\partial y^2}\right) \qquad \text{... (9.4)}$$

Species transport:

$$\frac{\partial C}{\partial t}+\frac{\partial\left(u_x C\right)}{\partial x}+\frac{\partial\left(u_y C\right)}{\partial y}=D\left(\frac{\partial^2 C}{\partial x^2}+\frac{\partial^2 C}{\partial y^2}\right)+S\left(C\right) \qquad \text{... (9.5)}$$

In Eq. 9.5, $S(C)$ is a source term. For the transport of an inert tracer, by convection-diffusion, $S(C) = 0$, while for the transport of a reagent species, $S(C) = -kC$ (assuming a first-order irreversible reaction).

Simulations were run with the commercial package Fluent 6.0. The fluid considered is water ($v = \mu/\rho = 10^{-6}$ m²s⁻¹), and a tracer solution was created in Fluent's database with identical properties of water, so that it does not affect reactor hydrodynamics. A molecular diffusivity of 5×10^{-10} m²s⁻¹ was considered, which is a typical value for liquids. A tracer step input was used, with uniform concentration across the entrance section, together with the parabolic velocity profile defined in Eq. 9.1. Fluent can simulate both the hydrodynamics and chemical reaction processes, and therefore the reservoirs previously considered can be used for modelling continuous-flow reactors (e.g. biological lagoons). In this case, simulations were run by defining a reactant and a product, both with properties identical to water. The reactor is initially full of water (inert), the laminar flow is established; after that (time $t = 0$) the reactant is fed to the reactor, similarly to the tracer step input.

Conversion of reactant is calculated based on the time evolution of species concentration at the reactor exit, obtained from a mass-weighted average formulation. We must point out that to achieve a high level of accuracy, all the simulation results presented in this section involved a detailed analysis of the numerical algorithms, the mesh employed, and the time step adopted (in transient simulations). For instance, the quick scheme of Leonard was selected for discretisation of the convective terms, a second-order implicit formulation was used during unsteady stimulation, and the computational grid contained typically 100×100 elements (note that for reservoirs with L/H $\gg$ 1, it is convenient to use a larger number of cells in the x-direction).

It is always a good practice to perform the calculations on several meshes with different levels of refinement in order to obtain mesh-independent results. A similar procedure should be adopted regarding the time step used in transient calculations. A control-volume approach is used by Fluent to numerically solve the governing equations.

Results and Discussion

Hydrodynamic characterisation

Figure 9.3 shows contour plots of the stream function within the reservoir, with L/H = 1, which illustrates the trajectories of the fluid elements (streamlines). For a given inlet velocity (or more broadly speaking, a given Reynolds number), the formation of a recirculation zone above the entrance of the reservoir, where velocity is small, is evident, thus suggesting formation of a stagnant region. It is noteworthy that the importance of such a region increases with Re, becoming particularly large for Re values around 100 where fluid trajectories are almost linear. Lower Re numbers lead to a smaller stagnant region and more curved streamlines. For Re < 1, inertia is negligible and the trajectories obtained are equivalent to creeping-flow conditions.

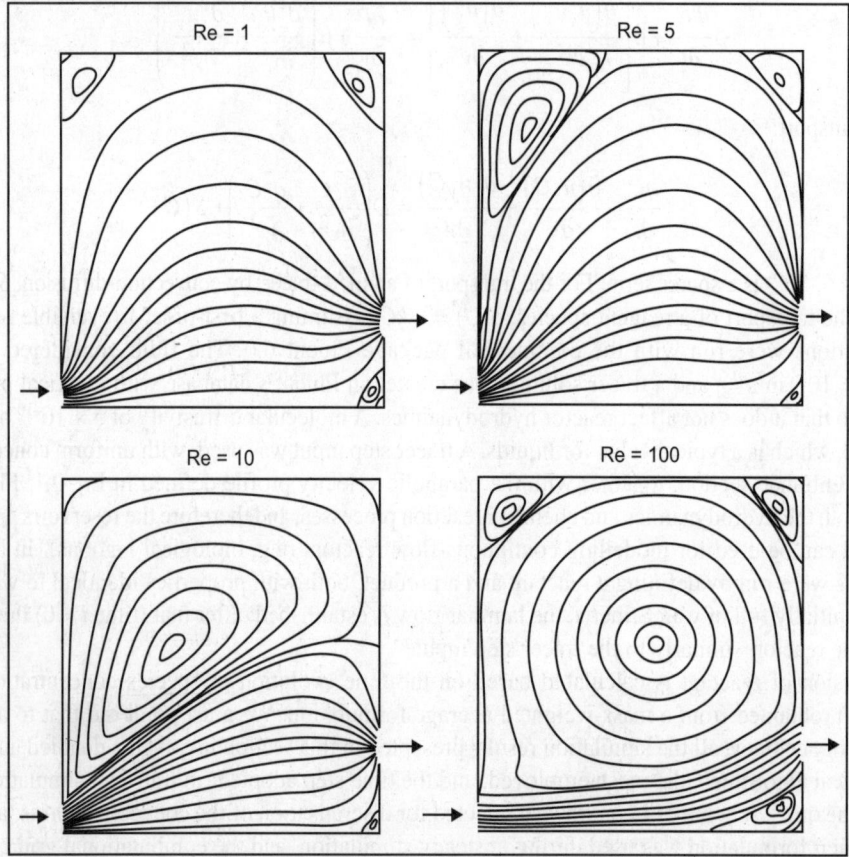

Fig. 9.3. Steady-state contours of the stream function for the reservoir with L/H = 1 as a function of the Reynolds number.

For longer reservoirs, the conclusions are similar, but now the importance of the recirculation zone decreases (for the same fluid inlet velocity). Indeed, the steady-state streamlines shown in Fig. 9.4, obtained for a reservoir with L/H = 5, show formation of a stagnant zone above the entrance, where the size increases with the Reynolds number. Comparison with the stream function contours of Fig. 9.3, however, shows that for the same Re, the fraction of dead volume decreases when the geometric ratio L/H increases.

RTD Determination from tracer experiments

After performing steady state simulations, students can proceed to transient runs, but they must first define a tracer step input at the inlet boundary condition. They must also be aware that for $t = 0$, no tracer exists within the reservoir and that the laminar flow is already established. They must first initialise the entire domain with a null tracer concentration and wait until the laminar regime is established before introducing the tracer step change at the inlet.

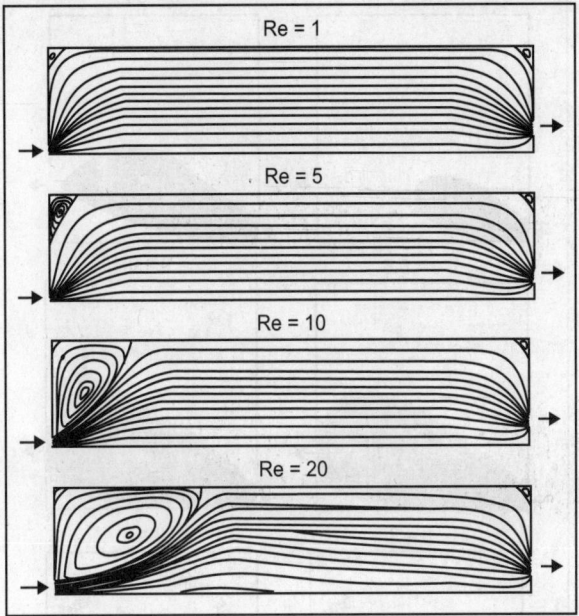

Fig. 9.4. Steady state contours of the stream function for the reservoir with L/H = 5 as a function of the Reynolds number.

After that, the CFD code solves the convection-diffusion equation that describes the tracer transport in the reservoir and the concentration field of tracer under transient regime is obtained. Particularly interesting is its concentration at the outflow boundary, $C_{out}(t)$. The contours of tracer concentration throughout the reservoir along time are also very interesting because they provide a good perspective on the evolution of concentration fronts. For a reservoir with L/H = 1, some frames were recorded at different times and are shown in Fig. 9.5. They show that only for $\theta = t/\tau$ around 0.22 can one start to 'see' tracer at the reservoir exit.

In addition, even for a very long time of operation (about five times the residence time), the reservoir is not completely full of tracer, due to the stagnant zone previously identified. To better illustrate this transient behaviour, it is possible to create an animation sequence with Fluent, using several frames obtained from the previous simulation.

With the data of transient tracer concentration at the outlet of the reservoir, which can be exported to an ASCII file, students can then compute the so-called Danckwerts' F curve, the normalised response of the reactor to a step input,

$$F(t) = \frac{C_{out}(t)}{C_{in}} \qquad \qquad ... (9.6)$$

Data can then be manipulated with a spreadsheet program such as Microsoft Excel. Results shown in Fig. 9.6 show the expected F curve, which only reaches the asymptotic value of 1 for very long times. It is worth mentioning the use of a logarithmic time scale, showing that tracer starts to exit the reservoir at around $\theta = t/\tau = 0.22$ due to its transport by convection, while fluid elements that enter the stagnant region only come out (by diffusion) much later (see detail of Fig. 9.6).

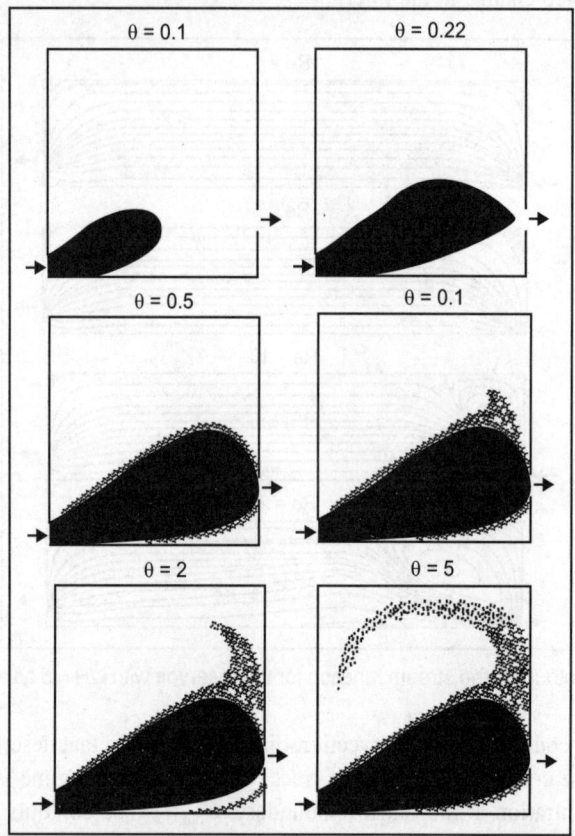

Fig. 9.5. Transient tracer concentration contours for the reservoir with L/H = 1 (Re =10).

With the response to a step input, the RTD function can now be computed by:

$$E(t) = \frac{dF(t)}{dt} \qquad \qquad \dots (9.7)$$

or, in terms of reduced time θ,

$$E(\theta) = \frac{dF(\theta)}{d\theta} = \tau E(t) \qquad \qquad \dots (9.8)$$

Figure 9.7 shows the RTD curves as a function of the Reynolds number, which gives an idea (for a given fluid) how space-time affects the RTD function. In all cases, the RTD curves evidence a very long tail and that a large fraction of fluid elements exits the reservoir with ages younger than the space-time, τ. Both features are indicative of the existence of large stagnant regions, together with slow recirculating flows near the inlet, as can be seen in Fig. 9.3. The fact that Re affects the RTD curve is also visible in Fig. 9.7. Higher Reynolds numbers imply that the fluid elements start to leave the reactor sooner and a higher fraction of fluid elements has a smaller residence time.

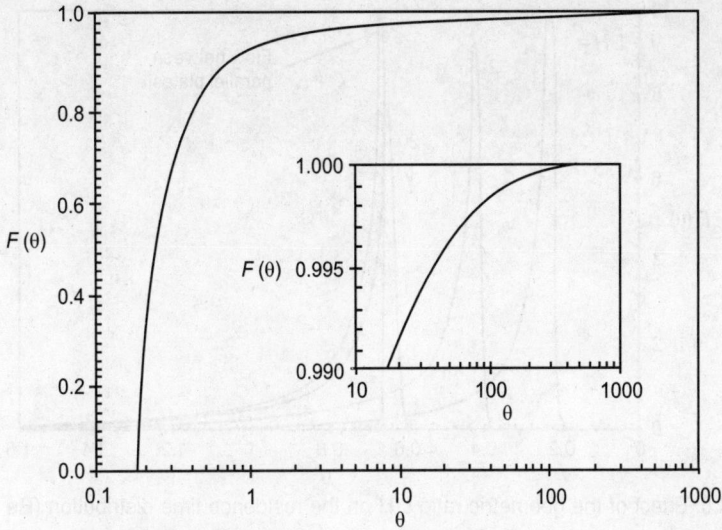

Fig. 9.6. Danckwerts' F curve for the reservoir, with L/H = 1 (Re = 10).

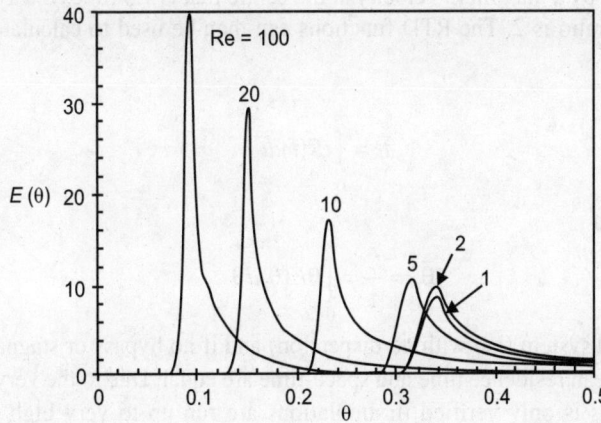

Fig. 9.7. Effect of the Reynolds number on the residence time distribution (reservoir with L/H = 1).

The effect of the reservoir geometry on the RTD is shown in Fig. 9.8. It can be seen that when L/H increases, the curves are shifted to the right, i.e. the mean residence time seems to increase because the importance of the recirculating zone decreases. This was previously found in the steady-state streamlines shown in Figs 9.3 and 9.4. It is particularly noteworthy that for very long reservoirs, one tends toward an asymptotic $E(\theta)$ curve, also shown in Fig. 9.8. This curve corresponds to the case of laminar flow between parallel plates, given by:

$$E(\theta) = \begin{cases} \dfrac{1}{30^3\sqrt{1-\dfrac{2}{30}}} & \theta \geq \dfrac{2}{3} \\[4mm] 0 & \theta < \dfrac{2}{3} \end{cases} \qquad \qquad ...(9.9)$$

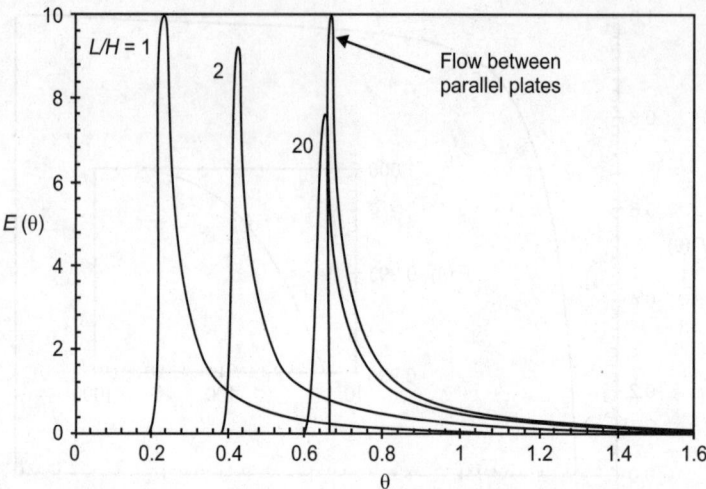

Fig. 9.8. Effect of the geometric ratio L/H on the residence time distribution (Re = 10).

It must be stressed that for this situation, i.e. laminar flow between parallel plates, the parabolic profile is characterised by a maximum velocity at the centre that is 1.5 times the average velocity, while for flow in pipes this ratio is 2. The RTD functions can then be used to calculate the mean residence time, defined as:

$$\bar{t}_r = \int_0^\infty t E(t)dt \qquad \text{... (9.10)}$$

or

$$\bar{\theta}_r = \frac{\bar{t}_r}{\tau} = \int_0^\infty \theta E(\theta)d\theta \qquad \text{... (9.11)}$$

For a closed-closed system (i.e. with no dispersion) and if no bypass or stagnant regions exist, it is well known that the mean residence time and space-time are equal. Due to the very long tail of the RTD function, however, this is only verified if simulations are run up to very high times, typically θ of $0(10^3)$. Otherwise, the normalisation condition:

$$\int_0^\infty E(t)dt = \int_0^\infty E(\theta)d\theta = 1$$

is not satisfied and the computed mean residence time is smaller than the real space-time.

Calculation of the mean residence time is very important for evaluation of malfunctions during the reactor's operation. Indeed, a straightforward way to diagnose the reactor's flow performance consists of comparing the computed value of the mean residence time from Eq. 9.10 with the space-time ($\tau = V/Q$, where V is the total volume of the reactor and Q is the volumetric flow rate), which is equivalent to comparing $\bar{\theta}_r$ with 1.

Disagreement between these two values may indicate the existence of bypasses, dead volumes, and odouriser. For instance, as indicated by Froment, if a region of the vessel retains a portion of the fluid for an order of magnitude greater than the mean residence time of the total fluid, then for all practical purposes, that portion is essentially at rest and the region is wasted space in the vessel.

The values obtained for the mean residence time shown in Tables 9.1 and 9.2 were calculated by integration of the RTD curves up to θ = 10. It is evident that they strongly depend on both the Reynolds number and the geometry of the reservoir. As Re decreases, the mean residence time increases (Table 9.1). As expected, when L/H increases, the mean residence time approaches the space-time value (Table 9.2).

Table 9.1. Influence of the Reynolds number on the mean residence time and fraction of dead volume (Reservoir with L/H = 1).

	Values calculated from RTD curves up to θ = 10	
	Mean residence time	Fraction of dead volume
Re	$\bar{\theta}_r = \bar{t}_r / \tau$	V_d/V
1	0.778	0.222
5	0.724	0.276
10	0.412	0.588
20	0.318	0.682
100	0.261	0.739

Table 9.2. Influence of reservoir geometry on the mean residence time and fraction of dead volume (for Re = 10).

	Values calculated from RTD curves up to θ = 10	
	Mean residence time	Fraction of dead volume
L/H	$\bar{\theta}_r = \bar{t}_r / \tau$	V_d/V
0.5	0.300	0.700
1	0.412	0.588
2	0.752	0.248
5	0.935	0.065
10	0.992	0.008
20	0.997	0.003

Because in all cases, $\bar{t}_r < \tau$ or $\bar{\theta}_r < 1$, we can conclude that a stagnant region exists, which in practice would be a dead volume, leading to a lower reactor performance. The fraction of the reactor volume occupied by the dead region is given by:

$$\frac{V_d}{V} = 1 - \frac{\bar{t}_r}{\tau} \qquad \text{... (9.12)}$$

The dead volume fractions obtained, shown in Tables 9.1 and 9.2, indicate that for high Re values (or high fluid velocities) and for geometries where L/H approaches 1, or even smaller, a large fraction of the reservoir will not be efficiently used for reaction purposes.

Prediction of conversion in the continuous-flow reactor

In a real reactor, the RTD function can be used to predict the limiting values of conversion under the two extremes of micromixing, using the well-known total segregation or maximum mixedness models. For first-order reactions (linear systems), however, the state of mixing does not affect conversion, and therefore the easy-to-use segregation approach can be applied to predict reactor performance. The total segregation model assumes that all fluid elements having the same age (residence time) 'travel together' in the reactor and do not mix with elements of different ages until they exit the reactor. Because there is no interchange of matter between fluid elements, each one acts as a batch reactor, so the mean steady-state conversion $\left(\bar{X}\right)$ in the real reactor is given by:

$$\bar{X} \int_0^\infty X_{batch}(t) = E(t)dt = \int_0^\infty X_{batch}(\theta)E(\theta)d\theta \qquad \text{... (9.13)}$$

where, $X_{batch}(t)$, for a first-order reaction, is given by:

$$X_{batch} = 1 - e^{(-kt)} = 1 - e^{(-Da\theta)} \qquad \text{... (9.14)}$$

and $Da = k\tau$ is the Damköhler number.

Steady-state conversion is then computed by using Eq. 9.13, with the RTD function previously determined. It is important to remark that the segregation model can also be used for prediction of the reactor transient behaviour. In this case, the upper integration limit in Eq. 9.13 must be set to t (or θ). This was done for our case study and the results, shown in Fig. 9.9, illustrate the reactant conversion in transient conditions, up to steady-state, for different Damköhler values. One must take care that the RTD used for prediction of reactor performance depends on its geometry and on the Reynolds number. In addition, because of the very long tail of the RTD function [as shown in Fig. 9.6 for the $F(\theta)$ curve], prediction of steady-state conversion requires RTD data up to very large times (note the logarithmic time scale).

This interesting feature is also evident in Fig. 9.9 — a non-negligible contribution to the overall reactor performance (in terms of fractional conversion), which is noticed at very long times. Such behaviour can be attributed to the stagnant region and to the different time scales for the involved phenomena: reaction, convection, and molecular diffusion.

As mentioned above, Fluent can also be used to simulate the system in the presence of a reaction, and so the data shown in Fig. 9.9 can be obtained either through the total segregation model or directly from CFD simulations (obtained curves coincide).

Asking students to compare results from both approaches is important because they feel more confident about the simulation results and calculations.

Simulation of the continuous flow reactor via CFD can also be used to evidence the contours of species concentration throughout the reactor, for instance at steady-state.

Finally, it is convenient to ask students to compare the steady-state conversion attained in the real reactor with those achieved with the ideal reactors that they learned in previous CRE courses: continuous stirred tank and plug flow reactors. For a first-order reaction, performance achieved by these reactors is given by:

$$X_{CSTR} = \frac{Da}{1+Da} \qquad \text{... (9.15)}$$

$$X_{PFR} = 1 - e^{(-Da)} \qquad \text{... (9.16)}$$

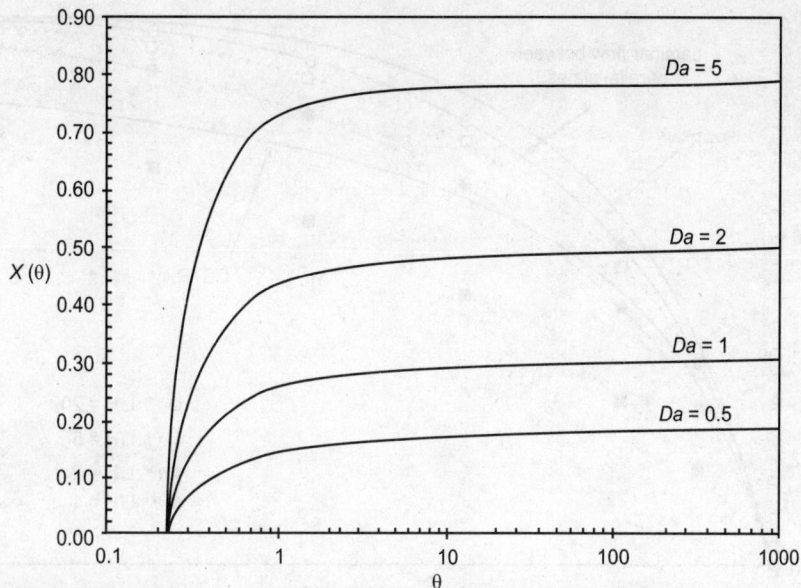

Fig. 9.9. Unsteady-state conversion obtained for the reactor with L/H = 1, predicted from the segregation model (Re = 10).

Data shown in Fig. 9.10 indicate that when dead regions or stagnant zones are negligible, i.e. for geometries where L/H is higher than about 5 (for Re = 10), the performance of the real reactor lies between that of the CSTR and PFR. When L/H is close to 1, or even smaller, such anomaly (dead volume) leads to a much lower performance of the nonideal reactor — even lower than that achieved with a perfectly mixed reactor. It is also noteworthy that for very long reactors (i.e. high L/H ratios), one approaches the theoretical behaviour of a laminar flow reactor (now between parallel plates) computed using the RTD function given in Eq. 9.9 and the segregation model.

Using reservoirs/reactors with different geometries and/or different operating conditions, students may be asked to:

1. Characterise the hydrodynamics.
2. Determine the residence time distribution from tracer experiments, which provides diagnosis of reactor operation.
3. Predict cunversion in a continuous-flow reactor (both steady-state and transient behaviour).

Our experience shows that use of the CFD code allows students to more easily understand some of the basic concepts taught in CRE curricula. Finally, comparison of numerical with analytical solutions known for laminar flow between parallel plates (i.e. for geometries with high U/H ratios) improves their self-reliance regarding CFD results.

In a survey sent a few years ago to chemical engineering departments spread all over the world, two of the main points addressed by the departments to a question relating to the future of CRE courses were: the increasing importance of computer applications and software packages, and putting more emphasis on nonideal reactors. With the case study herein proposed, both issues are dealt with. In addition, students learn the potential of CFD codes, which have been successfully used in practice to design commercial-size reactors, usually with complex flow processes.

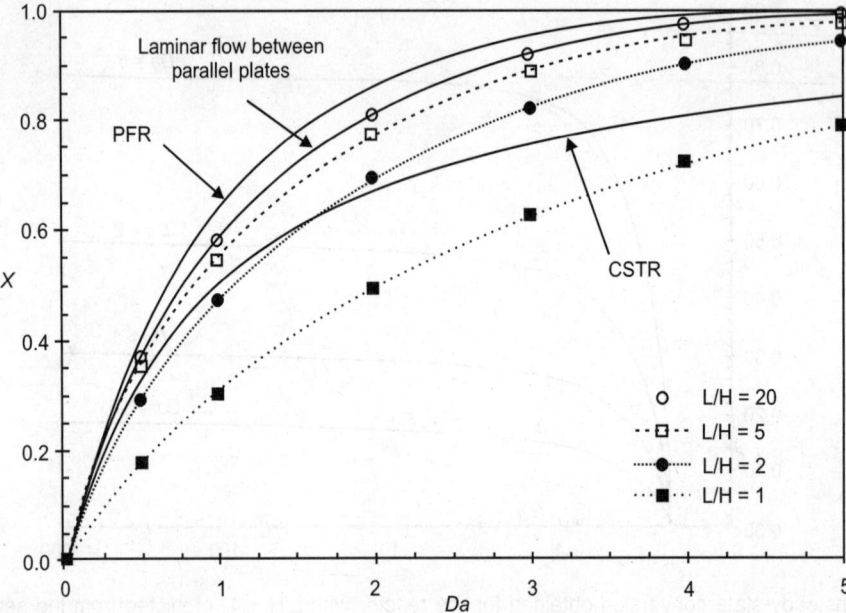

Fig. 9.10. Steady-state conversion versus Damköhler number for reactors with different geometries (Re = 10).

BOUNDARY CONDITIONS

In a numerical simulation, it is impossible and unnecessary to simulate the whole universe. Generally we choose a region of interest in which we conduct a simulation. The interesting region has a certain boundary with the surrounding environment. Numerical simulations also have to consider the physical processes in the boundary region. In most cases, the boundary conditions are very important for the simulation region's physical processes. Different boundary conditions may cause quite different simulation results. Improper sets of boundary conditions may introduce nonphysical influences on the simulation system, while a proper set of boundary conditions can avoid that. So arranging the boundary conditions for different problems becomes very important. While at the same time, different variables in the environment may have different boundary conditions according to certain physical problems. Commonly there are several different types of boundary conditions:

Fixed Boundary Condition

This kind of boundary condition fits for those environment values that do not change with time and physical processes well interior to the simulation region. If the physical process concentrates in the centre of the simulation region and causes very little influence on the boundary, and at the same time, the surrounding environment of the simulation region is stable, then we can set the boundary to fixed boundary conditions.

$$f(x_b) = f_0 \qquad \qquad ... (9.17)$$

where, $f(x_b)$ is the boundary value, f_0 is a fixed value.

Linear Boundary Condition

If the influence of the physical processes in the simulation region is large enough to reach boundary, then we have to consider the interaction of this influence and outer environment. At this time, the boundary will change according to the result of this interaction. A simple treatment of this case is to look on the boundary as a linear continuous boundary. Assuming that $f(x_b)$ is the boundary value, $f(x_b - \Delta x)$ the value of the inner point adjacent to x_b, $f(x_b - \Delta x)$ the value of the outer point adjacent to x_b. Then for a linear boundary condition we can have:

$$f(x_b - \Delta x) = 2f(x_b) - f(x_b - \Delta x) \qquad \qquad \text{... (9.18)}$$

Using the newly found $f(x_b - \Delta x)$ we can easily continue our simulation on the boundary.

Symmetric Boundary Condition

In some of the simulations, we can assume a symmetric state exists on the boundary. This treatment of the boundary condition corresponds to the physical assumption that, on the two sides of boundary, the same physical processes exist. The variable values at the same distance from the boundary at the two sides are the same. The function of such a boundary is that of a mirror that can reflect all the fluctuations generated by the simulation region. Assume $f(x_b)$, $f(x_b - \Delta x)$ are two adjacent boundary values, a typical set of symmetric boundary value is:

$$f(x_b + \Delta x) = f(x_b) \qquad \qquad \text{... (9.19)}$$

$$f(x_b + 2\Delta x) = f(x_b - \Delta x) \qquad \qquad \text{... (9.20)}$$

Time Varying Boundary Condition

Some simulations, such as global MHD simulation, space weather forecasting simulation, etc. combine observational data in the model and use it as a boundary condition. Then the boundary condition will change with time. The treatment of such boundary is more complicated as an additional interface between observation data and program is needed. At this time, the boundary condtion can be expressed as the following form:

$$f(x_b) = f(t) \qquad \qquad \text{... (9.21)}$$

where, $f(t)$ is decided by the observational data or other time varying data.

Despite its complexity, a time varing boundary condition is a more realistic way to do space numerial simulations.

Boundary Value Problem

In mathematics, in the field of differential equations, a boundary value problem is a differential equation together with a set of additional restraints, called the boundary conditions. A solution to a boundary value problem is a solution to the differential equation which also satisfies the boundary conditions (Fig. 9.11).

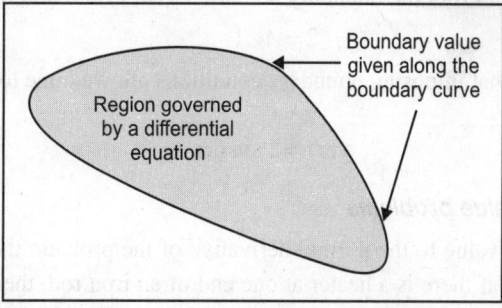

Fig. 9.11. Shows a region where a differential equation is valid and the associated boundary values.

Boundary value problems arise in several branches of physics as any physical differential equation will have them. Problems involving the wave equation, such as the determination of normal modes, are often stated as boundary value problems. A large class of important boundary value problems are the Sturm–Liouville problems. The analysis of these problems involves the eigenfunctions of a differential operator. To be useful in applications, a boundary value problem should be well posed. This means that given the input to the problem there exists a unique solution, which depends continuously on the input. Much theoretical work in the field of partial differential equations is devoted to proving that boundary value problems arising from scientific and engineering applications are in fact well-posed.

Among the earliest boundary value problems to be studied is the Dirichlet problem, of finding the harmonic functions (solutions to Laplace's equation); the solution was given by the Dirichlet's principle.

Initial Value Problem

A more mathematical way to picture the difference between an initial value problem and a boundary value problem is that an initial value problem has all of the conditions specified at the same value of the independent variable in the equation (and that value is at the lower boundary of the domain, thus the term 'initial' value). On the other hand, a boundary value problem has conditions specified at the extremes of the independent variable. For example, if the independent variable is time over the domain $[0,1]$ an initial value problem would specify a value of $y(t)$ and $y'(t)$ at time $t = 0$, while a boundary value problem would specify values for $y(t)$ at both $t = 0$ and $t = 1$.

If the problem is dependent on both space and time, then instead of specifying the value of the problem at a given point for all time the data could be given at a given time for all space. For example, the temperature of an iron bar with one end kept at absolute zero and the other end at the freezing point of water would be a boundary value problem.

Concretely, an example of a boundary value (in one spatial dimension) is the problem:

$$y'' (x) + y(x) = 0$$

to be solved for the unknown function $y(x)$ with the boundary conditions:

$$y(0) = 0, \; y(\pi/2) = 2.$$

Without the boundary conditions, the general solution to this equation is:

$$y(x) = A \sin (x) + B \cos(x).$$

From the boundary condition $y(0) = 0$ one obtains:

$$0 = A \cdot 0 + B \cdot 1$$

which implies that $B = 0$. From the boundary condition $y(\pi/2) = 2$ one finds:

$$2 = A \cdot 1$$

and so $A = 2$. One sees that imposing boundary conditions allowed one to determine a unique solution, which in this case is:

$$y(x) = 2 \sin (x).$$

Types of boundary value problems

If the boundary gives a value to the normal derivative of the problem then it is a Neumann boundary condition. For example, if there is a heater at one end of an iron rod, then energy would be added at a constant rate but the actual temperature would not be known (Fig. 9.12).

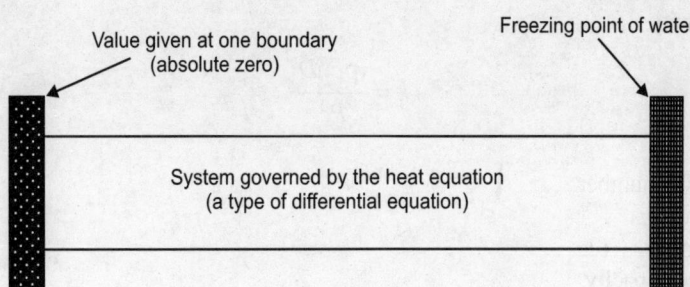

Fig. 9.12. The boundary value problem for an idealised 2D rod.

If the boundary gives a value to the problem then it is a Dirichlet boundary condition. For example, if one end of an iron rod is held at absolute zero, then the value of the problem would be known at that point in space. If the boundary has the form of a curve or surface that gives a value to the normal derivative and the problem itself then it is a Cauchy boundary condition.

Aside from the boundary condition, boundary value problems are also classified according to the type of differential operator involved. For an elliptic operator, one discusses elliptic boundary value problems. For an hyperbolic operator, one discusses hyperbolic boundary value problems. These categories are further subdivided into linear and various nonlinear types.

Peclet Number Calculator

Peclet number is a dimensionless number relating the rate of advection of a flow to its rate of diffusion, often thermal diffusion. Here we can calculate for Peclet number, velocity, density, heat capacity, characteristic length, thermal conductivity.
Paclet number:

$$Pe = \frac{v\rho c_p D}{k}$$

Velocity:

$$v = \frac{Pek}{\rho c_p D}$$

Density:

$$\rho = \frac{Pek}{v c_p D}$$

Heat capacity:

$$c_p = \frac{Pek}{v\rho D}$$

Characteristic length:

$$D = \frac{Pek}{v\rho c_p}$$

Thermal conductivity:

$$k = \frac{v\rho c_p D}{Pe}$$

where,

Pe = Peclet number.
v = Velocity.
ρ = Density.
c_p = Heat capacity.
D = Characteristic length.
k = Thermal conductivity.

Laminar Flow and Turbulent Flow of Fluids

Resistance to flow in a pipe

When a fluid flows through a pipe the internal roughness (e) of the pipe wall can create local eddy currents within the fluid adding a resistance to flow of the fluid. Pipes with smooth walls such as glass, copper, brass and polyethylene have only a small effect on the frictional resistance. Pipes with less smooth walls such as concrete, cast iron and steel will create larger eddy currents which will sometimes have a significant effect on the frictional resistance.

The velocity profile in a pipe will show that the fluid at the centre of the stream will move more quickly than the fluid towards the edge of the stream. Therefore, friction will occur between layers within the fluid. Fluids with a high viscosity will flow more slowly and will generally not support eddy currents and therefore the internal roughness of the pipe will have no effect on the frictional resistance. This condition is known as laminar flow.

Reynolds number

The Reynolds number (Re) of a flowing fluid is obtained by dividing the kinematic viscosity (viscous force per unit length) into the inertia force of the fluid (velocity × diameter)

$$\text{Kinematic viscosity} = \frac{\text{Dynamic viscosity}}{\text{Fluid density}}$$

$$\text{Reynolds number} = \frac{\text{Fluid velocity} \times \text{internal pipe diameter}}{\text{Kinematic viscosity}}$$

Laminar flow

Where the Reynolds number is less than 2300 laminar flow will occur and the resistance to flow will be independent of the pipe wall roughness. The friction factor for laminar flow can be calculated from 64/Re.

Turbulent flow: Turbulent flow occurs when the Reynolds number exceeds 4000 (Fig. 9.13).

Eddy currents are present within the flow and the ratio of the internal roughness of the pipe to the internal diameter of the pipe needs to be considered to be able to determine the friction factor. In large diameter pipes the overall effect of the eddy currents is less significant. In small diameter pipes the internal roughness can have a major influence on the friction factor. The 'relative roughness' of the pipe and the Reynolds number can be used to plot the friction factor on a friction factor chart. The friction factor can be used with the Darcy-Weisbach formula to calculate the frictional resistance in the pipe. Between the Laminar and Turbulent flow conditions (Re 2300 to Re 4000) the flow condition is known as critical. The flow is neither wholly laminar nor wholly turbulent. It may be considered as a combination of the two flow conditions.

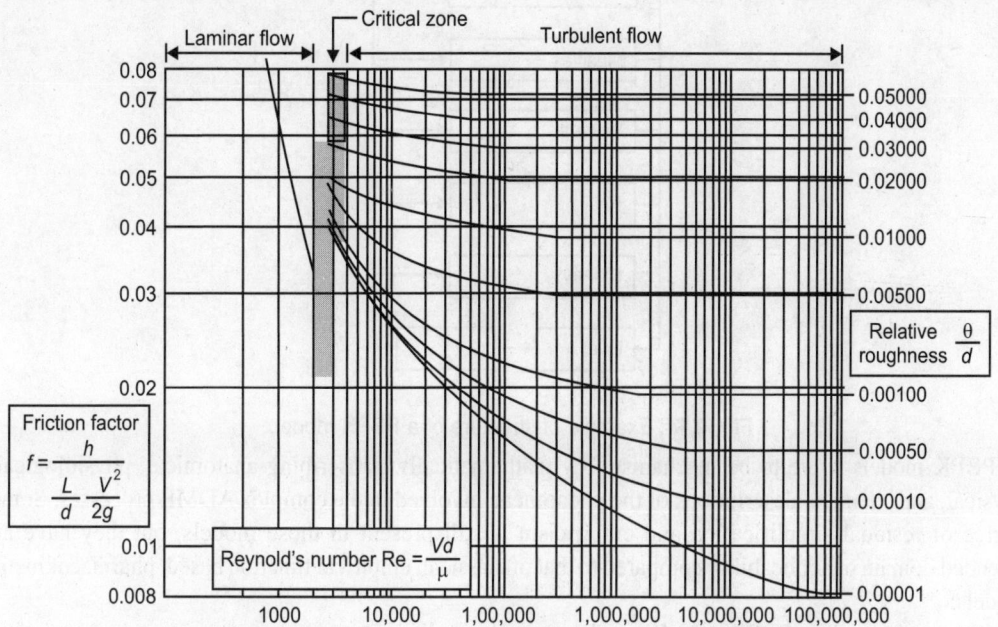

Fig. 9.13. Turbulent flow occurs when the Reynolds number exceeds 4000.

The friction factor for turbulent flow can be calculated from the Colebrook-White equation:

$$\frac{1}{\sqrt{f}} = 1.14 - 2\log_{10}\left(\frac{e}{D} + \frac{9.35}{Re\sqrt{f}}\right) \text{ for } Re > 4000$$

Internal roughness (e) of common pipe materials is given in Table 9.3.

Table 9.3. Internal roughness (e) of common pipe materials.

Cast iron (Asphalt dipped)	0.1220 mm	0.004800″
Cast iron	0.4000 mm	0.001575″
Concrete	0.3000 mm	0.011811″
Copper	0.0015 mm	0.000059″
PVC	0.0050 mm	0.000197″
Steel	0.0450 mm	0.001811″
Steel (galvanised)	0.1500 mm	0.005906″

PHYSIOLOGICALLY-BASED PHARMACOKINETIC MODELLING

Physiologically-based pharmacokinetic (PBPK) modelling is a mathematical modelling technique for predicting the absorption, distribution, metabolism and excretion (ADME) of a compound in humans and other animal species. PBPK modelling is used in pharmaceutical research and development, and in health risk assessment (Fig. 9.14).

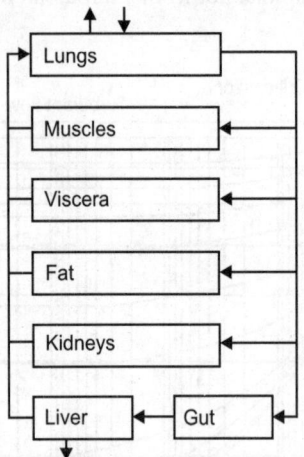

Fig. 9.14. Example of structure of a PBPK model.

PBPK models strive to be mechanistic by mathematically transcribing anatomical, physiological, physical, and chemical descriptions of the phenomena involved in the complex ADME processes. Some degree of residual simplification and empiricism is still present in those models, but they have an extended domain of applicability compared to that of classical, empirical function based, pharmacokinetic models.

Given that property, PBPK models may have purely predictive uses, but other uses, such as statistical inference, have been made possible by the development of Bayesian statistical tools able to deal with complex models. That is true for both toxicity risk assessment and therapeutic drug development.

PBPK models try to rely *a priori* on the anatomical and physiological structure of the body. These are usually also multi-compartment models, but the compartments correspond to predefined organs or tissues, for which the interconnections correspond to blood or lymph flows (more rarely to diffusions). A system of differential equations can still be written, but its parameters represent blood flows, pulmonary ventilation rate, organ volumes, etc. for which information is available in scientific publications. Indeed the description of the body is simplified and a balance needs to be struck between complexity and simplicity.

Besides the advantage of allowing the recruitment of *a priori* information about parameter values, these models also facilitate inter-species transpositions or extrapolation from one mode of administration to another (e.g. inhalation to oral).

It is interesting to note that the first pharmacokinetic model described in the scientific literature was in fact a PBPK model. It led, however, to computations intractable at that time. The focus shifted then to simpler models, for which analytical solutions could be obtained (such solutions were sums of exponential terms, which led to further simplifications). The availability of computers and numerical integration algorithms marked a renewed interest in physiological models in the early 1970s.

PBPK models are compartmental models like many others, but they have a few advantages over so-called 'classical' pharmacokinetic models, which are less grounded in physiology. PBPK models can first be used to abstract and eventually reconcile disparate data (from physico-chemical or biochemical experiments, *in vitro* or *in vivo* pharmacological or toxicological experiments, etc.). They give also access to internal body concentrations of chemicals or their metabolites, and in particular at the site of their effects, be it therapeutic or toxic. Finally they also help interpolation and extrapolation of knowledge between:

1. Doses: e.g. from the high concentrations typically used in laboratory experiments to those found in the environment.
2. Exposure duration: e.g. from continuous to discontinuous, or single to multiple exposures.
3. Routes of administration: e.g. from inhalation exposures to ingestion.
4. Species: e.g. transpositions from rodents to human, prior to giving a drug for the first time to subjects of a clinical trial, or when experiments on humans are deemed unethical, such as when the compound is toxic without therapeutic benefit.
5. Individuals: e.g. from males to females, from adults to children, from non-pregnant women to pregnant.

Some of these extrapolations are 'parametric': Only changes in input or parameter values are needed to achieve the extrapolation (this is usually the case for dose and time extrapolations). Others are 'nonparametric' in the sense that a change in the model structure itself is needed (e.g. when extrapolating to a pregnant female, equations for the fetus should be added).

Nomenclature

C	Concentration of tracer, reactant, or product (mol.m^3) or kg.m^{-3}).
D	Diffusivity (m^{2}s).
Da	Damköhler number, dimensionless.
$E(t)$	Resident-time distribution function (s^{-1}).
$E(\theta)$	Normalised RTD function, dimensionless
$F(t)$	Danckwerts' F curve, dimensionless.
H	Height of the reservoir/reactor (m).
k	Reaction rate constant of the first-order reaction (s^{-1}).
L	Length of the reservoir/reactor (m).
P	Pressure (*Pa*).
S	Source term (mol m^{-3}s^{-1} or kg m^{-3}s^{-1}).
Q	Volumetric flow rate (m^{3}s^{-1}).
Re	Reynolds number, dimensionless.
t	Time (s).
$\bar{t_r}$	Mean residence time (s).
U_{max}	Fluid velocity at the centre of the inlet boundary (ms^{-1}).
U_{mean}	Fluid velocity at the inlet boundary (ms^{-1}).
u_x	x-velocity (ms^{-1}).
u_y	y-velocity (ms^{-1}).
V	Volume of reservoir/reactor (m^3).
V_d	Dead volume in the reservoir/reactor (m^3).
X	Conversion, dimensionless.
x	Horizontal coordinate (m).
y	Vertical coordinate (m).

Subscripts

batch	Refers to batch reactor.
CSTR	Continuous stirred tank reactor.
in	Inlet conditions.
PFR	Plug flow reactor.
out	Outflow conditions.

Greek symbols

$\theta = t/\tau$	Reduced time, dimensionless.
$\bar{\theta}_r = \bar{t}_r / \tau$	Reduced mean residence time, dimensionless.
μ	Viscosity of the fluid (kg m^{-1}s^{-1}).
v	Kinematic viscosity of the fluid (m^{2}s^{-1}).
ρ	Density of the fluid (kg m^{-3}).
τ	Space-time (s).

Chapter 10

Chain Reactions and Combustion Reactors

INTRODUCTION

Combustion or burning is the sequence of exothermic chemical reactions between a fuel and an oxidant accompanied by the production of heat and conversion of chemical species. The release of heat can result in the production of light in the form of either glowing or a flame. Fuels of interest often include organic compounds (especially hydrocarbons) in the gas, liquid or solid phase.

Polymerisation also involves chain reactions, so some aspects related to polymerisation processes are also discussed.

CHAIN REACTIONS

In this section we will consider chain reactions in a homogeneous system. Chain reactions can be broken down into the following steps: initiation, chain propagation, inhibition, branching, and termination steps. The reactive species in a chain reaction are called the chain carriers which are reactive intermediates (normally free radicals) that are generated in an initiation step. In the propagation step the chain carriers react with reactants to produce products and regenerate the chain carriers. Chain carriers can also react with products to reform the reactants and the chain carrier. When this step occurs it is called the inhibition step which involves no reduction in the number of chain carriers. If there is a step in which two or more carriers are produced from a single chain carrier, the step is called chain branching. In the termination step chain carriers are consumed.

Free Radicals

Free radicals are typically produced by the homolytic cleavage of covalent bonds. There are numerous methods for producing free radicals. As an example consider the dissociation of chlorine gas Cl_2 which has an absorption band in the visible spectrum. The bond dissociation energy for Cl_2 is about 58 kcal/mol. Thus sunlight can provide sufficient energy for the cleavage of the chlorine-chlorine bond:

$$\text{Sunlight:} \qquad Cl-Cl \xrightarrow{hv} Cl^{\bullet} + Cl^{\bullet} \qquad \qquad ... (10.1)$$

The bond cleavage is called homolytic because each chlorine atom takes one electron from the bonding electron pair. The product from the bond cleavage $Cl^{\bullet}$ is called a neutral radical, or free radical, that has an unpaired electron in its outer shell. Recall each chlorine atom has 7 electrons in its outer shell. Free radicals can also be formed by thermal cracking, an important process in the refining of petroleum crude. At temperatures in excess of 500°C, high molecular weight alkanes break down into smaller alkane and alkene fragments through the homolysis of the C-C bond.

Peroxides are another source of free radicals. The weak O-O bond of peroxides are cleaved at temperatures around $80°-150°C$ to produce oxy radicals, as shown in this example involving the decomposition of tert-butyl peroxide:

$$(CH_3)_3C-O-O-C(CH_3)_3 \longrightarrow (CH_3)_3C-O^{\bullet} + O^{\bullet} - C(CH_3)_3 \xrightarrow{hv} Cl^{\bullet} + Cl^{\bullet} \quad ... (10.2)$$

Azobisisobutyronitrile (AIBN) is a common radical initiator from the azo compound family (R–N=N–R) that decomposes at $70°$ to $80°C$ to produce 2 isobutyronitrile radicals and nitrogen:

$$\underset{\substack{|\\CH_3}}{\overset{\substack{CN\\|}}{H_3C-C}}-N=N-\underset{\substack{|\\CH_3}}{\overset{\substack{CN\\|}}{C}}-CH_3 \rightarrow 2H_3C-\underset{\substack{|\\CH_3}}{\overset{\substack{CN\\|}}{C}} + N_2 \qquad ... (10.3)$$

Example 10.1: Decomposition of acetaldehyde

The classic example of a chain reaction is the decomposition of acetaldehyde to methane and carbon monoxide. The overall reaction is given by:

$$CH_3CHO \rightarrow CH_4 + CO \qquad ... (10.4)$$

In a chain reaction extremely reactive species (called chain carriers) are initially produced by thermal effects (from collisions) or photochemically in what is called an initiation step. The chain carriers are regenerated in a series of propagation steps and if the chain is not terminated one can achieve a runaway reaction leading to explosions. Chain carriers are normally neutral radicals such as $H^{\bullet}$, $CH_3^{\bullet}$ with unpaired electrons that make them highly reactive. Several mechanisms for the decomposition of acetaldehyde have been proposed. We will start our discussion with the following mechanism:

Initiation step: $\qquad CH_3CHO \xrightarrow{k_1} CH_3^{\bullet} + CHO^{\bullet}$

Propagation step I: $\qquad CH_3^{\bullet} + CH_3CHO \xrightarrow{k_4} CH_4 + CH_3CO^{\bullet}$

Propagation step II: $\qquad CH_3CO^{\bullet} \xrightarrow{k_5} CH_3^{\bullet} + CO \qquad\qquad ... (10.5)$

Termination step: $\qquad 2CH_3^{\bullet} \xrightarrow{k_6} C_2H_6$

The above mechanism does not predict the formation of a minor product H_2 which is known to occur. But for our initial discussion the above mechanism will suffice. We will examine a more complicated mechanism in the next section. First, note that the two propagations steps yield the overall decomposition reaction. Physically what this means is that the propagation steps occur much faster than the initiation and termination steps. Note the formation of ethane in the termination step. This is called a minor product of the chain reaction as it is not produced during the propagation steps. The free radicals involved in the propagation step are highly unstable and react rapidly. The notation $\bullet$ means that these species have an unpaired electron—the reason for their reactivity. We say the chain is fed by the initiation step (where the chain carriers are produced) and destroyed by the termination step which consumes the radical $CH_3^{\bullet}$. One can think of the propagation steps as a kinetic chain reaction: carriers are feed into the reaction sequence, and products are produced.

Note if we add the two propagation steps we get:

$$CH_3CHO + CH_3^{\bullet} \rightarrow CO + CH_4 + CH_3^{\bullet} \qquad ... (10.6)$$

Thus we can think of CH_3^* as the catalyst for the reaction. In fact mechanism (Eq. 10.6) is an autocatalytic reaction, because the initiation step generates the catalyst (CH_3^*), which is needed for the overall reaction. Figure 10.1 shows a sketch of the chain reaction and the catalytic loop.

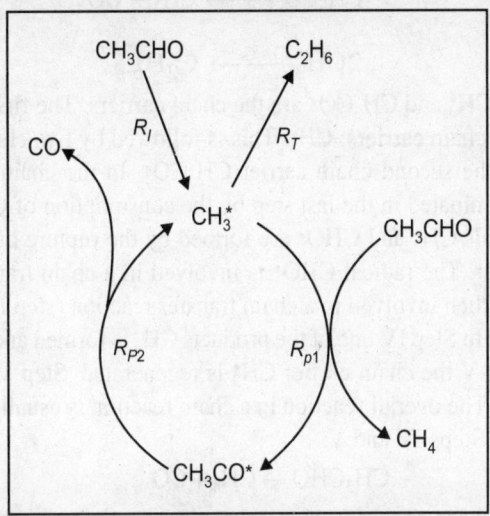

Fig. 10.1. Sketch of the chain reaction and the catalytic loop.

In the above sketch, R_I and R_T denote the rates of reaction for the initiator and termination steps:

$$R_I = k_I \, C_{CH_3CHO}$$

$$R_T = k_T C_{CH_3^*}^2 \qquad \qquad \text{... (10.7)}$$

The reaction rates for the propagations steps are:

$$R_{P1} = k_{P1} C_{CH_3^*} C_{CH_3CHO}$$

$$R_{P2} = k_{P2} C_{CH_3CO^\bullet} \qquad \qquad \text{... (10.8)}$$

The above figure show clearly how the chain carrier is fed into the autocatalytic loop by the initiation step and remove from the autocatalytic loop by the termination step. Note that the figure does not account for the radical species $CHO^\bullet$. Thus the diagram of the mechanism is incomplete. The reason is because we have not accounted for the minor products H_2. This is done in the next section.

Rice-Hersfeld Mechanism

In this section examine the free radical chain mechanism for the thermal decomposition of acetaldehyde to CH_4 and CO, that also produces the minor products H_2 and C_2H_6. The mechanism involves what is called transfer steps. The mechanism is due to Rice-Hersfeld. There are six elementary steps:

R_1 – Initiation step I: $\qquad\qquad CH_3CHO \xrightarrow{\ k_1\ } CH_3^* + CHO^\bullet$

R_2 – Chain transfer step I: $\qquad CHO^\bullet \xrightarrow{\ k_2\ } CO + H^\bullet$

R_3 – Chain transfer step II: $\qquad H^\bullet + CH_3CHO \xrightarrow{\ k_3\ } CH_3CHO^\bullet + H_2$

(contd ...)

R_4 – Propagation step I: $\qquad CH_3^{\cdot} + CH_3CHO \xrightarrow{k_4} CH_4 + CH_3CO^{\cdot}$

R_5 – Propagation step II: $\qquad CH_3CO^{\cdot} \xrightarrow{k_5} CH_3^{\cdot} + CO \qquad \qquad$ ... (10.9)

R_6 – Termination step I: $\qquad 2CH_3^{\cdot} \xrightarrow{k_6} C_2H_6$

In this reaction network $CH_3^{\cdot}$ and $CH_3CO^{\cdot}$ are the chain carriers. The first reaction step is the chain initiation step for one of the chain carriers: $CH_3^{\cdot}$. This is followed by two chain transfer steps, in which H_2 is produced, as well as the second chain carrier $CH_3CO^{\cdot}$. In the chain propagation steps $CH_3^{\cdot}$ is reproduced. The chain is terminated in the last step by the consumption of $CH_3^{\cdot}$.

Thus in Step I, the radicals $CH_3^{\cdot}$ and $CHO^{\cdot}$ are formed by the rupture of a C-C bond. Only $CH_3^{\cdot}$ is believed to be a chain carrier. The radical $CHO^{\cdot}$ is involved in a chain transfer step (II) in which $H^{\cdot}$ is formed. The $H^{\cdot}$ radical is then involved in a chain transfer reaction (step II) in which the other chain carrier $CH_3CO^{\cdot}$ is produced. In Step IV one of the products CH_4 is formed and the chain carrier $CH_3CO^{\cdot}$ is reproduced, while in Step V the chain carrier $CH_3^{\cdot}$ is regenerated. Step VI consumes CH_3 which in turn terminates the reaction. The overall reaction in a chain reaction is usually given by the propagation steps, which in this case are Steps IV and V:

Overall reaction: $\qquad \qquad CH_3CHO \rightarrow CH_4 + CO \qquad \qquad$ ... (10.10)

In this particular reaction sequence side reactions takes place involving the chain transfer and termination steps that produce C_2H_6 and H_2. We can determine a rate law for the overall reaction by finding a rate expression for R_{CH_4}. A schematic of the various pathways for the Rice-Hersfeld mechanism is shown in Fig. 10.2.

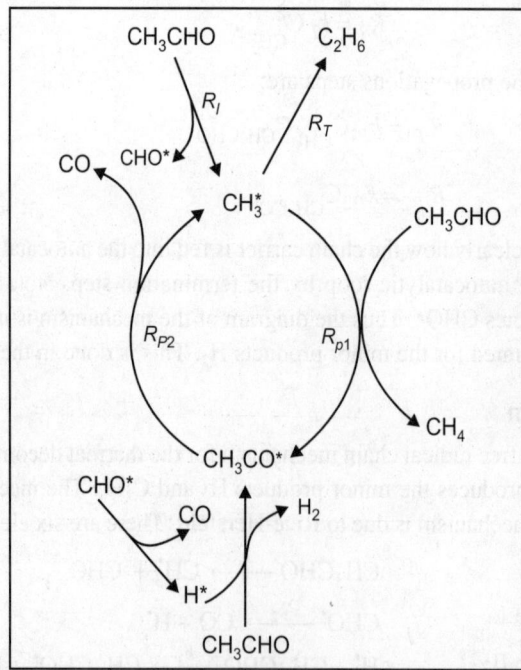

Fig. 10.2. A schematic diagram of the various pathways for the Rice-Hersfeld mechanism.

Let us construct the stoichiometric matrix S for this system:

$$S = \begin{bmatrix} -1 & 0 & -1 & -1 & 0 & 0 \\ 1 & 0 & 0 & -1 & 1 & -2 \\ 1 & -1 & 0 & 0 & 0 & 0 \\ 0 & 1 & 0 & 0 & 1 & 0 \\ 0 & 1 & -1 & 0 & 0 & 0 \\ 0 & 0 & 1 & 1 & -1 & 0 \\ 0 & 0 & 1 & 0 & 0 & 0 \\ 0 & 0 & 0 & 1 & 0 & 0 \\ 0 & 0 & 0 & 0 & 0 & 1 \end{bmatrix}$$ ... (10.11)

Here the rows represent the species $\{CH_3CHO, CH_3^\bullet, CHO^\bullet, CO, H^\bullet, CH_3CO^\bullet, H_2, CH_4, CH_3CO^\bullet, C_2H_6\}$ and the columns the reaction steps $R_1, R_2, ..., R_6$. It is a simple matter to check that the rank (S) = 6. Thus the reaction steps are linearly independent and hence the dimension of the null space of S is zero. Thus if this reaction were undertaken in a batch reactor there would be no cycles at steady-state. The reaction rate of each step would be zero. Note if we add up the two propagation steps we see that this system is autocatalytic

$$CH_3^\bullet + CH_3CHO \rightarrow CH_3^\bullet + CO \qquad ... (10.12)$$

where the radical $CH_3^\bullet$ plays the role of a catalyst.

MATHEMATICA ASSISTED ANALYSIS

For each elementary step we write down a rate expression based on mass action kinetics. We will use Mathematica to assist us with the algebra. To keep the Mathematica code readable, we will use the molecular formulas as subscripts on the production rate expressions and concentrations. Since we will not use the notations package, it is important to use appropriate symbols. To designate a radical we use the 'FilledSmallCircle' symbol $\bullet$.

This can be entered from the CompleteCharacters palette, or by the shortcut using the escape key: ESC fsci ESC. Note also we use $R_X[i]$, where X is a chemical formula for an individual species to denote the rate of production of species X in the ith step.

We define a function Reaction Rates that is a list of all the production rates for the species in the individual steps. The mechanism involves 6 steps, see Eq. 10.9.

ReactionRates $= \{R_{CH_3CHO}[1] = k_1\, C_{CH_3CHO},\ R_{CH_3^\bullet}[1] = k_1\, C_{CH_3CHO},\ R_{CHO^\bullet}[1] = k_1\, C_{CH_3CHO},$

$R_{CHO^\bullet}[2] = -k_2\, C_{CHO^\bullet},\ R_{CO}[2] = k_2\, C_{CHO^\bullet},\ R_{H^\bullet}[2] = k_2\, C_{CHO^\bullet},\ R_{H^\bullet}[3] = -k_3\, C_{H^\bullet} C_{CH_3CHO},$

$R_{CH_3CHO}[3] = -k_3\, C_{H^\bullet} C_{CH_3CHO},\ R_{H_2}[3] = k_3\, C_{H^\bullet} C_{CH_3CHO},\ R_{CH_3CO^\bullet}[3] = k_3\, C_{H^\bullet} C_{CH_3CHO},$

$R_{CH_3^\bullet}[4] = -k_4\, C_{CH_3^\bullet} C_{CH_3CHO},\ R_{CH_3CHO}[4] = -k_4\, C_{CH_3^\bullet} C_{CH_3CHO},\ R_{CH_4}[4] = k_4\, C_{CH_3^\bullet} C_{CH_3CHO},$

$R_{CH_3CO^\bullet}[4] = k_4\, C_{CH_3^\bullet} C_{CH_3CHO},\ R_{CH_3CO^\bullet}[5] = -k_5\, C_{CH_3CO^\bullet},\ R_{CH_3^\bullet}[5] = k_5\, C_{CH_3CO^\bullet},$

$R_{CO}[5] = k_5\, C_{CH_3CO^\bullet},\ R_{CH_3^\bullet}[6] = -2k_6\, C_{CH_3^\bullet}^2,\ R_{C_2H_6}[6] = k_6\, C_{CH_3^\bullet}^2\}$

$$\{-C_{CHOCH_3} \, k_1, C_{CHOCH_3} \, k_1, C_{CHOCH_3} \, k_1, -C_{CHO^\bullet} \, k_2, C_{CHO^\bullet} \, k_2, C_{CHO^\bullet} \, k_2,$$

$$-C_{H^\bullet} \, C_{CHOCH_3} \, k_3, -C_{H^\bullet} \, C_{CHOCH_3} \, k_3, C_{H^\bullet} \, C_{CHOCH_3} \, k_3, C_{H^\bullet} \, C_{CHOCH_3} \, k_3,$$

$$-C_{CHOCH_3} \, C_{CH_3^\bullet} k_4, -C_{CHOCH_3} \, C_{CH_3^\bullet} \, k_4, C_{CHOCH_3} \, C_{CH_3^\bullet} k_4, C_{CHOCH_3} \, C_{CH_3^\bullet} \, k_4,$$

$$-C_{CO^\bullet CH_3} \, k_5, C_{CO^\bullet CH_3} \, k_5, C_{CO^\bullet CH_3} \, k_5, -2C_{CH_3^\bullet}^2 \, k_6, C_{CH_3^\bullet}^2 \, k_6\}$$

Here is the rate of production for $CH_3^\bullet$ in step 4

$$R_{CH_3^\bullet}[4]$$

$$-C_{CHOCH_3} C_{CH_3^\bullet} k_4$$

We will make the Bodenstein approximation (or Pseudo-steady-state assumption) that the overall rate of production of the intermediates species (the radicals $CHO^\bullet$, $CH_3^\bullet$, $CH_3CO^\bullet$, $H^\bullet$) is approximately zero. This assumption is expressed by the following equations:

$$\text{Eqs.} = \{R_{CHO^\bullet}[1] + R_{CHO^\bullet}[2] = 0, R_{CH_3^\bullet}[6] + R_{CH_3^\bullet}[5] + R_{CH_3^\bullet}[4] + R_{CH_3^\bullet}[1] = 0,$$

$$R_{H^\bullet}[3] + R_{H^\bullet}[2] = 0, R_{CH_3CO^\bullet}[5] + R_{CH_3CO^\bullet}[4] + R_{CH_3CO^\bullet}[3] = 0\}$$

$$\{C_{CHOCH_3} \, k_1 - C_{CHO^\bullet} \, k_2 = 0, C_{CHOCH_3} \, k_1 - C_{CHOCH_3} C_{CH_3^\bullet} \, k_4 + C_{CO^\bullet CH_3} \, k_5 - 2 \, C_{CH_3^\bullet}^2 k_6 = 0,$$

$$C_{CHO^\bullet} \, k_2 - C_{H^\bullet} \, C_{CHOCH_3} \, k_3 = 0, C_{H^\bullet} \, C_{CHOCH_3} \, k_3 + C_{CHOCH_3} \, C_{CH_3^\bullet} \, k_4 - C_{CO^\bullet CH_3} \, k_5 = 0\}$$

Note that we have four algebraic equations for the 4 radical species (chain carriers). We can use *Mathematica*'s Solve routine to determine a solution

$$\text{soll} = \text{Solve}\left[\text{Eqs} \left\{C_{CHO^\bullet}, C_{CH_3^\bullet}, C_{H^\bullet}, C_{CO^\bullet CH_3}\right\}\right]$$

$$\{\{C_{CHO^\bullet} \to \frac{C_{CHOCH_3} \, k_1}{k_2}, C_{CO^\bullet CH_3} \to \frac{C_{CHOCH_3} \, k_1 - \dfrac{C_{CHOCH_3}^{3/2} \sqrt{k_1} k_4}{\sqrt{k_6}}}{k_5},$$

$$C_{H^\bullet} \to \frac{k_1}{k_3}, C_{CH_3^\bullet} \to \frac{\sqrt{C_{CHOCH_3}} \sqrt{k_1}}{\sqrt{k_6}}\}, \{C_{CHO^\bullet} \to \frac{C_{CHOCH_3} \, k_1}{k_2},$$

$$C_{CO^\bullet CH_3} \to \frac{C_{CHOCH_3} \, k_1 + \dfrac{C_{CHOCH_3}^{3/2} \sqrt{k_1} k_4}{\sqrt{k_6}}}{k_5}, C_{H^\bullet} \to \frac{k_1}{k_3}, C_{CH_3^\bullet} \to \frac{\sqrt{C_{CHOCH_3}} \sqrt{k_1}}{\sqrt{k_6}}\}\}$$

Note that we get two solutions. Since concentrations cannot be negative, we can discard the first solution. If we apply the appropriate solution to the expression for the rate of production $R_{CH_4}[4]$ we get:

$$R_{CH_4}[4]/\cdot \text{sol1}[(2)]$$

$$\frac{C_{CHOCH_3}^{3/2}\sqrt{k_1}k_4}{\sqrt{k_6}}$$

Thus the rate of production of CH_4 is given by:

$$R_{CH_4} = \frac{C_{CHOCH_3}^{2/3}k_4\sqrt{k_1}}{\sqrt{k_6}} \qquad \dots (10.13)$$

Note further that the overall reaction is based on steps IV and V (Eq. 10.10) and thus R_{CH_3CHO} is determined by:

$$R_{CH_3CHO}[4]/\cdot \text{sol1}[(2)]$$

$$-\frac{C_{CHOCH_3}^{3/2}\sqrt{k_1}k_4}{\sqrt{k_6}}$$

This result is consistent with Step 4 in Eq. 10.9. Finally, the overall rate of consumption of CH_3CHO (including side reactions) is given by:

$$(R_{CH_3CHO}[1] + R_{CH_3CHO}[3] + R_{CH_3CHO}[4])/\cdot \text{sol1}[(2)]$$

$$-2\,C_{CHOCH_3}\,k_1 - \frac{C_{CHOCH_3}^{3/2}\sqrt{k_1}k_4}{\sqrt{k_6}}$$

Thus, if we represent the reaction as the overall schema:

Overall reaction: $\qquad\qquad CH_3CHO \rightarrow CH_4 + CO \qquad\qquad \dots (10.14)$

then we can write for this schema, the overall rate of reaction as:

$$r_{overall} = -2\,C_{CHOCH_3}\,k_1 - \frac{C_{CHOCH_3}^{3/2}\sqrt{k_1}k_4}{\sqrt{k_6}} \qquad \dots (10.15)$$

ANALYSIS OF COMPOSITION MATRIX

For this system we have the following species:

$$CH_3CHO,\ CH_3^\bullet,\ CHO^\bullet,\ CO,\ H^\bullet,\ CH_3CO^\bullet,\ H_2,\ CH_4,\ C_2H_6$$

The composition matrix is:

$$\mathbb{C} = \begin{pmatrix} 2 & 1 & 1 & 1 & 0 & 2 & 0 & 1 & 2 \\ 4 & 3 & 1 & 0 & 1 & 3 & 2 & 4 & 6 \\ 1 & 0 & 1 & 1 & 0 & 1 & 0 & 0 & 0 \end{pmatrix} \qquad \dots (10.16)$$

Let us compute the NullSpace and rank of this matrix:

$$\mathbb{C} = \begin{pmatrix} \mathbf{2} & \mathbf{1} & \mathbf{1} & \mathbf{1} & \mathbf{0} & \mathbf{2} & \mathbf{0} & \mathbf{1} & \mathbf{2} \\ \mathbf{4} & \mathbf{3} & \mathbf{1} & \mathbf{0} & \mathbf{1} & \mathbf{3} & \mathbf{2} & \mathbf{4} & \mathbf{6} \\ \mathbf{1} & \mathbf{0} & \mathbf{1} & \mathbf{1} & \mathbf{0} & \mathbf{1} & \mathbf{0} & \mathbf{0} & \mathbf{0} \end{pmatrix};$$

RowReduce[ℂ] // MatrixForm

$$\begin{pmatrix} 1 & 0 & 1 & 0 & 1 & 0 & 2 & 1 & 0 \\ 0 & 1 & -1 & 0 & -1 & 1 & -2 & 0 & 2 \\ 0 & 0 & 0 & 1 & -1 & 1 & -2 & -1 & 0 \end{pmatrix}$$

Thus the rank is 3, and hence we have $N - 3 = 6$ basis vectors for the null space. These are:

NullSpace[ℂ] // MatrixForm

$$\begin{pmatrix} 0 & -2 & 0 & 0 & 0 & 0 & 0 & 0 & 1 \\ -1 & 0 & 0 & 1 & 0 & 0 & 0 & 1 & 0 \\ -2 & 2 & 0 & 2 & 0 & 0 & 1 & 0 & 0 \\ 0 & -1 & 0 & -1 & 0 & 1 & 0 & 0 & 0 \\ -1 & 1 & 0 & 1 & 1 & 0 & 0 & 0 & 0 \\ -1 & 1 & 1 & 0 & 0 & 0 & 0 & 0 & 0 \end{pmatrix}$$

The stoichiometric schema for this null space is:

Schema I: $2CH_3^* \rightarrow C_2H_6$

Schema II: $CH_3CHO \rightarrow CO + CH_4$

Schema III: $2CH_3CHO \rightarrow 2CH_3^* + 2CO + H_2$

Schema IV: $CH_3^* + CO \rightarrow CH_3CO^{\bullet}$ ... (10.17)

Schema V: $CH_3CHO \rightarrow CH_3^* + CO + H^{\bullet}$

Schema VI: $CH_3CHO \rightarrow CH_3^* + CHO^{\bullet}$

These schema can be used to construct the elementary steps described in the previous section.

Example 10.2: The hydrogen bromide reaction

The mechanism suggested for the reaction of hydrogen with bromine to produce hydrogen bromide is

Initiation step $Br_2 + M \xrightarrow{k_1} 2Br^{\bullet} + M$ $q_1 = k_1 \, C_{Br_2} \, C_M$

Propagation step I $Br^{\bullet} + H_2 \xrightarrow{k_2} HBr + H^{\bullet}$ $q_2 = k_2 \, C_{Br^{\bullet}} \, C_{H_2}$

Propagation step II $H^{\bullet} + Br_2 \xrightarrow{k_3} HBr + Br^{\bullet}$ $q_3 = k_3 \, C_{H^{\bullet}} \, C_{Br_2}$ ... (10.18)

Inhibition step $H^{\bullet} + HBr \xrightarrow{k_4} H_2 + Br^{\bullet}$ $q_4 = k_4 \, C_{H^{\bullet}} \, C_{HBr}$

Termination step $2Br^{\bullet} + M \xrightarrow{k_5} Br_2 + M^{\bullet}$ $q_5 = k_5 \, C_{Br^{\bullet}}^2 \, C_M$

Here q_i are progress (reaction) rates for the individual steps. The quantity M is called a collision partner that provides collision energy to cause bond cleavage. The concentration of M depends on the concentration of inerts (such as N_2), as well as Br_2, H_2 and HBr. Note that in this chain reaction there is an inhibition step. A schematic of the hydrogen bromide cycle is shown in Fig. 10.3.

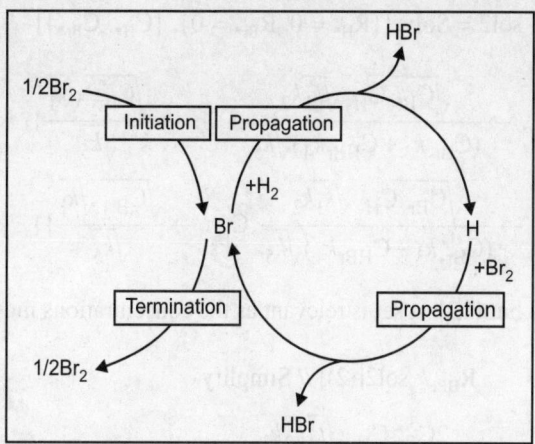

Fig. 10.3. A schematic diagram of the hydrogen bromide cycle.

The reaction rate constants k_i at 500 K are estimated to be:

$$
\left.
\begin{aligned}
k_1 &= 3.8 \times 10^{-8}\ C_M\ \mathrm{dm^3\ mol^{-1}\ s^{-1}} \\
k_2 &= 380\ \mathrm{dm^3\ mol^{-1}\ s^{-1}} \\
k_3 &= 9.6 \times 10^{10}\ \mathrm{dm^3\ mol^{-1}\ s^{-1}} \\
k_4 &= 7.2 \times 10^{9}\ \mathrm{dm^3\ mol^{-1}\ s^{-1}} \\
k_5 &= 4.2 \times 10^{-13}\ C_M\ \mathrm{dm^3\ mol^{-1}\ s^{-1}}
\end{aligned}
\right\} \qquad \ldots (10.19)
$$

An estimate of the reaction rates q_i are given below:

$$q_1 \approx 1\ \mathrm{mol/s},\ q_2 \approx 100\ \mathrm{mol/s},\ q_3 \approx 100\ \mathrm{mol/s},\ q_4 \approx 0.1\ \mathrm{mol/s},\ q_5 \approx 1\ \mathrm{mol/s} \qquad \ldots (10.20)$$

Now the rates of production of the species are:

$$
\left.
\begin{aligned}
R_{HBr} &= q_2 + q_3 - q_4 \approx 200 - 0.1 \approx 200\ \mathrm{mol/s} \\
R_{H\bullet} &= q_2 - q_3 - q_4 \approx -0.1\ \mathrm{mol/s} \\
R_{Br\bullet} &= 2q_1 - q_2 + q_3 + q_4 - 2q_5 = 0.1\ \mathrm{mol/s} \\
R_{Br_2} &= -q_1 + q_5 \approx 0 \\
R_{H_2} &= -q_2 + q_4 \approx -100\ \mathrm{mol/s}
\end{aligned}
\right\} \qquad \ldots (10.21)
$$

Applying the PSS assumption to the radical species gives:

$$q_2 - q_3 - q_4 = 0,\ 2q_1 - q_2 + q_3 + q_4 - 2q_5 = 0 \qquad \ldots (10.22)$$

We can use these two equations to solve for the concentrations of the intermediate species $C_{H\bullet}$ and $C_{Br\bullet}$. Here is the Mathematica code to do these calculations:

ReactionRates =

$(q_1 = k_1\ C_{Br_2}\ C_M,\ q_2 = k_2\ C_{Br\bullet}\ C_{H_2},\ q_3 = k_3\ C_{H\bullet}\ C_{Br_2},\ q_4 = k_4\ C_{H\bullet}\ C_{HBr},\ q_5 = k_5\ C_{Br\bullet}^2\ C_M\}$

$\{C_M\ C_{Br_2}\ k_1,\ C_{Br\bullet}\ C_{H_2}\ k_2,\ C_{H\bullet}\ C_{Br_2}\ k_3,\ C_{H\bullet}\ C_{HBr}\ k_4,\ C_{Br\bullet}^2\ C_M\ k_5\}$

ProductionRates =

$\{R_{HBr} = q_2 + q_3 - q_4,\ R_{H\bullet} = q_2 - q_3 - q_4,\ R_{Br\bullet} = 2q_1 - q_2 + q_3 + q_4 - 2q_5,\ R_{Br_2} = -q_1 + q_5,\ R_{H_2} = -q_2 + q_4\}$

$\{C_{Br\bullet}\ C_{H_2}\ k_2 + C_{H\bullet}\ C_{Br_2}\ k_3 - C_{H\bullet}\ C_{HBr}\ k_4,\ C_{Br\bullet}\ C_{H_2}\ k_2 - C_{H\bullet}\ C_{Br_2}\ k_3 - C_{H\bullet}\ C_{HBr}\ k_4,\ 2C_M\ C_{Br_2}\ k_1 -$
$C_{Br\bullet}\ C_{H_2}\ k_2 + C_{H\bullet}\ C_{Br_2}\ k_3 + C_{H\bullet}\ C_{HBr}\ k_4 - 2C_{Br\bullet}^2\ 2C_M\ k_5,\ -C_M\ C_{Br_2}\ k_1 + 2C_{Br\bullet}^2\ C_M\ k_5,\ -C_{Br\bullet}\ C_{H_2}\ k_2 +$
$C_{H\bullet}\ C_{HBr}\ k_4\}$

$$sol2 = Solve[\{R_{H\bullet} = 0, R_{Br\bullet} = 0\}, \{C_{H\bullet}, C_{Br\bullet}\}]$$

$$\left\{\left\{C_{H\bullet} \rightarrow \frac{\sqrt{C_{Br_2}} C_{H_2} \sqrt{k_1 k_2}}{(C_{Br_2} k_3 + C_{HBr} k_4)\sqrt{k_5}}, C_{Br\bullet} \rightarrow \frac{\sqrt{C_{Br_2}} \sqrt{k_1}}{\sqrt{k_5}}\right\},\right.$$

$$\left.\left\{C_{H\bullet} \rightarrow \frac{\sqrt{C_{Br_2}} C_{H_2} \sqrt{k_1 k_2}}{(C_{Br_2} k_3 + C_{HBr} k_4)\sqrt{k_5}}, C_{Br\bullet} \rightarrow \frac{\sqrt{C_{Br_2}} \sqrt{k_1}}{\sqrt{k_5}}\right\}\right\}$$

We have two solutions but the last set is relevant as the concentrations must be positive. The rate of production of HBr is then

$$R_{HBr}/\cdot sol2[(2)] \text{ // Simplify}$$

$$\frac{2 C_{Br_2}^{3/2} C_{H_2} \sqrt{k_1 k_2 k_3}}{(C_{Br_2} k_3 + C_{HBr} k_4)\sqrt{k_5}}$$

Dividing through by $k_3 C_{Br_2}$, gives:

$$R_{HBr} = \frac{2\sqrt{k_1}\, k_2\, C_{Br_2}^{1/2} C_{H_2}}{\left(1 + \dfrac{C_{HBr} k_4}{C_{Br_2} k_3}\right)\sqrt{k_5}} = \frac{K\, C_{Br_2}^{1/2} C_{H_2}}{1 + K' \dfrac{C_{HBr}}{C_{Br_2}}} \qquad \ldots (10.23)$$

Which is identical to what Bodenstein and Lind determined empirically.

AUTOXIDATION

Autoxidation is any oxidation that occurs in open air or in presence of oxygen and/or UV radiation and forms peroxides and hydroperoxides. A classic example of autoxidation is that of simple ethers like diethyl ether, whose peroxides can be dangerously explosive. It can be considered to be a slow, flameless combustion of materials by reaction with oxygen. Autoxidation is important because it is a useful reaction for converting compounds to oxygenated derivatives, and also because it occurs in situations where it is not desired (as in the destructive cracking of the rubber in automobile tyres).

Although virtually all types of organic materials can undergo air oxidation, certain types are particularly prone to autoxidation, including unsaturated compounds that have allylic hydrogens or benzylic hydrogens; these materials are converted to hydroperoxides by autoxidation.

Mechanism

Autoxidation is a free radical chain process. Such reactions can be divided into three stages: chain initiation, propagation, and termination. In the initiation process, some event causes free radicals to be formed. For example, free radicals can be produced purposefully by the decomposition of a radical initiator, such as benzoyl peroxide. In some cases, initiation occurs by a process that is not well understood but is thought to be the spontaneous reaction of oxygen with a material with a readily abstractable hydrogen. Destructive autoxidation processes also are initiated by pollutants such as those in smog.

Once free radicals are formed, they react in a chain to convert the material to a hydroperoxide. The chain is ended by termination reactions in which free radicals collide and combine their odd electrons to form a new bond.

Chain initiation

$$ROOH + RH \xrightarrow{\text{Energy}} RO\bullet + \bullet OH + RH \rightarrow RH\bullet + H_2O + R\bullet$$

$$RO\bullet + RH \xrightarrow{\text{H-abstraction}} R\bullet + ROH$$

Chain propagation

$$R^\bullet + O_2 \xrightarrow{\text{Fast}} ROO^\bullet$$

$$ROO^\bullet + RH \xrightarrow{\text{H-abstraction}} ROOH + {}^\bullet R$$

Chain termination

$$2ROO^\bullet \rightarrow 2RO^\bullet + O_2 \rightarrow ROH + QO + O_2$$

Source of alcohol and ketone

$$ROOH \rightarrow ROO^\bullet \rightarrow ROOH + Q^\bullet OOH \rightarrow ROOH + QO + OH$$

Reaction rate

In steady-state, the concentration of chain-carrying radicals is constant, thus the rate of initiation equals the rate of termination.

$$r_{\text{init}} = k_{\text{init}} \bullet [ROOH] = k_{\text{term}} \bullet [ROO^\bullet]^2$$

$$r_{\text{prop}} = k_{\text{prop}} \bullet [RH] \bullet [ROO^\bullet] = k_{\text{prop}} \bullet [RH] \bullet \sqrt{\frac{k_{\text{init}}}{k_{\text{term}}}} \bullet \sqrt{[ROOH]}$$

Autoxidations in industry

Autoxidation is a process of enormous economic impact, since all foods, plastics, gasolines, oils, rubber, and other materials that must be exposed to air undergo continuous destructive reactions of this type. All plastics and rubber and most processed foods contain antioxidants to protect them against the attack of oxygen. In the chemical industry many chemicals are produced by autoxidation:

1. In the cumene process phenol and acetone are made from benzene and propylene.
2. The autoxidation of cyclohexane yields cyclohexanol and cyclohexanone.
3. *p*-Xylene is oxidised to terephthalic acid.
4. Ethylbenzene is oxidised to ethylbenzene hydroperoxide, an epoxidising agent in the propylene oxide/styrene process POSM.

Autoxidation in food

It is well known that fats become rancid, even when kept at low temperatures. The complex mixture of compounds in wines including polyphenols, polysaccharides, proteins can undergo autoxidation during the ageing process. Simple polyphenols can lead to the formation of B-type procyanidins in wines or in model solutions.

This is correlated to the browning colour change characteristic of this process. This phenomenon is also observed in carrot puree.

Spoilage of Food

Food spoilage means the original nutritional value, texture, flavour of the food are damaged, the food become harmful to people and unsuitable to eat.

There are three types of micro-organisms that cause food spoilage—yeasts, moulds and bacteria.

1. Yeasts growth causes fermentation which is the result of yeast metabolism. There are two types of yeasts: true yeast and false yeast. True yeast metabolises sugar producing alcohol and carbon dioxide gas. This is known as fermentation. False yeast grows as a dry film on a food surface, such as on pickle brine. False yeast occurs in foods that have a high sugar or high acid environment.

2. Moulds grow in filaments forming a tough mass which is visible as 'mould growth'. Moulds form spores which, when dry, float through the air to find suitable conditions where they can start the growth cycle again.

3. Mould can cause illness, especially if the person is allergic to moulds. Usually though, the main symptoms from eating mouldy food will be nausea or vomiting from the bad taste and smell of the mouldy food.

4. Both yeasts and moulds can thrive in high acid foods like fruit, tomatoes, jams, jellies and pickles. Both are easily destroyed by heat. Processing high acid foods at a temperature of $100°C$ ($212°F$) in a boiling water canner for the appropriate length of time destroys yeasts and moulds.

Antioxidant

An antioxidant is a molecule capable of inhibiting the oxidation of other molecules. Oxidation is a chemical reaction that transfers electrons from a substance to an oxidising agent. Oxidation reactions can produce free radicals. In turn, these radicals can start chain reactions that damage cells. Antioxidants terminate these chain reactions by removing free radical intermediates, and inhibit other oxidation reactions. They do this by being oxidised themselves, so antioxidants are often reducing agents such as thiols, ascorbic acid or polyphenols.

Although oxidation reactions are crucial for life, they can also be damaging; hence, plants and animals maintain complex systems of multiple types of antioxidants, such as glutathione, vitamin C, and vitamin E as well as enzymes such as catalase, superoxide dismutase and various peroxidases. Low levels of antioxidants, or inhibition of the antioxidant enzymes, cause oxidative stress and may damage or kill cells. As oxidative stress might be an important part of many human diseases, the use of antioxidants in pharmacology is intensively studied, particularly as treatments for stroke and neurodegenerative diseases. However, it is unknown whether oxidative stress is the cause or the consequence of disease.

Antioxidants are widely used as ingredients in dietary supplements in the hope of maintaining health and preventing diseases such as cancer, coronary heart disease and even altitude sickness. Although initial studies suggested that antioxidant supplements might promote health, later large clinical trials did not detect any benefit and suggested instead that excess supplementation may be harmful. In addition to these uses of natural antioxidants in medicine, these compounds have many industrial uses, such as preservatives in food and cosmetics and preventing the degradation of rubber and gasoline.

CHEMICAL SYNTHESIS BY AUTOXIDATION

Now we consider some very positive examples of this type of reaction sequence. Some organic molecules have weak C–H bonds that are easily broken. This fact has been exploited for some key industrial reactions. The tertiary C–H bond in isobutane is fairly weak (91 kcal/mole) compared to other bonds in the molecule, and this can be exploited to make two very important chemicals.

Propylene Oxide and Isobutylene

Propylene oxide

Propylene oxide is an organic compound with the molecular formula CH_3CHCH_2O. This colourless volatile liquid is produced on a large scale industrially, its major application being its use for the production of polyether polyols for use in making polyurethane plastics. It is chiral epoxide, although it is commonly used as a racemic mixture. Industrial production of propylene oxide starts from propylene. Two general approaches are employed, one involving hydrochlorination and the other involving oxidation.

Hydrochlorination route: The traditional route proceeds via the conversion of propylene to chloropropanols:

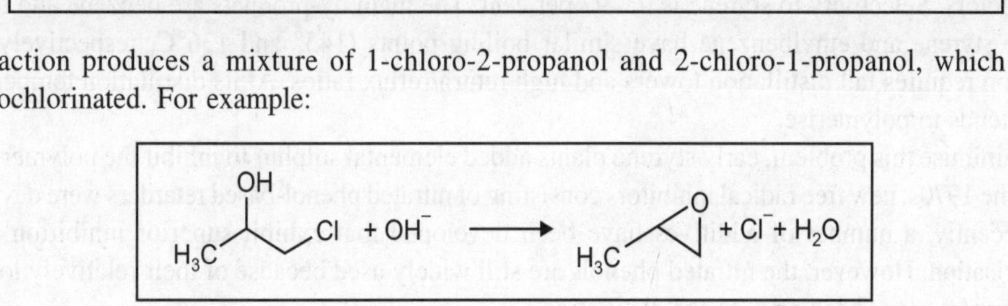

The reaction produces a mixture of 1-chloro-2-propanol and 2-chloro-1-propanol, which is then dehydrochlorinated. For example:

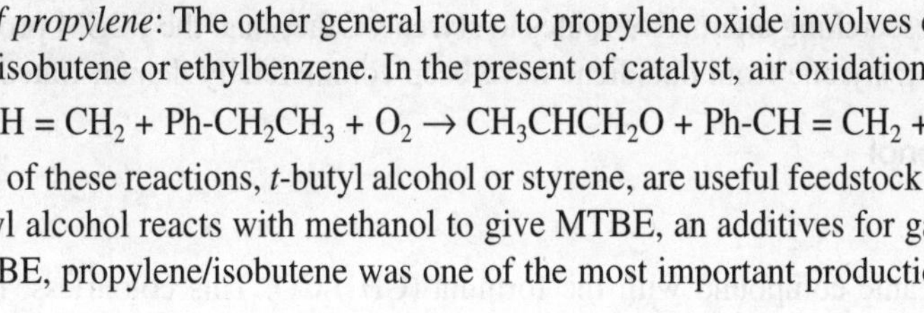

Lime is often used as a chlorine absorber.

Co-oxidation of propylene: The other general route to propylene oxide involves co-oxidation of the organic chemicals isobutene or ethylbenzene. In the present of catalyst, air oxidation occurs as follows:

$$CH_3CH = CH_2 + Ph\text{-}CH_2CH_3 + O_2 \rightarrow CH_3CHCH_2O + Ph\text{-}CH = CH_2 + H_2O$$

The coproducts of these reactions, *t*-butyl alcohol or styrene, are useful feedstock for other products. For example *t*-butyl alcohol reacts with methanol to give MTBE, an additives for gasoline. Before the current ban of MTBE, propylene/isobutene was one of the most important production process.

Isobutylene

Isobutylene (or 2-methylpropene) is a hydrocarbon of significant industrial importance. It is a four-carbon branched alkene (olefin), one of the four isomers of butylene. At standard temperature and pressure it is a colourless flammable gas.

Isobutylene can be isolated from refinery streams by reaction with sulphuric acid, but the most common industrial method for its production is by catalytic dehydrogenation of isobutane.

Styrene

Styrene, also known as vinyl benzene, is an organic compound with the chemical formula $C_6H_5CH = CH_2$. This cyclic hydrocarbon is a colourless oily liquid that evaporates easily and has a sweet smell, although high concentrations confer a less pleasant odour. Styrene is the precursor to polystyrene and several copolymers.

Styrene is most commonly produced by the catalytic dehydrogenation of ethylbenzene. Ethylbenzene is mixed in the gas phase with 10–15 times its volume in high-temperature steam, and passed over a solid catalyst bed. Most ethylbenzene dehydrogenation catalysts are based on iron(III) oxide, promoted by several per cent potassium oxide or potassium carbonate.

Steam serves several roles in this reaction. It is the source of heat for powering the endothermic reaction, and it removes coke that tends to form on the iron oxide catalyst through the water gas shift reaction. The potassium promoter enhances this decoking reaction. The steam also dilutes the reactant and products, shifting the position of chemical equilibrium towards products. A typical styrene plant consists of two or three reactors in series, which operate under vacuum to enhance the conversion and selectivity. Typical per-pass conversions are ca. 65 per cent for two reactors and 70–75 per cent for three reactors. Selectivity to styrene is 93–97 per cent. The main by-products are benzene and toluene. Because styrene and ethylbenzene have similar boiling points (145° and 136°C, respectively), their separation requires tall distillation towers and high return/reflux ratios. At its distillation temperatures, styrene tends to polymerise.

To minimise this problem, early styrene plants added elemental sulphur to inhibit the polymerisation. During the 1970s, new free radical inhibitors consisting of nitrated phenol-based retarders were developed. More recently, a number of additives have been developed that exhibit superior inhibition against polymerisation. However, the nitrated phenols are still widely used because of their relatively low cost. These reagents are added prior to the distillation.

Improving conversion and so reducing the amount of ethylbenzene that must be separated is the chief impetus for researching alternative routes to styrene. Other than the POSM process, none of these routes like obtaining styrene from butadiene have been commercially demonstrated.

Acetone and Phenol

Acetone

Acetone is the organic compound with the formula $(CH_3)_2CO$. This colourless, mobile, flammable liquid is the simplest example of the ketones. Owing to the fact that acetone is miscible with water it serves as an important solvent in its own right, typically as the solvent of choice for cleaning purposes in the laboratory.

Acetone is produced directly or indirectly from propylene. Most commonly, in the cumene process, benzene is alkylated with propene and the resulting cumene (isopropylbenzene) is oxidised to give phenol and acetone:

$$C_6H_5CH(CH_3)_2 + O_2 \rightarrow C_6H_5OH + (CH_3)_2CO$$

This conversion entails the intermediacy of cumene hydroperoxide, $C_6H_5C(OOH)(CH_3)_2$. Acetone is also produced by the direct oxidation of propene with a Pd(II)/Cu(II) catalyst, akin to the Wacker process. Small amounts of acetone are produced in the body by the decarboxylation of ketone bodies.

Phenol

Phenol, also known as carbolic acid, is an organic compound with the chemical formula C_6H_5OH. It is a white, crystalline solid. This functional group consists of a phenyl, bonded to a hydroxyl (–OH). It is produced on a large scale as a precursor to many materials and useful compounds. It is a mildly acidic compound that requires careful handling.

Phenol can be made from the partial oxidation of benzene, by the cumene process, or by the Raschig-Hooker process. It can also be found as a product of coal oxidation. The dominant method starts from cumene (isopropylbenzene):

$$C_6H_5CH(CH_3)_2 + O_2 \rightarrow C_6H_5OH + (CH_3)_2CO$$

Adipic Acid

Adipic acid is the organic compound with the formula $(CH_2)_4(COOH)_2$. From the industrial perspective, it is the most important dicarboxylic acid: About 2.5 billion kilograms of this white crystalline powder are produced annually, mainly as a precursor for the production of nylon. Adipic acid otherwise rarely occurs in nature.

Historically, adipic acid was prepared from various fats using oxidation. Currently adipic acid is produced from a mixture of cyclohexanol and cyclohexanone called 'KA oil', the abbreviation of 'ketone-alcohol oil'. The KA oil is oxidised with nitric acid to give adipic acid, via a multistep pathway. Early in the reaction the cyclohexanol is converted to the ketone, releasing nitrous acid:

$$HOC_6H_{11} + HNO_3 \rightarrow OC_6H_{10} + HNO_2 + H_2O$$

Among its many reactions, the cyclohexanone is nitrosated, setting the stage for the scission of the C-C bond:

$$HNO_2 + HNO_3 \rightarrow NO^+NO_3^- + H_2O$$
$$OC_6H_{10} + NO^+ \rightarrow OC_6H_9\text{-}2\text{-}NO + H^+$$

Side products of the method include glutaric and succinic acids. Related processes start from cyclohexanol, which is obtained from the hydrogenation of phenol.

COMBUSTION

Combustion or burning is the sequence of exothermic chemical reactions between a fuel and an oxidant accompanied by the production of heat and conversion of chemical species. The release of heat can result in the production of light in the form of either glowing or a flame. Fuels of interest often include organic compounds (especially hydrocarbons) in the gas, liquid or solid phase.

In a complete combustion reaction, a compound reacts with an oxidising element, such as oxygen or fluorine, and the products are compounds of each element in the fuel with the oxidising element. For example:

$$CH_4 + 2O_2 \rightarrow CO_2 + 2H_2O + energy$$
$$CH_2S + 6F_2 \rightarrow CF_4 + 2HF + SF_6$$

A simple example can be seen in the combustion of hydrogen and oxygen, which is a commonly used reaction in rocket engines:

$$2H_2 + O_2 \rightarrow 2H_2O(g) + heat$$

The result is water vapour.

In the large majority of industrial applications of combustion and in fires, air is the source of oxygen (O_2), in air, each kg (lbm) of oxygen is mixed with approximately 3.76 kg (lbm) of nitrogen. The resultant flue gas from the combustion will contain nitrogen:

$$CH_4 + 2O_2 + 7.52N_2 \rightarrow CO_2 + 2H_2O + 7.52N_2 + \text{heat}$$

Complete combustion is almost impossible to achieve. In reality, as actual combustion reactions come to equilibrium, a wide variety of major and minor species will be present such as carbon monoxide and pure carbon (soot or ash). Additionally, any combustion in air, which is 78 per cent nitrogen, will also create several forms of nitrogen oxides.

In complete combustion, the reactant burns in oxygen, producing a limited number of products. When a hydrocarbon burns in oxygen, the reaction will only yield carbon dioxide and water. When elements are burned, the products are primarily the most common oxides. Carbon will yield carbon dioxide, nitrogen will yield nitrogen dioxide, sulphur will yield sulphur dioxide, and iron will yield iron(III) oxide.

In most industrial applications and in fires, air is the source of oxygen (O_2). In air, each kg (lbm) of oxygen is mixed with approximately 3.76 kg (lbm) of nitrogen. Nitrogen does not take part in combustion, but at high temperatures, some nitrogen will be converted to NO_x, usually between 1 per cent and 0.002 per cent (2 ppm). A more complete air combustion reaction is therefore:

$$2CH_4 + xO_2 + N_2 \rightarrow CO_2 + 4H_2O + CO + 2NO_x + \text{heat}$$

Incomplete combustion occurs when there isn't enough oxygen to allow the fuel to react completely to produce carbon dioxide and water. It also happens when the combustion is quenched by a heat sink such as a solid surface or flame trap.

The quality of combustion can be improved by design of combustion devices, such as burners and internal combustion engines. Smouldering is the slow, low-temperature, flameless form of combustion, sustained by the heat evolved when oxygen directly attacks the surface of a condensed-phase fuel. It is a typically incomplete combustion reaction.

Rapid combustion is a form of combustion, otherwise known as a fire, in which large amounts of heat and light energy are released, which often results in a flame. This is used in a form of machinery such as internal combustion engines and in thermobaric weapons. Combustion resulting in a turbulent flame is the most used for industrial application (e.g. gas turbines, gasoline engines, etc.) because the turbulence helps the mixing process between the fuel and oxidiser.

HYDROGEN OXIDATION

Hydrogen is the chemical element with atomic number 1. It is represented by the symbol H. With an atomic weight of 1.00794 u (1.007825 u for hydrogen-1), hydrogen is the lightest and most abundant chemical element, constituting roughly 75 per cent of the Universe's elemental mass. Stars in the main sequence are mainly composed of hydrogen in its plasma state. Naturally occurring elemental hydrogen is relatively rare on earth.

The most common isotope of hydrogen is protium (name rarely used, symbol H) with a single proton and no neutrons. In ionic compounds it can take a negative charge (an anion known as a hydride and written as H^-), or as a positively charged species H^+. The latter cation is written as though composed of a bare proton, but in reality, hydrogen cations in ionic compounds always occur as more complex species. Hydrogen forms compounds with most elements and is present in water and most organic compounds. It plays a particularly important role in acid-base chemistry with many reactions exchanging protons between soluble molecules. As the simplest atom known, the hydrogen atom has been of theoretical use. For example, as the only neutral atom with an analytic solution to the Schrödinger equation, the study of the energetics and bonding of the hydrogen atom played a key role in the development of quantum mechanics.

Industrial production is mainly from the steam reforming of natural gas, and less often from more energy-intensive hydrogen production methods like the electrolysis of water. Most hydrogen is employed near its production site, with the two largest uses being fossil fuel processing (e.g. hydrocracking) and ammonia production, mostly for the fertiliser market. Hydrogen is a concern in metallurgy as it can embrittle many metals, complicating the design of pipelines and storage tanks.

Hydrogen gas (dihydrogen or molecular hydrogen) is highly flammable and will burn in air at a very wide range of concentrations between 4 and 75 per cent by volume. The enthalpy of combustion for hydrogen is –286 kJ/mol:

$$2H_2(g) + O_2(g) \rightarrow 2H_2O(l) + 572kJ \ (286 \ kJ/mol)$$

Hydrogen gas forms explosive mixtures with air in the concentration range 4–74 per cent (volume per cent of hydrogen in air) and with chlorine in the range 5–95 per cent. The mixtures spontaneously detonate by spark, heat or sunlight. The hydrogen autoignition temperature, the temperature of spontaneous ignition in air, is 500°C (932°F). Pure hydrogen-oxygen flames emit ultraviolet light and are nearly invisible to the naked eye, as illustrated by the faint plume of the Space Shuttle main engine compared to the highly visible plume of a Space Shuttle Solid Rocket Booster. The detection of a burning hydrogen leak may require a flame detector; such leaks can be very dangerous. The destruction of the Hindenburg airship was an infamous example of hydrogen combustion; the cause is debated, but the visible flames were the result of combustible materials in the ship's skin. Because hydrogen is buoyant in air, hydrogen flames tend to ascend rapidly and cause less damage than hydrocarbon fires. Two-thirds of the Hindenburg passengers survived the fire, and many deaths were instead the result of falls or burning diesel fuel.

Hydrogen can be prepared in several different ways, but economically the most important processes involve removal of hydrogen from hydrocarbons. Commercial bulk hydrogen is usually produced by the steam reforming of natural gas. At high temperatures (1000–1400 K, 700°–1100°C or 1300°–2000°F), steam (water vapour) reacts with methane to yield carbon monoxide and H_2.

$$CH_4 + H_2O \rightarrow CO + 3H_2$$

This reaction is favoured at low pressures but is nonetheless conducted at high pressures (2.0 MPa, 20 atm or 600 in Hg). This is because high-pressure H_2 is the most marketable product and pressure swing adsorption (PSA) purification systems work better at higher pressures. The product mixture is known as 'synthesis gas' because it is often used directly for the production of methanol and related compounds. Hydrocarbons other than methane can be used to produce synthesis gas with varying product ratios. One of the many complications to this highly optimised technology is the formation of coke or carbon:

$$CH_4 \rightarrow C + 2H_2$$

Consequently, steam reforming typically employs an excess of H_2O. Additional hydrogen can be recovered from the steam by use of carbon monoxide through the water gas shift reaction, especially with an iron oxide catalyst. This reaction is also a common industrial source of carbon dioxide:

$$CO + H_2O \rightarrow CO_2 + H_2$$

Other important methods for H_2 production include partial oxidation of hydrocarbons:
$$2CH_4 + O_2 \rightarrow 2CO + 4H_2$$
and the coal reaction, which can serve as a prelude to the shift reaction above:
$$C + H_2O \rightarrow CO + H_2$$

Hydrogen is sometimes produced and consumed in the same industrial process, without being separated. In the Haber process for the production of ammonia, hydrogen is generated from natural gas. Electrolysis of brine to yield chlorine also produces hydrogen as a co-product.

Hydrogen is not an energy resource, except in the hypothetical context of commercial nuclear fusion power plants using deuterium or tritium, a technology presently far from development. The sun's energy comes from nuclear fusion of hydrogen, but this process is difficult to achieve controllably on earth. Elemental hydrogen from solar, biological, or electrical sources require more energy to make it than is obtained by burning it, so in these cases hydrogen functions as an energy carrier, like a battery. Hydrogen may be obtained from fossil sources (such as methane), but these sources are unsustainable.

H_2 is a product of some types of anaerobic metabolism and is produced by several micro-organisms, usually via reactions catalysed by iron- or nickel-containing enzymes called hydrogenases. These enzymes catalyse the reversible redox reaction between H_2 and its component two protons and two electrons. Creation of hydrogen gas occurs in the transfer of reducing equivalents produced during pyruvate fermentation to water.

Water splitting, in which water is decomposed into its component protons, electrons, and oxygen, occurs in the light reactions in all photosynthetic organisms. Some such organisms, including the alga *Chlamydomonas reinhardtii* and cyanobacteria, have evolved a second step in the dark reactions in which protons and electrons are reduced to form H_2 gas by specialised hydrogenases in the chloroplast. Efforts have been undertaken to genetically modify cyanobacterial hydrogenases to efficiently synthesise H_2 gas even in the presence of oxygen. Efforts have also been undertaken with genetically modified alga in a bioreactor. Hydrogen poses a number of hazards to human safety, from potential detonations and fires when mixed with air to being an asphyxant in its pure, oxygen-free form.

Alkane Oxidation

A method for making a composition comprising alkanes includes dehydrating a feedstock solution comprising a carbohydrate, in the presence of an acid catalyst, to yield at least one furan derivative compound, in a reaction vessel containing a biphasic reaction medium: an aqueous reaction solution and a substantially immiscible organic extraction solution. The furan derivative compound is then subjected to a self-aldol condensation reaction or a crossed-aldol condensation reaction with another carbonyl compound to yield a beta-hydroxy carbonyl compound and/or an alpha-beta unsaturated carbonyl compound. The beta-hydroxy carbonyl and/or alpha-beta unsaturated compounds are then hydrogenated to yield a saturated or partially saturated compound, followed by hydrodeoxygenation (e.g. dehydrating and hydrogenating) of the saturated or partially saturated compound to yield a composition of matter comprising alkanes.

Reactions with oxygen (combustion reaction): All alkanes react with oxygen in a combustion reaction, although they become increasingly difficult to ignite as the number of carbon atoms increases. The general equation for complete combustion is:
$$C_nH_{2n+2} + (1.5n+0.5)O_2 \rightarrow (n+1)H_2O + nCO_2$$

In the absence of sufficient oxygen, carbon monoxide or even soot can be formed, as shown below:

$$C_nH_{(2n+2)} + \tfrac{1}{2}nO_2 \rightarrow (n+1)H_2O + nCO$$

For example methane:

$$2CH_4 + 3O_2 \rightarrow 2CO + 4H_2O$$

$$CH_4 + 1.5O_2 \rightarrow CO + 2H_2O$$

The standard enthalpy change of combustion, $\Delta_c H°$, for alkanes increases by about 650 kJ/mol per CH_2 group. Branched-chain alkanes have lower values of $\Delta_c H°$ than straight-chain alkanes of the same number of carbon atoms, and so can be seen to be somewhat more stable.

Isomerisation and reformation: Isomerisation and reformation are processes in which straight-chain alkanes are heated in the presence of a platinum catalyst. In isomerisation, the alkanes become branched-chain isomers. In reformation, the alkanes become cycloalkanes or aromatic hydrocarbons, giving off hydrogen as a by-product. Both of these processes raise the octane number of the substance.

Other reactions: Alkanes will react with steam in the presence of a nickel catalyst to give hydrogen. Alkanes can be chlorosulphonated and nitrated, although both reactions require special conditions. The fermentation of alkanes to carboxylic acids is of some technical importance. In the Reed reaction, sulphur dioxide, chlorine and light convert hydrocarbons to sulphonyl chlorides.

AUTOCATALYSIS

A single chemical reaction is said to have undergone autocatalysis, or be autocatalytic, if the reaction product is itself the catalyst for that reaction. A set of chemical reactions can be said to be 'collectively autocatalytic' if a number of those reactions produce, as reaction products, catalysts for enough of the other reactions that the entire set of chemical reactions is self sustaining given an input of energy and food molecules.

Rate Law in Autocatalytic Reactions

The rate law for the second order autocatalytic reaction:

$$A + B \rightarrow 2B \text{ is } v = k[A][B].$$

The concentrations of A and B vary in time according to:

$$[A] = \frac{[A]_0 + [B]_0}{1 + \dfrac{[B]_0}{[A]_0} e^{([A]_0 + [B]_0)kt}}$$

and

$$[B] = \frac{[A]_0 + [B]_0}{1 + \dfrac{[A]_0}{[B]_0} e^{-([A]_0 + [B]_0)kt}}$$

The graph for these equations is a sigmoid curve, which is typical for autocatalytic reactions: these chemical reactions proceed slowly at the start because there is little catalyst present, the rate of reaction increases progressively as the reaction proceeds as the amount of catalyst increases and then it again slows down as the reactant concentration decreases. If the concentration of a reactant or product in an experiment follows a sigmoid curve, the reaction is likely to be autocatalytic (Fig. 10.4).

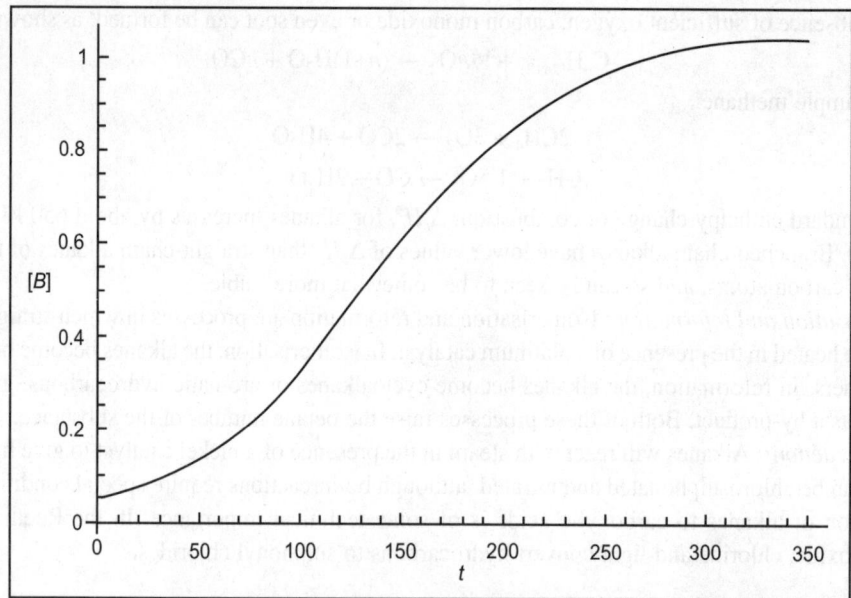

Fig. 10.4. Sigmoid variation of product concentration in autocatalytic reactions.

PREMIXED FLAMES

A premixed flame is a flame in which the oxidiser has been mixed with the fuel before it reaches the flame front. This creates a thin flame front as all of the reactants are readily available. If the mixture is rich, a diffusion flame will generally be found farther downstream.

If the flow of the fuel–oxidiser mixture is laminar, the flame speed of premixed flames is dominated by the chemistry. If the flow rate is below the flame speed, the flame will move upstream until the fuel is consumed or until it encounters a flame holder. If the flow rate is equal to the flame speed, we would expect a stationary flat flame front normal to the flow direction. If the flow rate is above the flame speed, the flame front will become conical such that the component of the velocity vector normal to the flame front is equal to the flame speed. As a result, the flame front of most premixed flames in daily life are roughly conical.

Premix Burner (In Flame Spectroscopy)

A burner in which fuel and oxidant are thoroughly mixed inside the burner housing before they leave the burner ports and enter the primary-combustion or inner zone of the flame. This type of burner usually produces and approximately laminar flame, and is commonly combined with a separate unit for nebulising the sample.

Leminar Diffusion Flames

In many cases, the fuel and air in a flame are often initially unmixed. This leads to a completely different type of flame termed the diffusion or non-premixed flame. In the simplest case, gas and air are injected in parallel and ignited. At the interface, molecular diffusion occurs resulting in a very region where combustion is possible where the flame sits. As the flame progresses downstream, the mixing layer increases in thickness until eventually a uniform mixture becomes of products is formed.

In these flames, it is solely the diffusion rate of fresh mixture into the flame zone which sustains combustion as opposed to the chemical kinetics of the system, hence the name. The reaction progresses in a very complex manner. As the pure fuel migrates towards the flame zone it is heated as before. However, because there is no oxygen in this case, the fuel is pyrolysed or broken down to smaller molecules and radicals. This also results in the formation of soot (carbon) which is strongly luminous and gives these flames their distinctive yellow colour. As the flame zone is approached, more oxygen is present to allow the products of the pyrolysis to react until finally they pass through the stoichiometric contour and reaction is completed. The resulting flame zone is substantially thicker than premixed flames and the combustion rate is entirely limited by mass diffusion of fresh mixture to the flame zone as opposed to the chemistry of the combustion as with premixed flames.

Turbulence and Recirculation

All practical combustion systems are turbulent in nature and often have some form of recirculation region to stabilise the combustion. Laminar flames are impractical since they rely on either thermal (premix) or molecular (non-premixed) diffusion to occur, both of which are slow processes. Turbulence has the effect of greatly enhancing the mixing in systems and thus greatly enhances the local combustion intensity. It said that to have a successful system you need the three T's of combustion:

1. Time (could also be viewed as volume).
2. Temperature.
3. Turbulence.

Studies of turbulence show how the unstable mixing layer between the fuel and oxidant or hot products allows entrainment and hence increases the mixing and/or ignition of fresh mixture. Premixed flame speeds are therefore greatly enhanced. For turbulent jet diffusion flames, the result of increased jet speed and hence turbulence actually results in a reduction rather than increase in flame length.

Often turbulence alone is still not enough for a stable flame, depending on the combustion intensity required. Recirculation is used as a method of increasing this yet higher. Recirculation may be achieved by one or a combination of three methods.

ENERGY GENERATION

All electricity generation systems have a 'carbon footprint', that is, at some points during their construction and operation carbon dioxide (CO_2) is emitted. There is some debate about how large these footprints are, especially for 'low carbon' technologies such as wind and nuclear.

All electricity generation technologies generate carbon dioxide (CO_2) and other greenhouse gas emissions. To compare the impacts of these different technologies accurately, the total CO_2 amounts emitted throughout a system's life must be calculated. Emissions can be both direct—arising during operation of the power plant, and indirect—arising during other nonoperational phases of the life cycle. Fossil fuelled technologies (coal, oil, gas) have the largest carbon footprints, because they burn these fuels during operation. Non-fossil fuel based technologies such as wind, photovoltaics (solar), hydro, biomass, wave/tidal and nuclear are often referred to as 'low carbon' or 'carbon neutral' because they do not emit CO_2 during their operation. However, they are not 'carbon free' forms of generation since CO_2 emissions do arise in other phases of their life cycle such as during extraction, construction, maintenance and decommissioning.

A 'carbon footprint' is the total amount of CO_2 and other greenhouse gases, emitted over the full life cycle of a process or product. It is expressed as grams of CO_2 equivalent per kilowatt hour of generation (gCO_2eq/kWh), which accounts for the different global warming effects of other greenhouse gases. This POSTnote deals only with life cycle CO_2eq emissions from electricity generation. All other emissions are outside the scope of this study.

Calculating Carbon Footprints

Carbon footprints are calculated using a method called life cycle assessment (LCA), and is also referred to as the 'cradle-to-grave' approach. This method is used to analyse the cumulative environmental impacts of a process or product through all the stages of its life. It takes into account energy inputs and emission outputs throughout the whole production chain from exploration and extraction of raw materials to processing, transport and final use. The LCA method is internationally accredited by ISO 14000 standards. The robustness of the method means that although carbon footprints vary between individual power plants, the ranking of electricity generation technologies does not change with different sources of data.

Life cycle assessment (LCA)

A complete LCA consists of four phases: (i) goal and scope (boundary) definition, (ii) inventory analysis (LCI), (iii) impact assessment (LCIA), and (iv) interpretation/improvement. The 'carbon footprint' is just one output from the life cycle inventory (LCI) step.

Life cycle inventory (LCI)

The inventory table is the most objective result of a LCA study, referring mainly to measures of mass and energy, i.e. raw materials and energy consumption, and the emission of solid, liquid and gaseous wastes. However, it does not say anything about the environmental impact of a particular emitted substance.

Life cycle impact assessment (LCIA)

In this phase all the LCI phase outputs are analysed to determine their impacts. The outputs contribute to impact categories such as global warming potential.

Use of other evidence-based tools

Life cycle assessment cannot replace the decision-making process itself. The information needs to be complemented by other considerations such as social and economic aspects. For example, despite the small carbon footprint that LCA of nuclear shows, these analyses are not always sufficient to answer criticisms of the safety and security of nuclear power.

Carbon Footprints

Fossil fuelled technologies

The carbon footprint of fossil fuelled power plants is dominated by emissions during their operation. Indirect emissions during other life cycle phases such as raw material extraction and plant construction are relatively minor. Coal burning power systems have the largest carbon footprint of all the electricity generation systems analysed here. Conventional coal combustion systems result in emissions of the order of >1000 gCO_2eq/kWh.

Lower emissions can be achieved using newer gasification plants (< 800 gCO$_2$eq/kWh), but this is still an emerging technology so is not as widespread as proven combustion technologies. Current gas powered electricity generation has a carbon footprint around half that of coal (~500 gCO$_2$eq/kWh), because gas has a lower carbon content than coal. Like coal fired plants, gas plants could co-fire biomass to reduce carbon emissions in the future.

Low carbon technologies

In contrast to fossil fuelled power generation, the common feature of renewable and nuclear energy systems is that emissions of greenhouse gases and other atmospheric pollutants are 'indirect', that is, they arise from stages of the life cycle other than power generation.

Biomass

Biomass is obtained from organic matter, either directly from dedicated energy crops like short-rotation coppice willow and grasses such as straw and miscanthus, or indirectly from industrial and agricultural by-products such as wood-chips. The use of biomass is generally classed as 'carbon neutral' because the CO$_2$ released by burning is equivalent to the CO$_2$ absorbed by the plants during their growth. However, other life cycle energy inputs affect this 'carbon neutral' balance, for example emissions arise from fertiliser production, harvesting, drying and transportation.

Biomass fuels are much lower in energy and density than fossil fuels. This means that large quantities of biomass must be grown and harvested to produce enough feedstock for combustion in a power station. Transporting large amounts of feedstock increases life cycle CO$_2$ emissions, so biomass electricity generation is most suited to small-scale local generation facilities, or operating as combined heat and power (CHP) plants. The range of carbon footprints for biomass is related to the type of organic matter and the way it is burned. Combustion of low density miscanthus results in higher life cycle emissions (93 gCO$_2$eq/kWh), than gasification of higher density wood-chip (25 gCO$_2$eq/kWh). Biomass can also be 'cofired' with fossil fuels in conventional power stations. Replacing a component of the fossil fuel with 'carbon neutral' biomass reduces the overall CO$_2$ emissions from these power stations.

Photovoltaics (PV)

Photovoltaics (PV), also known as solar cells, are made of crystalline silicon, a semi-conducting material which converts sunlight into electricity. The silicon required for PV modules is extracted from quartz sand at high temperatures. This is the most energy intensive phase of PV module production, accounting for 60 per cent of the total energy requirement.

Marine technologies (wave and tidal)

There are two types of marine energy devices; wave energy converters and tidal (stream and barrage) devices. Marine based electricity generation is still an emerging technology and is not yet operating on a commercial scale.

Hydropower

Hydropower converts the energy from flowing water, via turbines and generators, into electricity. There are two main types of hydroelectric schemes; storage and run-of-river. Storage schemes require dams. In run-of-river schemes, turbines are placed in the natural flow of a river. Once in operation, hydro schemes emit very little CO$_2$, although some methane emissions do arise due to decomposition of flooded vegetation. Storage schemes have a higher footprint, (~10–30 gCO$_2$eq/kWh), than run-of-river schemes as they require large amounts of raw materials (steel and concrete) to construct the dam. Run-of-river schemes have very small reservoirs (those with weirs) or none at all so do not give rise to significant emissions during their operation. Carbon footprints for this type of hydro scheme are some of the lowest of all electricity generation technologies (<5 g CO_2 eq/kWh).

Wind

Electricity generated from wind energy has one of the lowest carbon footprints. As with other low carbon technologies, nearly all the emissions occur during the manufacturing and construction phases, arising from the production of steel for the tower, concrete for the foundations and epoxy/fibre glass for the rotor blades.

These account for 98 per cent of the total life cycle CO_2 emissions. Emissions generated during operation of wind turbines arise from routine maintenance inspection trips.

Nuclear

Nuclear power generation has a relatively small carbon footprint (~5 gCO_2eq/kWh). Since there is no combustion, (heat is generated by fission of uranium or plutonium), operational CO_2 emissions account for <1 per cent of the total. Most emissions occur during uranium mining, enrichment and fuel fabrication. Decommissioning accounts for 35 per cent of the lifetime CO_2 emissions, and includes emissions arising from dismantling the nuclear plant and the construction and maintenance of waste storage facilities. The most energy intensive phase of the nuclear cycle is uranium extraction, which accounts for 40 per cent of the total CO_2 emissions.

Some commentators have suggested that if global nuclear generation capacity increases, higher grade uranium ore deposits would be depleted, requiring use of lower grade ores. This has raised concerns that the carbon footprint of nuclear generation may increase in the future.

Flammability Limit

Flammability limits, also called flammable limits, or explosive limits give the proportion of combustible gases in a mixture, between which limits this mixture is flammable. Gas mixtures consisting of combustible, oxidising, and inert gases are only flammable under certain conditions. The lower flammable limit (LFL) (lower explosive limit) describes the leanest mixture that is still flammable, i.e. the mixture with the smallest fraction of combustible gas, while the upper flammable limit (UFL) (upper explosive limit) gives the richest flammable mixture. Increasing the fraction of inert gases in a mixture raises the LFL and decreases UFL. A deflagration is a propagation of a combustion zone at a velocity less than the speed of sound in the unreacted medium. A detonation is a propagation of a combustion zone at a velocity greater than the speed of sound in the unreacted medium.

Flammability limits of mixtures of several combustible gases can be calculated using Le Chatelier's mixing rule for combustible volume fractions x_i:

$$LFL_{mix} = \frac{1}{\sum \frac{x_i}{LFL_i}}$$

and similar for UEL.

Temperature and pressure also influences flammability limits. Higher temperature results in lower LFL and higher UFL, while greater pressure increases both values. The effect of pressure is very small at pressures below 10 millibar and difficult to predict, since it has hardly been studied.

COMBUSTION OF LIQUID AND SOLID FUELS

Liquid Fuels

Combustion of a liquid fuel in an oxidising atmosphere actually happens in the gas phase. It is the vapour that burns, not the liquid. Therefore, a liquid will normally catch fire only above a certain temperature: its flash point. The flash point of a liquid fuel is the lowest temperature at which it can form an ignitable mix with air. It is also the minimum temperature at which there is enough evaporated fuel in the air to start combustion.

Solid Fuels

The act of combustion consists of three relatively distinct but overlapping phases:

1. Preheating phase, when the unburned fuel is heated up to its flash point and then fire point. Flammable gases start being evolved in a process similar to dry distillation.
2. Distillation phase or gaseous phase, when the mix of evolved flammable gases with oxygen is ignited. Energy is produced in the form of heat and light. Flames are often visible. Heat transfer from the combustion to the solid maintains the evolution of flammable vapours.
3. Charcoal phase or solid phase, when the output of flammable gases from the material is too low for persistent presence of flame and the charred fuel does not burn rapidly anymore but just glows and later only smoulders.

A general scheme of polymer combustion is shown in Fig. 10.5.

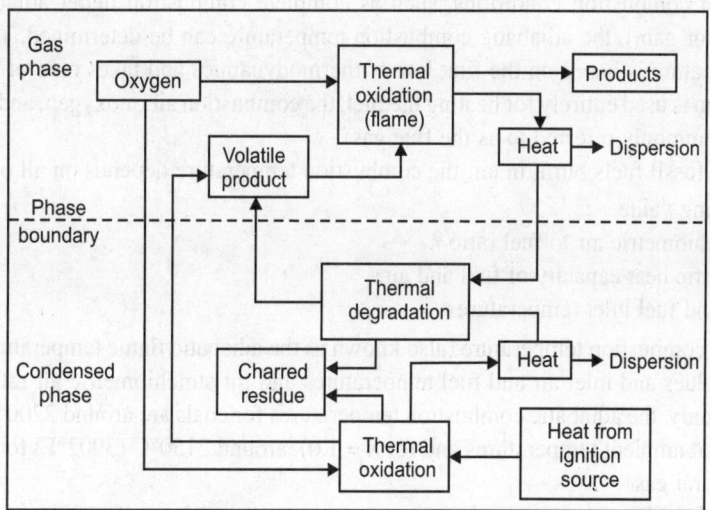

Fig. 10.5. A general scheme of polymer combustion.

Reaction mechanism

Combustion in oxygen is a radical chain reaction where many distinct radical intermediates participate. The high energy required for initiation is explained by the unusual structure of the dioxygen molecule. The lowest-energy configuration of the dioxygen molecule is a stable, relatively unreactive diradical in a triplet spin state. Bonding can be described with three bonding electron pairs and two antibonding electrons, whose spins are aligned, such that the molecule has nonzero total angular momentum. Most fuels, on the other hand, are in a singlet state, with paired spins and zero total angular momentum. Interaction between the two is quantum mechanically a 'forbidden transition', i.e. possible with a very low probability. To initiate combustion, energy is required to force dioxygen into a spin-paired state, or singlet oxygen. This intermediate is extremely reactive. The energy is supplied as heat. The reaction produces heat, which keeps it going.

Combustion of hydrocarbons is thought to be initiated by hydrogen atom abstraction (not proton abstraction) from the fuel to oxygen, to give a hydroperoxide radical (HOO). This reacts further to give hydroperoxides, which break up to give hydroxyl radicals. There are a great variety of these processes that produce fuel radicals and oxidising radicals. Oxidising species include singlet oxygen, hydroxyl, monatomic oxygen, and hydroperoxyl. Such intermediates are short-lived and cannot be isolated. However, non-radical intermediates are stable and are produced in incomplete combustion. An example is acetaldehyde produced in the combustion of ethanol. An intermediate in the combustion of carbon and hydrocarbons, carbon monoxide, is of special importance because it is a poisonous gas, but also economically useful for the production of syngas.

Solid fuels also undergo a great number of pyrolysis reactions that give more easily oxidised, gaseous fuels. These reactions are endothermic and require constant energy input from the combustion reactions. A lack of oxygen or other poorly designed conditions result in these noxious and carcinogenic pyrolysis products being emitted as thick, black smoke.

Temperature

Assuming perfect combustion conditions, such as complete combustion under adiabatic conditions (i.e. no heat loss or gain), the adiabatic combustion temperature can be determined. The formula that yields this temperature is based on the first law of thermodynamics and takes note of the fact that the heat of combustion is used entirely for heating the fuel, the combustion air or oxygen, and the combustion product gases (commonly referred to as the flue gas).

In the case of fossil fuels burnt in air, the combustion temperature depends on all of the following:

1. The heating value.
2. The stoichiometric air to fuel ratio λ.
3. The specific heat capacity of fuel and air.
4. The air and fuel inlet temperatures.

The adiabatic combustion temperature (also known as the adiabatic flame temperature) increases for higher heating values and inlet air and fuel temperatures and for stoichiometric air ratios approaching one. Most commonly, the adiabatic combustion temperatures for coals are around 2200°C (3992°F) (for inlet air and fuel at ambient temperatures and for $\lambda = 1.0$), around 2150°C (3902°F) for oil and 2000°C (3632°F) for natural gas.

In industrial fired heaters, power plant steam generators, and large gas-fired turbines, the more common way of expressing the usage of more than the stoichiometric combustion air is per cent excess combustion air. For example, excess combustion air of 15 per cent means that 15 per cent more than the required stoichiometric air is being used.

Instabilities

Combustion instabilities are typically violent pressure oscillations in a combustion chamber. These pressure oscillations can be as high as 180 dB, and long-term exposure to these cyclic pressure and thermal loads reduces the life of engine components. In rockets, such as the F1 used in the Saturn V program, instabilities led to massive damage of the combustion chamber and surrounding components. This problem was solved by redesigning the fuel injector. In liquid jet engines the droplet size and distribution can be used to attenuate the instabilities. Combustion instabilities are a major concern in ground-based gas turbine engines because of NO_x emissions. The tendency is to run lean, an equivalence ratio less than 1, to reduce the combustion temperature and thus reduce the NO_x emissions; however, running the combustion lean makes it very susceptible to combustion instabilities.

The Rayleigh Criterion is the basis for analysis of thermoacoustic combustion instabilities and is evaluated using the Rayleigh Index over one cycle of instability:

$$G(x) = \frac{1}{T}\int_T q'(x,t)p'(x,t)dt$$

where, q' is the heat release rate and p' is the pressure fluctuation. When the heat release oscillations are in phase with the pressure oscillations, the Rayleigh Index is positive and the magnitude of the thermo acoustic instability increases. On the other hand, if the Rayleigh Index is negative, then thermoacoustic damping occurs. The Rayleigh Criterion implies that a thermoacoustic instability can be optimally controlled by having heat release oscillations 180 degrees out of phase with pressure oscillations at the same frequency. This minimises the Rayleigh Index.

Charcoal

Charcoal is the dark grey residue consisting of impure carbon obtained by removing water and other volatile constituents from animal and vegetation substances. Charcoal is usually produced by slow pyrolysis, the heating of wood, sugar, bone char, or other substances in the absence of oxygen. The resulting soft, brittle, lightweight, black, porous material resembles coal and is 50 to 95 per cent carbon with the remainder consisting of volatile chemicals and ash.

Pyrolysis

Pyrolysis is a form of incineration that chemically decomposes organic materials by heat in the absence of oxygen. Pyrolysis typically occurs under pressure and at operating temperatures above 430°C (800°F). In practice, it is not possible to achieve a completely oxygen-free atmosphere. Because some oxygen is present in any pyrolysis system, a small amount of oxidation occurs (Fig. 10.6).

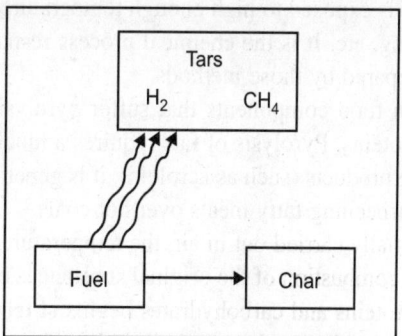

Fig. 10.6. Pyrolysis chemistry.

Pyrolysis is a special case of thermolysis, and is most commonly used for organic materials, being then one of the processes involved in charring. The pyrolysis of wood, which starts at 200°–300°C

(390°–570°F), occurs for example in fires or when vegetation comes into contact with lava in volcanic eruptions. In general, pyrolysis of organic substances produces gas and liquid products and leaves a solid residue richer in carbon content. Extreme pyrolysis, which leaves mostly carbon as the residue, is called carbonisation.

This chemical process is heavily used in the chemical industry, for example, to produce charcoal, activated carbon, methanol and other chemicals from wood, to convert ethylene dichloride into vinyl chloride to make PVC, to produce coke from coal, to convert biomass into syngas, to turn waste into safely disposable substances, and for transforming medium-weight hydrocarbons from oil into lighter ones like gasoline. These specialised uses of pyrolysis may be called various names, such as dry distillation, destructive distillation, or cracking.

Pyrolysis also plays an important role in several cooking procedures, such as baking, frying, grilling, and caramelising. And it is a tool of chemical analysis, for example in mass spectrometry and in carbon-14 dating. Indeed, many important chemical substances, such as phosphorus and sulphuric acid, were first obtained by this process. Pyrolysis has been assumed to take place during catagenesis, the conversion of buried organic matter to fossil fuels. It is also the basis of pyrography.

Pyrolysis differs from other high-temperature processes like combustion and hydrolysis in that it does not involve reactions with oxygen, water, or any other reagents. However, the term has also been applied to the decomposition of organic material in the presence of superheated water or steam (hydrous pyrolysis), for example in the steam cracking of oil.

Occurrence and uses

Fire

Pyrolysis is usually the first chemical reaction that occurs in the burning of many solid organic fuels, like wood, cloth, and paper, and also of some kinds of plastic. In a wood fire, the visible flames are not due to combustion of the wood itself, but rather of the gases released by its pyrolysis; whereas the flameless burning of embers is the combustion of the solid residue (charcoal) left behind by it. Thus, the pyrolysis of common materials like wood, plastic, and clothing is extremely important for fire safety and fire-fighting.

Cooking

Pyrolysis occurs whenever food is exposed to high enough temperatures in a dry environment, such as roasting, baking, toasting, grilling, etc. It is the chemical process responsible for the formation of the golden-brown crust in foods prepared by those methods.

In normal cooking, the main food components that suffer pyrolysis are carbohydrates (including sugars, starch, and fibre) and proteins. Pyrolysis of fats requires a much higher temperature, and since it produces toxic and flammable products (such as acrolein), it is generally avoided in normal cooking. It may occur, however, when barbecuing fatty meats over hot coals.

Even though cooking is normally carried out in air, the temperatures and environmental conditions are such that there is little or no combustion of the original substances or their decomposition products. In particular, the pyrolysis of proteins and carbohydrates begins at temperatures much lower than the ignition temperature of the solid residue, and the volatile subproducts are too diluted in air to ignite. (In flambé dishes, the flame is due mostly to combustion of the alcohol, while the crust is formed by pyrolysis as in baking.) Pyrolysis of carbohydrates and proteins require temperatures substantially higher than 100°C (212°F), so pyrolysis does not occur as long as free water is present, e.g. in boiling food—

not even in a pressure cooker. When heated in the presence of water, carbohydrates and proteins suffer gradual hydrolysis rather than pyrolysis. Indeed, for most foods, pyrolysis is usually confined to the outer layers of food, and only begins after those layers have dried out.

Food pyrolysis temperatures are however lower than the boiling point of lipids, so pyrolysis occurs when frying in vegetable oil or suet, or basting meat in its own fat. Pyrolysis also plays an essential role in the production of barley, tea, coffee, and roasted nuts such as peanuts and almonds. As these consist mostly of dry materials, the process of pyrolysis is not limited to the outermost layers but extends throughout the materials. In all these cases, pyrolysis creates or releases many of the substances that contribute to the flavour, colour, and biological properties of the final product. It may also destroy some substances that are toxic, unpleasant in taste, or those that may contribute to spoilage.

Controlled pyrolysis of sugars starting at 170°C (338°F) produces caramel, a beige to brown water-soluble product which is widely used in confectionery and (in the form of caramel colouring) as a colouring agent for soft drinks and other industrialised food products.

Solid residue from the pyrolysis of spilled and splattered food creates the brown-black encrustation often seen on cooking vessels, stove tops, and the interior surfaces of ovens.

Charcoal

Pyrolysis has been used since ancient times for turning wood into charcoal in an industrial scale. Besides wood, the process can also use sawdust and other wood waste products. Charcoal is obtained by heating wood until its complete pyrolysis (carbonisation) occurs, leaving only carbon and inorganic ash. In many parts of the world, charcoal is still produced semi-industrially, by burning a pile of wood that has been mostly covered with mud or bricks. The heat generated by burning part of the wood and the volatile by-products pyrolyses the rest of the pile. The limited supply of oxygen prevents the charcoal from burning too. A more modern alternative is to heat the wood in an airtight metal vessel, which is much less polluting and allows the volatile products to be condensed. The original vascular structure of the wood and the pores created by escaping gases combine to produce a light and porous material. By starting with dense wood-like material, such as nutshells or peach stones, one obtains a form of charcoal with particularly fine pores (and hence a much larger pore surface area), called activated carbon, which is used as an adsorbent for a wide range of chemical substances.

Biochar

Residues of incomplete organic pyrolysis, e.g. from cooking fires, are thought to be the key component of the terra preta soils associated with ancient indigenous communities of the Amazon basin. Terra preta is much sought by local farmers for its superior fertility compared to the natural red soil of the region. Efforts are underway to recreate these soils through biochar, the solid residue of pyrolysis of various materials, mostly organic waste. Biochar improves the soil texture and ecology, increasing its ability to retain fertilisers and release them slowly. It naturally contains many of the micronutrients needed by plants, such as selenium. It is also safer than other 'natural' fertilisers such as manure or sewage since it has been disinfected at high temperature, and since it releases its nutrients at a slow rate, it greatly reduces the risk of water table contamination. Biochar is also being considered for carbon sequestration, with the aim of mitigation of global warming.

Coke

Pyrolysis is used on a massive scale to turn coal into coke for metallurgy, especially steel-making. Coke can also be produced from the solid residue left from petroleum refining.

Those starting materials typically contain hydrogen, nitrogen or oxygen atoms combined with carbon into molecules of medium to high molecular weight. The coke-making or 'coking' process consists in heating the material in closed vessels to very high temperatures (up to 2000°C or 3600°F), so that those molecules are broken down into lighter volatile substances, which leave the vessel, and a porous but hard residue that is mostly carbon and inorganic ash. The amount of volatiles varies with the source material, but is typically 25–30 per cent of it by weight.

Carbon fibre

Carbon fibres are filaments of carbon that can be used to make very strong yarns and textiles. Carbon fibre items are often produced by spinning and weaving the desired item from fibres of a suitable polymer, and then pyrolysing the material at a high temperature (from 1500°–3000°C or 2730°–5430°F).

The first carbon fibres were made from rayon, but polyacrylonitrile has become the most common starting material. For their first workable electric lamps, Joseph Wilson Swan and Thomas Edison used carbon filaments made by pyrolysis of cotton yarns and bamboo splinters, respectively.

Biofuel

Pyrolysis is the basis of several methods that are being developed for producing fuel from biomass, which may include either crops grown for the purpose or biological waste products from other industries.

Although synthetic diesel fuel cannot yet be produced directly by pyrolysis of organic materials, there is a way to produce similar liquid (bio-oil) that can be used as a fuel, after the removal of valuable bio-chemicals that can be used as food additives or pharmaceuticals. Higher efficiency is achieved by the so-called flash pyrolysis where finely divided feedstock is quickly heated to between 350° and 500°C (660° and 930°F) for less than 2 seconds.

Fuel bio-oil resembling light crude oil can also be produced by hydrous pyrolysis from many kinds of feedstock, including waste from pig and turkey farming, by a process called thermal depolymerisation (which may however include other reactions besides pyrolysis).

Plastic waste disposal

Anhydrous pyrolysis can also be used to produce liquid fuel similar to diesel from plastic waste.

Processes

In many industrial applications, the process is done under pressure and at operating temperatures above 430°C (806°F). For agricultural waste, for example, typical temperatures are 450° to 550°C (840° to 1000°F).

Vacuum pyrolysis

In vacuum pyrolysis, organic material is heated in a vacuum in order to decrease boiling point and avoid adverse chemical reactions. It is used in organic chemistry as a synthetic tool. In flash vacuum thermolysis or FVT, the residence time of the substrate at the working temperature is limited as much as possible, again in order to minimise secondary reactions.

Processes for biomass pyrolysis

Since pyrolysis is endothermic, various methods have been proposed to provide heat to the reacting biomass particles:

1. Partial combustion of the biomass products through air injection. This results in poor-quality products.

2. Direct heat transfer with a hot gas, ideally product gas that is reheated and recycled. The problem is to provide enough heat with reasonable gas flow-rates.

3. Indirect heat transfer with exchange surfaces (wall, tubes). It is difficult to achieve good heat transfer on both sides of the heat exchange surface.

4. Direct heat transfer with circulating solids: Solids transfer heat between a burner and a pyrolysis reactor. This is an effective but complex technology.

For flash pyrolysis the biomass must be ground into fine particles and the insulating char layer that forms at the surface of the reacting particles must be continuously removed. The following technologies have been proposed for biomass pyrolysis:

1. Fixed beds were used for the traditional production of charcoal. Poor, slow heat transfer resulted in very low liquid yields.

2. Augers: This technology is adapted from a Lurgi process for coal gasification. Hot sand and biomass particles are fed at one end of a screw. The screw mixes the sand and biomass and conveys them along. It provides a good control of the biomass residence time. It does not dilute the pyrolysis products with a carrier or fluidising gas. However, sand must be reheated in a separate vessel, and mechanical reliability is a concern. There is no large-scale commercial implementation.

3. Ablative processes: Biomass particles are moved at high speed against a hot metal surface. Ablation of any char forming at the particles surface maintains a high rate of heat transfer. This can be achieved by using a metal surface spinning at high speed within a bed of biomass particles, which may present mechanical reliability problems but prevents any dilution of the products. As an alternative, the particles may be suspended in a carrier gas and introduced at high speed through a cyclone whose wall is heated; the products are diluted with the carrier gas. A problem shared with all ablative processes is that scale-up is made difficult since the ratio of the wall surface to the reactor volume decreases as the reactor size is increased. There is no large-scale commercial implementation.

4. Rotating cone: Pre-heated hot sand and biomass particles are introduced into a rotating cone. Due to the rotation of the cone, the mixture of sand and biomass is transported across the cone surface by centrifugal force. Like other shallow transported-bed reactors relatively fine particles are required to obtain a good liquid yield. There is no large scale commercial implementation.

5. Fluidised beds: Biomass particles are introduced into a bed of hot sand fluidised by a gas, which is usually a recirculated product gas. High heat transfer rates from fluidised sand result in rapid heating of biomass particles. There is some ablation by attrition with the sand particles, but it is not as effective as in the ablative processes. Heat is usually provided by heat exchanger tubes through which hot combustion gas flows. There is some dilution of the products, which makes it more difficult to condense and then remove the bio-oil mist from the gas exiting the condensers. This process has been scaled up by companies such as Dynamotive and Agri-Therm. The main challenges are in improving the quality and consistency of the bio-oil.

6. Circulating fluidised beds: Biomass particles are introduced into a circulating fluidised bed of hot sand. Gas, sand and biomass particles move together, with the transport gas usually being a recirculated product gas, although it may also be a combustion gas. High heat transfer rates

from sand ensure rapid heating of biomass particles and ablation is stronger than with regular fluidised beds. A fast separator separates the product gases and vapours from the sand and char particles. The sand particles are reheated in fluidised burner vessel and recycled to the reactor. Although this process can be easily scaled up, it is rather complex and the products are much diluted, which greatly complicates the recovery of the liquid products.

Coke and Charcoal

Coke

Coke is the solid carbonaceous material derived from destructive distillation of low-ash, low-sulphur bituminous coal. Cokes from coal are grey, hard, and porous. Coke is usually produced from coal; the process is called coking.

Volatile constituents of the coal—including water, coal-gas, and coal-tar—are driven off by baking in an airless furnace or oven at temperatures as high as 2000°C. This fuses together the fixed carbon and residual ash. Most modern facilities have 'by-product' coking ovens. Today, the volatile hydrocarbons are mainly used, after purification, in a separate combustion process to generate energy. Non by-product coking furnaces or coke furnaces (ovens) burn the hydrocarbon gases produced by the coke-making process to drive the carbonisation process. Bituminous coal must meet a set of criteria for use as coking coal, determined by particular coal assay techniques. These include moisture content, ash content, sulphur content, volatile content, tar, and plasticity.

The greater the volatile matter in coal, the more by-product can be produced, but too low or too high a level of volatile matter in the coal results in inferior coke produced in respect to coke quality properties. It is generally considered that levels of 26–29 per cent of volatile matter in the coal blend is good for coking purposes. Thus different types of coal are proportionally blended to reach acceptable levels of volatility before the coking process begins.

Charcoal grill

Charcoal grills use either charcoal briquets or all-natural lump charcoal as their fuel source. The charcoal, when burned, will transform into embers radiating the heat necessary to cook food.

There is contention among grilling enthusiasts on what type of charcoal is best for grilling. Users of charcoal briquets emphasise the uniformity in size, burn rate, heat creation, and quality exemplified by briquets. Users of all-natural lump charcoal emphasise the reasons they prefer it: subtle smoky aromas, high heat production, and lack of binders and fillers often present in briquets.

There are many different charcoal grill configurations. Some grills are square, round, or rectangular, some have lids while others do not, and some may or may not have a venting system for heat control.

EXPLOSIVES

An explosive is defined as a material (chemical or nuclear) that can be initiated to undergo very rapid, self-propagating decomposition that results in the formation of more stable material, the liberation of heat, or the development of a sudden pressure effect through the action of heat on produced or adjacent gases. All of these outcomes produce energy; a weapon's effectiveness is measured by the quantity of energy or damage potential—it delivers to the target.

Modern weapons use both kinetic and potential energy to achieve maximum lethality. Kinetic energy systems rely on the conversion of kinetic energy to work, while potential energy systems use explosive energy directly in the form of heat and blast, or by accelerating metal as a shaped charge, EFP or case fragments to increase their kinetic energy and damage volume.

Energy may be broadly classified as potential or kinetic. Potential energy is energy of configuration or position, or the capacity to perform work. For example, the relatively unstable chemical bonds among the atoms that comprise trinitrotoluene (TNT) possess chemical potential energy. Potential energy can, under suitable conditions, be transformed into kinetic energy, which is energy of motion. When a conventional explosive such as TNT is detonated, the relatively unstable chemical bonds are converted into bonds that are more stable, producing kinetic energy in the form of blast and thermal energies. This process of transforming a chemical system's bonds from lesser to greater stability is exothermic (there is a net production of energy).

A chemical explosive is a compound or a mixture of compounds which, when subjected to heat, impact, friction, or shock, undergoes very rapid, self-propagating, heat-producing decomposition. This decomposition produces gases that exert tremendous pressures as they expand at the high temperature of the reaction. The work done by an explosive depends primarily on the amount of heat given off during the explosion. The term detonation indicates that the reaction is moving through the explosive faster than the speed of sound in the unreacted explosive; whereas, deflagration indicates a slower reaction (rapid burning). A high explosive will detonate; a low explosive will deflagrate. All commercial explosives except black powder are high explosives.

Low-order explosives (LE) create a subsonic explosion (below 3300 feet per second) and lack HE's over-pressurisation wave. Examples of LE include pipe bombs, gunpowder, and most pure petroleum-based bombs such as Molotov cocktails or aircraft improvised as guided missiles. A high explosive (HE) is a compound or mixture which, when initiated, is capable of sustaining a detonation shockwave to produce a powerful blast effect. A detonation is the powerful explosive effect caused by the propagation of a high-speed shockwave through a high explosive compound or mixture. During the process of detonation, the high explosive is largely decomposed into hot, rapidly expanding gas.

The most important single property in rating an explosive is detonation velocity, which may be expressed for either confined or unconfined conditions. It is the speed at which the detonation wave travels through the explosive. Since explosives in boreholes are confined to some degree, the confined value is the more significant. The detonation velocity of an explosive is dependent on the density, ingredients, particle size, charge diameter, and degree of confinement. Decreased particle size, increased charge diameter, and increased confinement all tend to increase the detonation velocity. Unconfined velocities are generally 70 to 80 per cent of confined velocities.

The confined detonation velocity of commercial explosives varies from 4000 to 25,000 fps. With cartridge explosives the confined velocity is seldom attained. Some explosives and blasting agents are sensitive to diameter changes. As diameter is reduced, the velocity is reduced until at some critical diameter, propagation is no longer assured and misfires are likely.

Relative effectiveness factor (RE factor) is a measurement of an explosive's power for military demolitions purposes. It measures the detonating velocity relative to that of TNT, which has an RE factor of 1.00. TNT equivalent is a measure of the energy released from the detonation of a nuclear weapon, or from the explosion of a given quantity of fissionable material, in terms of the amount of TNT (trinitrotoluene) which could release the same amount of energy when exploded. The twelve-kiloton Hiroshima atomic bomb had a blast effect alone equivalent to some twenty-five million pounds of TNT-that's million. Denser explosives usually give higher detonation velocities and pressures. A dense explosive may be desirable for difficult blasting conditions or where fine fragmentation is required.

Low-density explosives will suffice in easily fragmented or closely jointed rocks and are preferred for quarrying coarse material.

Energetic materials are made in two ways. The first is by physically mixing solid oxidisers and fuels, a process that, in its basics, has remained virtually unchanged for centuries. Such a process results in a composite energetic material such as black powder. The second process involves creating a monomolecular energetic material, such as TNT, in which each molecule contains an oxidising component and a fuel component. For the composites, the total energy can be much greater than that of monomolecular materials. However, the rate at which this energy is released is relatively slow when compared to the release rate of monomolecular materials. Monomolecular materials such as TNT work fast and thus have greater power than composites, but they have only moderate energy densities — commonly half those of composites. Greater energy densities versus greater power — that's been the traditional trade-off. Ingredients of high explosives are classified as explosive bases, combustibles, oxygen carriers, antacids, and absorbents. Some ingredients perform more than one function. An explosive base is a solid or liquid which, upon the application of sufficient heat or shock, decomposes to gases with an accompanying release of considerable heat. A combustible combines with excess oxygen to prevent the formation of nitrogen oxides. An oxygen carrier assures complete oxidation of the carbon to prevent the formation of carbon monoxide. The formation of nitrogen oxides or carbon monoxide, in addition to being undesirable from the standpoint of fumes, results in lower heat of explosion and efficiency than when carbon dioxide and nitrogen are formed. Antacids increase stability in storage, and absorbents absorb liquid explosive bases.

Explosives are classified as primary or secondary based on their susceptibility to initiation. Primary explosives, which include lead azide and lead styphnate, are highly susceptible to initiation. Primary explosives often are referred to as initiating explosives because they can be used to ignite secondary explosives. Secondary explosives, which include nitroaromatics and nitramines are much more prevalent at military sites than are primary explosives. Because they are formulated to detonate only under specific circumstances, secondary explosives often are used as main charge or bolstering explosives.

Secondary explosives can be loosely categorised into melt-pour explosives, which are based on nitroaromatics such as TNT, and plastic-bonded explosives which are based on a binder and crystalline explosive such as RDX.

Propellants include both rocket and gun propellants. Most rocket propellants are composites based on a rubber binder, ammonium perchlorate oxidiser, and a powdered aluminium fuel; or composites based on a nitrate esters, usually nitroglycerine or nitrocellulose and nitramines. If a binder is used, it usually is an isocyanate-cured polyester or polyether. Some propellants also contain combustion modifiers, such as lead oxide. One group of gun propellants are called 'single base' (principally nitrocellulose), 'double base' (nitrocellulose and nitroglycerine), or 'triple base' (nitrocellulose, nitroglycerine, and nitroguanidine). Some of the newer, lower vulnerability gun propellants contain polymer binders and crystalline nitramines.

Pyrotechnics include illuminating flares, signalling flares, coloured and white smoke generators, tracers, incendiary delays, fuses, and photo-flash compounds. Pyrotechnics usually are composed of an inorganic oxidiser and metal powder in a binder. Illuminating flares contain sodium nitrate, magnesium, and a binder. Signalling flares contain barium, strontium, or other metal nitrates.

Explosive and incendiary (fire) bombs are further characterised based on their source. 'Manufactured' implies standard military-issued, mass produced, and quality-tested weapons. 'Improvised' describes weapons produced in small quantities, or use of a device outside its intended purpose, such as converting a commercial aircraft into a guided missile. Manufactured (military) explosive weapons are exclusively HE-based. Terrorists will use whatever is available—illegally obtained manufactured weapons or improvised explosive devices (also known as 'IEDs') that may be composed of HE, LE, or both. Manufactured and improvised bombs cause markedly different injuries. Plastic explosive means an explosive material in flexible or elastic sheet form formulated with one or more high explosives which in their pure form has a vapour pressure less than 10^{-4} Pa at a temperature of 25°C, is formulated with a binder material, and is as a mixture malleable or flexible at normal room temperature.

The energetic materials used by the military as propellants and explosives are mostly organic compounds containing nitro (NO_2) groups. The three major classes of these energetic materials are nitroaromatics (e.g. trinitrotoluene or TNT), nitramines (e.g. hexahydro-1,3,5 trinitroazine or RDX), and nitrate esters (e.g. nitrocellulose and nitroglycerine).

Since the invention of the cannon, the explosive fills used to drive lethal mechanisms have been the subject of ever increasing interest and study. Traditionally, munitions designers have used such explosives as Comp-B, TNT, or LX-14, depending upon the particular application.

During the 1920s and into the 1940s, the Army's Picatinny Arsenal was instrumental in designing, modelling and evaluating such high explosive material as TNT, RDX, and Haleite. This work greatly influenced battlefield lethality during World War II where explosives exhibiting a higher brisance, or shattering effect, than TNT were in great demand.

The 1960s brought new explosives such as HMX that was chemically analogous to RDX, but even more powerful to give soldiers greater lethality capability. Picatinny laboratories also developed precision warheads for several missile systems, including the DRAGON-MAW, a Medium Antiarmour Weapon.

The Army uses Research Department Explosive (RDX) and High Melt Explosive (HMX) as basic explosives for munitions and tactical missiles as well as propellants for strategic missiles rather than TNT because of their superior energy.

Most modern explosives are reasonably stable and require percussive shock or other triggering devices for detonation. Energetic materials are especially vulnerable to elevated temperature, with possible consequences ranging from mild decomposition to vigorous deflagration or detonation. Energetic materials can also be initiated by mechanical work through friction, impact, or electricity (e.g. current flow, spark, electrostatic discharge, or electromagnetic radiation). Other stimuli (e.g. focused laser light or chemical incompatibility) can have consequences ranging from mild decomposition to detonation.

Explosives may be toxic, with exposure pathways being inhalation of dust or vapour, ingestion, or skin contact. Most explosives are not highly toxic, but improper handling can result in systemic poisoning, usually affecting the bone marrow (i.e. the blood cell-producing system) and the liver. Some explosives are vasodilators, which cause headaches, low blood pressure, chest pains, and possible heart attacks. Some explosives may irritate the skin.

Some detonation or combustion products from explosives are toxic. Such products can be respiratory and skin irritants and lead to systemic effects following short-term exposure to high levels. Soot from detonated explosives is not mutagenic; however, soot from burned gun propellants may be mutagenic and is therefore treated as a mutagen.

Fortunately, contamination usually occurs in dilute, aqueous solutions or in relatively low concentrations in the soil and present no explosion hazard. Masses of pure crystalline explosive material have, however, been encountered in soils associated with waste-water lagoons, leach pits, burn pits, and firing ranges. These materials remain hazardous for long periods of time and great care must be used during the investigation and remediation process. Molecular weights are moderate, of the order of a few hundreds of grams per mole. The molecular structure, particularly the types and positions of subsidiary functional groups, controls environmental behaviour.

All of the common explosives are solid at normal environmental temperatures and pressures. Melting point temperatures for explosives solids are moderate (50°–205°C). Melting points are of little direct value in predicting environmental fate and transport, but several parameter estimation relations for solids incorporate the influence of molecular crystal bonding by including a term dependent on the melting point. Melting points are not available for many of the breakdown products. Most of the explosives and associated contaminants have very low volatility, with vapour pressures estimated to be less than 6×10^{-4} torr. Henry's law constants (KH) range from 10^{-4} to 10^{-11} atm·m-2·mole-1. Only those with KH greater than 10^{-5} volatilise significantly from aqueous solution 12. Though explosives compounds may not be volatile, some of the transformation products, other key reactants, or products may be volatile to semivolatile.

EXPLOSIONS AND DETONATIONS

Explosions

An explosion is a rapid increase in volume and release of energy in an extreme manner, usually with the generation of high temperatures and the release of gases. An explosion creates a shock wave. If the shock wave is a supersonic detonation, then the source of the blast is called a 'high explosive'. Subsonic shock waves are created by low explosives through the slower burning process known as deflagration.

Explosions can occur in nature. Most natural explosions arise from volcanic processes of various sorts. Explosive volcanic eruptions occur when magma rising from below has much dissolved gas in it; the reduction of pressure as the magma rises causes the gas to bubble out of solution, resulting in a rapid increase in volume. Explosions also occur as a result of impact events. Explosions can also occur outside of earth in the universe in events such as supernova. Explosions frequently occur during Bushfires in Eucalyptus forests where the volatile oils in the tree tops suddenly combust.

Chemical

The most common artificial explosives are chemical explosives, usually involving a rapid and violent oxidation reaction that produces large amounts of hot gas. Gunpowder was the first explosive to be discovered and put to use.

Nuclear

A nuclear weapon is a type of explosive weapon that derives its destructive force from the nuclear reaction of fission or from a combination of fission and fusion. As a result, even a nuclear weapon with a small yield is significantly more powerful than the largest conventional explosives available, with a single weapon capable of completely destroying an entire city.

Electrical

A high current electrical fault can create an electrical explosion by forming a high energy electrical arc which rapidly vapourises metal and insulation material. This arc flash hazard is a danger to persons working on energised switchgear. Also, excessive magnetic pressure within an ultra-strong electromagnet can cause a magnetic explosion.

Vapour

Boiling liquid expanding vapour explosions are a type of explosion that can occur when a vessel containing a pressurised liquid is ruptured, causing a rapid increase in volume as the liquid evaporates.

Astronomical

Among the largest known explosions in the universe are supernova, which result from stars exploding, and gamma ray bursts, whose nature is still in some dispute. Solar flares are an example of explosion common on the sun, and presumably on most other stars as well. The energy source for solar flare activity comes from the tangling of magnetic field lines resulting from the rotation of the sun's conductive plasma. Another type of large astronomical explosion occurs when a very large meteoroid or an asteroid impacts the surface of another object, such as a planet.

Mechanical

Strictly a physical process, as opposed to chemical or nuclear, e.g. the bursting of a sealed or partially-sealed container under internal pressure is often referred to as a 'mechanical explosion'. Examples include an overheated boiler or a simple tin can of beans tossed into a fire.

Detonation

Detonation involves an exothermic front accelerating through a medium that eventually drives a shock front propagating directly in front of it. They are observed in both conventional solid and liquid explosives, as well as in reactive gases. The velocity of detonations in solid and liquid explosives is much higher than that in gaseous ones, which allows far clearer resolution of the wave system in the latter.

Gaseous detonations normally occur in confined systems but are occasionally observed in large vapour clouds. Again, they are often associated with a gaseous mixture of fuel and oxidant of a composition, somewhat below conventional flammability limits. There is an extraordinary variety of fuels that may be present as gases, as droplet fogs and as dust suspensions. Other materials, such as acetylene, ozone and hydrogen peroxide are detonable in the absence of oxygen. Oxidants include halogens, ozone, hydrogen peroxide and oxides of nitrogen and chlorine.

In terms of external damage, it is important to distinguish between detonations and deflagrations where the exothermic wave is subsonic and maximum pressures are at most a quarter of those generated by the former.

REACTOR SAFETY

The objective of reactor safety is that reactors will be built and operated to pose no undue risk to public health and safety. Since the early days of nuclear reactors, it has been recognised that a society can gain the benefits of nuclear technology only if that society generally understands and accepts the risks from nuclear reactors. Reactor safety, therefore, is an essential prerequisite of reactor operation and must be placed on an equal footing with electricity production and other benefits of nuclear technology. It is further important that nuclear safety experts continue to improve their understanding of the risks from nuclear reactors and communicate that information to the public.

The essential attributes of reactor safety are the following:

1. A solid foundation of scientific and technological knowledge.
2. A robust design that uses established codes and standards and embodies margins, qualified materials, and redundant and diverse safety systems.

3. Construction and testing in accordance with the applicable design specifications and safety analyses.
4. A comprehensive organisational safety culture.
5. Qualified operational and maintenance personnel that have a profound respect for the reactor core and radioactive materials, and any supporting systems.
6. Technical specifications that define and control the safety operating envelope.
7. A strong engineering function that provides support to operations and maintenance.
8. Adherence to a defense-in-depth safety philosophy to maintain multiple barriers, both physical and procedural, to protect people.
9. Risk insights derived from analysis and experience.
10. Effective quality assurance, self-assessment, and corrective action programs.
11. Emergency plans protecting both onsite workers and off-site populations.
12. Access to a continuing program of nuclear safety research.
13. A strong management and fiscal organisation.
14. Safety regulatory authorities that are responsible for independently assuring operational safety.
15. Newer reactor designs that continue to incorporate enhanced safety features.

SOLVED EXAMPLES

Example 10.3: The pyrolysis of ethane proceeds with an activation energy of about 75,000 Cal. How much faster is the decomposition at 650°C than at 500°C?

Solution: At 500°C (773 K):

Rate of decomposition of ethane, $\quad -r_A = K\, C_A$

$$-r_{A1} = K_1\, C_A = K_0 \times e^{-E/RT} \times C_A$$
$$= K_0 \times e^{-75,000/(1.987)(773)} \times C_A$$
$$= 6.216 \times 10^{-22}\, K_0\, C_A$$

At 650°C (923 K):

$$-r_{A2} = K_2\, C_A = K_0 \times e^{-E/RT} \times C_A$$
$$= K_0 \times e^{-75000/(1.987)(973)} \times C_A$$
$$= 1.737 \times 10^{-18}\, K_0\, C_A$$

$$\therefore \qquad \frac{-r_{A2}}{-r_{A1}} = 2794$$

∴ Decomposition at 650°C is 2794 times faster than at 500°C.

Example 10.4: Pyrolysis of methyl acetoxy propionate at temperature near 500°C has been studied.

$$CH_3\, COOH\, (CH)_3\, COOH_3 \rightarrow CH_3\, COOH + CH_2 = CHCOO\, CH_3$$

Below 565°C the pyrolysis reaction is essentially 1st order with a rate constant given by $K = 7.8 \times 10^9 \times e^{-19220/T}$ sec^{-1}, where, T is expressed in K. It is desired to design a plug flow reactor to operate isothermally at 580°C.

What length of 16 cm diameter pipe is required to convert 90 per cent of the raw feed stack to methyl acrylate ($CH_2 = COOCH_3$). The feed enters at 5 atm at a flow rate of 230 kg/hr. Ideal gas flow may be assumed.

Solution:

Data: 1st order, gas phase, PFR

$$A \rightarrow Product$$

$$T = 500°C = 773°K$$
$$K = 7.8 \times 10^9 \times e^{-19220/773} = 0.1241/sec.$$
$$P_{Ao} = 5 \text{ atm}, X_A = 0.9$$

$$F_{Ao} = 230 \text{ kg/hr} = \frac{230}{170 \times 3600} = 3.758 \times 10^{-4} \text{ Kmol/s}$$

For 1st order irreversible, variable volume system:

$$-\ln (1 - X_A) (1 + \varepsilon_A) - \varepsilon_A X_A = K \tau \qquad \qquad ... (10.24)$$

$$C_{Ao} = \frac{P_{Ao}}{RT} = \frac{5}{0.082 \times 773} = 0.0788 \text{ Kmol/m}^3$$

$$\tau = \frac{C_{Ao}V}{F_{Ao}} = \frac{0.0788 \, V}{3.758 \times 10^{-4}} = 205.67 \, V$$

$$\varepsilon = \frac{2-1}{1} = 1$$

Now putting all these values in Eq. 10.23:

$$-\ln (1 - 0.9) (1 + 1) - (1 \times 0.9) = 0.1241 \times 205.67 \, V$$

$$\therefore \qquad V = 0.14238 \text{ m}^3$$

$$V = (\pi \, d^2 L)/4 = 0.14238$$

$$= (\pi/4) (0.16)^2 \, L = 0.14238$$

$$\therefore \qquad L = 7.081 \text{ m}$$

Example 10.5: Under appropriate conditions acetaldehyde vapour react to give methane and carbon monoxide by the reaction:

$$CH_3CHO \xrightarrow{K} CH_4 + CO$$

0.1 kg/s of acetaldehyde vapour is to be decomposed at 520°C and 1 atm in a tubular reactor. The reaction is 2nd order and irreversible with respect to CH_3CHO. $K = 0.43$ m^3/(kmol.s). What will be the volume of reactor for 35 and 90 per cent decomposition of acetaldehyde.

Solution:

Data: Order of reaction = 2

$$A \rightarrow B + C, \text{ Gas phase}$$

$$T = 520°C = 793°K, P_{Ao} = 1 \text{ atm.}$$
$$F_{Ao} = 0.1/44 = 2.272 \times 10^{-3} \text{ kmol/s}$$
$$K = 0.43 \text{ m}^3/\text{kmol}$$

$$C_{Ao} = \frac{P_{Ao}}{RT} = 0.01537$$

$$X_A = 0.35 \text{ and } 0.9$$

$$\varepsilon_A = (2-1)/1 = 1$$

$$\tau = \frac{C_{Ao}V}{F_{Ao}} = 6.768 \ V$$

For second order irreversible reaction,

$$\tau = C_{Ao} \int_0^{X_A} \frac{dX_A}{-r_A} = C_{Ao} \int_0^{X_A} \frac{dX_A}{KC_A^2}, = C_A \frac{C_{Ao}(1-X_A)}{1+\varepsilon X_A}$$

$$\Rightarrow 2\varepsilon_A(1+\varepsilon_A)\ln(1-X_A) + \varepsilon_A^2 X_A + (\varepsilon_A+1)^2 \frac{X_A}{1-X_A} = C_{Ao}K\tau$$

At $X_A = 0.35$:

$$2 \times 1 \times (1+1)\ln(1-35) + (\text{At } X_A = 0.35) + 4 \times \frac{0.35}{1-0.35}$$

$$= 0.01537 \times 0.43 \times 6.768 \ V$$

∴ $$V = 17.448 \ m^3$$

At $$X_A = 0.90; \ V = 619 \ m^3.$$

Chapter 11

Polymerisation Reactions and Reactors

INTRODUCTION

Polymers are an example of 'products-by-process', where the final product properties are mostly determined during manufacture, in the reactor. An understanding of processes occurring in the polymerisation reactor is, therefore, crucial to achieving efficient, consistent, safe and environmentally friendly production of polymeric materials.

Polymer reaction engineering provides the link between the fundamentals of polymerisation kinetics and polymer microstructure achieved in the reactor.

Polymerisation is a process of reacting monomer molecules together in a chemical reaction to form three-dimensional networks or polymer chains. There are many forms of polymerisation and different systems exist to categorise them.

In chemical compounds, polymerisation occurs via a variety of reaction mechanisms that vary in complexity due to functional groups present in reacting compounds and their inherent steric effects explained by VSEPR Theory. In more straightforward polymerisation, alkenes, which are relatively stable due to σ bonding between carbon atoms form polymers through relatively simple radical reactions; in contrast, more complex reactions such as those that involve substitution at the carbonyl group require more complex synthesis due to the way in which reacting molecules polymerise.

As alkenes can be formed in somewhat straightforward reaction mechanisms, they form useful compounds such as polyethylene and polyvinyl chloride (PVC) when undergoing radical reactions, which are produced in high tonnages each year due to their usefulness in manufacturing processes of commercial products, such as piping, insulation and packaging. Polymers such as PVC are generally referred to as 'homopolymers' as they consist of repeated long chains or structures of the same monomer unit, whereas polymers that consist of more than one molecule are referred to as 'copolymers'.

Other monomer units, such as formaldehyde hydrates or simple aldehydes, are able to polymerise themselves at quite low temperatures ($> -80°C$) to form trimers; molecules consisting of 3 monomer units which can cyclise to form ring cyclic structures, or undergo further reactions to form tetramers, or 4 monomer-unit compounds. Further compounds either being referred to as oligomers in smaller molecules. Generally, because formaldehyde is an exceptionally reactive electrophile it allows nucleophillic addition of hemiacetal intermediates, which are generally short-lived and relatively unstable 'mid stage' compounds which react with other molecules present to form more stable polymeric compounds. Polymerisation that is not sufficiently moderated and proceeds at a fast rate can be very hazardous. This phenomenon is known as hazardous polymerisation and can cause fires and explosions.

MECHANISM OF POLYMERISATION

Mechanism of polymerisation of single monomer is given below. If M represents the monomer and $Init_2$ represents a free radical initiator then we can write,

$$Init_2 \xrightarrow{\;k_i\;} 2\,Init \qquad\qquad \text{Chain initiation}$$

$$\left.\begin{array}{l} Init_2 + M \xrightarrow{\;k_p\;} Init\,M \\ \quad\cdots \qquad \cdots \qquad \cdots \\ Init\,M_n + M \xrightarrow{\;k_p\;} Init\,M_{n+1} \end{array}\right\} \qquad \text{Chain propagation}$$

$$\left.\begin{array}{l} Init\,M_n + Init\,M_m \xrightarrow{\;k_t\;} Init\,M_{m+n}\,Init \\ \quad \text{or} \\ \qquad\qquad \xrightarrow{\;k_t\;} Init\,M_n\!+\!Init\,M_m \end{array}\right\} \qquad \text{Chain termination}$$

Using a steady-state approximation on the concentration of all types of radicals $[R*]$ we obtain,

$$[R*] = \left(\frac{k_i}{k_t}\right)^{1/2} [Init]^{1/2}$$

and rate of propagation
$$= k_p[R*][M]$$
$$= k_p\,(k_i/k_t)^{1/2}\,[Init_2]^{1/2}\,[M]$$
$$= k\,[Init_2]^{1/2}\,[M]$$

The above mechanism is also consistent with the kinetics of copolymerisation.

Step-growth

Step-growth polymers are defined as polymers formed by the stepwise reaction between functional groups of monomers. Most step-growth polymers are also classified as condensation polymers, but not all step-growth polymers (like polyurethanes formed from isocyanate and alcohol bifunctional monomers) release condensates. Step-growth polymers increase in molecular weight at a very slow rate at lower conversions and reach moderately high molecular weights only at very high conversion (i.e. > 95 per cent).

To alleviate inconsistencies in these naming methods, adjusted definitions for condensation and addition polymers have been developed. A condensation polymer is defined as a polymer that involves elimination of small molecules during its synthesis, or contains functional groups as part of its backbone chain, or its repeat unit does not contain all the atoms present in the hypothetical monomer to which it can be degraded.

Chain-growth

Chain-growth polymerisation (or addition polymerisation) involves the linking together of molecules incorporating double or triple chemical bonds. These unsaturated monomers (the identical molecules that make up the polymers) have extra internal bonds that are able to break and link up with other monomers to form the repeating chain. Chain-growth polymerisation is involved in the manufacture of polymers such as polyethylene, polypropylene, and polyvinyl chloride (PVC). A special case of chain-growth polymerisation leads to living polymerisation. In the radical polymerisation of ethylene, its pi bond is broken, and the two electrons rearrange to create a new propagating centre like the one that attacked it. The form this propagating centre takes depends on the specific type of addition mechanism. There are several mechanisms through which this can be initiated. The free radical mechanism was one of the first methods to be used. Free radicals are very reactive atoms or molecules that have unpaired electrons. Taking the polymerisation of ethylene as an example, the free radical mechanism can be divided into three stages: chain initiation, chain propagation, and chain termination (Fig. 11.1).

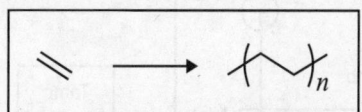

Fig. 11.1. Polymerisation of ethylene.

Free radical addition polymerisation of ethylene must take place at high temperatures and pressures, approximately 300°C and 2000 atm. While most other free radical polymerisations do not require such extreme temperatures and pressures, they do tend to lack control. One effect of this lack of control is a high degree of branching. Also, as termination occurs randomly, when two chains collide, it is impossible to control the length of individual chains. A newer method of polymerisation similar to free radical, but allowing more control involves the Ziegler-Natta catalyst, especially with respect to polymer branching.

Other forms of chain growth polymerisation include cationic addition polymerisation and anionic addition polymerisation. While not used to a large extent in industry yet due to stringent reaction conditions such as lack of water and oxygen, these methods provide ways to polymerise some monomers that cannot be polymerised by free radical methods such as polypropylene. Cationic and anionic mechanisms are also more ideally suited for living polymerisations, although free radical living polymerisations have also been developed.

GENERALISED PREDICTIVE CONTROL OF BATCH POLYMERISATION REACTOR

Polymerisation reactors play a key role in polymer engineering and the importance of their effective control is well-recognised in the polymerisation literature. A major characteristic of polymerisation reactors is their complex nonlinear behaviour.

During the eighties, significant advances were made in the area of nonlinear control, primarily within the differential geometric framework. Not only the system theoretic properties of nonlinear system are now well-understood, but also controller design technique are available, like the Globally Linearising Control (GLC) method. The GLC method with a PI linear controller has been used in for nonlinear control of a batch polymerisation reactor. In this section we use GLC and MPC methods to control the temperature of a batch polymerisation reactor in which solution of poly-methyl-methacrylate (PMMA) polymerisation takes place. The rest of this section is as follows: The description of Mathematical model, after a brief review of the GLC design method, followed by a brief review of MPC method and finally application control method to batch reactor system.

Mathematical Model

The GLC is a model-based control method; therefore, a mathematical description of the process (in state-space form) is needed to synthesise the control law. Figure 11.2 shows the schematic diagram of batch reactor.

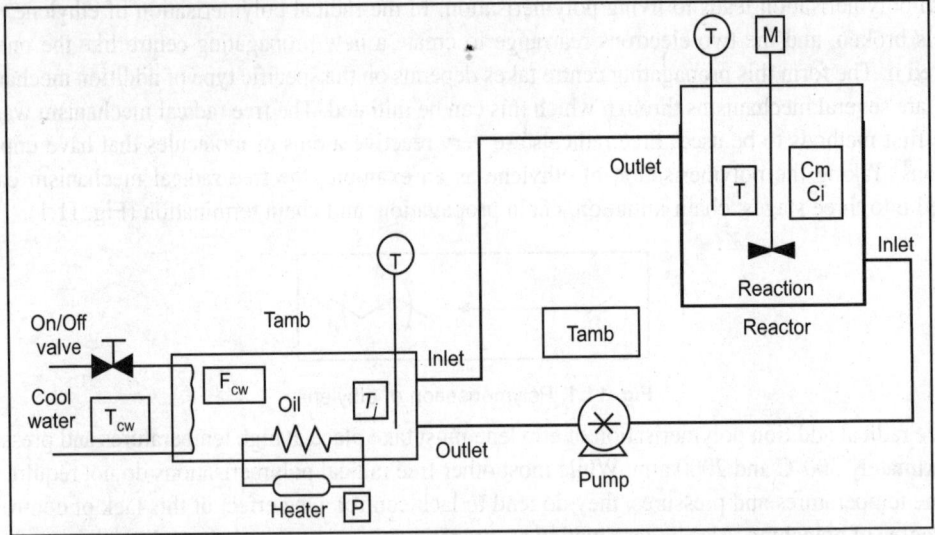

Fig. 11.2. Schematic diagram of the batch polymerisation reactor.

Reactor and jacket dynamics

Species balances for the monomer, initiator, solvent and dead polymer for the reactor and energy balances for the reactor and jacket, under standard assumptions, the reactor and jacket dynamics are given as follow:

$$
\begin{cases}
\dfrac{dC_m}{dt} = f_1(C_m, C_i, T) \\[2mm]
\dfrac{dC_i}{dt} = f_2(C_m, C_i, T) \\[2mm]
\dfrac{dT}{dt} = f_3(C_m, C_i, T, T_j) \\[2mm]
\dfrac{dT_j}{dt} = f_4(C_m, T, T_j) + \alpha_4 u
\end{cases}
\qquad \text{... (11.1)}
$$

where,

$$
u = P - \frac{\alpha_5(F_{cw})}{\alpha_4} \cdot (T_j - T_{cw}) \qquad \text{... (11.2)}
$$

u is the net heat input to the jacket by the heater and inlet cooling water. f_1, f_2, f_3, f_4 are scalar functions; for brevity, these are not given here. α_4 is a constant parameter.

Control problem

The control problem is to force the reactor to follow the optimal temperature profile file $T^*(t)$ (Fig. 11.3) by using the nonlinear control method and model predictive controller.

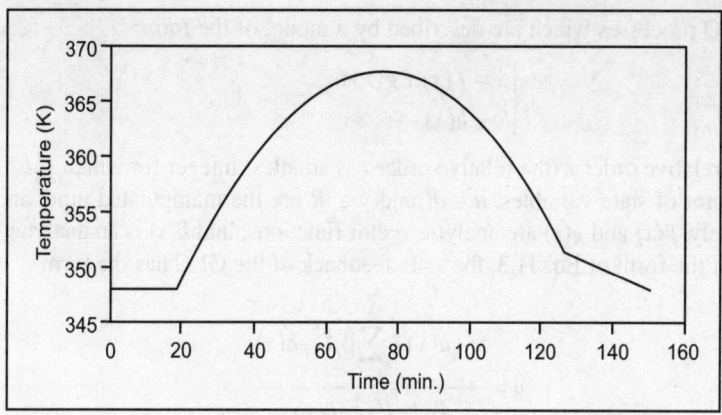

Fig. 11.3. Temperature profile.

GLC Method

Brief review

The GLC structure (shown in Fig. 11.4) consists of (i) a static state feedback (in inner loop), under which the closedloop input/output system is exactly linear, and (ii) an external linear controller (in outer loop) to ensure offset-less tracking of set point in the presence of modelling errors and process disturbances.

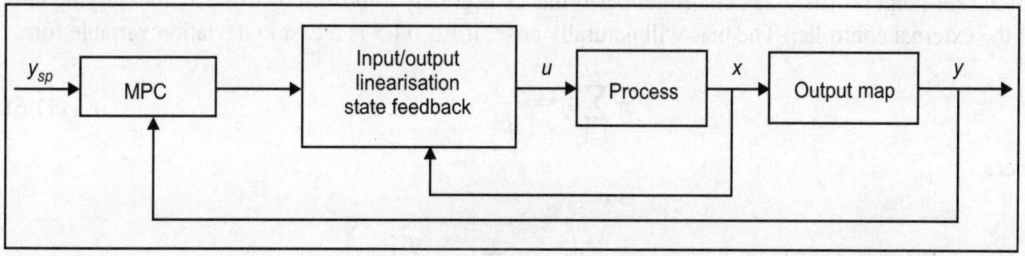

Fig. 11.4. The GLC structure.

In the case the state variables are not measured on-line, they should be reconstructed by using state observers (Fig. 11.5). In what follows, a brief review of the GLC synthesis approach will be provided.

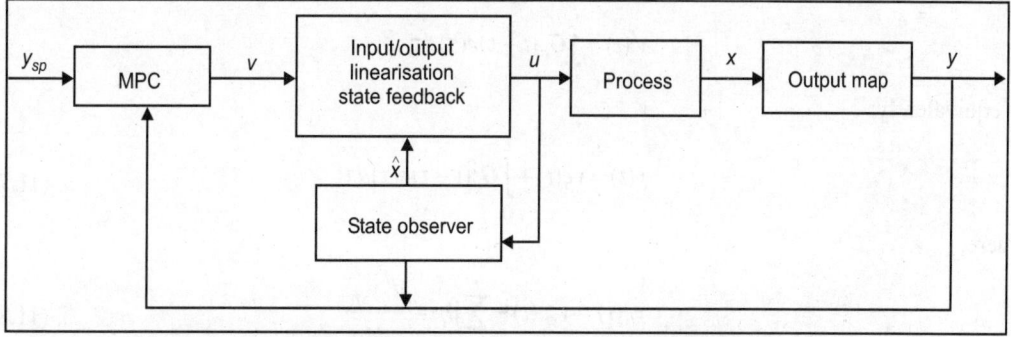

Fig. 11.5. The GLC-output feedback structure.

Consider SISO processes which are described by a model of the form:

$$\begin{cases} \dot{x} = f(x) + g(x)u \\ y = h(x) \end{cases} \qquad \text{... (11.3)}$$

With a finite relative order r (the relative order r is smallest integer for which $L_g L_f^{r-1} h(X) \neq 0$. Here $x \in R^n$ is the vector of state variables, $u \in R$ and $y \in R$ are the manipulated input and the controlled output, respectively. $f(x)$ and $g(x)$ are analytic vector functions, and $h(x)$ is an analytic scalar function. For the system of the form of Eq. 11.3, the state feedback of the GLC has the form:

$$u = \frac{v - h(x) - \sum_{l=1}^{r} \beta_l L_f^l h(x)}{\beta_r L_g L_f^{r-1} h(x)} \qquad \text{... (11.4)}$$

where, β_ls are tunable parameters. And v is the output of the external controller. Under the state feedback of Eq. 11.4, the input/output behaviour of the closed-loop $(v - y)$ system is linear without zeros:

$$y + \sum_{l=1}^{r} \beta_l \cdot \frac{d^l y}{dt^l} = v \qquad \text{... (11.5)}$$

For control problems that involve constant set points, the bias of the external controller is normally taken to be constant. In batch processes, however, where the objective is to track an a priori known smooth time-varying set-point profile $y_{sp}(t)$, controller performance is greatly improved by using a time-varying bias for the external controller. The bias will naturally arise, if Eq. 11.5 is recast in deviation variable form:

$$y' + \sum_{l=1}^{r} \beta_l \cdot \frac{d^l y'}{dt^l} = v' \qquad \text{... (11.6)}$$

where,

$$y' = y - y_{sp}$$

$$v' = v - \left(y_{sp} + \sum_{l=1}^{r} \beta_l \cdot \frac{d^l y_{sp}}{dt^l} \right)$$

and then an external bias-free error feedback controller [e.g. with transfer function $G_e(s)$ such that $1/G_e(0) = 0$] is used:

$$v'(t) = \int_0^t G_e(t - \tau) e(\tau) d\tau$$

or equivalently:

$$v(t) = v_b(t) + \int_0^t G_e(t - \tau) e(\tau) d\tau \qquad \text{... (11.7)}$$

where,

$$v_b(t) = y_{sp}(t) + \sum_{l=1}^{r} \beta_l \cdot \frac{d^l y_{sp}}{dt^l} \qquad \text{... (11.8)}$$

is the external controller bias.

Synthesis of the control law

The nonlinear control law is synthesised by following the steps of the GLC method, i.e.:

1. Recasting the model described by Eq. 11.1 in the standard state-space form of Eq. 11.3:

$$\frac{d}{dt}\begin{Bmatrix} C_m \\ C_i \\ T \\ T_j \end{Bmatrix} = \begin{bmatrix} f_1(C_m,C_i,T) \\ f_2(C_m,C_i,T) \\ f_3(C_m,C_i,T,T_j) \\ f_4(C_m,T,T_j) \end{bmatrix} + \begin{bmatrix} 0 \\ 0 \\ 0 \\ \alpha_4 \end{bmatrix} u$$

here, $x = [C_m \ C_i \ T \ T_j]^T \in R^4$.

2. Calculating the relative order:

$$r = 2 \ (L_g L_f h = \alpha_1 \cdot \alpha_4 \neq 0, \ L_g h = 0)$$

3. Calculating the state feedback:

$$u = \frac{v - \beta_2 \sum_{i=1}^{4} \frac{\partial f_3(x)}{\partial X_i} f_i(x) - \beta_1 \cdot f_3(x) - T}{\beta_2 \cdot \alpha_1 \cdot \alpha_4} \qquad \qquad \text{... (11.9)}$$

where, β_1 and β_2 are tunable parameter.

4. As an external linear controller, using GPC controller.

Once the value of u calculated by the control law the corresponding values of the two actual manipulated inputs (P, F_{cw}), are calculated according to the same coordination rules.

$$P = \begin{cases} u & 0 < u < P_{max} \\ P_{max} & u > P_{max} \\ u + \dfrac{\alpha_5}{\alpha_4}(T_j - T_{cw}) & -\dfrac{\alpha_5}{\alpha_4}(T_j - T_{cw}) < u < 0 \\ 0 & u < -\dfrac{\alpha_5}{\alpha_4}(T_j - T_{cw}) \end{cases}$$

$$\alpha_5(F_{cw}) = \begin{cases} \alpha_5 & u < 0 \\ 0 & u > 0 \end{cases}$$

Reduced order state observer

In the state feedback (Eq. 11.9), u is a function of the four states C_m, C_i, T and T_j. From these four states, C_m, C_i are not measured on-line, therefore, they should be estimated. A reduced-order observer is used to estimate the concentrations of the monomer and initiator. For the model of Eq. 11.1 this involves on-line integration of the first two differential equations of the model, i.e.:

$$\begin{cases} \dfrac{d\hat{C}_m}{dt} = f_1(\hat{C}_m, \hat{C}_i, T), & \hat{C}_m(0) = C_m \\[2mm] \dfrac{d\hat{C}_i}{dt} = f_2(\hat{C}_m, \hat{C}_i, T), & \hat{C}_i(0) = C_i \end{cases}$$

where, the $\hat{C}_m$ and $\hat{C}_i$ denote the estimates of the concentrations C_m and C_i, using the measured reactor temperature as input.

Model Predictive Control

Model predictive control (MPC) originated in the late seventies and has developed considerably since then. MPC is an optimisation-based multivariable control strategy that uses a mathematical model, incorporated into a control system, to predict in real-time the control action to be taken on the process. The predictive model represents the relationship between the process inputs and the process outputs. The MPC has the ability to predict process behaviour and proactively take measures to optimise control.

Predictive control determines future values of the manipulated variable by optimising a cost function, which expresses the control objectives and constraints. As it were, at the present time t, the present and future control inputs on the control horizon N_u, $u(t)$, $u(t+1)$,..., $u(t+M-1)$ u and predicted outputs over the prediction horizon N, $\hat{y}(t)$, $\hat{y}(t+1)$, ..., $\hat{y}(t+N)$ are obtained by solving an optimisation problem represented by a specified objective function. Among these solutions, only the first input $u_d(t)$ is implemented for time $(t, t+1)$. At the next time step, new values of the measured output are acquired, the control and prediction horizons are shifted forward by one step and the same calculations are repeated. To compensate the modelling error, new measurements have to be done at each time step.

A general objective function is the following quadratic form, mostly referred to as generalised predictive control (GPC):

$$J(N_1,N_2,N_u) = \sum_{j=N_1}^{N_2} \delta(j)[\hat{y}(t+j|t)-...w(t+j)]^2 + \sum_{j=1}^{N_u} \lambda(j)[\Delta u(t+j-1)]^2 \quad ...(11.10)$$

where, $\hat{y}(t+j/t)$ is an optimum j step ahead prediction of the system output on data up to time t, N_1 and N_2 are the minimum and maximum costing horizons, N_u is the control horizon, $\delta(j)$ and $\lambda(j)$ are weighting sequences and $w(t+j)$ is the future reference trajectory. For simplicity we can assume that $\lambda(j) = \lambda$, $\delta(j) = 1$.

The objective of predictive control is to compute the future control sequence $u(t)$, $u(t+1)$, ..., $u(t+N_u)$ in such a way that the future plant output $y(t+j)$ is driven closed to $w(t+j)$. This is accomplished by minimising $J(N_1, N_2, N_u)$.

The process to be controlled is described by following model:

$$A(q^{-1})y(t) = B(q^{-1})u(t-1) + e(t) \quad ...(11.11)$$

This model is CARIMA model. In this model $y(t)$ is reactor temperature, $u(t)$ is the net heat input to the jacket, $e(t)$ a random noise sequence that models the error between output plant and output model and q^{-1} is the backward shift operator. $A(q^{-1})$ and $B(q^{-1})$ are polynomials as a function of the backward shift operator, q^{-1}. Their order are n_a and n_b, respectively.

We use this model to predict output over prediction horizon. To derive j-step-ahead predictor of $\hat{y}(t+j)$ based on the Eq. 11.11, consider the Diophantine equation. Then a set of N_2 j-step-ahead outputs prediction over prediction horizon expressed as:

$$\hat{Y} = G\Delta U + F \quad ...(11.12)$$

where,

$$\hat{Y} = [\hat{y}(t+1/t)\,\hat{y}(t+2/t)...\hat{y}(t+N/t)]^T$$

$$\Delta U = [\Delta u(t)\Delta u(t+1)...\Delta u(t+N-1)]^T$$

$F = [f_1 f_2 \ldots f_N]^T$ is free response of system.

$$G = \begin{bmatrix} g_0 & 0 & \cdots & 0 \\ g_1 & g_0 & \cdots & 0 \\ \vdots & \vdots & \ddots & \vdots \\ g_{N-1} & g_{N-2} & \cdots & g_0 \end{bmatrix}$$

The first column of the matrix G is step response coefficient of system. If control horizon is less than prediction horizon ($N_u < N$), in other words the control signal is kept constant after control horizon, ($\Delta u (t + j - 1) = 0$ for $j > N_u$), the set of predictions which effect the objective function can be expressed as:

$$\hat{Y} = G_s \Delta U_s + F \qquad \ldots (11.13)$$

where,

$$G_s = \begin{bmatrix} g_0 & 0 & \cdots & 0 \\ g_1 & g_0 & \cdots & 0 \\ \vdots & \vdots & \ddots & \vdots \\ g_{N-1} & g_{N-2} & & g_{N-Nu} \end{bmatrix}$$

$$\Delta U_s = [\Delta u_s(t) \Delta u_s(t+1) \ldots \Delta u_s(t + N_u - 1)]^T$$

Now, the objective function of Eq. 11.12 can be rewritten as the following form:

$$J = (G\Delta U + F - W)^T (G\Delta U + F - W) + \ldots \lambda U^T U \qquad \ldots (11.14)$$

where,

$W = [w(t + 1), w(t + 2), \ldots, w(t + N)]^T$ is the reference trajectory.

Equation 11.14 can be written as:

$$J = 1/2 \Delta U^T H \Delta U + b^T \Delta U + F_0 \qquad \ldots (11.15)$$

where,

$$\left. \begin{array}{l} H = 2(G^T G + \lambda I) \\ b^T = 2(F - W)^T G \\ F_0 = (F - W)^T (F - W) \end{array} \right] \qquad \ldots (11.16)$$

The minimum of J, assuming there are no constraints on the control signals, can be found by making the gradient of J equal to zero, which leads to:

$$\Delta U = -H^{-1}b = (G^T G + \lambda I)^{-1} G^T (W - F) \qquad \ldots (11.17)$$

Note that the control signal that is actually sent to the process is the first element of vector ΔU, given by:

$$\Delta u(t) = K(W - F) \qquad \ldots (11.18)$$

where, K is the first row of matrix $(G^T G + \lambda I)^{-1} G^T$. This has a clear meaning that can easily be derived from Fig. 11.6. If there are no future predicted errors, that is, if $(W - F) = 0$, then there is no control move, since the objective will be fulfilled with the free evolution of the process. However in the other case, there will be an increment in the control action proportional (with a factor K) to that future error. Notice that the action is taken with respect to future errors, not past errors, as is the case in conventional feedback controllers. Notice that only the first element of ΔU is applied and the procedure is repeated at next sampling time.

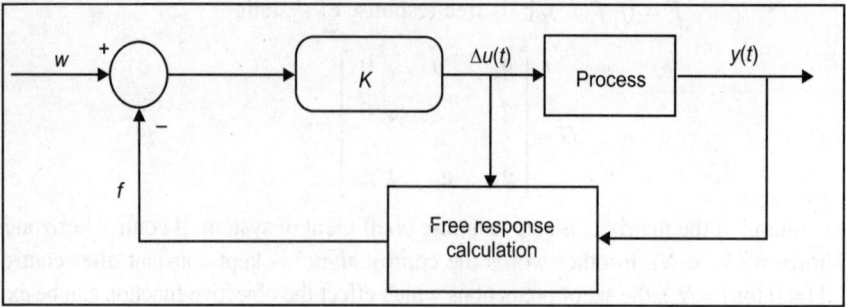

Fig. 11.6. The GPC control law.

Application of Batch Reactor Temperature Control

In this section the proposal algorithm is applied to batch polymerisation reactor. Assume that the tuning parameter is:

$$\beta_1 = 550, \ \beta_2 = 2.5 \times 10^4$$

Since the batch polymerisation reactor is a slow system, the sampling time was chosen to be 20 s. The time is needed for calculation by a 2.5 Ghz cpu is 16 ms that is suitable for computations.

The temperature profile has been shown in Fig. 11.3. Since GLC method create a linear model for system we can applied set point to linear model and show that the system need to a linear controller to have good performance. Figure 11.7 shows output of system when we do not use linear controller. It shows that the GLC method cannot track set point and system has not good performance.

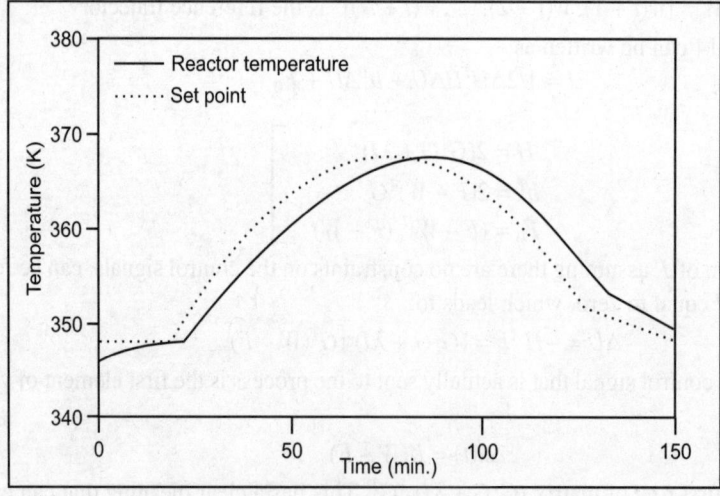

Fig. 11.7. The GLC linear model output.

In Fig. 11.8 we applied MPC controller to system. It shows the ability of MPC controller to output track set point and system has good performance.

The optimum values of parameter are as follows:

$$N_2 = 8, \ N_u = 6, \ \lambda = 0.6$$

The performance of proposed controller has been shown in Fig. 11.9.

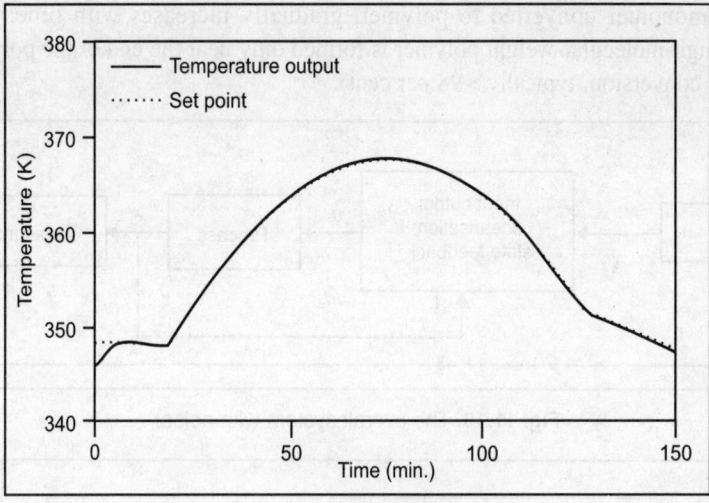

Fig. 11.8. The MPC controller response.

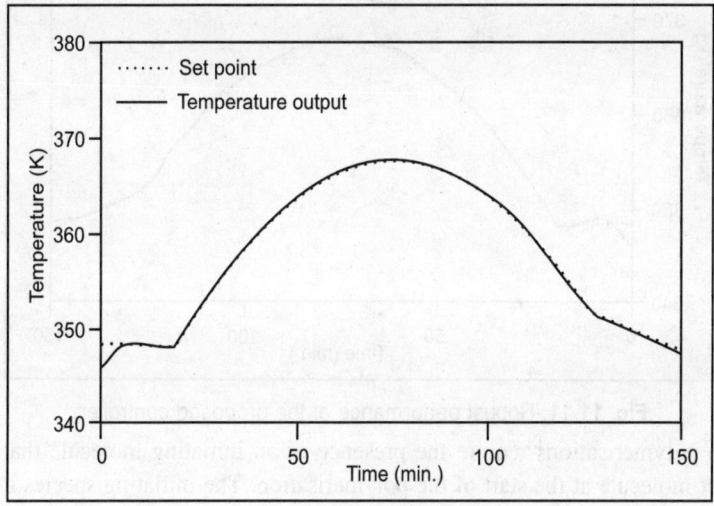

Fig. 11.9. Performance of the proposed controller.

Now, we assume that there is a uniform distribute noise in output (Fig. 11.10) then, by $N_2 = 10$, $N_u = 8$, $\lambda = 5$ the response of system is shown in Fig. 11.11. It shows that MPC controller has good performance when system effected by noise.

CHAIN-GROWTH POLYMERISATION

More recently, another classification scheme based on polymerisation kinetics has been adopted over the more traditional addition and condensation categories. According to this scheme, all polymerisation mechanisms are classified as either step growth or chain-growth. Most condensation polymers are step growth, while most addition polymers are chain-growth. During chain-growth polymerisation, high-molecular-weight polymer is formed early during the polymerisation, and the polymerisation yield, or the per cent of monomer converted to polymer, gradually increases with time. In step-growth polymerisation, high-molecular-weight polymer is formed only near the end of the polymerisation (i.e. at high monomer conversion, typically >98 per cent).

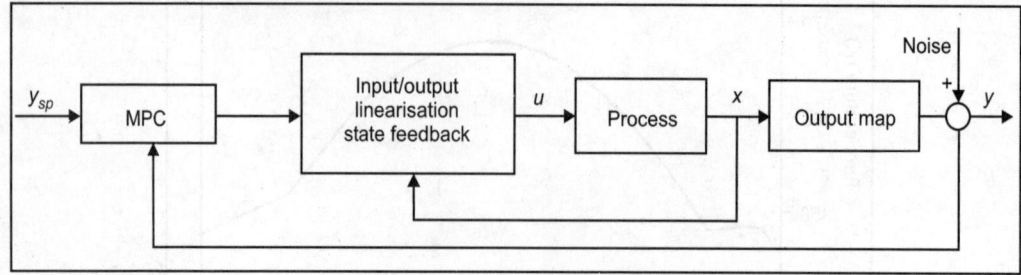

Fig. 11.10. The overall system with noise.

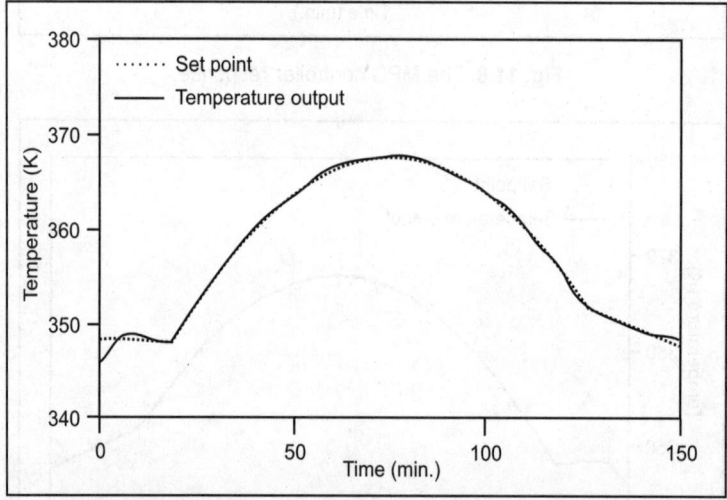

Fig. 11.11. Robust performance of the proposed controller.

Chain-growth polymerisations require the presence of an initiating molecule that can be used to attach a monomer molecule at the start of the polymerisation. The initiating species may be a radical, anion, or cation, as discussed in the following sections. Free-radical, anionic, and cationic chain-growth polymerisations share three common steps — initiation, propagation, and termination. Whether the polymerisation of a particular monomer can occur by one or more mechanisms (i.e. free radical, anionic, or cationic) depends, in part, on the chemical nature of the substituent group. Monomers with an electron-withdrawing group can polymerise by an anionic pathway, while those with an electron-donating group follow a cationic pathway. Some polymers with a resonance-stabilised substituent-group such as a phenyl ring may be polymerised by more than one pathway. For example, polysterene can be polymerised by both free-radical and anionic methods.

POLYMER MOLECULAR WEIGHT

Polymer molecular weight is important because it determines many physical properties. Some examples include the temperatures for transitions from liquids to waxes to rubbers to solids and mechanical properties such as stiffness, strength, viscoelasticity, toughness, and viscosity. If molecular weight is too low, the transition temperatures and the mechanical properties will generally be too low for the polymer material to have any useful commercial applications. For a polymer to be useful it must have transition temperatures to waxes or liquids that are above room temperatures and it must have mechanical properties sufficient to bear design loads.

For example, consider the property of tensile strength. Figure 11.12 shows a typical plot of strength as a function of molecular weight. At low molecular weight, the strength is too low for the polymer material to be useful. At high molecular weight, the strength increases eventually saturating to the infinite molecular weight result of S_∞. The strength-molecular weight relation can be approximated by the inverse relation:

$$S = S_\infty - \frac{A}{M} \qquad \qquad \text{... (11.19)}$$

where, A is a constant and M is the molecular weight. Many properties have similar molecular weight dependencies. They start at a low value and eventually saturate at a high value that is characteristic for infinite or very large molecular weight. Unlike small molecules, however, the molecular weight of a polymer is not one unique value. Rather, a given polymer will have a distribution of molecular weights.

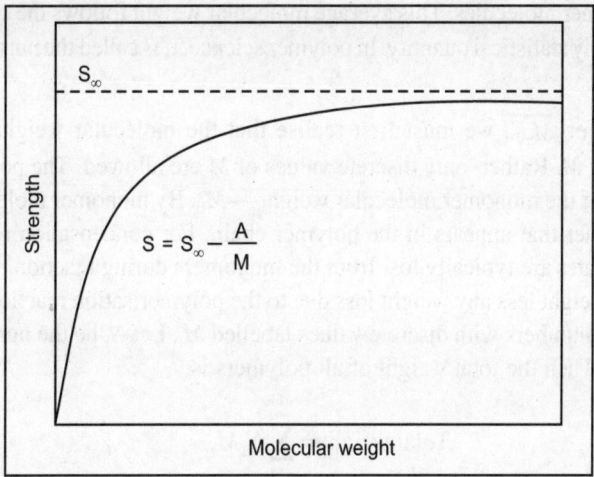

Fig. 11.12. A typical plot of tensile strength as a function of molecular weight.

The distribution will depend on the way the polymer is produced. For polymers we should not speak of a molecular weight, but rather of the distribution of molecular weight, $P(M)$, or of the average molecular weight, (M). Polymer physical properties will be functions of the molecular weight distribution function as in:

$$S = S_\infty - \frac{A}{F[P(M)]} \qquad \qquad \text{... (11.20)}$$

where, $F[P(M)]$ is some function of the complete molecular weight distribution function. For some properties, $F[P(M)]$ my reduce to simply an average molecular weight. The property will thus be a function of the average molecular weight, (M), and insensitive to other the details of the molecular weight distribution function:

$$S = S_\infty - \frac{A}{(M)} \qquad \qquad \text{... (11.21)}$$

There are many ways, however, to calculate an average molecular weight. The question, therefore, is how do you define the average molecular weight for a given distribution of molecular weights. The answer is that the type of property being studied will determine the desired type of average molecular weight. For example, strength properties may be influenced more by high molecular weight molecules than by low molecular weight molecules and thus the average molecular weight for strength properties should be weighted to emphasise the presence of high molecular weight polymer. In this section we will consider several ways of calculating molecular weights. We also consider the meanings of those averages. Finally, we consider typical distributions of molecular weights.

Number Average Molecular Weight

Consider a property which is only sensitive to the number of molecules present—a property that is not influenced by the size of any particle in the mixture. The best example of such properties are the colligative properties of solutions such as boiling point elevation, freezing point depression, and osmotic pressure. For such properties, the most relevant average molecular weight is the total weight of polymer divided by the number of polymer molecules. This average molecular weight follows the conventional definition for the mean value of any statistical quantity. In polymer science, it is called the number average molecular weight—$\overline{M_N}$.

To get a formula for $\overline{M_N}$, we must first realise that the molecular weight distribution is not a continuous function of M. Rather, only discrete values of M are allowed. The possible values of M are the various multiples of the monomer molecular weight—M_0. By monomer molecular weight we mean the weight per monomer that appears in the polymer chain. For condensation reactions, for example, where molecules of water are typically lost from the monomers during reaction, we will take M_0 as the monomer molecular weight less any weight loss due to the polymerisation reaction. The possible values of M make up a set of numbers with discrete values labelled M_i. Let N_i be the number of polymers with molecular weight M_i. Then the total weight of all polymers is:

$$\text{Total weight} = \sum_{i=1}^{\infty} N_i M_i \qquad \qquad \text{... (11.22)}$$

and the total number of polymer molecules is:

$$\text{Total number} = \sum_{i=1}^{\infty} N_i \qquad \qquad \text{... (11.23)}$$

As discussed above, the number average molecular weight is:

$$\overline{M_N} = \frac{\sum_{i=1}^{\infty} N_i M_i}{\sum_{i=1}^{\infty} N_i} = \frac{\text{Total weight}}{\text{Number of polymers}} = \frac{\text{Weight}}{\text{Polymer}} \qquad \text{... (11.24)}$$

The term $N_i/\Sigma N_i$ is physically the number fraction of polymers with molecular weight M_i. If we denote number fraction as X_i (i.e. mole fraction) the number average molecular weight is:

$$\overline{M_N} = \sum_{i=1}^{\infty} X_i M_i \qquad \qquad ... (11.25)$$

In lab experiments it is more common to measure out certain weights of a polymer rather than certain numbers of moles of a polymer. It is thus useful to derive an alternate form for $\overline{M_N}$ in terms or weight fraction of polymers with molecular weight M_i denoted as w_i. First we note that the concentration of polymer species i is (in weight per unit volume):

$$c_i = \frac{N_i M_i}{V} \qquad \qquad ... (11.26)$$

Inserting c_i for $N_i M_i$ and expressing N_i in terms of c_i results in:

$$\overline{M_N} = \frac{\sum_{i=1}^{\infty} c_i}{\sum_{i=1}^{\infty} \frac{c_i}{M_i}} \qquad \qquad ... (11.27)$$

Dividing numerator and denominator by Σc_i results in:

$$\overline{M_N} = \frac{1}{\sum_{i=1}^{\infty} \frac{w_i}{M_i}} \qquad \qquad ... (11.28)$$

where, w_i is the weight fraction of polymer i or the weight of polymer i divided by the total polymer weight:

$$w_i = \frac{N_i M_i}{\sum_{i=1}^{\infty} N_i M_i} = \frac{c_i}{\sum_{i=1}^{\infty} c_i} \qquad \qquad ... (11.29)$$

Weight Average Molecular Weight

Consider of polymer property which depends not just on the number of polymer molecules but on the size or weight of each polymer molecule. A classic example is light scattering. For such a property we need a weight average molecular weight. To derive the weight average molecular weight, replace the appearance of the number of polymers of molecular weight i or N_i in the number average molecular weight formula with the weight of polymer having molecular weight i or $N_i M_i$. The result is:

$$\overline{M_W} = \frac{\sum_{i=1}^{\infty} N_i M_i^2}{\sum_{i=1}^{\infty} N_i M_i} \qquad \qquad ... (11.30)$$

By noting that $N_i M_i / \Sigma N_i M_i$ is the weight fraction of polymer with molecular weight i, w_i, an alternative form for weight average molecular weight in terms of weight fractions

$$\overline{M_W} = \sum_{i=1}^{\infty} w_i M_i \qquad \qquad ... (11.31)$$

Comparing this expression to the expression for number average molecular weight in terms of number fraction (Eq. 11.25) we see that $\overline{M_N}$ is the average M_i weighted according to number fractions and that

$\overline{M_W}$ is the average M_i weighted according to weight fractions. The meanings of their names are thus apparent.

Other Average Molecular Weights

To get $\overline{M_W}$ from $\overline{M_N}$ we replaced N_i by N_iM_i. We can generalise this process and replace N_i by $N_iM_i^k$ to get an average molecular weight denoted as $\overline{M_k}$:

$$\overline{M_k} = \frac{\sum_{i=1}^{\infty} N_i M_i^{k+1}}{\sum_{i=1}^{\infty} N_i M_i^k} \qquad \qquad \text{... (11.32)}$$

Thus $\overline{M_0} = \overline{M_N}$, and $\overline{M_1} = \overline{M_W}$. Several other $\overline{M_k}$ forms appear in experiments. Two examples are $\overline{M_2} = \overline{M_z}$ and $\overline{M_3} = \overline{M_{z+1}}$ which are used in analysis of ultracentrifugation experiments.

One average molecular weight which does not fit into the mould of $\overline{M_k}$ is the viscosity average molecular weight or $\overline{M_v}$. It is defined by:

$$\overline{M_v} = \left(\frac{\sum_{i=1}^{\infty} N_i M_i^{1+a}}{\sum_{i=1}^{\infty} N_i M_i} \right)^{1/a} \qquad \qquad \text{... (11.33)}$$

where, a is a constant that depends on the polymer/solvent pair used in the viscosity experiments.

Schematically, a typical molecular weight distribution might appear as in Fig. 11.13.

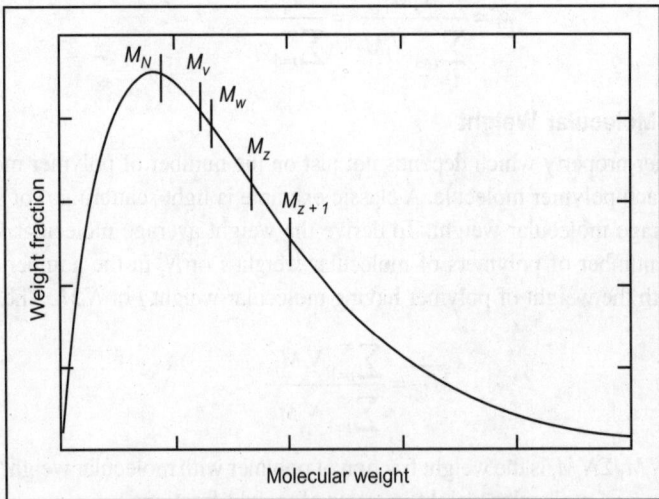

Fig. 11.13. A schematic plot of a distribution of molecular weights along with the rankings of the various average molecular weights.

For any molecular weight distribution, the various average molecular weights always rank in the order given below:

$$\overline{M_N} \leq \overline{M_v} \leq \overline{M_W} \leq \overline{M_z} \leq \overline{M_{z+1}} \leq \overline{M_4} \leq ... \qquad \qquad \text{... (11.34)}$$

The equalities hold only when the polymer is monodisperse, i.e. only when all molecules have the same molecular weight. For monodisperse polymers all molecular weight averages are the same and equal to the one molecular weight. For polydisperse polymers, the average molecular weights will all be different and will rank in the above order. Historically this fact was not always recognised thus it was sometimes difficult to reconcile conflicting experimental results. Say two scientists measured average molecular weight, but one used a colligative property which yields $\overline{M_N}$ and the other used light scattering which yields $\overline{M_W}$. Until it was recognises that $\overline{M_N} \neq \overline{M_W}$, it was difficult to explain differing experimental results on the same polymer solution.

POLYOLEFIN

A polyolefin is a polymer produced from a simple olefin (also called an alkene with the general formula C_nH_{2n}) as a monomer. For example, polyethylene is the polyolefin produced by polymerising the olefin ethylene. An equivalent term is polyalkene; this is a more modern term, although polyolefin is still used in the petrochemical industry. Polypropylene is another common polyolefin which is made from the olefin propylene. Polyolefins are impossible to join by solvent cementing because they have excellent chemical resistance and can only be adhesively bonded after surface treatment because they have very low surface energies. They are also extremely inert chemically and exhibit decreased strength at lower temperatures. A more specific type of olefin is a poly-alpha-olefin (or poly-α-olefin, sometimes abbreviated as PAO), a polymer made by polymerising an alpha-olefin. An alpha-olefin (or α-olefin) is an alkene where the carbon-carbon double bond starts at the α-carbon atom, i.e. the double bond is between the #1 and #2 carbons in the molecule. Common alpha-olefins used as co-monomers to give a polymer alkyl branching groups are similar to 1-hexene or may be longer (Fig. 11.14).

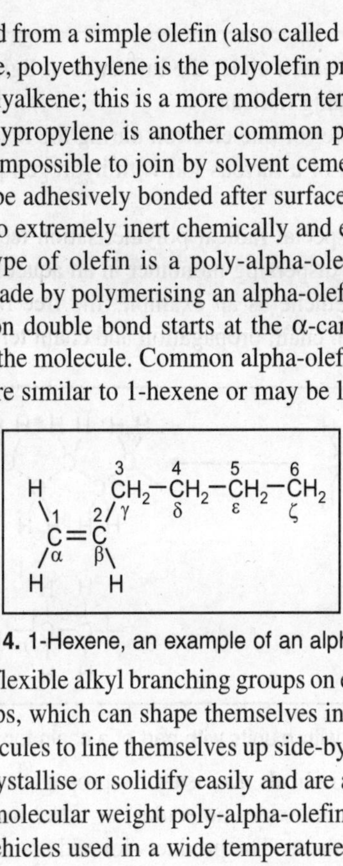

Fig. 11.14. 1-Hexene, an example of an alpha-olefin.

Many poly-alpha-olefins have flexible alkyl branching groups on every other carbon of their polymer backbone chain. These alkyl groups, which can shape themselves in numerous conformations, make it very difficult for the polymer molecules to line themselves up side-by-side in an orderly way. Therefore, many poly-alpha-olefins do not crystallise or solidify easily and are able to remain oily, viscous liquids even at lower temperatures. Low molecular weight poly-alpha-olefins are useful as synthetic lubricants such as synthetic motor oils for vehicles used in a wide temperature range.

Even polyethylenes copolymerised with a small amount of alpha-olefins (such as 1-hexene, 1-octene, or longer) are more flexible than simple straight chain high density polyethylene, which has no branching. The methyl branch groups on a polypropylene polymer are not long enough to make typical commercial polypropylene more flexible than polyethylene.

RADICAL POLYMERISATION

Radical polymerisation is a type of polymerisation in which the reactive centre of a polymer chain consists of a radical. The polymerisation reaction is initiated by three classes of free-radical initiators.

1. Certain compounds that can be broken down in two radicals at temperatures just above room temperature. Such compounds include organic peroxides such as benzoyl peroxide and certain azo compounds such as AIBN.

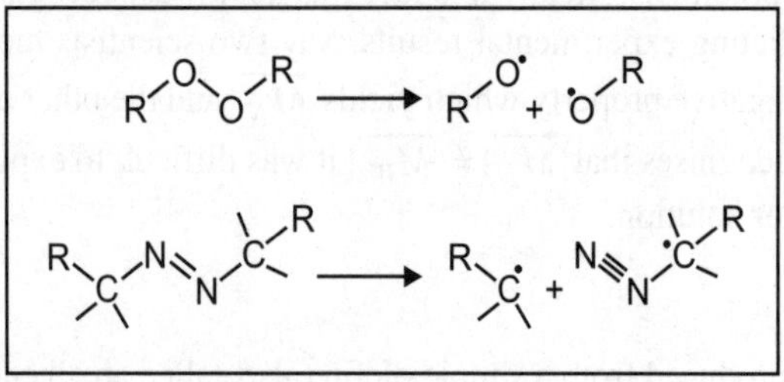

2. Photosensitive molecules, which under the influence of light, get into an excited state or react with other molecules, forming radicals.
3. A redox-system with transfer of one electron during the reaction. This often involves a metal-ion such as in the reaction of a ferrous ion with hydrogen peroxide to a ferric ion in which a hydroxyl radical is formed.

Emulsion polymerisation is a special radical polymerisation technique in which reactive sites are kept separated from each other by dispersing monomer in an aqueous medium.

Taking the polymerisation of ethene as an example, the free radical reaction mechanism can be divided into three stages: initiation, chain propagation and chain termination (Fig. 11.15).

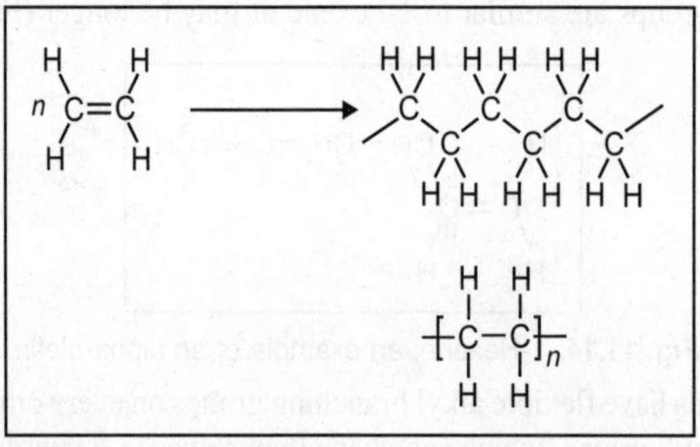

Fig. 11.15. Ethylene polymerisation with n units with part of a chain top and repeating units in chain bottom.

Initiation is the creation of free radicals necessary for propagation. The radicals can be created from radical initiators, such as organic peroxide molecules, or other molecules containing an O-O single bond or by reacting oxygen with ethene. The products formed are unstable and easily break down into two radicals. In an ethene monomer, one electron pair is held securely between the two carbons in a sigma bond. The other is more loosely held in a pi bond. The free radical uses on electron from the pi bond to form a more stable bond with the carbon atom. The other electron returns to the second carbon atom, turning the whole molecule into another radical.

Propagation is the rapid reaction of this radicalised ethene molecule with another ethene monomer, and the subsequent repetition to create the repeating chain.

Termination occurs when a radical reacts in a way that prevents further propagation. The most common method of termination is by coupling where two radical species react with each other forming a single molecule. Another, less common method of termination is chain disproportionation where two radicals meet, but instead of coupling, they exchange a proton, which gives two terminated chains, one saturated and the other with a terminal double bond.

Termination is suppressed in emulsion polymerisation because the radical concentration is low. A chain transfer reaction is also a side-reaction in radical polymerisation and serves to reduce the average chain length.

Free radical addition polymerisation of ethylene must take place at high temperatures and pressures — approximately 300°C and 2000 At. While most other free radical polymerisations do not require such extreme temperatures and pressures (for instance styrene will polymerise at 80°C in benzene or toluene), they do tend to lack stereocontrol.

Another lack of control is a high degree of branching, this is due to the rearrangement of the free radical to cause branching, this is why free radical polymerised ethylene forms low density polyethylene (LDPE) which has very different properties to high density polyethylene (HDPE) which is made using a Ziegler-Natta catalyst.

As termination occurs randomly when two chains collide, it is impossible to control the length of individual chains. Electron rich alkenes when used as monomers tend to form radicals which are more able to react with electron poor alkenes and vice versa, hence mixtures of electron poor and electron rich alkenes then to copolymerise forming polymers where the two monomers alternate. Classic examples of these pairs include ethylene and tetrafluoroethylene, and maleic anhydride and styrene. These combinations of monomers form polymers which are of great industrial importance.

For free radical polymerisation, there must be free radicals which will add on monomer and form reactive chains. However, there are different types of radical initiators: (i) benzoyl peroxide, (ii) AIBN, (iii) UV light, and (iv) hydrogen peroxide.

However, these tend to form side reactions in the polymerisation process, therefore each radical initiator has its efficiency. The rate of dissociation of the radical is given by:

$$R = 2K_d f(\text{I})$$

The propagation step is where the active polymer keeps adding on the monomer.

$$R = (K_p (\text{M}^{\cdot})(\text{M})$$

The termination is where two active ends combine to give a dead chain.

$$R = K_t (\text{M}^{\cdot})^2$$

Moreover, we do have inhibitors which are radical scavengers. Examples are: oxygen, benzoquinone, 1,4 benzene-diol. Therefore, in the reacting vessel we should ensure the absence of these compounds.

Transfer agents seem to be the same as inhibitors but they are two distinct components. Transfer agents allow to have transfer of protons from the solvent to the active polymer chains. But we do not have formation of dead chain, that is the proton can be abstracted again by the solvent. An example of transfer agent is RSH, where the S-H bond is very labile.

Chain Transfer

Chain transfer is a polymerisation reaction by which the activity of a growing polymer chain is transferred to another molecule.

$$\text{P}^{\cdot} + \text{XR}' \rightarrow \text{PX} + \text{R}'\cdot$$

Chain transfer reactions reduce the average molecular weight of the final polymer. Chain transfer can be either introduced deliberately into a polymerisation (by use of a chain transfer agent) or it may be an unavoidable side-reaction with various components of the polymerisation. Chain transfer reactions occur in most forms of addition polymerisation including radical polymerisation, ring-opening polymerisation, coordination polymerisation, and cationic addition polymerisation.

Chain transfer reactions are usually categorised by the nature of the molecule that reacts with the growing chain.

1. Transfer to chain transfer agent: Chain transfer agents have at least one weak chemical bond, which therefore facilitates the chain transfer reaction. Common chain transfer agents include thiols, especially DDM, and halocarbons such as carbon tetrachloride. Chain transfer agents are sometimes called modifiers or regulators.

2. Transfer to monomer: Chain transfer to monomer may take place in which the growing polymer chain abstracts an atom from unreacted monomer existing in the reaction medium. Because, by definition, polymerisation reactions only take place in the presence of monomer, chain transfer to monomer determines the theoretical maximum molecular weight that can be achieved by a given monomer. Chain transfer to monomer is especially significant in cationic addition polymerisation and ring-opening polymerisation.

3. Transfer to polymer: Chain transfer may take place with an already existing polymer chain, especially under conditions in which much polymer is present. This often occurs at the end of a radical polymerisation when almost all monomer has been consumed. Branched polymers are formed as monomer adds to the new radical site which is located along the polymer backbone. The properties of low-density polyethylene are critically determined by the amount of chain transfer to polymer that takes place.

4. Transfer to solvent: In solution polymerisation, the solvent can act as a chain transfer agent. Unless the solvent is chosen to be inert, very low molecular weight polymers (oligomers) can result.

CATALYTIC POLYMERISATION

Free-radical polyolefin reactions form polymers with many 'mistakes' in addition to the ideal long-chain alkanes because of chain-branching and chain-termination steps. This produces a fairly heterogeneous set of polymer molecules with a broad molecular-weight distribution, and these molecules do not crystallise when cooled but rather form amorphous polymers, which are called low-density polyethylene. It was discovered by Ziegler in Germany and Natta in Italy in the 1950s that metal alkyls were very efficient catalysts to promote ethylene polymerisation at low pressures and low temperatures, where free-radical polymerisation is very slow.

They further found that the polymer they produced had fewer side chains because there were fewer growth mistakes caused by chain transfer and radical recombination. Therefore, this polymer was more crystalline and had a higher density than polymer prepared by free-radical processes. Thus were discovered linear and high-density polymers.

Ziegler and Natta found that Ti alkyls promoted with chloride gave good performance, and scientists from Phillips Petroleum found that Cr alkyls were also effective. The mechanisms is thought to involve alkyl ligands R-bonded to the Ti or Cr atom to form a species such as R-Ti and the subsequent bonding and insertion of ethylene to form $R-C_2H_4-Ti$ which adds successive ethylenes to create linear polymer of higher molecular weight. A Ziegler–Natta catalyst is a catalyst used in the production of polymers of 1-alkenes (α-olefins). Ziegler–Natta catalysts are typically based on titanium compounds and organometallic aluminium compounds, such as the undefined methylaluminoxane (MAO) or well defined triethylaluminium, $(C_2H_5)_3Al$. Ziegler-Natta catalysts are used to polymerise terminal 1-alkenes.

$$nCH_2 = CHR \rightarrow -[CH_2-CHR]_n-$$

Karl Ziegler prepared linear polyethylene with the catalyst he discovered. Giulio Natta used similar catalysts to polymerise 1-alkenese. Poly (1-alkenes)s can be isotactic, syndiotactic, or atactic, depending on the relative orientation of the alkyl groups in polymer chains consisting of units $-[CH_2-CHR]-$. In isotactic polymers, all chiral centres CHR share the same stereochemistry. Chiral centres in syndiotactic polymers alternate their relative steriochemistry. Atactic polymers lack regular steriochemistry. The sterioregularity of the polymer depends on the type of catalyst used to prepare it, and once prepared, the polymer's stereochemistry does not change.

The Ziegler–Natta catalysts represented a major breakthrough in polymerisation chemistry because they produce a variety of commercially important polymers and can be highly stereoselective. Previously known radical polymerisation reactions result in the formation of atactic polymers. $TiCl_4$-derived catalyst systems convert propylene, and many other 1-alkenese, to isotactic polymers such as polypropylene. Related systems employing VCl_4 yield syndiotactic polymers.

The first Ziegler–Natta catalyst was produced by treating crystalline α-$TiCl_3$ with $[AlCl(C_2H_5)_2]_2$. Polymerisation reactions of any alkene occur at special Ti centres located on the exterior of the crystallites. Most titanium ions in these crystallites are surrounded by six chloride ligands to give an octahedral structure. At the surface, however, 'defects' occur where some Ti centres lack their full complement of chloride ligands. The alkene molecule binds at these 'vacancies'. In ways the are still not fully clear, the alkene converts to an alkyl ligand group. The most probable pathway of this reaction is the insertion of the $C = C$ bond of the alkene molecule into the Ti–C bond:

$$L_nTi-CH_2-CHR-Polymer + CH_2 = CHR \rightarrow L_nTi-CH_2-CHR-CH_2-CHR-Polymer$$

The coordination sphere of the Ti atom restricts the approach of incoming alkene molecules, thereby imposing stereoregularity on the growing polymer chain. The Cossee–Arlman mechanism describes the growth of stereospecific polymers.

Many thousands of alkene insertion reactions occur at each active centre resulting in the formation of long polymer chains attached to the centre. On occasion, the polymer chains are disengaged from the active centres in the reaction:

$$L_nTi-CH_2-CHR-Polymer + CH_2 = CHR \rightarrow L_nTi-CH_2-CH_2R-CH_2-CR-Polymer$$

This reaction occurs quite rarely and the formed polymers have a too high molecular weight to be of commercial use. To reduce the molecular weight, hydrogen is added to the polymerisation reaction:

$$L_nTi-CH_2-CHR-Polymer + H_2 \rightarrow L_nTi-H + CH_3-CHR-Polymer$$

During the past 40 years, a large number of different supported Ziegler–Natta catalysts were developed which afford a much higher activity in alkene polymerisation reactions and much higher contents of crystalline isotactic fractions in the polymers they produce, up to 97–99 per cent. The principal source of Ti in all these catalysts is $TiCl_4$, and the principal support is $MgCl_2$. In order to maintain the high selectivity for an isotactic polymer product, a variety of catalyst modifiers, Lewis bases, must be used. To form these catalysts, several techniques were developed for combining $TiCl_4$, $MgCl_2$, and the Lewis base in a single solid precatalyst. The final catalyst system is prepared by combining this solid powder with $AlEt_3$ and another Lewis base compound.

It should be noted that titanium(IV) chloride, all solid Ziegler–Natta catalysts and alkyl aluminium compounds are unstable in air, and the alkylaluminium compounds are pyrophoric. The catalysts, therefore, must be prepared and handled under an inert atmosphere.

Activity depends on the nature of the metal. Ti, Zr, Hf form highly active catalysts. It is theorised that these catalysts feature d^0 species. Without any d-electrons, the titanium-alkene bond is not stabilised by pi backbonding, so the barrier for alkene binding is decreased.

The length of a polymer chain is determined by two competing rate constants, the rate of chain propagation (transferring the alkene to the growing polymer chain) versus the rate of termination. Termination usually occurs by β-H elimination. By tuning, one can effectively 'dial in' the molecular weight of the polymer product. For example, 'half-sandwich' zirconium species, tend to give low molecular weight polymers because of their enhanced tendency to undergo β-hydride elimination.

Significant effort has been dedicated to developing other catalysts that effectively polymerise a number of branched alkenes. In addition, there has been an interest in developing homogeneous Ziegler–Natta catalysts (that do not require the aluminium cocatalyst); these species are cationic and become active in solution by losing a labile ligand. One such catalyst is the agostic complex $[Cp_2Zr(CH_3)$ $CH_3B(C_6F_5)_3]$. The borate anion dissociates, leaving a vacant active site to bind alkene, allowing polymerisation to commence. Developments have built upon advances in non-coordinating anions. In addition to those based on cyclopentadienyl ligands, catalysts are increasingly designed using nitrogen-based ligands.

Polymers prepared by Ziegler–Natta catalysts are: (i) polyethylene, (ii) polypropylene, (iii) amorphous poly-alpha-olefins (APAO), (iv) polyvinyl alcohol, and (v) polyacetylene.

Coordination (Ziegler–Natta) Polymerisation

Early work: Insertion of aluminium alkyls into olefins was studied by Ziegler:

Oligomers with an even number of carbons

Important discovery: R3Al + Lewis acids:

Another important discovery: tacticity control:

Overall scheme of coordination polymerisation

1. Limited to ethylene and other α-olefins like propylene. (Actually, it is the only good way to polymerise these monomers.)
2. Produces linear polymer, with very few branches (e.g. high density polyethylene, HDPE).
3. Capable of producing homo-tactic polymers.
4. Most commercial initiators are insoluble complexes or supported on insoluble carriers.
5. Very complex mechanism, still poorly understood for the heterogeneous systems.
6. Termination is almost exclusively by chain transfer.
7. Modern 'high mileage' initiators produce up to 1000's of kg per g initiator.
8. Initiators are often called 'catalysts' even though they are consumed by the process. Many chains are started per molecule of initiator.

Mechanism of coordination polymerisation

The mechanism is poorly understood because it takes place on the surface of an insoluble particle, a difficult situation to probe experimentally. The mechanism shown in Fig. 11.16 is one of several models proposed to at least partially explain the action of the Ziegler–Natta systems, but it is only an approximation of the more complex process that actually occurs.

FLORY–HUGGINS THEORY

In the early 1940s, Paul Flory and Maurice Huggins, working independently, developed a theory based upon a simple lattice model that could be used to understand the nonideal nature of polymer solutions. In the Flory–Huggins model, the lattice sites, or holes, are chosen to be the size of the solvent molecule. As the simple example, consider the mixing of a low-molecular-weight solvent (component 1) with a low-molecular-weight solute (component 2). The solute molecule is assumed to have the same size as a solvent molecule, and therefore only one solute or one solvent molecule, and therefore only one solute or one solvent molecule can occupy a single lattice site at a given time. A representation of the lattice model of this case is illustrated in Fig. 11.17.

The increase in entropy due to mixing of a solvent and solute, ΔS_m, may be obtained from the Boltzmann relation

$$\Delta S_m = k \ln \Omega \qquad \qquad \dots (11.35)$$

where, k is Boltzmann's constant (1.38×10^{-23} J K^{-1}) and Ω gives the total number of ways of arranging n_1 solvent molecules and n_2 solute molecules, where, $N = n_1 + n_2$ is the total number lattices sites. The probability function is given as:

$$\Omega = \frac{N!}{n_1! n_2!}$$

... (11.36)

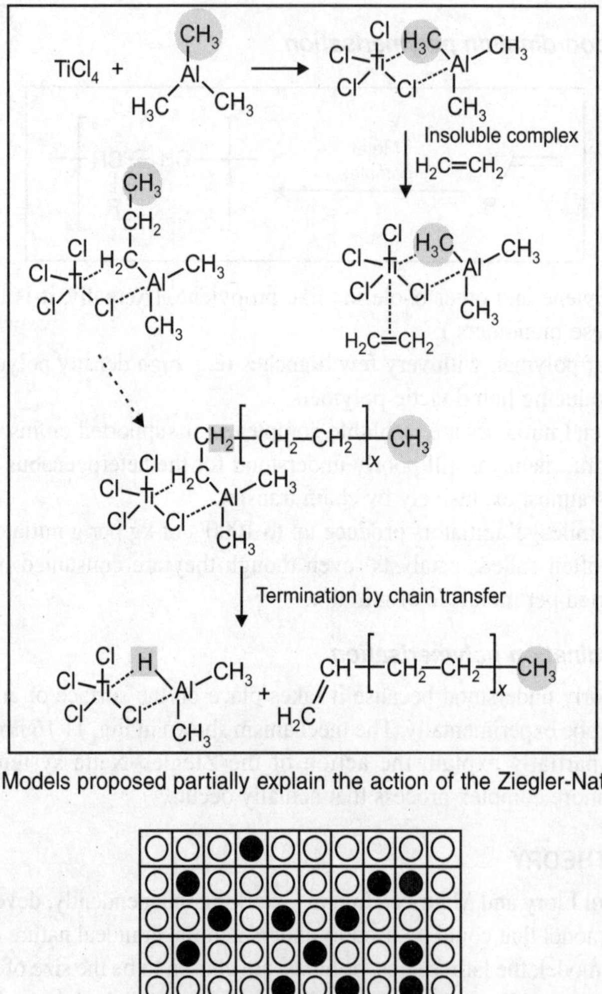

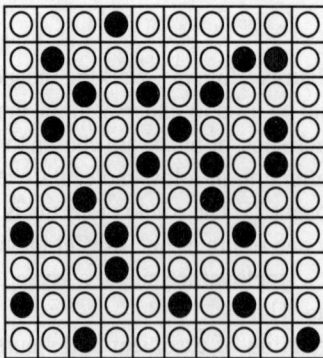

Fig. 11.16. Models proposed partially explain the action of the Ziegler-Natta systems.

Fig. 11.17. Representation of two-dimensional Flory–Huggins lattice containing solvent molecules (O) and a low-molecular-weight solute (●).

Use of Stirling's approximation:

$$\ln n! = n \ln n - n \qquad \qquad \text{... (11.37)}$$

leads to the expression for the entropy of mixing as:

$$\Delta S_m = -k(n_1 \ln x_1 + n_2 \ln x_2) \qquad \qquad \text{... (11.38a)}$$

or

$$\Delta S_m = -R(x_1 \ln x_1 + x_2 \ln x_2) \qquad \qquad \text{... (11.38b)}$$

where, R is the ideal gas constant and x_1 is the mole fraction of the solvent given as:

$$x_1 = \frac{n_1}{n_1 + n_2} \qquad \qquad \text{... (11.39)}$$

Equation 11.38a is the well-known relation for the entropy change due to mixing of an ideal mixture, which can also be obtained from classical thermodynamics of an ideal solution following the Lewis–Randal law. Equation 11.38b can be written for a multicomponent system having N components as:

$$\Delta S_m^{id} = -R \sum_{i=1}^{N} x_i \ln x_i \qquad \qquad \text{... (11.40)}$$

The entropy of mixing a low-molecular-weight solvent with a high-molecular-weight polymer is small than given by Eq. 11.40 for a low-molecular-weight mixture. This is due to the loss in conformation entropy resulting from the linkage of individual repeating units along a polymer chain compared to the less ordered case of unassociated low-molecular-weight solute-molecules dispersed in a low-molecular-weight solvent. In the development of an expression for ΔS_m for a high-molecular-weight polymer in a solvent, the lattice is established by dividing the polymer chain into r segments, each the size of a solvent molecule, where r is the ratio of polymer volume to solvent volume. For n_2 polymer molecules, the total number of lattice sites is then $N = n_1 + r n_2$. A lattice containing low-molecular weight solvent molecules and a single polymer-chain is illustrated in Fig. 11.18.

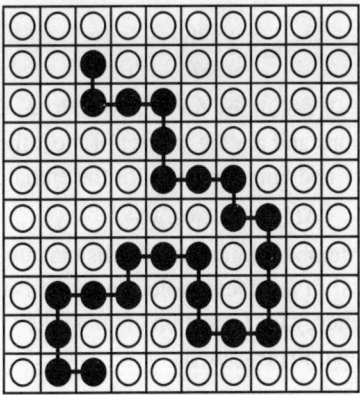

Fig. 11.18. Lattice model for a polymer chain in solution. Symbols represent solvent molecules (O) and polymer-chain segments (●).

Without going into details of the derivations, the expression for the entropy change due to mixing obtained by Flory and Huggins is given as:

$$\Delta S_m = -k(n_1 \ln \phi_1 + n_2 \ln \phi_2) \qquad \qquad \text{... (11.41)}$$

where, ϕ_1 and ϕ_2 are the lattice volume fractions of solute (component 1) or polymer (component 2), respectively. These are given as:

$$\phi_1 = \frac{n_1}{n_1 + rn_2} \qquad \qquad \text{... (11.42a)}$$

and

$$\phi_2 = \frac{rn_2}{n_1 + rn_2} \qquad \qquad \text{... (11.42b)}$$

For a polydisperse polymer, Eq. 11.41 may be modified as:

$$\Delta S_m = -k\left(n_1 \ln\phi_1 + \sum_{i=2}^{N} n_i \ln\phi_i \right) \qquad \qquad \text{... (11.43)}$$

where the summation is over all polymer chains (N) in the molecular-weight distribution. For simplicity, the most commonly used form of the entropy expression, Eq. 11.41, will be used in further discussion. Equation 11.41 provides the entropy term in the expression for the Gibbs free energy of mixing, ΔG_m, of a polymer solution given as:

$$\Delta G_m = \Delta H_m - T\Delta S_m \qquad \qquad \text{... (11.44)}$$

Once an expression for the enthalpy of mixing, ΔH_m, is known, expressions for the chemical potential and activity of the solvent can be obtained as:

$$\Delta\mu_1 = \mu_1 - \mu_1^{\,\circ} = \overline{\Delta G}_m = \left(\frac{\partial \Delta G_m}{\partial n_1} \right)_{T,p} \qquad \qquad \text{... (11.45)}$$

where, $\overline{\Delta G}_m$ is the partial-molar Gibbs free energy of mixing and the activity is related to the chemical potential as:

$$\ln a = \frac{\Delta\mu_1}{kT} \qquad \qquad \text{... (11.46)}$$

For an ideal solutions, $\Delta H_m = 0$. Solutions for which $\Delta H_m \neq 0$ but for which ΔS_m is given by Eq. 11.38 are termed regular solutions and are the subject of most thermodynamic models for polymer mixtures. The expression that Flory and Huggins gave for the enthalpy of mixing is:

$$\Delta H_m = zn_1 r_1 \phi_2 \Delta\omega_{12} \qquad \qquad \text{... (11.47)}$$

where, z is the lattice coordination number or number of cells that are first neighbours to a given cell, r_1 represents the number of 'segments' in a solvent molecule for consideration of the most general case, and $\Delta\omega_{12}$ is the change in internal energy for formation of an unlike molecular pair (solvent–polymer or 1–2) contacts given by the mean-field expression as:

$$\Delta\omega_{12} = \omega_{12} - \frac{1}{2}(\omega_{11} + \omega_{22}) \qquad \qquad \text{... (11.48)}$$

where, ω_{ij} is the energy of i–j contacts. It is clear from Eqs 11.47 and 11.48 that an ideal solution ($\Delta H_m = 0$) is one for which the energies of 1–1, 1–2, and 2–2 molecular interactions are equal.

Since z and ω_{12} have the character of empirical parameters, it is convenient to define a single energy parameter called the Flory interaction parameter, χ_{12}, given as:

$$\chi_{12} = \frac{z r_1 \Delta \omega_{12}}{kT} \qquad \text{... (11.49)}$$

The interaction parameter is a dimensionless quantity that characterises the interaction energy per solvent molecule (having r_1 segments) divided by kT. As Eq. 11.49 indicates χ_{12} is inversely related to temperature but is independent of concentration.

The expression for the enthalpy of mixing may then be written by combining Eqs 11.47 and 11.49 as:

$$\Delta H_m = kT\chi_{12} n_1 \phi_2 \qquad \text{... (11.50)}$$

Combining the expression for the entropy (Eq. 11.41) and enthalpy (Eq. 11.50) of mixing gives the well-known Flory–Huggins expression for the Gibbs free energy of mixing

$$\Delta G_m = kT(n_1 \ln \phi_1 + n_2 \ln \phi_2 + \chi_{12} n_1 \phi_2) \qquad \text{... (11.51)}$$

From this relationship, the activity of the solvent (Eq. 11.46) can be obtained from Eq. 11.51 as:

$$\ln a_1 = \ln(1 - \phi_2) + \left(1 - \frac{1}{r}\right)\phi_2 + \chi_{12}\phi_2^2 \qquad \text{... (11.52)}$$

In the case of high-molecular-weight polymers for which the number of solvent-equivalent segments, r, is large, the $1/r$ term within parentheses on the right-hand side of Eq. 11.52 can be neglected to give:

$$\ln a_1 = \ln (1 - \phi_2) + \phi_2 + \chi_{12}\phi_2^2 \qquad \text{... (11.53)}$$

The Flory–Huggins equation is still widely used and has been largely successful in describing the thermodynamics of polymer solutions; however, there are a number of important limitations of the original expression that should be emphasised. The most important are the following:

1. Applicability only to solutions that are sufficiently concentrated that they have uniform segment-density.
2. There is no volume change of mixing (whereas favourable interactions between polymer and solvent molecules should result in a negative volume change).
3. There are no energetically-preferred arrangements of polymer segments and solvent molecules in the lattice.

There have been a number of subsequent developments to extend the applicability of the original Flory–Huggins theory and to improve agreement between theoretical and experimental results. For example, Flory and Krigbaum have developed a thermodynamic theory for dilute polymer solutions, which was given in Flory's original text. Koningsveld and others have improved the agreement of the original Flory–Huggins theory with experimental data by an empirical modification of χ_{12} to include a composition dependence and to account for polymer polydispersity. Both of these approaches are presented briefly in the following section. More recent approaches employ equation-of-state theories such as those developed by Flory and others for which a volume change of mixing can be incorporated.

Flory–Krigbaum and Modified Flory–Huggins Theory

Flory–Krigbaum theory

Flory and Krigbaum have provided a model to describe the thermodynamics of a dilute polymer solution in which individual polymer chains are isolated and surrounded by regions of solvent molecules. In contrast to the case of a semidilute solution addressed by the Flory–Higgins theory, segmental density can no longer be considered to be uniform. In their development, Flory and Krigbaum viewed the dilute solution as a dispersion of clouds consisting of polymer segments surrounded by regions of pure solvent. For a dilute solution, the expression for solvent activity was given as:

$$\ln a_1 = (\kappa_1 - \psi_1)\phi_2^2 \qquad \qquad \text{... (11.54)}$$

where, κ_1 and ψ_1 are heat and entropy parameters, respectively. They defined an 'ideal' or theta (θ) temperature as:

$$\theta = \frac{k_1 T}{\psi_1} \qquad \qquad \text{... (11.55)}$$

from which Eq. 11.54 can be written as:

$$\ln a_1 = -\psi_1\left(1 - \frac{\theta}{T}\right)\phi_2^2 \qquad \qquad \text{... (11.56)}$$

It follows from Eq. 11.56 that solvent activity approaches unity as temperature approaches the θ temperature. At the θ temperature, the dimensions of a polymer chain collapse to unperturbed dimensions (i.e. in the absence of excluded-volume effects).

Modified Flory–Huggins

In the original lattice theory, χ_{12} was given an inverse dependence upon temperature (Eq. 11.49) but there was no provision for concentration dependence which experimental studies has shown to be important. Koningsveld and others have introduced an eimpirical dependence to improve the agreement with experimental data by casting the Flory–Huggins expression in the general form:

$$\Delta G_m = RT(\phi_1 \ln \phi_1 + \phi_2 \ln \phi_2 + g\phi_1\phi_2) \qquad \qquad \text{... (11.57)}$$

In Eq. 11.57, g is an interaction energy term for which the concentration dependence can be given as a power series in ϕ_2 as:

$$g = g_0 + g_1\phi_2 + g_2\phi_2^2 + ... \qquad \qquad \text{... (11.58)}$$

where each g term, g_k ($k = 1, 2, 3...$), has a temperature dependence that can be expressed in the form:

$$g_k = -g_{k,1} + \frac{g_{k,2}}{T} \qquad \qquad \text{... (11.59)}$$

Elastomer

Elastomers are amorphous polymers existing above their glass transition temperature, so the considerable segmental motion is possible. At ambient temperatures rubbers are thus relatively soft (E~3 MPa) and deformable.

Elastomers are usually thermosets (requiring vulcanisation) but may also be thermoplastic. The long polymer chains cross-link during curing, i.e. vulcanising. The molecular structure of elastomers can be imagined as 'spaghetti and meatball' structure, with the meatballs signifying cross-links. The elasticity is derived from the ability of the long chains to reconfigure themselves to distribute and applied stress. The covalent cross-linkages ensure that the elastomer will return to its original configuration when the stress is removed. As a result of this extreme flexibility, elastomers can reversibly extend from 5–700 per cent, depending on the specific material. Without the cross-linkages or with short, uneasily reconfigured chains, the applied stress would result in a permanent deformation. Temperature effects are also present in the demonstrated elasticity of a polymer. Elastomers that have cooled to a glassy or crystalline phase will have less mobile chains, and consequentially less elasticity, than those manipulated at temperatures higher than the glass transition temperature of the polymer. It is also possible for a polymer to exhibit elasticity that is not due to covalent cross-links, but instead for thermodynamic reasons.

CONDENSATION POLYMERISATION

Condensation polymers are any kind of polymers formed through a condensation reaction, releasing small molecules as by-products such as water or methanol, as opposed to addition polymers which involve the reaction of unsaturated monomers. Types of condensation polymers include polyamides, polyacetals and polyesters. Condensation polymerisation, a form of step-growth polymerisation, is a process by which two molecules join together, resulting loss of small molecules which is often water. The types of end product resulting from a condensation polymerisation is dependent on the number of functional end groups of the monomer which can react.

Monomers with only one reactive group terminate a growing chain, and thus give and products with a lower molecular weight. Linear polymers are created using monomers with two reactive end groups and monomers with more than two end groups give three-dimensional polymers which are cross-linked.

Dehydration synthesis often involves joining monomers with an –OH (hydroxyl) group and a freely ionised –H on either end (such as a hydrogen from the $-NH_2$ in nylon or proteins). Normally, two or more different monomers are used in the reaction. The bonds between the hydroxyl group, the hydrogen atom and their respective atoms break forming water from the hydroxyl and hydrogen, and the polymer.

Polyester is created through ester linkages between monomers, which involve the functional groups carboxyl and hydroxyl (an organic acid and an alcohol monomer). Nylon is another common condensation polymer. It can be manufactured by reacting di-amines with carboxyl derivatives. In this example the derivative is a di-carboxylic acid, but di-acyl chlorides are also used. Another approach used is the reaction of di-functional monomers, with one amine and one carboxylic acid group on the same molecule:

General chemical structure of one type of condensation polymer

The carboxylic acids and amines link to form peptide bonds, also know as amide groups. Proteins are condensation polymers made from amino acid monomers. Carbohydrates are also condensation polymers made from sugar monomers such as glucose and galactose. Condensation polymerisation is occasionally used to form simple hydrocarbons. This method, however, is expensive and inefficient, so the addition polymer of ethene (polyethylene) is generally used.

Condensation polymers, unlike addition polymers, may be biodegradable. The peptide or ester bonds between monomers can be hydrolysed by acid catalysts or bacterial enzymes breaking the polymer chain into smaller pieces. The most commonly known condensation polymers are proteins, fabrics such as nylon, silk, or polyester.

Mechanism of Condensation Polymerisation

As we know that monomers that are joined by condensation polymerisation have two functional groups. Also a carboxylic acid and an amine can form an amide linkage, and a carboxylic acid and an alcohol can form an ester linkage. Since each monomer has two reactive sites, they can form long-chain polymers by making many amide or ester links. Let's look at two examples of common polymers made from the monomers we have studied.

Example 1: A carboxylic acid monomer and an amine monomer can join in an amide linkage.

Adipic acid   1,6-Diaminohexane   Water

As before, a water molecule is removed, and an amide linkage is formed. Notice that an acid group remains on one end of the chain, which can react with another amine monomer. Similarly, an amine group remains on the other end of the chain, which can react with another acid monomer.

Thus, monomers can continue to join by amide linkages to form a long chain. Because of the type of bond that links the monomers, this polymer is called a polyamide. The polymer made from these two six-carbon monomers is known as nylon-6,6. (Nylon products include hosiery, parachutes, and ropes.)

Example 2: A carboxylic acid monomer and an alcohol monomer can join in an ester linkage.

Terephthalic acid   Ethylene glycol   Water

A water molecule is removed as the ester linkage is formed. Notice the acid and the alcohol groups that are still available for bonding.

Because the monomers above are all joined by ester linkages, the polymer chain is a polyester. This one is called PET, which stands for poly(ethylene terephthalate). (PET is used to make soft-drink bottles, magnetic tape, and many other plastic products).

Let's summarise: As difunctional monomers join with amide and ester linkages, polyamides and polyesters are formed, respectively. We have seen the formation of the polyamide nylon-6,6 and the polyester PET. There are numerous other examples.

The above process is called condensation polymerisation because a molecule is removed during the joining of the monomers. This molecule is frequently water.

Nylon

Nylon is a thermoplastic silky material, first used commercially in a nylon-bristled toothbrush, followed more famously by women's stockings. It is made of repeating units linked by peptide bonds (another name for amide bonds) and is frequently referred to as polyamide (PA). Nylon was the first commercially successful synthetic polymer. There are two common methods of making nylon for fibre applications. In one approach, molecules with an acid (COOH) group on each end are reacted with molecules containing amine (NH_2) groups and each end. The resulting nylon is named on the basis of the number of carbon atoms separating the two acid groups and the two amines. These are formed into monomers of intermediate molecular weight, which are then reacted to form long polymer chains.

Nylons are condensation copolymers formed by reacting equal parts of a diamine and a dicarboxylic acid, so that peptide bonds form at both ends of each monomer is a process analogous to polypeptide biopolymers. Chemical elements included are carbon, hydrogen, nitrogen, and oxygen. The numerical suffix specifies the numbers of carbons donated by the monomers; the diamine first and the diacid second. The most common variant is nylon 6-6 which refers to the fact that the diamine (hexamethylene diamine, IUPAC name: 1,6-diamenohexane) and the diacid (adipic acid, IUPAC name: hexane-1,6-diacarboxylic acid) each donate 6 carbons to the polymer chain. As with other regular copolymers like polyesters and polyurethanes, the 'repeating unit' consists of one of each monomer, so that they alternate in the chain. Since each monomer in this copolymer has the same reactive group on both ends, the direction of the amide bond reverses between each monomer, unlike natural polyamide proteins which have overall directionality: C terminal $\rightarrow$ N terminal. In the laboratory, nylon 6-6 can also be made using adipoyl chloride instead of adipic. It is difficult to get the proportions exactly correct, and deviations can lead to chain termination at molecular weights less than a desirable 10,000 daltons (u). To overcome this problem, a crystalline, solid 'nylon salt' can be formed at room temperature, using an exact 1:1 ratio of the acid and the base to neutralise each other. Heated to 285°C, the salt reacts to form nylon polymer. Above 20,000 daltons, it is impossible to spin the chains into yarn, so to combat this, some acetic acid is added to react with a free amine end group during polymer elongation to limit the molecular weight. In practice, and especially for 6,6, the monomers are often combined in a water solution. The water used to make the solution is evaporated under controlled conditions, and the increasing concentration of 'salt' is polymerised to the final molecular weight.

DuPont patented nylon 6,6, so in order to compete, other companies (particularly the German BASF) developed the homopolymer nylon 6, or polycaprolactam — not a condensation polymer, but formed by a ring-opening polymerisation (alternatively made by polymerising aminocaproic acid). The peptide bond within the caprolactam is broken with the exposed active groups on each side being incorporated into two new bonds as the monomer becomes part of the polymer backbone. In this case, all amide bonds lie in the same direction, but the properties of nylon 6 are sometimes indistinguishable from those of nylon 6,6 — except for melt temperature (N6 is lower) and some fibre properties in products like carpets and textiles. There is also nylon 9.

Protein

Proteins (also know as polypeptides) are organic compounds adjacent amino acid residues. The sequence of amino acids in a protein is defined by the sequence of a gene, which is encoded in the genetic code. In general, the genetic code specifies 20 standard amino acids; however, in certain organisms the genetic code can include elenocysteine — and in certain archaea — pyrrolysine. Shortly after or even during synthesis, the residues in a protein are often chemically modified by post-translational modification, which alters the physical and chemical properties, folding, stability, activity, and ultimately, the function of the proteins. Proteins can also work together to achieve a particular function, and they often associate to form stable complexes.

Like other biological macromolecules such as polysaccharides and nucleic acids, proteins are essential parts of organisms and participate in virtually every process within cells. Many proteins are enzymes that catalyse biochemical reactions and are vital to metabolism. Proteins also have structural or mechanical functions, such as actin and myosin in muscle and proteins in the cytoskeleton, which form a system of scaffolding that maintains cell shape. Other proteins are important in cell signalling, immune responses, cell adhesion, and the cell cycle. Proteins are also necessary in animals' diets, since animals cannot synthesise all the amino acids they need and must obtain essential amino acids from food. Through the process of digestion, animals breakdown ingested protein into free amino acids that are then used in metabolism.

Proteins may be purified from other cellular components using a variety of techniques such as ultracentrifugation, precipitation, electrophoresis, and chromatography; the advent of genetic engineering has made possible a number of method to facilitate purification. Methods commonly used to study protein structure and function include immunohistochemistry, site-directed mutagenesis, and mass spectrometry.

FISCHER–TROPSCH PROCESS

The Fischer–Tropsch process is a catalysed chemical reaction in which synthesis gas, a mixture of carbon monoxide and hydrogen, is converted into liquid hydrocarbons of various forms. The most common catalysts are based on iron and cobalt, although nickel and ruthenium have also been used. The principal purpose of this process is to produce a synthetic petroleum substitute, typically from coal, natural gas or biomass, for use as synthetic lubrication oil or as synthetic fuel. This synthetic fuel runs trucks, cars, and some aircraft engines. The use of diesel is increasing in recent years.

Combination of biomass gasification (BG) and Fischer–Tropsch (FT) synthesis is a possible route to produce renewable transportation fuels (biofuels).

The Fischer–Tropsch process involves a variety of competing chemical reactions, which lead to a series of desirable products and undesirable by-products. The most important reactions are those resulting in the formation of alkanes. These can be described by chemical equations of the form:

$$(2n+1)H_2 + nCO \rightarrow C_nH_{(2n+2)} + nH_2O$$

where, n is a positive integer. The simplest of these ($n=1$), results in formation of methane, which is generally considered an unwanted by-product (particularly when methane is the primary feedstock used to produce the synthesis gas). Process conditions and catalyst composition are usually chosen to favour higher order reactions ($n > 1$) and thus minimise methane formation. Most of the alkanes produced tend to be straight-chained, although some branched alkanes are also formed. In addition to alkane formation, competing reactions result in the formation of alkenes, as well as alcohols and other oxygenated hydrocarbons. Usually, only relatively small quantities of these non-alkane products are formed, although catalysts favouring some of these products have been developed. Another important reaction is the water gas shift reaction:

$$H_2O + CO \rightarrow H_2 + CO_2$$

Although this reaction results in formation of unwanted CO_2, it can be used to shift the H_2:CO ratio of the incoming synthesis gas. This is especially important for synthesis gas derived from coal, which tends to have a ratio of ~0.7 compared to the ideal ratio of ~2. It should be noted that, according to published data on the current commercial implementations of the coal-based Fischer–Tropsch process, these plants can produce as much as 7 tonnes of CO_2 per tonne of liquid hydrocarbon products (excluding the reaction water product). This is due in part to the high energy demands required by the gasification process, and in part by the design of the process as implemented.

Process Conditions

Generally, the Fischer–Tropsch is operated in the temperature range of $150°-300°C$ ($302°-572°F$). Higher temperatures lead to faster reactions and higher conversion rates, but also tend to favour methane production. As a result the temperature is usually maintained at the low to middle part of the range. Increasing the pressure leads to higher conversion rates and also favours formation of long-chained alkanes both of which are desirable. Typical pressures are in the range of one to several tens of atmospheres. Chemically, even higher pressures would be favourable, but the benefits may not justify the additional costs of high-pressure equipment.

A variety of synthesis gas compositions can be used. For cobalt-based catalysts the optimal H_2:CO ratio is around 1.8–2.1. Iron-based catalysts promote the water-gas-shift reaction and thus can tolerate significantly lower ratios. This can be important for synthesis gas derived from coal or biomass, which tend to have relatively low H_2:CO ratios (<1).

Product Distribution

In general the product distribution of hydrocarbons formed during the Fischer–Tropsch process follows an Anderson–Schulz–Flory distribution, which can be expressed as:

$$W_n/n = (1-\alpha)^2\alpha^{n-1}$$

where, W_n is the weight fraction of hydrocarbon molecules containing n carbon atoms. α is the chain growth probability or the probability that a molecule will continue reacting to form a longer chain. In general, α is largely determined by the catalyst and the specific process conditions.

Examination of the above equation reveals that methane will always be the largest single product; however by increasing α close to one, the total amount of methane formed can be minimised compared to the sum of all of the various long-chained products. Increasing α increases the formation of long-chained hydrocarbons. The very long-chained hydrocarbons are waxes, which are solid at room temperature. Therefore, for production of liquid transportation fuels it may be necessary to crack some of the Fischer–Tropsch products. In order to avoid this, some researchers have proposed using zeolites or other catalyst substrates with fixed sized pores that can restrict the formation of hydrocarbons longer than some characteristic size (usually $n < 10$). This way they can drive the reaction so as to minimise methane formation without producing lots of long-chained hydrocarbons. So far, such efforts have had only limited success.

Fischer–Tropsch Catalysts

A variety of catalysts can be used for the Fischer–Tropsch process, but the most common are the transition metals cobalt, iron, and ruthenium. Nickel can also be used, but tends to favour methane formation. Cobalt seems to be the most active catalyst, although iron also performs well and can be more suitable for low-hydrogen-content synthesis gases such as those derived from coal due to its promotion of the water-gas-shift reaction. In addition to the active metal the catalysts typically contain a number of promoters, including potassium and copper, as well as high-surface-area binders/supports such as silica, alumina, or zeolites. Unlike the other metals used for this process (Co, Ni, Ru) which remain in the metallic state during synthesis, iron catalysts tend to form a number of chemical phases, including various iron oxides and iron carbides during the reaction. Control of these phase transformations can be important in maintaining catalytic activity and preventing breakdown of the catalyst particles.

The Fishcer–Tropsch catalysts are notoriously sensitive to the presence of sulphur-containing compounds among other poisons. The sensitivity of the catalyst to sulphur is higher for cobalt-based catalysts than for their iron counterparts.

Cobalt catalysts are preferred for Fischer–Tropsch synthesis when the feedstock is natural gas due to the higher activity of the cobalt catalyst. Natural gas has a high hydrogen to carbon ratio, so the water-gas-shift is not needed for cobalt catalysts. Iron catalysts are preferred for lower quality feedstocks such as coal or biomass.

While iron catalysts are also susceptible to sulphur poisoning from coal with high sulphur content, the lower cost of iron makes sacrificial catalyst at the front of a reactor bed economical. Also, as mentioned earlier, iron can catalyse the water-gas-shift to increase the hydrogen to carbon ratio to make the reaction more favourably selective.

Synthesis Gas Production

The initial reactants (synthesis gases) used in the Fischer–Tropsch process are hydrogen gas (H_2) and carbon monoxide (CO). These chemicals are usually produced by one of two methods:

1. The partial combustion of a hydrocarbon:
$$C_nH_{(2n+2)} + \tfrac{1}{2}nO_2 \rightarrow (n+1)H_2 + nCO$$
When $n = 1$ (methane), the equation becomes $2CH_4 + O_2 \rightarrow 4H_2 + 2CO$

2. The gasification of coal, biomass, or natural gas:
$$CH_x + H_2O \rightarrow (1 + 0.5x)H_2 + CO$$
The value of x depends on the type of fuel. For example, natural gas has a greater hydrogen content (from $x = 4$ to $x \approx 2.5$) than coal ($x < 1$).

The energy needed for this endothermic reaction is usually provided by the (exothermic) combustion of the hydrocarbon source with oxygen.

The mixture of carbon monoxide and hydrogen is called synthesis gas or syngas. The resulting hydrocarbon products are refined to produce the desired synthetic fuel.

The carbon dioxide and carbon monoxide is generated by partial oxidation of coal and wood-based fuels. The utility of the process is primarily in its role in producing fluid hydrocarbons from a solid feedstock, such as coal or solid carbon-containing wastes of various types. Non-oxidative pyrolysis of the solid material produces syngas which can be used directly as a fuel without being taken through Fischer–Tropsch transformations. If liquid petroleum-like fuel, lubricant, or wax is required, the Fischer–Tropsch process can be applied.

CRYSTALLISATION KINETICS

For a given polymer, the extent of crystallisation attained during melt processing depends upon the rate of crystallisation and the time during which melt temperatures are maintained. Above T_m, some polymers that have low rates of crystallisation, such as poly(ethylene terephthalate) and polycaprolactone, can be quenched rapidly enough to achieve an amorphous state. Other polymers having much higher rates of crystallisation, such as polyethylene, cannot be quenched quickly enough to prevent crystallisation. For a given polymer, the rate of crystallisation depends upon the crystallisation temperature, as illustrated by Fig. 11.19, which shows the effect of temperature on the rate of spherulite growth in poly(ethylene terephthalate) (PET).

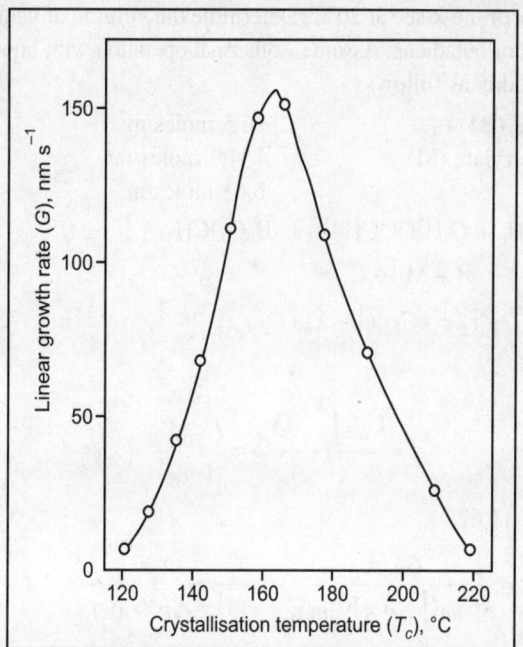

Fig. 11.19. Plot of linear growth rate of spherulites in poly(ethylene terephthalate) (PET) as a function of temperature and at a pressure of 1 bar. The maximum growth rate is observed near 178°C. Values of T_g and T_m for PET are approximately 69° and 265°C, respectively.

At T_m, the crystalline lamellae are destroyed as fast as they are formed from the melt and, therefore, the net rate of crystallisation is zero. Since the large-scale segmental mobility required for chain folding ceases at T_g, the crystallisation rate is again zero. At some intermediate temperature, T_{max}, an optimum balance is reached between chain mobility and lamellae growth. The temperature at which the crystallisation rate is maximum is independent of molecular weight; however, the maximum crystallisation rate decreases as the molecular weight increases.

The rate of crystallisation can be followed by a variety of techniques, such as dilatometric measurement of volume changes, infrared spectroscopy, and optical-microscopic measurement of the growth of spherulite radii with time (e.g. Fig. 11.19). During the crystallisation process, the fractional crystallinity, ϕ, at time t may be approximated by the Avrami equation.

$$\phi = 1 - \exp\left(-kt^n\right) \qquad \qquad \text{... (11.60)}$$

where, k is a temperature-dependent growth-rate parameter and n is a temperature-independent nucleation index. Typically, n varies between 1 and 4 depending on the nature of nucleation and growth processes. For example, in the case of sporadically nucleating spherulites, as may result during quiescent melt crystallisation near T_m, the nucleation index is approximately 4. The fractional crystallinity of a polymer can be determined by a variety of techniques, including infrared spectroscopy, X-ray analysis and density measurements and by calorimetric methods.

SOLVED EXAMPLE

Example 11.1. The condensation of butadiene and methyl acrylate in benzene solution was studied. Reactant butadiene is the limiting step. It is reported that butadiene follows second order reaction with rate constant 1.15×10^{-3} m³/mole-k sec at 20°C. Determine the volume of back mixed reactor needed to give 40 per cent conversion of butadiene. Assume isothermal operation with liquid feed rate of 500 m³/sec. The composition of the feed is as follows:

Butadiene (B)	96.5 moles/m³
Methyl acrylate (M)	184.0 moles/m³
AlCl₃	6.63 moles/m³

$$C_4H_6 + C_3H_3OOCH_3 \rightarrow C_7H_9OOCH_3$$

$$B + M \rightarrow C.$$

Solution:

$$-r_B = kC_{B0}(1 - X_B)\, C_{AlCl_3,0} \qquad \qquad ... (11.61)$$

The design equation:

$$\tau = \frac{C_{B0}\int_0^{X_B} dX_B}{(-r_B)} = \frac{C_{B0}X_B}{(-r_B)} \qquad \qquad ... (11.62)$$

Combining Eqs 11.61 and 11.62:

$$\tau = \frac{C_{B0}X_B}{kC_{B0}(1-X_B)C_{AlCl_3,0}} = \frac{X_B}{k(1-X_B)C_{AlCl_3,0}} \qquad \qquad ... (11.63)$$

Substituting the numerical values gives:

$$\tau = \frac{0.4}{(1.15\times10^{-3})(1-0.4)6.63}$$

$$= 87.4 \text{ k sec} \approx 24.3 \text{ hrs}$$

Volume of reactor, $V = \tau Q$

$$= 87.4 \times 0.5 = 43.7 \text{ m}^3.$$

Chapter 12

Biological Reactions Engineering

INTRODUCTION

A biological process is a process of a living organism. Biological processes are made up of any number of chemical reactions or other events that results in a transformation. Regulation of biological processes occurs where any process is modulated in its frequency, rate or extent. Biological processes are regulated by many means; examples include the control of gene expression, protein modification or interaction with a protein or substrate molecule. Dioxygen (O_2) plays an important role in the energy metabolism of living organisms. Free oxygen is produced in the biosphere through photolysis (light-driven oxidation and splitting) of water during photosynthesis in cyanobacteria, green algae, and plants. During oxidative phosphorylation in cellular respiration, oxygen is reduced to water, thus closing the biological water-oxygen redox cycle. In nature, free oxygen is produced by the light-driven splitting of water during oxygenic photosynthesis. Green algae and cyanobacteria in marine environments provide about 70 per cent of the free oxygen produced on earth. The remainder is produced by terrestrial plants, although for example, almost all oxygen produced in tropical forests is consumed by organisms living there. A simplified overall formula for photosynthesis is:

$$6CO_2 + 6H_2O + photons \rightarrow C_6H_{12}O_6 + 6O_2$$

(or simply carbon dioxide + water + sunlight $\rightarrow$ glucose + oxygen).

ENZYME KINETICS

The kinetic behaviour of enzymes has been studied in detail for a century, beginning with the classic work of Henri and Michaelis and Menten. The objectives have been three-fold: to gain an understanding of the mechanisms of enzyme action; to illuminate the physiological roles of enzyme-catalysed reactions; and, mainly in recent years, to manipulate enzyme properties for biotechnological ends. Experimental practice has been overwhelmingly dominated by the first of these aims; most experiments have been designed as if shedding light on the mechanism was the principal, or even the only, objective. However, although much valuable information has been obtained in this way, there are some important aspects of enzyme function that are obscured when working mainly with isolated enzymes in conditions far removed from those that exist in the cell.

Michaelis–Menten Kinetics

Michaelis–Menten kinetics (occasionally also referred to as Michaelis–Menten–Henri kinetics) approximately describes the kinetics of many enzymes. It is named after Leonor Michaelis and Maud

Menten. This kinetic model is relevant to situations where very simple kinetics can be assumed, (i.e. there is no intermediate or product inhibition, and there is no allostericity or cooperativity). More complex models exist for the cases where the assumptions of Michaelis–Menten kinetics are no longer appropriate.

The starting point for any discussion of enzyme kinetics is the Michaelis–Menten equation, which expresses the initial rate v of a reaction at a concentration a of the substrate transformed in a reaction catalysed by an enzyme at total concentration e_0:

$$v = \frac{k_0 e_0 a}{K_m + a} = \frac{Va}{K_m + a} \qquad \qquad \text{... (12.1)}$$

The parameters are k_0, the catalytic constant, and K_m, the Michaelis constant. The form shown in the middle is more fundamental than that on the right, but the second form, in which $k_0 e_0$ is written as the limiting rate V, is often used because the enzyme concentration in meaningful units is often not known. V is the limit that v approaches as the enzyme becomes saturated (that is, when a becomes very large) and K_m is the value of a at which $v = 0.5V$ (that is, at which the rate is half-maximal). The ratio k_0/K_m is called the specificity constant and given the symbol k_A: It is a more fundamental constant than K_m in the analysis of enzyme mechanisms (i.e. it has a simpler mechanistic meaning), but the equation is usually written in terms of V (or k_0) and K_m nonetheless.

The principal assumption implied by Eq. 12.1 is that the rate of the reverse reaction is negligible: This may be because the reaction is irreversible for practical purposes, but even with reversible reactions the reverse reaction can be made negligible by measuring the rate in the absence of products and by extrapolating the rate back to zero time, that is, by estimating the initial rate at a time when no products have accumulated. Even if there is no significant reverse reaction, products can still affect the rate of the forward reaction because product inhibition is a common phenomenon.

Another assumption is that the reaction is allowed sufficient time to reach a steady state because all reactions pass through an initial acceleration phase known as the transient state. This phase is normally very brief (a few milliseconds) and in practice enzymes are usually studied under steady-state conditions. Note, however, that this is made experimentally possible by working with extremely small enzyme concentrations compared with those that may exist in the cell.

The use of very low enzyme concentrations has two important consequences: First, it normally means that the enzyme concentration can be neglected in comparison with the substrate concentration, and second it makes the steady-state rate sufficiently slow to be easily measured and the steady-state phase long enough to be meaningful.

If high enzyme concentrations were used, the transient state would be as brief as before, but the steady state would also be very brief, so that there would be no period in which one could adequately treat the rate as constant.

Equation 12.1 may be derived from the following model:

$$\text{A + E} \underset{k_{-1}}{\overset{k_1}{\rightleftharpoons}} \text{EA} \underset{k_{-2}}{\overset{k_2}{\rightleftharpoons}} \text{EP} \underset{k_{-3}}{\overset{k_3}{\rightleftharpoons}} \text{E + P} \qquad \qquad \text{... (12.2)}$$

which assumes that the reaction passes through an enzyme–substrate complex EA, which undergoes catalytic transformation to an enzyme–product complex EP, which then breaks down to form products. Although real enzyme mechanisms may be more complicated than this, every reaction passes through steps of substrate binding, chemical transformation, and product release. In simple introductory treatments the second and third steps are often treated as a single step, but conceptually they are clearly distinct. In reactions of more than one substrate, the steps do not necessarily occur in the order one might guess; that is, some products may be released before all substrates have bound, but the general principle that any reaction involves the same three kinds of step remains valid. The Michaelis–Menten parameters can be defined as follows in terms of the rate constants shown in Eq. 12.2:

$$k_0 = \frac{k_2 k_3}{k_{-2} + k_2 + k_3}; k_A = \frac{k_1 k_2 k_3}{k_{-1}k_{-2} + k_{-1}k_3 + k_2 k_3}; K_m = \frac{k_{-1}k_{-2} + k_{-1}k_3 + k_2 k_3}{k_1(k_{-2} + k_2 + k_3)} \quad \text{... (12.3)}$$

Note that none of the three parameters has a simple transparent meaning. The interpretations commonly attributed to them depend on additional simplifying assumptions that are not always correct. For example, K_m is often said to be equal to the equilibrium constant k_{-1}/k_1 for dissociation of A from EA, from the expression in Eq. 12.3 does not take this form unless k_2 is very small.

As there is no good reason for k_2 to be small, and indeed ideas of evolutionary optimisation of enzyme function lead one to expect the opposite when the enzyme is acting on its natural physiological substrate, m follows that K_m should not, in general, be regarded as a measure of the equilibrium dissociation constant.

Despite these difficulties in providing a detailed mechanistic meaning to K_m, it does provide a measure of the tightness of substrates binding in the steady-state, as it is quite correct to take K_m as equal to the ratio of the sum of concentrations of all enzyme complexes (i.e. both EA and EP) over the concentration of free enzyme. Similarly k_0 provides a valid measure of the capacity of the enzyme–substrate complex to mean to give products, even if it cannot be interpreted as the rate constant for a unique step in the mechanism.

The reason for the term specificity constant for k_A, that is, the relationship to enzyme specificity, will become clear once inhibition has been considered.

Graphical Analysis

Equation 12.1 defines a metabolic dependence of rate on substrate concentration, as illustrated in Fig. 12.1. The initial steep rise in v as a measures from zero is rapidly transformed into the phenomenon of saturation, whereby further increases in a procure smaller and smaller increases in v, which approaches but does not reach or exceed, the limiting rate V. The rectangular hyperbola makes this type of plot inconvenient for estimating the values of the kinetic parameters (because the line does not approach the saturation limit closely enough at reasonable values of a to allow direct measurement of V). For this purpose, therefore, it is usual to transform Eq. 12.1 into one of the following three forms, which underlie the three straight-line plots illustrated in Figs 12.2 through 12.4:

$$\frac{1}{v} = \frac{1}{V} + \frac{K_m}{V}\frac{1}{a} \quad \text{... (12.4)}$$

$$\frac{a}{v} = \frac{K_m}{V} + \frac{1}{V}a \quad \text{... (12.5)}$$

$$v = V - K_m\frac{v}{a} \quad \text{... (12.6)}$$

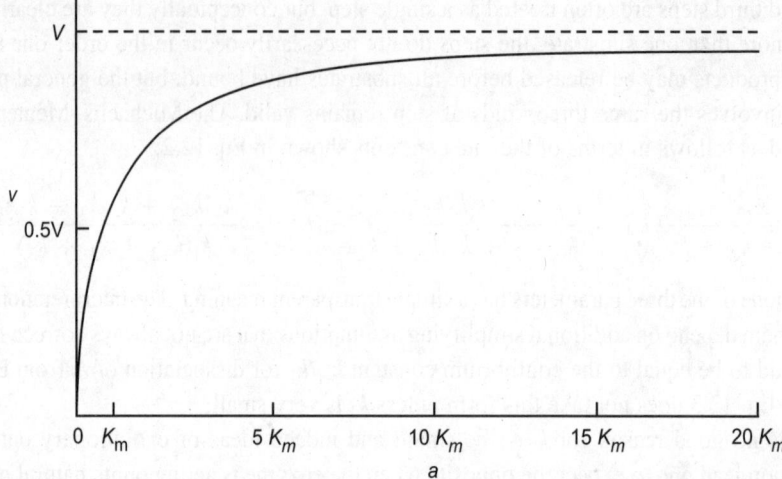

Fig. 12.1. Michaelis–Menten dependence of rate on substrate concentration. The curve is a rectangular hyperbola through the origin, approaching a limit of $v = V$ at saturation. The rate is 0.5V at a substrate concentration equal to the Michaelis constant, K_m, but note that the approach to the limit is slow, so that, for example, even at $a = 10\ K_m$ the rate is still nearly 10 per cent less than V.

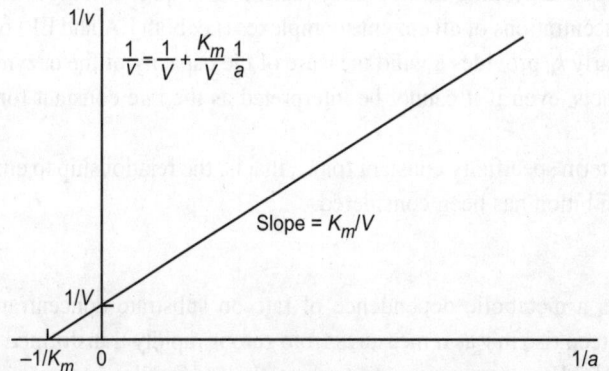

Fig. 12.2. The double-reciprocal plot. This is the most widely used method of plotting the Michaelis–Menten equation as a straight-line. However, the severe distortion of any experimental errors in the original data causes it to give a misleading impression.

The double-reciprocal plot, illustrated in Fig. 12.2 and based on Eq. 12.4, is the mostly widely used, but it is also the least satisfactory because it distorts the effect of experimental error to such an extent that it is difficult to form any visual judgement of where the best line should be drawn. The other two plots are better, and the plot of v against v/a (Fig. 12.4, Eq. 12.6) has the particular advantage that the entire observable range of v values, from 0 to V, is mapped onto a finite range of graph; this makes it easy to judge by eye if an experiment has been well designed. On the other hand, it has the disadvantage that v, normally the less reliable measurement, contributes to both coordinates, and errors in v cause deviations along lines through the origin, rather than parallel with one or the other axis.

In modern practice it is usually best to regard these plots as for illustration purposes only, and to use suitable computer programs for the actual parameter estimation.

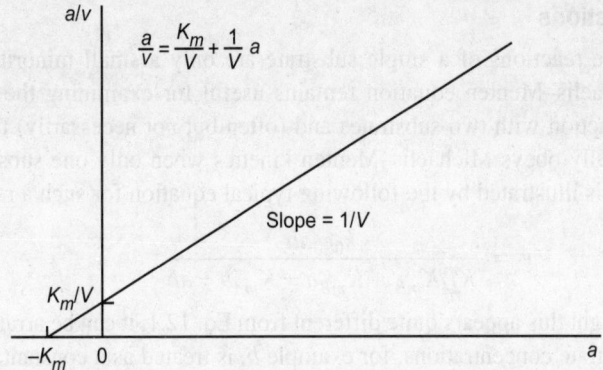

Fig. 12.3. The plot of a/v against a. This alternative to the plot shown in Fig. 12.2 produces much less distortion of the experimental error.

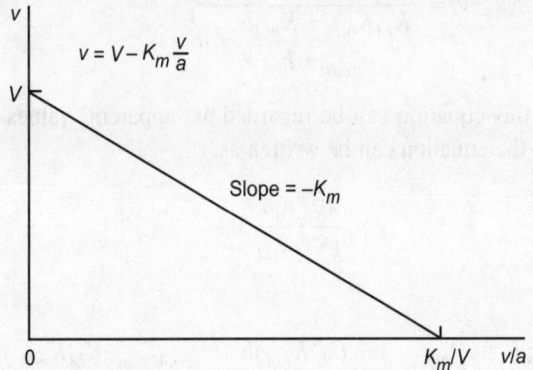

Fig. 12.4. The plot of 'v' against v/a. The third way of plotting the Michaelis–Menten equation as a straight-line also avoids the error-distorting property of the plot shown in Fig. 12.2, and maps the entire range of observable rates (from 0 to V) onto a finite range of paper. This is a desirable property because it makes it impossible to disguise deficiencies in the experimental design.

For this purpose, it is not sufficient just to apply unweighted linear regression to the straight-line plots, as this suffers from the same statistical distortions as the plots themselves. Full treatment would require more space than is available here.

The following two equations for calculating best-fit values of K_m and V give satisfactory results if the v values have uniform coefficient of variation (uniform standard deviation expressed as a percentage), as is usually at least approximately correct:

$$K_m = \frac{\sum v^2 \sum (v/a) - \sum (v^2/a) \sum v}{\sum (v^2/a^2) \sum v - \sum (v^2/a) \sum (v/a)} \qquad \text{... (12.7)}$$

$$V = \frac{\sum (v^2/a^2) \sum v^2 - [\sum (v^2/a)]^2}{\sum (v^2/a^2) \sum v - \sum (v^2/a) \sum (v/a)} \qquad \text{... (12.8)}$$

Each summation is made over all observations.

Two-Substrate Reactions

Enzymes that catalyse reactions of a single substrate are only a small minority of all the enzymes known, but the Michaelis–Menten equation remains useful for examining the kinetics of the more common case of a reaction with two substrates and (often but not necessarily) two products, because such a reaction normally obeys Michaelis–Menten kinetics when only one substrate concentration is varied at a time. This is illustrated by the following typical equation for such a reaction:

$$v = \frac{k_0 e_0 ab}{K_{iA}K_{mB} + K_{mB}a + K_{mA}b + ab} \quad \text{... (12.9)}$$

Although at first sight this appears quite different from Eq. 12.1, it can be arranged in the same form if one of the two substrate concentrations, for example b, is treated as a constant:

$$v = \frac{\left(\dfrac{k_0 b}{K_{mB} + b}\right) e_0 a}{\dfrac{K_{iA}K_{mB} + K_{mA}b}{K_{mB} + b} + a} \quad \text{... (12.10)}$$

The two fractions in this equation can be regarded as 'apparent' values of the Michaelis–Menten parameters for A, that is, the equation can be written as:

$$v = \frac{k_0^{app} e_0 a}{k_{mA}^{app} + a} \quad \text{... (12.11)}$$

with

$$k_0^{app} = \frac{k_0 b}{k_{mB} + b}; \; k_A^{app} = \frac{(k_0/K_{mA})b}{(K_{iA}K_{mB}/K_{mA}) + b}; \; k_{mA}^{app} = \frac{K_{iA}K_{mB} + K_{mA}b}{K_{mB} + b} \quad \text{... (12.12)}$$

Notice that the expressions for the apparent values of k_0 and k_A are both individually of Michaelis–Menten form with respect to b, whereas that for the apparent value of K_m is more complicated. This behaviour is quite typical and is one of the reasons why k_0 and k_A should regarded as more fundamental parameters than K_m. More generally, the concept of apparent parameters pervades the analysis of simple cases in steady-state enzyme kinetics, being important for the study of reactions with multiple substrates and inhibition.

Limitations

The first source of limitations for the Michaelis–Menten kinetics is that it is an approximation of the kinetics derived by the law of mass action. The second limitation is that Michaelis–Menten kinetics relies upon the law of mass action which is derived from the assumptions of free (Fickian) diffusion and thermodynamically-driven random collision. However, many biochemical or cellular processes deviate significantly from such conditions. For example, the cytoplasm inside a cell behaves more like a gel than a freely flowable or watery liquid, due to the very high concentration of protein (up to ~400 mg/ml) and other solutes, which can severely limit molecular movements (e.g. diffusion or collision). This causes macromolecular crowding, which can alter reaction rates and dissociation constants.

For heterogeneous enzymatic reactions, such as those of membrane enzymes, molecular mobility of the enzyme or substrates can also be severely restricted, due to the immobilisation or phase-separation of the reactants. For some homogeneous enzymatic reactions, the mobility of the enzyme or substrate may also be limited, such as the case of DNA polymerase where the enzyme moves along a chained substrate, rather than having a three-dimensional freedom. The limitation on molecular mobility (as well as other 'non-ideal' conditions) demands modifications on the conventional mass-action laws, and Michaelis–Menten kinetics, to better reflect certain real world situations. Although it has been shown that the law of mass action can be valid in heterogeneous environments.

INHIBITION AND ACTIVATION

Inhibition

For most enzyme-catalysed reactions, molecules exist that resemble the substrate closely enough to bind to the enzyme, but not closely enough to undergo a chemical reaction. Such a molecule is known as a competitive inhibitor and causes competitive inhibition, characterised by a rate equation of the following form:

$$v = \frac{k_0 e_0 a}{k_m(1 + i/K_{ic}) + a} \qquad \text{... (12.13)}$$

in which i is the concentration of the inhibitor and K_{ic} is the competitive inhibition constant. (The qualification competitive and the second subscript c are usually omitted if only this simplest kind of inhibition is being considered.)

Inhibitors can interfere with catalysis as well as with substrate binding. In the simplest case, an inhibitory term affects the variable term in the denominator of the Michaelis–Menten equation, instead of the constant term:

$$v = \frac{k_0 e_0 a}{k_m + a(1 + i/K_{iu})} \qquad \text{... (12.14)}$$

This is called uncompetitive inhibition, and the inhibition constant K_{iu} is the uncompetitive inhibition constant. This is important as a limiting case of inhibition, but in its pure form it is not at all common. Much more often one has mixed inhibition, when both competitive and uncompetitive effects occur simultaneously:

$$v = \frac{k_0 e_0 a}{k_m(1 + i/K_{ic}) + a(1 + i/K_{iu})} \qquad \text{... (12.15)}$$

There is no particular reason for the two inhibition constants K_{ic} and K_{iu} to be equal, and most of the mechanisms one might propose to account for mixed inhibition lead one to expect them to be different, yet the case where $K_{ic} = K_{iu}$ is often given an undeserved prominence in discussions of inhibition, largely because experiments done many years ago suggested that it was a more common phenomenon than it is. This is called non-competitive inhibition and its rate equation is the same as Eq. 12.15, but with both K_{ic} and K_{iu} written simply as K_i.

All of these kinds of inhibition are conveniently discussed in terms of apparent Michaelis–Menten parameters. In the general case (Eq. 12.15), these are as follows:

$$k_0^{app} = \frac{k_0}{1 + i/K_{iu}}; \quad k_A^{app} = \frac{k_A}{1 + i/K_{ic}}; \quad k_m^{app} = \frac{K_m(1 + i/K_{ic})}{1 + i/K_{iu}} \qquad \text{... (12.16)}$$

Note that the first two expressions have the same form, and both simplify to independence of i in the event that one or other inhibition term is negligible. The expression for the apparent value of K_m is more complicated, especially when one considers how it varies with the different types of inhibition. It increases with the concentration of a competitive inhibitor, it decreases as the concentration of an uncompetitive inhibitor increases, it may change in either direction as the concentration of a mixed inhibitor increases, or it is independent of inhibitor concentration if the inhibition is non-competitive. In general, it is simplest to regard k_A as the parameter affected by competitive inhibition, negligibly so when the competitive component is negligible, k_0 as the parameter affected by uncompetitive inhibition, negligibly so when the uncompetitive component is negligible, and K_m just as the ratio of the two, so $K_m = k_0/k_A$. The effects of the different kinds of inhibition on the common plots as illustrated in Figs 12.2 through 12.4 follows naturally from Eq. 12.16.

Any competitive effect affects the apparent value of k_A, hence, it increases the slope of the plot of $1/v$ against $1/a$ (Fig. 12.2), it increases the ordinate intercept of the plot of a/v against a (Fig. 12.3), and it decreases the abscissa intercept of the plot of v against v/a (Fig. 12.4). Conversely, any uncompetitive effect increases the ordinate intercept of the plot of $1/v$ against $1/a$, increases the slope of the plot of a/v against a, and decreases the ordinate intercept of the plot of v against v/a. When both components of the inhibition are present, both kinds of effects occur. As an illustration we may consider just one example, the effect of competitive inhibition on the plot of $1/v$ against $1/a$: Plots made at various different inhibitor concentrations produce a family of straight-lines intersecting on the ordinate axis, as shown in Fig. 12.5, the lack of effect on the ordinate intercept being a direct consequence of the lack of effect on the apparent value of V.

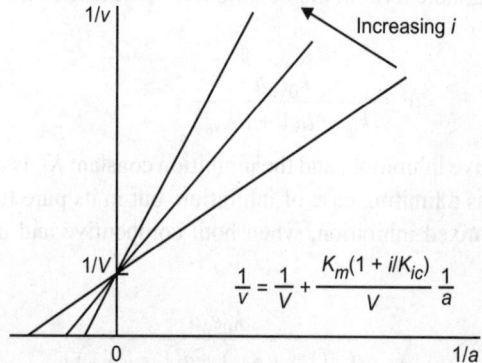

Fig. 12.5. Effect of competitive inhibition on the double-reciprocal plot.

Specificity

Specificity is the most fundamentally important property of enzymes. Although one is often impressed by the catalytic effectiveness of enzymes, accelerating a reaction is not, in reality, difficult: Heating the reaction mixture in a sealed tube is an efficient way of accelerating virtually any reaction, essentially without limit. What is difficult is to accelerate one selected reaction without at the same time accelerating a great mass of unwanted reactions. What is important about an enzyme, therefore, is not that it is an excellent catalyst for a small set of reactions, but that it is an extremely bad catalyst—virtually without any activity—for all other reactions. In other words, the essential properties of enzymes are that they act under very mild conditions and are highly specific.

In the past, specificity was often assessed by comparing the kinetic parameters for different reactions measured in isolation from one another, which led to sterile arguments as to whether specificity was best measured in terms of k_0, K_m, or some combination of the two. This type of argument was resolved once it was realised that the only meaningful way of defining specificity is as a property of an enzyme that allows it to discriminate between substrates that are mixed together. The simplest way to consider this is with a model similar to that for competitive inhibition, except that one assumes that both molecules are capable of reacting. The equation for reaction of one substrate A in the presence of a competing substrate A' follows an equation similar to that for competitive inhibition (Eq. 12.13):

$$v = \frac{k_0 e_0 a}{K_m(1 + a'/K'_m) + a} \qquad \qquad \text{... (12.17)}$$

with the inhibitor concentration replaced by a', the concentration of A', and the inhibition constant by K'_m, the Michaelis constant for the reaction of A' considered in isolation. The rate of reaction of A' is given by the same equation with an obvious transposition of symbols:

$$v' = \frac{k'_0 e_0 a'}{K'_m(1 + a/K_m) + a'} \qquad \qquad \text{... (12.18)}$$

It can then be seen that the ratio of rates is the ratio of substrate concentrations multiplied by the ratio of specificity constants:

$$\frac{v}{v'} = \frac{k_0/K_m}{k'_0/K'_m} \frac{a}{a'} = \frac{k_A a}{k'_A a'} \qquad \qquad \text{... (12.19)}$$

This result, which is still less well known than its importance merits, is the reason for the term specificity constant. Note that although inspection of Eq. 12.1 suggests that k_A is no more than the parameter that defines the rate at very low substrate concentrations, no assumption about the magnitudes of the concentrations was made in arriving at Eq. 12.19. It is thus valid at all concentrations, and the specificity constant measures specificity at all concentrations, not just low ones.

Activation

Activation is the opposite from inhibition, in which a reaction proceeds more rapidly in the presence of a particular molecule than in its absence. It is less common than inhibition, and discussion is complicated by the fact that a variety of quite different phenomena have been termed 'activation'. The most important of these is a confusion between true activation and the case where the activator is really a component of the substrate. Numerous ATP-handling enzymes are said to be activated by magnesium ions, when in reality the complex MgATP is the true substrate, that is, the species that reacts with the enzyme. Other metal ions, such as the zinc in a number of enzymes, may be true activators as they bind to the enzyme itself and confer catalytic properties on it.

Another misuse of the term activation relates to coenzymes such as NAD in many dehydrogenases: Alcohol dehydrogenase, for example, may be said to be activated by NAD, but although consideration of the metabolic pathway in which the reaction occurs may make it convenient to regard ethanol as the substrate and NAD as the coenzyme, this is meaningless when the reaction is considered in isolation. So far as alcohol dehydrogenase is concerned, it catalyses a reaction that requires two substrates, ethanol and oxidised NAD; the reaction will not proceed unless both are present, and neither has any more reason to be called the substrate than the other.

When such improper uses of the term are excluded, there remain a number of enzymes for which the true inverse of inhibition occurs. In the simplest cases the equations are just the inverse of inhibition equations, with terms of the form i/K_i replaced by ones of the form K_x/x (for an activator X with activation constant K_x). However, the simplest cases constitute a smaller proportion of the whole than they do for inhibition. This is because whereas most inhibitors inhibit completely, in the sense that enzyme species with inhibitor bound retain no activity as long as the inhibitor remains bound, many enzymes subject to activation retain some activity in the absence of the activator. As a result, full analysis of activation is often more complicated than it is for inhibition, but this will not be discussed further here.

Irreversible Inhibition

The types of inhibition considered so far are examples of reversible inhibition; the inhibitor binds reversibly and catalytic activity returns when the inhibitor is released. Irreversible inhibition also occurs, in which the inhibitor either binds so tightly that for practical purposes it cannot be removed, or reacts with the enzyme and converts it irreversibly to a form that has no catalytic activity. These two cases are conceptually different, and the former is more correctly called tight-binding inhibition, rather than irreversible inhibition. However, they are not easy to distinguish in practice, and have similar practical effects and, hence, similar practical uses.

Although irreversible inhibition has played a smaller part than reversible inhibition in the academic study of enzyme mechanisms, it has far greater industrial and pharmacological importance. This is because competitive inhibitors, the most common kind of reversible inhibitors, are almost completely ineffective in complete physiological systems, for reasons to be considered shortly. By contrast, whenever irreversible (or tight-binding) inhibition occurs in a physiological system, it can be expected to have profound effects. Many toxic and pharmacologically active substances owe their effects to irreversible inhibition.

Both tight-binding and irreversible inhibition manifest themselves in ways that allow them to be confused with non-competitive inhibition, as in Eq. 12.15 with the two inhibition constants equal. This is because the practical effect of irreversible inhibition is not on any of the kinetic parameters in the Michaelis–Menten equation, but on e_0, the total enzyme concentration. However, a decrease in e_0 can be confused with a decrease in the apparent value of k_0, as they occur as a product in Eq. 12.1. Although uncompetitive inhibition affects k_0, it does so without affecting k_A, and thus also changes K_m. Decreasing k_0 without affecting K_m, similar to what one would observe if e_0 decreases, is a definition of noncompetitive inhibition.

Inhibitory Effects in Metabolic Systems

Competitive and uncompetitive inhibition are sufficiently similar in their effects in artificial experiments on isolated enzymes that they are often not distinguished, and an uncompetitive component in mixed inhibition often passes unnoticed.

Many inhibitors are described in the literature as competitive inhibitors in the absence of any real evidence of the type of inhibition. This sort of confusion can easily lead to the entirely false idea that they are similar in their effects in systems where the inhibited enzyme is mixed with other enzymes and catalyses a step in the middle of a pathway.

In a typical experiment *in vitro*, one decides the concentrations of the various components in advance and measures the rate that results; however, this is very different from what happens in the cell. To a first approximation, an enzyme catalysing a step in the middle of a pathway must transform its substrate at the rate at which it arrives, that is, within certain limits it has little or no effect on the rate of its reaction, but instead determines the concentrations of the metabolites around it. (This is an oversimplification: but is useful for discussion.)

It is useful therefore to transform Eqs 12.13 and 12.14 into expressions for a in terms of i:

$$a = \frac{vK_m(1 + i/K_{ic})}{k_0e_0 - v} \qquad \qquad \cdots (12.20)$$

$$a = \frac{vK_m}{k_0e_0 - v(1 + i/K_{iu})} \qquad \qquad \cdots (12.21)$$

However, similar Eqs 12.13 and 12.14 may seem, their transformed versions are drastically different. Equation 12.20 shows a linear dependence of a on i, which means that increasing i can never result in uncontrolled increases in a. This is illustrated in Fig. 12.6(a). Even at an inhibition concentration equal to the inhibition constant, the substrate concentration is only doubled. By contrast, the curve defined by Eq. 12.21 is a rectangular hyperbola [Fig. 12.6(b)] that produces a steep and uncontrolled rise in substrate concentration at quite moderate inhibitor concentrations. The point is that in competitive inhibition, rises in substrate and inhibitor concentrations oppose one another—not only does the inhibitor compete with the substrate, but equally, the substrate competes with the inhibitor. In uncompetitive inhibition, however, these effects potentiate one another.

It follows that although it is relatively easy to find molecules that will act as competitive inhibitors, it is also largely useless as a strategy for designing pesticides or drugs because it is correspondingly easy for the organism to counteract the effect of the inhibition. To produce major metabolic effects one needs uncompetitive inhibitors, irreversible inhibitors, or tight-binding inhibitors: None of these are as easy to produce as weakly binding competitive inhibitors, but they are far more effective.

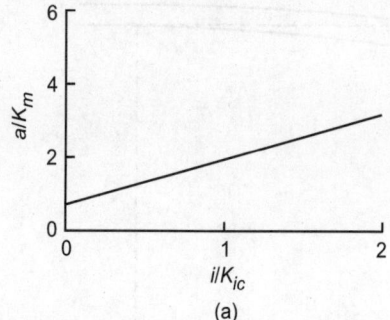
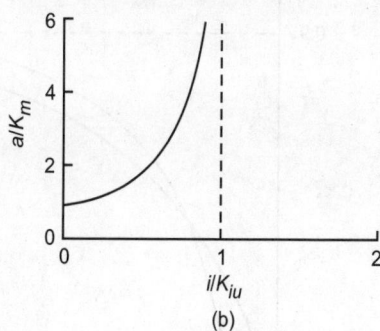

Fig. 12.6. Effects of (a) competitive and (b) uncompetitive inhibition on the concentration of substrate in a constant-rate system. Both curves are drawn for the case of $a = K_m$ in the absence of inhibitor. Note that both kinds of inhibitor have quantitatively equal effects at very low concentrations, but the initial slope is maintained indefinitely if the inhibition is competitive, whereas it rapidly becomes infinite if the inhibition is uncompetitive.

NON-MICHAELIS–MENTEN BEHAVIOUR

All of the cases considered so far can be regarded as generalisations of the Michaelis–Menten equation (Eq. 12.1). However, although many enzymes do behave in this way, at least as a first approximation, there are some important exceptions. It is simple to calculate from Eq. 12.1 that if $a = K_m/9$ then $v = 0.1V$ and if $a = 9\,K_m$ then $v = 0.9V$; in other words, spanning the 10–90 per cent range of available rates requires an 81-fold range of substrate concentrations, almost two orders of magnitude. Similar calculations may be done with any of the equations of the Michaelis–Menten type for additional substrates, inhibitors, or activators. Their implication is that as long as enzymes follow Michaelis–Menten kinetics, their rates cannot be adequately varied by manipulating concentrations of substrates, for example, because effective regulation will often require sensitivity to small changes in signals—certainly changes much smaller than two orders of magnitude. A second difficulty arises from the fact that inhibition of the types considered commonly derives from structural similarities between inhibitors and substrates or products, whereas there is no reason to expect the molecules needed for metabolic signals to resemble the substrates or products of the enzymes that need to respond to the signals. In reality, the concentration of the end product of a pathway often serves as such a signal: too low, and the pathway needs to be activated; too high, and it needs to be inhibited. It is often found, therefore, that the enzyme that catalyses the first committed step of a pathway, that is, the first step after a branch point, in the branch that leads to the end product in question, is inhibited by that end product. For inhibition of this kind to be possible, the enzyme must have a specific binding site for the end product, independent of the binding sites for substrates and products. Such a site is called an allosteric site, and the phenomenon is called allosteric inhibition. Because the need for it often coincides with the need for higher sensitivity than is provided by Michaelis–Menten kinetics and the common kinds of inhibition, allosteric inhibition is often cooperative. This means that the equations that define it are more complicated than those considered above, allowing, for example, a change from 10 per cent to 90 per cent inhibited over a concentration range much smaller than 81-fold, and typically less than 10-fold (though rarely, if ever, less than 3-fold) (Fig. 12.7).

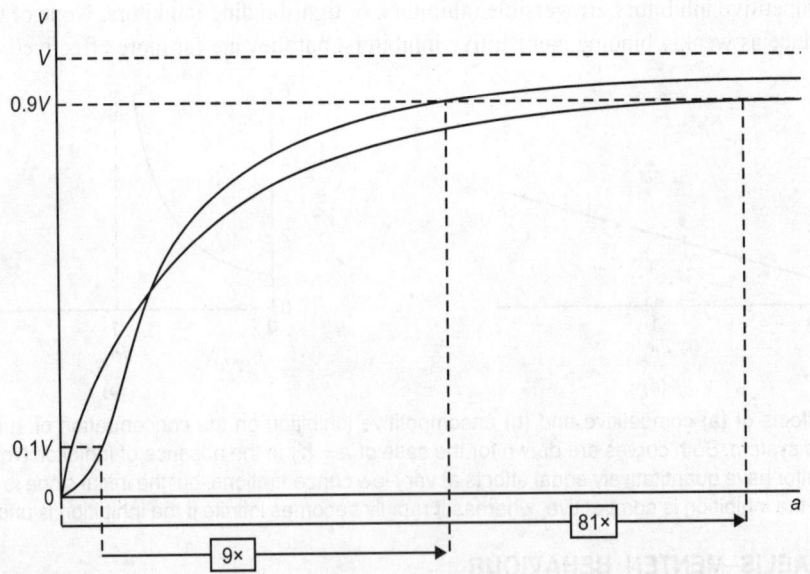

Fig. 12.7. Non-Michaelis–Menten kinetics. For an enzyme obeying the Michaelis–Menten equation (Fig. 12.1), an 81-fold increase in substrate concentration is needed to bring the rate from 10 per cent to 90 per cent of V. If the enzyme shows positive cooperativity the curve typically becomes sigmoid (S-shaped), and this range of substrate concentrations is decreased (to nine-fold in the example, but in strongly cooperative cases it can be as small as three-fold).

FERMENTATION METHODS AND SYSTEMS

Fermentation is classically defined as the microbiological conversion, in the absence of oxygen, of sugars to alcohol or lactic acid. By the enzymatic conversion of sugars to alcohol, the micro-organisms gain biochemical energy (ATP). Fermentation is used in many commercial food processes, including the making of sauerkraut, yogurt, wines and beers. Certain bacteria and fungi are capable of causing fermentation to occur in the absence of oxygen. Some industrial processes are called fermentation which are aerobic (conducted in the presence of oxygen) and would be more correctly be termed oxidative processes. Fruits, grains, milk, and other organic substances naturally undergo fermentation during spoilage, a phenomenon which cavemen no doubt observed.

There are two main types of industrial fermentation processes, batch fermentations and continuous fermentations. In batch fermentations, sterile growth medium is inoculated with the micro-organisms and no additional growth medium is added. In continuous fermentations, growth medium is added to the fermenting medium to sustain the fermentation process. Fermentation occurs by the production of cellular enzymic reactions instead of chemical reactions aided by inanimate catalysts, sometimes operating at elevated temperature and pressure.

Growth Kinetics

Growth kinetics, i.e. the relationship between specific growth rate and the concentration of a substrate, is one of the basic tools in microbiology. However, despite more than half a century of research, many fundamental questions about the validity and application of growth kinetics as observed in the laboratory to environmental growth conditions are still unanswered. For pure cultures growing with single substrates, enormous inconsistencies exist in the growth kinetic data reported.

Microbial growth kinetics, i.e. the relationship between the specific growth rate (μ) of a microbial population and the substrate concentration (s), is an indispensable tool in all fields of microbiology, be it physiology, genetics, ecology, or biotechnology, and therefore it is an important part of the basic teaching of microbiology.

Batch Fermentation Process

A tank of fermenter is filled with the prepared mash of raw materials to be fermented. The temperature and pH for microbial fermentation is properly adjusted, and occasionally nutritive supplements are added to the prepared mash. The mash is steam sterilised in a pure culture process. The inoculum of a pure culture is added to the fermenter, from a separate pure culture vessel.

Fermentation proceeds, and after the proper time the contents of the fermenter, are taken out for further processing. The fermenter is cleaned and the process is repeated. Thus each fermentation is a discontinuous process divided into batches.

In autecological studies, bacterial growth in batch culture can be modelled with four different phases: (i) lag phase, (ii) exponential or log phase, (iii) stationary phase, and (iv) death phase.

1. During lag phase, bacteria adapt themselves to growth conditions. It is the period where the individual bacteria are maturing and not yet able to divide. During the lag phase of the bacterial growth cycle, synthesis of RNA, enzymes and other molecules occurs.

2. Exponential phase (sometimes called the log phase) is a period characterised by cell doubling. The number of new bacteria appearing per unit time is proportional to the present population. If growth is not limited, doubling will continue at a constant rate so both the number of cells and the rate of population increase doubles with each consecutive time period. For this type of

exponential growth, plotting the natural logarithm of cell number against time produces a straight line. The slope of this line is the specific growth rate of the organism, which is a measure of the number of divisions per cell per unit time. The actual rate of this growth (i.e. the slope of the line in the Fig. 12.8) depends upon the growth conditions, which affect the frequency of cell division events and the probability of both daughter cells surviving. Exponential growth cannot continue indefinitely, however, because the medium is soon depleted of nutrients and enriched with wastes.

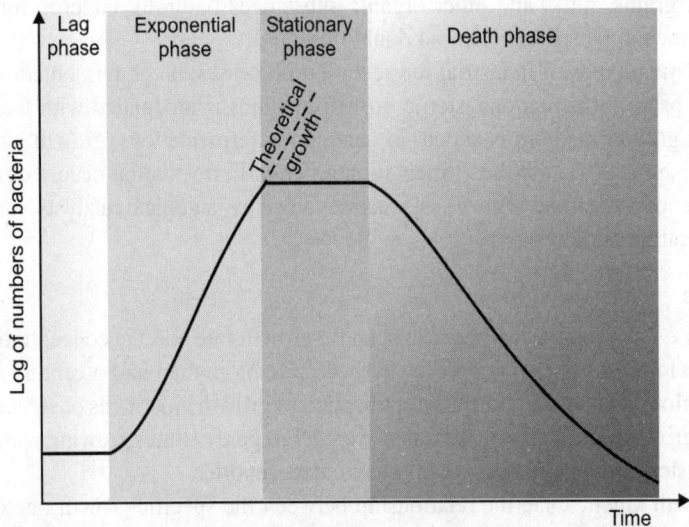

Fig. 12.8. Bacterial growth curve.

3. During stationary phase, the growth rate slows as a result of nutrient depletion and accumulation of toxic products. This phase is reached as the bacteria begin to exhaust the resources that are available to them. This phase is a constant value as the rate of bacterial growth is equal to the rate of bacterial death.

4. At death phase, bacteria run out of nutrients and die.

Fed-batch

A 'fed-batch' is a biotechnological batch process which is based on feeding of a growth limiting nutrient substrate to a culture. The fed-batch strategy is typically used in bioindustrial processes to reach a high cell density in the bioreactor. Mostly the feed solution is highly concentrated to avoid dilution of the bioreactor. The controlled addition of the nutrient directly affects the growth rate of the culture and allows to avoid overflow metabolism (formation of side metabolites, such as acetate for *Escherichia coli*, lactic acid in cell cultures, ethanol in *Saccharomyces cerevisiae*), oxygen limitation (anaerobiosis). In most cases the growth-limiting nutrient is glucose which is fed to the culture as a highly concentrated glucose syrup (600–850 g/l). Two basic approaches to the fed-batch fermentation can be used: the constant volume fed-batch culture — fixed volume fed-batch — and the variable volume fed-batch.

Fixed volume fed-batch

In this type of fed-batch, the limiting substrate is fed without diluting the culture. The culture volume can also be maintained practically constant by feeding the growth limiting substrate in undiluted form, for example, as a very concentrated liquid or gas (e.g. oxygen). Alternatively, the substrate can be added by dialysis or, in a photosynthetic culture, radiation can be the growth limiting factor without affecting the culture volume. A certain type of extended fed-batch—the cyclic fed-batch culture for fixed volume systems — refers to a periodic withdrawal of a portion of the culture and use of the residual culture as the starting point for a further fed-batch process. Basically, once the fermentation reaches a certain stage, (for example, when aerobic conditions cannot be maintained any more), the culture is removed and the biomass is diluted to the original volume with sterile water or medium containing the feed substrate. The dilution decreases the biomass concentration and result in an increase in the specific growth rate. Subsequently, as feeding continues, the growth rate will decline gradually as biomass increases and approaches the maximum sustainable in the vessel once more, at which point the culture may be diluted again.

Variable volume fed-batch

As the name implies, a variable volume fed-batch is one in which the volume changes with the fermentation time due to the substrate feed. The way this volume changes is dependent on the requirements, limitations and objectives of the operator.

The feed can be provided according to one of the following options:
1. The same medium used in the batch mode is added.
2. A solution of the limiting substrate at the same concentration as that in the initial medium is added.
3. A very concentrated solution of the limiting substrate is added at a rate less than (1), (2) and (3).

This type of fed-batch can still be further classified as repeated fed-batch process or cyclic fed-batch culture, and single fed-batch process. The former means that once the fermentation reached a certain stage after which is not effective any more, a quantity of culture is removed from the vessel and replaced by fresh nutrient medium. The decrease in volume results in a increase in the specific growth rate, followed by a gradual decrease as the quasi-steady-state is established. The latter type refers to a type of fed-batch in which supplementary growth medium is added during the fermentation, but no culture is removed until the end of the batch. This system presents a disadvantage over the fixed volume fed-batch and the repeated fed-batch process: much of the fermentor volume is not utilised until the end of the batch and consequently, the duration of the batch is limited by the fermentor volume.

No special piece of equipment is required over the equipment required for batch. However, some considerations should be made over the equipment used for a fed-batch fermentation. The vessels, particularly those used for the acid and base control, must be constructed from a nontoxic, corrosion-resistant material which is capable of withstanding repeated sterilisation cycles. There are two types of pumps which are suitable for the aseptic pumping of small volumes of culture media: the peristaltic pump and the diaphragm-dosing pump.

Control techniques for fed-batch fermentation

Adaptive control is the name given to a control system in which the controller learns about the process by acquiring data from a certain process and keeps on updating a control model. A parameter estimator monitors the process and estimates the process dynamics in terms of the parameters of a previously defined mathematical model of the process. A control design algorithm is then used to generate controller coefficients from those estimates, and a controller sets up the required control signals to the devices controlling the process. An extremely important feature of an adaptive controller is the structure of the model used by the parameter estimator to analyse estimates of process dynamics. The process can be described by a set of mass balance equations, whose quantities can be measured directly or indirectly. Figure 12.9 describes schematically the concept.

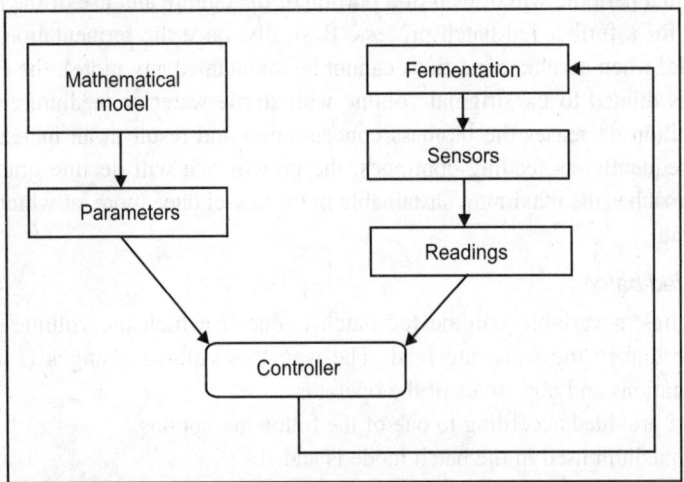

Fig. 12.9. Adaptive control: the controller compares the estimates from a mathematical model applied to the system to the readings obtained from the fermentation process. The controller then sends the signal to the device controlling the fermentation, for example, by increasing or decreasing a flow rate.

The optimal strategy for the fed-batch fermentation of most organisms is to feed the growth-limiting substrate at the same rate that the organism utilises the substrate, this is, to match the feed rate with demand for the substrate.

Four basic approaches have been used in attempts to balance substrate feed with demand (listed in order of increasing accuracy and/or complexity):

1. Open-loop control schemes in which feed is added according to historical data or predicted data.
2. Indirect control of substrate feed based on nonfeed source parameters such as pH, offgas analysis, dissolved O_2 or concentrations of organic products.
3. Indirect control schemes based on mass balance equations, the values of which are calculated from data obtained by sensors.
4. Direct control schemes based on direct, on-line measurements substrates.

Better and more flexible control may be obtained when there is direct measurement of substrate or an excreted metabolite in the medium, which can be used to influence feeding rates to the fermentation. This can be done off-line or semi-on-line, but on-line measurements are more useful because of:

1. The shorter analysis required.
2. Lower personnel requirement.
3. A reduced chance of fermentor contamination.

Regardless the type of control, the design is strongly influenced by both mathematical model availabilities and measurement possibilities.

Control and optimisation of bioreactors is strongly influenced by the quality of the sensors available for crucial response variables. Of primary importance is the ratio of the dynamic parameters of the sensor to those of the process. When these variables cannot be measured easily or quickly enough, a mathematical model must be used in some way in place of feedback information.

When an exact mathematical model is at disposal, an open-loop process control can be proposed which generally proves to be insufficient. The advantage of a feedback control is that a response to unforeseen and unexpected conditions during the fermentation is achieved and the process is controlled within the desired limits.

An indirect feedback control utilises an observable parameter, such as dissolved oxygen, pH, respiratory quotient, partial pressure of CO_2, culture fluorescence or by-product formation, which is closely related to the course of microbial fermentation. As examples of fed-batch systems using this concept, one can mention the pH-stat-a system in which the feed is provided depending on the pH, and the DO-stat-a system in which the feed is provided depending on the reading of the dissolved oxygen. A direct feedback controller uses the concentration of limiting substrate in the culture medium as a feedback feed-related parameter for control. A direct feedback control can have the disadvantage of not being very feasible due to the difficulty associated with obtaining accurate on-line measurements of substrate concentrations or even by the absence of on-line sensors for the important compound to control. The advantage of a feedback control is that a response to unforeseen and unexpected conditions during the fermentation is achieved and the process is controlled within the desired limits.

A feedback control can be implemented accordingly to not only a single measurement, but also to obtain a finer control action in a dual-level system. Turner, describes a control method applied to a fed-batch culture of recombinant *Escherichia coli* in which a two-level control was preferred because it provided much greater flexibility and better control over the substrate concentration in the medium and the production of by-products.

As compared with the batch fermentation, two more parameters need to be specified to determine the operating conditions of a fed-batch fermentation: feed and initial feeding time. These parameters are usually process and/or micro-organism specific and the parameters commonly used to define them.

Continuous fermentation process

Growth of micro-organisms during batch fermentation confirms to the characteristic growth curve, with a lag phase followed by a logarithmic phase. This, in turn, is terminated by progressive decrements in the rate of growth until the stationary phase is reached. This is because of limitation of one or more of the essential nutrients.

In continuous fermentation, the substrate is added to the fermenter continuously at a fixed rate. This maintains the organisms in the logarithmic growth phase. The fermentation products are taken out continuously. The design and arrangements for continuous fermentation, are somewhat complex.

Homogeneous bioreactor systems suspension cultures

The classical system is the suspension culture using a stirred tank with different impeller types and installations, equipped with or without a spin filter. In large-scale bioreactors, slight modifications of several internal parts of bioreactors used for bacterial fermentation are made in order to adopt them for culturing animal cells. The modifications are in the agitation system. Marine type impeller, vibromixer or rotating flexible sheets replace the turbine type impeller widely used in microbial fermentation. Perfusion systems were also developed for submerged cultivation of animal cells.

This is run as either a chemostat or turbidostat. A chemostat is a bioreactor to which fresh medium is continuously added, while culture liquid is continuously removed to keep the culture volume constant.

By changing the rate with which medium is added to the bioreactor the growth rate of the micro-organism can be easily controlled.

Steady-state: One of the most important features of chemostats is that micro-organisms can be grown in a physiological steady-state. In steady-state, growth occurs at a constant rate and all culture parameters remain constant (culture volume, dissolved oxygen concentration, nutrient and product concentrations, pH, cell density, etc.). In addition environmental conditions can be controlled by the experimenter. Micro-organisms grown in chemostats naturally strive to steady-state: if a low amount of cells are present in the bioreactor, the cells can grow at growth rates higher than the dilution rate, as growth is not limited by the addition of the limiting nutrient. The limiting nutrient is a nutrient essential for growth, present in the media at a limiting concentration (all other nutrients are usually supplied in surplus). However, if the cell concentration becomes too high, the amount of cells that are removed from the reactor cannot be replenished by growth as the addition of the limiting nutrient is insufficient. This results in an equilibrium situation (steady-state), where the rate of cell growth is equal to the rate of cell removal. Because obtaining a steady-state requires at least 5 volume changes, chemostats require large nutrient and waste reservoirs.

Dilution rate: At steady-state the specific growth rate (μ) of the micro-organism is equal to the dilution rate (D). The dilution rate is defined as the rate of flow of medium over the volume of culture in the bioreactor:

$$D = \frac{\text{Medium flow rate}}{\text{Culture volume}} = \frac{F}{V}$$

Maximal growth rate: Each micro-organism growing on a particular substrate has a maximum specific growth rate (μ_{max}) (the rate of growth observed if none of the nutrients are limiting). If a dilution rate is chosen that is higher than μ_{max}, the culture will not be able to sustain itself in the bioreactor, and will wash out.

Plug flow reactor model: The plug flow reactor (PFR) model is used to describe chemical reactions in continuous, flowing systems. The PFR model is used to predict the behaviour of chemical reactors, so that key reactor variables, such as the dimensions of the reactor, can be estimated. PFRs are also sometimes called continuous tubular reactors (CTRs) as shown in Fig. 12.10.

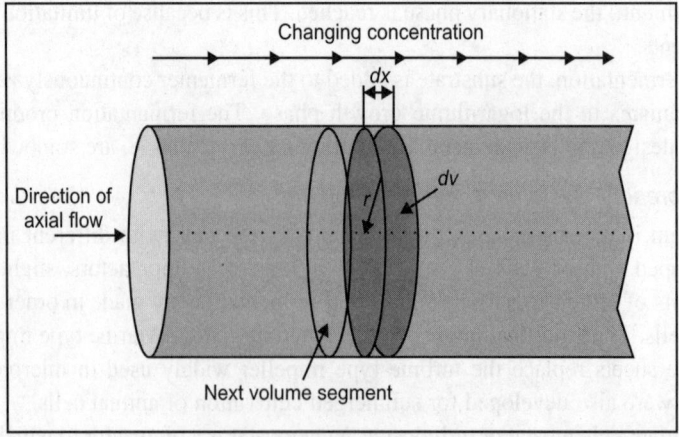

Fig. 12.10. Schematic diagram of a plug flow reactor (PFR).

Fluid going through a PFR may be modelled as flowing through the reactor as a series of infinitely thin coherent 'plugs', each with a uniform composition, travelling in the axial direction of the reactor, with each plug having a different composition from the ones before and after it. The key assumption is that as a plug flows through a PFR, the fluid is perfectly mixed in the radial direction but not in the axial direction (forwards or backwards). Each plug of differential volume is considered as a separate entity, effectively an infinitesimally small batch reactor, limiting to zero volume. As it flows down the tubular PFR, the residence time (τ) of the plug is a function of its position in the reactor. In the ideal PFR, the residence time distribution is therefore a 'Dirac delta' function with a value equal to τ.

PFR modelling: PFRs are frequently referred to as piston flow reactors, or sometimes as continuous tubular reactors. They are governed by ordinary differential equations, the solution for which can be calculated provided the appropriate boundary conditions are known.

The PFR model works well for many fluids, liquids, gases, and slurries: Although turbulent flow and axial diffusion cause a degree of mixing in the axial direction in real reactors, the PFR model is appropriate when these effects are sufficiently small that they can be ignored.

Operation and uses: PFRs are used to model the chemical transformation of compounds as they are transported in systems resembling 'pipes'. The pipe can represent a variety of engineered or natural conduits through which liquids or gases flow (e.g. rivers, pipelines, regions between two mountains, etc.).

An ideal plug flow reactor has a fixed residence time: Any fluid (plug) that enters the reactor at time t will exit the reactor at time $t + \tau$, where τ is the residence time of the reactor. The residence time distribution function is therefore a 'Dirac delta' function at τ. A real plug flow reactor has a residence time distribution that is a narrow pulse around the mean residence time distribution.

A typical plug flow reactor could be a tube packed with some solid material (frequently a catalyst). Typically these types of reactors are called packed bed reactors or PBR's. Sometimes the tube will be a tube in a shell and tube heat exchanger.

Continuous stirred-tank reactor

The continuous stirred-tank reactor (CSTR), also known as vat- or backmix reactor, is a common ideal reactor type in chemical engineering. A CSTR often refers to a model used to estimate the key unit operation variables when using a continuous agitated-tank reactor to reach a specified output. The mathematical model works for all fluids: liquids, gases, and slurries.

Baffle: Baffle may refer to a separator in a shell and tube heat exchanger used to support the tubes and direct fluid flow.

Reynolds number: In fluid mechanics, the Reynolds number (Re) is a dimensionless number that gives a measure of the ratio of inertial forces ($\rho V^2/L$) to viscous forces ($\mu V/L^2$) and consequently quantifies the relative importance of these two types of forces for given flow conditions. Reynolds numbers frequently arise when performing dimensional analysis of fluid dynamics problems, and as such can be used to determine dynamic similitude between different experimental cases. They are also used to characterise different flow regimes, such as laminar or turbulent flow: laminar flow occurs at low Reynolds numbers, where viscous forces are dominant, and is characterised by smooth, constant fluid motion, while turbulent flow occurs at high Reynolds numbers and is dominated by inertial forces, which tend to produce random eddies, vortices and other flow instabilities. Reynolds number can be defined for a number of different situations where a fluid is in relative motion to a surface (the definition of the Reynolds number is not to be confused with the Reynolds Equation or lubrication equation). These definitions generally include the fluid properties of density and viscosity, plus a velocity and a characteristic length or characteristic dimension. This dimension is a matter of convention — for example a radius or diameter are equally valid for spheres or circles, but one is chosen by convention. For aircraft or ships, the length or width can be used. For flow in a pipe or a sphere moving in a fluid the internal diameter is generally used today. Other shapes (such as rectangular pipes or nonspherical objects) have an equivalent diameter defined. For fluids of variable density (e.g. compressible gases) or variable viscosity (non-Newtonian fluids) special rules apply. The velocity may also be a matter of convention in some circumstances, notably stirred vessels.

$$\text{Re} = \frac{\rho V L}{\mu} = \frac{VL}{v} = \frac{QL}{vA}$$

where,

V is the mean fluid velocity (SI units: m/s)

L is a characteristic linear dimension (travelled length of fluid, or hydraulic radius when dealing with river systems) (m)

μ is the dynamic viscosity of the fluid (Pa·s or N·s/m² or kg/m·s)

v is the kinematic viscosity ($v = \mu/\rho$) (m²/s)

ρ is the density of the fluid (kg/m³)

Q is the volumetric flow rate (m³/s)

A is the pipe cross-sectional area (m²).

Note that this is equal to the ratio between $\rho V^2/L$, which is the drag (up to a numerical factor, half the drag coefficient), and $\mu V/L^2$, which is the force due to viscosity (up to a numerical factor depending on the form of the flow).

Power number: The power number N_p (also known as Newton number) is a commonly-used dimensionless number relating the resistance force to the inertia force. The power-number has different specifications according to the field of application, e.g. for stirrers the power number is defined as:

$$N_p = \frac{P}{\rho n^3 d^5}$$

where,

P = power

ρ = fluid density

n = rotational speed

d = diameter of stirrer.

Gas exchange and mass transfer

Mass transfer is the transfer of mass from high concentration to low concentration. The phrase is commonly used in engineering for physical processes that involve molecular and convective transport of atoms and molecules within physical systems. Mass transfer includes both fluid flow and separation unit operations.

Some common examples of mass transfer processes are the evaporation of water from a pond to the atmosphere; the diffusion of chemical impurities in lakes, rivers, and oceans from natural or artificial point sources; mass transfer is also responsible for the separation of components in an apparatus such as a distillation column. In HVAC examples of a heat and mass exchangers are cooling towers and evaporative coolers where evaporation of water cools that portion which remains as a liquid, as well as cooling and humidifying the air passing through. The driving force for mass transfer is a difference in concentration; the random motion of molecules causes a net transfer of mass from an area of high concentration to an area of low concentration. The amount of mass transfer can be quantified through the calculation and application of mass transfer coefficients. Mass transfer finds extensive application in chemical engineering problems, where material balance on components is performed.

Henry's law: Henry's law states that—'At a constant temperature, the amount of a given gas dissolved in a given type and volume of liquid is directly proportional to the partial pressure of that gas in equilibrium with that liquid.' An equivalent way of stating the law is that the solubility of a gas in a liquid at a particular temperature is proportional to the pressure of that gas above the liquid. Henry's law has since been shown to apply for a wide range of dilute solutions, not merely those of gases. An everyday example of Henry's law is given by carbonated soft drinks. Before the bottle or can is opened, the gas above the drink is almost pure carbon dioxide at a pressure slightly higher than atmospheric pressure. The drink itself contains dissolved carbon dioxide. When the bottle or can is opened, some of this gas escapes, giving the characteristic hiss (or pop in the case of a champagne bottle). Because the pressure above the liquid is now lower, some of the dissolved carbon dioxide comes out of solution as bubbles. If a glass of the drink is left in the open, the concentration of carbon dioxide in solution will come into equilibrium with the carbon dioxide in the air, and the drink will go flat.

Formula and the Henry's law constant: Henry's law can be put into mathematical terms (at constant temperature) as:

$$p = k_H c$$

where, p is the partial pressure of the solute in the gas above the solution, c is the concentration of the solute and k_H is a constant with the dimensions of pressure divided by concentration. The constant, known as the Henry's law constant, depends on the solute, the solvent and the temperature.

Some values for k_H for gases dissolved in water at 298 kelvins include:

1. Oxygen (O_2): 769.2 L·atm/mol.
2. Carbon dioxide (CO_2): 29.4 L·atm/mol.
3. Hydrogen (H_2): 1282.1 L·atm/mol.

There are other forms of Henry's law, each of which defines the constant k_H differently and requires different dimensional units. In particular, the 'concentration' of the solute in solution may also be expressed as a mole fraction or as a molality.

Oxygen transfer in bioreactors

Oxygen is needed by cells for respiration. Oxygen used by cells in suspension must be available as dissolved oxygen. Since oxygen solubility is quite small, about 6 to 7 mg/l under normal cultivation conditions, metabolic oxygen requirement is supplied on needed basis by continuous aeration of culture medium.

Actively respiring yeast requires about 0.15 g O^2 (g cell)$^{-1}$ hr. At a cell concentration of 10 g l^{-1}, medium saturated with air can support less than 30 seconds worth of metabolic oxygen. That is, a continuous supply of oxygen must be maintained in any viable aerobic manufacturing process. In this section, we will first get a quantitative appreciation for metabolic oxygen demand, followed by methods used in calculating rates at which oxygen is transferred from sparged air. We will then examine methods useful in characterising oxygen mass transfer coefficient. Finally we will evaluate bioreactor operation and design based on oxygen transfer capability.

Metabolic oxygen demand

Metabolic oxygen demand of an organism depends on the biochemical nature of the cell and cultivation conditions. Oxygen need is usually satisfied in most cells if the dissolved oxygen concentration in the medium is kept at about 1 mg/l. If the oxygen level is allowed to fall far below this value, oxygen consumption rate decreases with concomitant decrease in biochemical energy production, and as a result cell growth rate also decreases.

Volumetric oxygen mass transfer coefficient

In a typical aeration system, oxygen from the air bubble is transferred through the gas-liquid interface followed by liquid phase diffusion/bulk transport to the cells. Although this is a multi-step serial transport, in a well dispersed systems, the major resistance to oxygen transfer is in the liquid film surrounding the gas bubble.

Bioreactor oxygen balance

Let us now consider the case of oxygen balance within a bioreactor in which cells are growing and in the process consuming oxygen. There is a continuous inflow of air at a constant volumetric flow rate. The liquid broth is agitated by a Rushton agitator (flat blade stirrer). Let the metabolic oxygen uptake rate be q_{O_2} and cell concentration is X. Let us examine the reactor system over a sufficiently short period that we can treat X as a constant. Consider oxygen balance over the liquid phase of the bioreactor.

O_2 transferred from gas phase – O_2 consumed by cells = Accumulation

$$\left[k_{La} \left(C_{DO}^* - C_{DO} \right) \right] V - q_{O_2} XV = \frac{d(V\, C_{DO})}{dt} \qquad \text{... (12.22)}$$

For constant liquid phase volume, the above can be simplified to:

$$\frac{d(C_{DO})}{dt} = k_{La} \left(C_{DO}^* - C_{DO} \right) - q_{O_2} X \qquad \text{... (12.23)}$$

The concentration, C_{DO} is readily measured using a dissolved oxygen electrode. A later segment of the course on biosensors, will deal with principle of measurement and construction of DO electrodes. If oxygen being supplied is in exact balance with the oxygen consumed by the cells, we expect the dissolved oxygen concentration to remain constant; that is, the derivative in Eq. 12.23 will vanish. That is,

$$q_{O_2} X = k_{La} \left(C_{DO}^* - C_{DO} \right)$$

One useful application of the above is in estimating the maximum cell concentration a particular bioreactor is capable of supporting in terms of oxygen supply.

Factors affecting k_{La}

The mass transfer coefficient is strongly affected by agitation speed and air flow rate. In general,

$$k_{La} = k\, (P_Q/V_R)^{0.4}\, (V_S)^{0.5}\, (N)^{0.5}$$

where, k is a constant, P_Q is the power required for aerated bioreactor, V_R is the bioreactor volume, V_S is air flow rate, N is agitator speed. Note that the mass transfer coefficient increases with agitation speed and air flow rate.

Measurement of k_{La}

Most common method of measuring k_{La} is to conduct experiments in the bioreactor when cells are absent, or cell concentration is low so that consumption by cells can be neglected.

The latter condition is present immediately after inoculating the bioreactor. Consider Eq. 12.23 under these conditions:

$$\frac{d(C_{DO})}{dt} = k_{La}\left(C^*_{DO} - C_{DO}\right)$$

If we allow steady-state to occur, the dissolved oxygen concentration will reach saturation value, C^*_{DO} and the concentration-time profile will be flat, as shown in the Fig. 12.11.

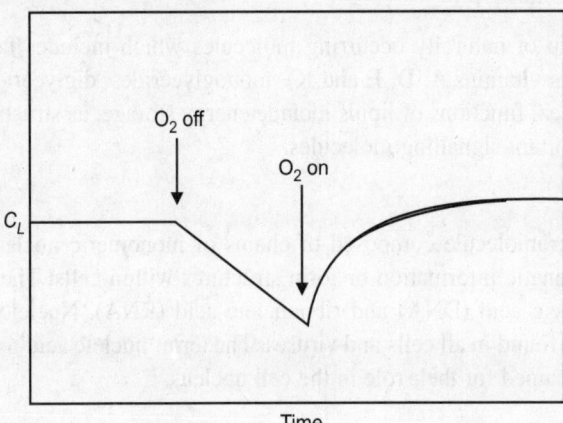

Fig. 12.11. Oxygen profile during a transient. The responses will be exponential, rather than straight lines.

If the oxygen source (air) is replaced by nitrogen, the resulting response of the system is described by the above equation with the term, C^*_{DO} set to zero. That is,

$$\frac{d(C_{DO})}{dt} = k_{La}\left(0 - C_{DO}\right) \text{ and } C_{DO}(t=0) = C^*_{DO}$$

The solution to the above is:

$$C_{DO} = C^*_{DO} \text{ Exp }(-k_{La}t)$$

If one plots the response on a semi-log plot, the slope will equal to the negative of mass transfer coefficient. It is relatively a simple experiment and the data analysis is also easy to do. When other type of transient mass transfer experiments are conducted, the above equations should be suitably modified. For example for the case of nitrogen to air switch, we should suitably modify the solution because the initial condition is now different.

Biological Molecules

Carbohydrate

A carbohydrate is an organic compound with the general formula $C_m(H_2O)_n$, that is, consists only of carbon, hydrogen and oxygen, with the last two in the 2:1 atom ratio. Carbohydrates can be viewed as hydrates of carbon, hence their name. Structurally however, it is more accurate to view them as polyhydroxy aldehydes and ketones.

In food science and in many informal contexts, the term carbohydrate often means any food that is particularly rich in starch (such as cereals, bread and pasta) or sugar (such as candy, jams and desserts).

Protein

Proteins (also known as polypeptides) are organic compounds made of amino acids arranged in a linear chain and folded into a globular form. The amino acids in a polymer are joined together by the peptide bonds between the carboxyl and amino groups of adjacent amino acid residues.

Like other biological macromolecules such as polysaccharides and nucleic acids, proteins are essential parts of organisms and participate in virtually every process within cells.

Lipid

Lipids are a broad group of naturally occurring molecules which includes fats, waxes, sterols, fat-soluble vitamins (such as vitamins A, D, E and K), monoglycerides, diglycerides, phospholipids, and others. The main biological functions of lipids include energy storage, as structural components of cell membranes, and as important signalling molecules.

Nucleic acid

A nucleic acid is a macromolecule composed of chains of monomeric nucleotides. In biochemistry these molecules carry genetic information or form structures within cells. The most common nucleic acids are deoxyribonucleic acid (DNA) and ribonucleic acid (RNA). Nucleic acids are universal in living things, as they are found in all cells and viruses. The term 'nucleic acid' is the generic name for a family of biopolymers, named for their role in the cell nucleus.

Fat

Fats consist of a wide group of compounds that are generally soluble in organic solvents and largely insoluble in water. Chemically, fats are generally triesters of glycerol and fatty acids. Fats may be either solid or liquid at room temperature, depending on their structure and composition.

Fats form a category of lipid, distinguished from other lipids by their chemical structure and physical properties. This category of molecules is important for many forms of life, serving both structural and metabolic functions. They are an important part of the diet of most heterotrophs (including humans). Fats or lipids are broken down in the body by enzymes called lipases produced in the pancreas.

Phosphate

A phosphate, an inorganic chemical, is a salt of phosphoric acid. In organic chemistry, a phosphate, or organophosphate, is an ester of phosphoric acid. Organic phosphates are important in biochemistry and biogeochemistry or ecology. Inorganic phosphates are mined to obtain phosphorus for use in agriculture and industry. At elevated temperatures in the solid state, phosphates can condense to form pyrophosphates.

CELLS

The cell is the functional basic unit of life. It was discovered by Robert Hooke and is the functional unit of all known living organisms. It is the smallest unit of life that is classified as a living thing, and is often called the building block of life. Some organisms, such as most bacteria, are unicellular (consist of a single cell). Other organisms, such as humans, are multicellular. (Humans have about 100 trillion or 10^{14} cells; a typical cell size is 10 µm; a typical cell mass is 1 nanogram.) The largest cells are about 135 µm in the anterior horn in the spinal cord while granule cells in the cerebellum, the smallest, can be some 4 µm and the longest cell can reach from the toe to the lower brain stem (Pseudounipolar cells). The largest known cells are unfertilised ostrich egg cells which weigh 3.3 pounds.

Components of Cell

All cells, whether prokaryotic or eukaryotic, have a membrane that envelops the cell, separates its interior from its environment, regulates what moves in and out (selectively permeable), and maintains the electric potential of the cell. Inside the membrane, a salty cytoplasm takes up most of the cell volume. All cells possess DNA, the hereditary material of genes, and RNA, containing the information necessary to build various proteins such as enzymes, the cell's primary machinery. There are also other kinds of biomolecules in cells.

Cell membrane: A cell's defining boundary

The cytoplasm of a cell is surrounded by a cell membrane or plasma membrane. The plasma membrane in plants and prokaryotes is usually covered by a cell wall. This membrane serves to separate and protect a cell from its surrounding environment and is made mostly from a double layer of lipids (hydrophobic fat-like molecules) and hydrophilic phosphorus molecules. Hence, the layer is called a phospholipid bilayer. It may also be called a fluid mosaic membrane.

Cytoskeleton: A cell's scaffold

The cytoskeleton acts to organise and maintain the cell's shape; anchors organelles in place; helps during endocytosis, the uptake of external materials by a cell, and cytokinesis, the separation of daughter cells after cell division; and moves parts of the cell in processes of growth and mobility. The eukaryotic cytoskeleton is composed of microfilaments, intermediate filaments and microtubules.

Genetic material

Two different kinds of genetic material exist: deoxyribonucleic acid (DNA) and ribonucleic acid (RNA). Most organisms use DNA for their long-term information storage, but some viruses (e.g. retroviruses) have RNA as their genetic material. The biological information contained in an organism is encoded in its DNA or RNA sequence. RNA is also used for information transport (e.g. mRNA) and enzymatic functions (e.g. ribosomal RNA) in organisms that use DNA for the genetic code itself. Transfer RNA (tRNA) molecules are used to add amino acids during protein translation.

Organelles

The human body contains many different organs, such as the heart, lung, and kidney, with each organ performing a different function. Cells also have a set of 'little organs', called organelles, that are adapted and/or specialised for carrying out one or more vital functions. There are several types of organelles within an animal cell. Some (such as the nucleus and golgi apparatus) are typically solitary, while others (such as mitochondria, peroxisomes and lysosomes) can be numerous (hundreds to thousands). The cytosol is the gelatinous fluid that fills the cell and surrounds the organelles.

Mitochondria and chloroplasts

Mitochondria are self-replicating organelles that occur in various numbers, shapes, and sizes in the cytoplasm of all eukaryotic cells. Mitochondria play a critical role in generating energy in the eukaryotic cell.

Ribosomes

Ribosome is a large complex of RNA and protein molecules. They each consist of two subunits, and act as an assembly line where RNA from the nucleus is used to synthesise proteins from amino acids.

Cell nucleus

The cell nucleus is the most conspicuous organelle found in a eukaryotic cell. It houses the cell's chromosomes, and is the place where almost all DNA replication and RNA synthesis (transcription) occur.

Endoplasmic reticulum

The endoplasmic reticulum (ER) is the transport network for molecules targeted for certain modifications and specific destinations, as compared to molecules that will float freely in the cytoplasm.

Golgi apparatus

The primary function of the Golgi apparatus is to process and package the macromolecules such as proteins and lipids that are synthesised by the cell. It is particularly important in the processing of proteins for secretion. The Golgi apparatus forms a part of the endomembrane system of eukaryotic cells.

Lysosomes and peroxisomes

Lysosomes contain digestive enzymes (acid hydrolases). They digest excess or worn-out organelles, food particles, and engulfed viruses or bacteria. Peroxisomes have enzymes that rid the cell of toxic peroxides.

Centrosome

The centrosome produces the microtubules of a cell — a key component of the cytoskeleton. It directs the transport through the ER and the Golgi apparatus.

Vacuoles

Vacuoles store food and waste. Some vacuoles store extra water. They are often described as liquid filled space and are surrounded by a membrane.

Cell Functions

Cell growth and metabolism

Between successive cell divisions, cells grow through the functioning of cellular metabolism. Cell metabolism is the process by which individual cells process nutrient molecules. Metabolism has two distinct divisions: catabolism, in which the cell breaks down complex molecules to produce energy and reducing power, and anabolism, in which the cell uses energy and reducing power to construct complex molecules and perform other biological functions.

Creation of new cells

Cell division involves a single cell (called a mother cell) dividing into two daughter cells. This leads to growth in multicellular organisms (the growth of tissue) and to procreation (vegetative reproduction) in unicellular organisms.

Prokaryotic cells divide by binary fission. Eukaryotic cells usually undergo a process of nuclear division, called mitosis, followed by division of the cell, called cytokinesis.

Protein synthesis

Cells are capable of synthesising new proteins, which are essential for the modulation and maintenance of cellular activities. This process involves the formation of new protein molecules from amino acid building blocks based on information encoded in DNA/RNA. Protein synthesis generally consists of two major steps: transcription and translation.

BIOENERGY AND METABOLIC PATHWAYS

Bioenergy is renewable energy made available from materials derived from biological sources. In its most narrow sense it is a synonym to biofuel, which is fuel derived from biological sources. In its broader sense it includes biomass, the biological material used as a biofuel, as well as the social, economic, scientific and technical fields associated with using biological sources for energy. This is a common misconception, as bioenergy is the energy extracted from the biomass, as the biomass is the fuel and the bioenergy is the energy contained in the fuel.

Biomass is any organic material which has stored sunlight in the form of chemical energy. As a fuel it may include wood, wood waste, straw, manure, sugar cane, and many other by-products from a variety of agricultural processes.

Solid Biomass

Biomass is material derived from recently living organisms, which includes plants, animals and their by-products. Manure, garden waste and crop residues are all sources of biomass. It is a renewable energy source based on the carbon cycle, unlike other natural resources such as petroleum, coal, and nuclear fuels. Animal waste is a persistent and unavoidable pollutant produced primarily by the animals housed in industrial-sized farms.

There are also agricultural products being grown for biofuel production. These include corn, and soyabeans and to some extend willow and switchgrass on a pre-commercial research level, primarily in the United States; rapeseed, wheat, sugar beet, and willow (15,000 ha in Sweden) primarily in Europe; sugar cane in Brazil; palm oil and miscanthus in Southeast Asia; sorghum and cassava in China; and jatropha in India. Hemp has also been proven to work as a biofuel. Biodegradable outputs from industry, agriculture, forestry and households can be used for biofuel production, using, e.g. anaerobic digestion to produce biogas, gasification to produce syngas or by direct combustion. Examples of biodegradable wastes include straw, timber, manure, rice husks, sewage, and food waste. The use of biomass fuels can therefore contribute to waste management as well as fuel security and help to prevent or slow down climate change, although alone they are not a comprehensive solution to these problems.

Metabolic Pathway

In biochemistry, metabolic pathways are series of chemical reactions occurring within a cell. In each pathway, a principal chemical is modified by chemical reactions. Enzymes catalyse these reactions, and often require dietary minerals, vitamins, and other cofactors in order to function properly. Because of the many chemicals that may be involved, pathways can be quite elaborate. In addition, many pathways can exist within a cell. This collection of pathways is called the metabolic network. Pathways are important to the maintenance of homoeostasis within an organism.

Metabolism is a step-by-step modification of the initial molecule to shape it into another product. The result can be used in one of three ways:

1. To be stored by the cell.
2. To be used immediately, as a metabolic product.
3. To initiate another metabolic pathway, called a flux generating step.

A molecule called a substrate enters a metabolic pathway depending on the needs of the cell and the availability of the substrate. An increase in concentration of anabolic and catabolic end-products would slow the metabolic rate for that particular pathway.

Each metabolic pathway is composed of a series of biochemical reactions that are connected by their intermediates: The reactants (or substrates) of one reaction are the products of the previous one, and so on. Metabolic pathways are usually considered in one direction (although all reactions are chemically reversible, conditions in the cell are such that it is thermodynamically more favourable for flux to be in one of the directions).

1. Glycolysis was the first metabolic pathway discovered:
    (a) As glucose enters a cell, it is immediately phosphorylated by ATP to glucose 6-phosphate in the irreversible first step. This is to prevent the glucose from leaving the cell.
    (b) In times of excess lipid or protein energy sources, glycolysis may run in reverse (gluconeogenesis) in order to produce glucose 6-phosphate for storage as glycogen or starch.
2. Metabolic pathways are often regulated by feedback inhibition, or by a cycle wherein one of the products in the cycle starts the reaction again, such as the Krebs cycle.
3. Anabolic and catabolic pathways in eukaryotes are separated either by compartmentation or by the use of different enzymes and cofactors.

Oxygen Transport

The oxygen that we obligate aerobes need for survival is transported from the lungs to peripheral tissues by the haemoglobin that is densely packed in our red blood cells (erythrocytes). Haemoglobin is the most intensively studied protein in the world, and its structure is known in intimate detail. It is made up of four protein subunits. Nestled deep in each of these protein's folds, is a planar structure called a porphyrin, which binds in its centre a single atom of iron, most commonly in the 2+ valency state. The iron-porphyrin group is called heme. In the lung, oxygen diffuses across the alveolar membrane, and then the red cell membrane in lung capillaries. When it encounters a molecule of haemoglobin, it wedges itself between the iron atom and a nitrogen attached to the globin chain, which helps to hold the heme group in place in the protein.

One molecule of haemoglobin with its four heme groups is capable of binding four molecules of diatomic oxygen, O_2. The loaded pigment is called oxyhaemoglobin, and it is a brilliant red colour as in arterial blood. Pressure from dissolved oxygen in plasma and in the surroundings in the red cell helps to keep the oxygen on its binding site.

As the blood circulates to the periphery, the small amount of dissolved oxygen is consumed first by cells in organs and tissues. This release in pressure makes available the much larger reservoir of heme-bound oxygen, which begins a sequential unloading of its four oxygen molecules. At the most, under normal circumstances only 3 molecules of oxygen are unloaded. Partially or fully unloaded haemoglobin is called deoxyhaemoglobin. It is a dark blue to purplish colour as in venous blood.

During oxygen unloading, the haemoglobin tetramer undergoes subtle intramolecular conformational changes called cooperativity. As a result of cooperativity, once the first oxygen has been unloaded, the unloading of the second oxygen is facilitated. The second oxygen can dissociate after a much smaller change in oxygen pressure than was needed to unload the first. Another conformational change facilitates dissociation of the third oxygen. Cooperativity is an important phenomenon that permits the loading and unloading of large amounts of oxygen at physiologically relevant oxygen pressures. Chemicals that interfere with oxygen transport and/or cooperativity are potentially lethal.

Photosynthesis

Photosynthesis is a process that converts carbon dioxide into organic compounds, especially sugars, using the energy from sunlight. Photosynthesis occurs in plants, algae, and many species of bacteria, but not in archaea. Photosynthetic organisms are called photoautotrophs, since they can create their own food. In plants, algae, and cyanobacteria, photosynthesis uses carbon dioxide and water, releasing oxygen as a waste product. Photosynthesis is vital for life on earth. As well as maintaining the normal level of oxygen in the atmosphere, nearly all life either depends on it directly as a source of energy, or indirectly as the ultimate source of the energy in their food (the exceptions are chemoautotrophs that live in rocks or around deep sea hydrothermal vents). The amount of energy trapped by photosynthesis is immense, approximately 100 terawatts: which is about six times larger than the power consumption of human civilisation (Fig. 12.12).

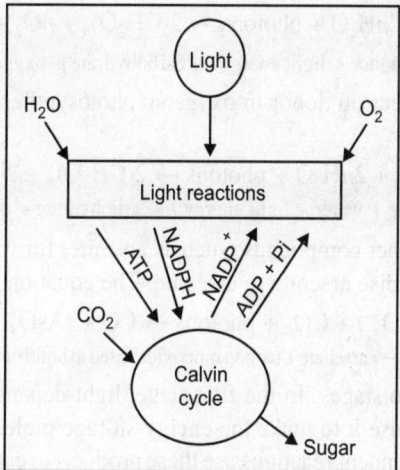

Fig. 12.12. Photosynthesis changes the energy from the sun into chemical energy and splits water to liberate O_2 and fixes CO_2 into sugar.

Although photosynthesis can happen in different ways in different species, some features are always the same. For example, the process always begins when energy from light is absorbed by proteins called photosynthetic reaction centres that contain chlorophylls. In plants, these proteins are held inside organelles called chloroplasts, while in bacteria they are embedded in the plasma membrane. Some of the light energy gathered by chlorophylls is stored in the form of adenosine triphosphate (ATP). The rest of the energy is used to remove electrons from a substance such as water. These electrons are then used in the reactions that turn carbon dioxide into organic compounds.

In plants, algae and cyanobacteria, this is done by a sequence of reactions called the Calvin cycle, but different sets of reactions are found in some bacteria, such as the reverse Krebs cycle in *Chlorobium*. Many photosynthetic organisms have adaptations that concentrate or store carbon dioxide. This helps reduce a wasteful process called photorespiration that can consume part of the sugar produced during photosynthesis.

Photosynthetic organisms are photoautotrophs, which means that they are able to synthesise food directly from carbon dioxide using energy from light. However, not all organisms that use light as a source of energy carry out photosynthesis, since photoheterotrophs use organic compounds, rather than carbon dioxide, as a source of carbon. In plants, algae and cyanobacteria, photosynthesis releases oxygen. This is called oxygenic photosynthesis. Although there are some differences between oxygenic photosynthesis in plants, algae and cyanobacteria, the overall process is quite similar in these organisms. However, there are some types of bacteria that carry out anoxygenic photosynthesis, which consumes carbon dioxide but does not release oxygen.

Carbon dioxide is converted into sugars in a process called carbon fixation. Carbon fixation is a redox reaction, so photosynthesis needs to supply both a source of energy to drive this process, and the electrons needed to convert carbon dioxide into carbohydrate, which is a reduction reaction. In general outline, photosynthesis is the opposite of cellular respiration, where glucose and other compounds are oxidised to produce carbon dioxide, water, and release chemical energy. However, the two processes take place through a different sequence of chemical reactions and in different cellular compartments.

The general equation for photosynthesis is therefore:

$$2nCO_2 + 2nH_2O + \text{photons} \rightarrow 2(CH_2O)_n + nO_2 + 2nA$$

Carbon dioxide + electron donor + light energy → carbohydrate + oxygen + oxidised electron donor

Since water is used as the electron donor in oxygenic photosynthesis, the equation for this process is:

$$2nCO_2 + 2nH_2O + \text{photons} \rightarrow 2(CH_2O)_n + 2nO_2$$

Carbon dioxide + water + light energy → carbohydrate + oxygen

Other processes substitute other compounds (such as arsenite) for water in the electron-supply role; the microbes use sunlight to oxidise arsenite to arsenate: The equation for this reaction is:

$$(AsO_3^{3-}) + CO_2 + \text{photons} \rightarrow CO + (AsO_4^{3-})$$

Carbon dioxide + arsenite + light energy → arsenate + carbon monoxide (used to build other compounds in subsequent reactions)

Photosynthesis occurs in two stages. In the first stage, light-dependent reactions or light reactions capture the energy of light and use it to make the energy-storage molecules ATP and NADPH. During the second stage, the light-independent reactions use these products to capture and reduce carbon dioxide.

Genetic Code

The genetic code is the set of rules by which information encoded in genetic material (DNA or mRNA sequences) is translated into proteins (amino acid sequences) by living cells. The code defines a mapping between trinucleotide sequences, called codons, and amino acids. With some exceptions, a triplet codon in a nucleic acid sequence specifies a single amino acid. Because the vast majority of genes are encoded with exactly the same code, this particular code is often referred to as the canonical or standard genetic code, or simply the genetic code, though in fact there are many variant codes. For example, protein synthesis in human mitochondria relies on a genetic code that differs from the standard genetic code.

Not all genetic information is stored using the genetic code. All organisms' DNA contains regulatory sequences, intergenic segments, and chromosomal structural areas that can contribute greatly to phenotype. Those elements operate under sets of rules that are distinct from the codon-to-amino acid paradigm underlying the genetic code.

POLYMERASE CHAIN REACTION

The polymerase chain reaction (PCR) is a technique in molecular biology to amplify a single or few copies of a piece of DNA across several orders of magnitude, generating thousands to millions of copies of a particular DNA sequence. The method relies on thermal cycling, consisting of cycles of repeated heating and cooling of the reaction for DNA melting and enzymatic replication of the DNA. Primers (short DNA fragments) containing sequences complementary to the target region along with a DNA polymerase (after which the method is named) are key components to enable selective and repeated amplification.

As PCR progresses, the DNA generated is itself used as a template for replication, setting in motion a chain reaction in which the DNA template is exponentially amplified. PCR can be extensively modified to perform a wide array of genetic manipulations.

Almost all PCR applications employ a heat-stable DNA polymerase, such as Taq polymerase, an enzyme originally isolated from the bacterium *Thermus aquaticus*. This DNA polymerase enzymatically assembles a new DNA strand from DNA building blocks, the nucleotides, by using single-stranded DNA as a template and DNA oligonucleotides (also called DNA primers), which are required for initiation of DNA synthesis. The vast majority of PCR methods use thermal cycling, i.e. alternately heating and cooling the PCR sample to a defined series of temperature steps. These thermal cycling steps are necessary first to physically separate the two strands in a DNA double helix at a high temperature in a process called DNA melting. At a lower temperature, each strand is then used as the template in DNA synthesis by the DNA polymerase to selectively amplify the target DNA. The selectivity of PCR results from the use of primers that are complementary to the DNA region targeted for amplification under specific thermal cycling conditions.

PCR is used to amplify a specific region of a DNA strand (the DNA target). Most PCR methods typically amplify DNA fragments of up to ~10 kilo base pairs (kb), although some techniques allow for amplification of fragments up to 40 kb in size.

The PCR is commonly carried out in a reaction volume of 10–200 μl in small reaction tubes (0.2–0.5 ml volumes) in a thermal cycler. The thermal cycler heats and cools the reaction tubes to achieve the temperatures required at each step of the reaction. Many modern thermal cyclers make use of the Peltier effect which permits both heating and cooling of the block holding the PCR tubes simply by reversing the electric current.

Thin-walled reaction tubes permit favourable thermal conductivity to allow for rapid thermal equilibration. Most thermal cyclers have heated lids to prevent condensation at the top of the reaction tube. Older thermocyclers lacking a heated lid require a layer of oil on top of the reaction mixture or a ball of wax inside the tube.

Steady-state Mass Transfer in Biological Systems

The transfer of a particular chemical species from one phase to another and the movement of such species along gradients within phases are fundamentally important processes in any system involving more than one chemical component. Forces that can promote such transfer include: free energy, pressure, electrical charge, temperature, and concentration. These driving forces are physical quantities of importance not only in biological systems but also in other systems encountered in engineering practice. The transport of chemical species as a result of such driving forces is commonly called mass transfer. Mass transfer subdivides into molecular mass transfer and convective mass transfer. The term diffusion is often used for the former. Diffusion deals with random molecular migration of matter through a medium, whereas convective mass transfer involves the migration of matter into a moving fluid or a stream of gas.

For example, if an open container of acetic acid is placed in one corner of a room, the odour will soon be detected throughout the entire room. In this case, there is a combination of molecular diffusion and convection. If a lump of sugar is placed in a glass of water, the sugar will eventually dissolve and spread in all parts of the volume without stirring. In this case convection will play a minor role. In solids, there can be no convection and all movements are by molecular diffusion. The general equation for all types of mass transfer is:

$$\text{Mass transfer rate} = \text{Drying force/Resistance} \qquad \qquad ... (12.24)$$

where the resistance is a function of the properties of the medium through which the matter is transferred. In practice, mass transfer always begins with an unsteady-state condition. Steady-state conditions may be established only after some time has elapsed. In steady-state conditions, the driving force (e.g. concentration), the resistance, and the transfer rate are constant over time. In unsteady-state conditions, all these properties change with time.

Chapter 13

Environmental Reactions Engineering

INTRODUCTION

The natural environment, encompasses all living and nonliving things occurring naturally on earth or some region thereof. The concept of the natural environment can be distinguished by components:

1. Complete ecological units that function as natural systems without massive human intervention, including all vegetation, micro-organisms, soil, rocks, atmosphere and natural phenomena that occur within their boundaries.
2. Universal natural resources and physical phenomena that lack clear-cut boundaries, such as air, water, and climate, as well as energy, radiation, electric charge, and magnetism, not originating from human activity.

The natural environment is contrasted with the built environment, which comprises the areas and components that are strongly influenced by humans.

Over-consumption, waste and pollution are a few of the rapidly growing environmental problems facing our society. Yet what happens to all of our waste? How come we are lucky enough to live in a country that provides safe drinking water? These are issues that environmental chemical engineers address. We take for granted such 'luxuries' because we have grown accustomed to these products of chemical and environmental engineering.

Environmental chemical engineers design systems, processes and equipment for air, water and soil quality control, solid waste disposal, and the remediation of contaminated soil, air and water. They develop strategies to reduce pollution at the source and treat wastes that cannot be eliminated. Applying chemistry theories, they calculate the impact of human activity on the environment and seek to design methods of environmental sustainability, conservation and protective efforts and reparations if necessary.

Environmental chemical engineers provide economical answers to clean up yesterday's waste and prevent tomorrow's pollution. Catalytic converters, reformulated gasoline and smoke stack scrubbers all help keep the world cleaner. Additionally, environmental chemical engineers help reduce the strain on natural materials through synthetic replacements, more efficient processing, and new recycling technologies. It is common for environmental chemical engineers to work with environmental scientists, planners, hazardous waste management technicians and other engineering specialists as well as lawyers and bankers.

Environmental chemical engineers work on both large and small scales in terms of the issues they work with. For instance some conduct hazardous-waste management studies and design municipal sewage systems while others deal with worldwide issues such as minimising the effects of acid rain, global warming, automobile emissions, and the protection of wildlife. Due to the growing number of environmental problems that seem to arise each day, the need for environmental chemical engineers is growing and new challenges present themselves to these professionals.

Environmental chemical engineers create plans on computers that test and predict possible environmental problems and in their research, they generate solutions. Most environmental chemical engineers travel to laboratories and sites to see their work in progress (if it is a structural project) and otherwise spend their days researching and formulating new ideas on environmental sustainability. They evaluate each project to find the most cost-effective solutions to problems while still maintaining recognised engineering and governmental standards.

Environmental chemical engineers are required to constantly update their skills and knowledge in order to keep up with technological advancements in this quickly changing field.

Environmental chemical engineers are consciously concerned about the decline of our environment therefore seeking ways to improve and promote a safe and clean environment. They should be knowledgeable about the implications of environmental legislation and the effects of human consumption on our environment. They should keep updated on a new technologies and changes in the environmental climate.

Environmental chemical engineers must be safety conscious and practical in decision-making. They possess good communication skills because they work closely with architects, lawyers and environmental activists. Environmental chemical engineers can analyse data, review calculations and prepare cost estimates and have the ability to visualise three-dimensional objects from two-dimensional drawings. They must be passionately dedicated to their projects, be creative in their designs and be as knowledgeable as possible in both the chemical engineering and environmental fields. Finally, they should enjoy being innovative, doing work that requires precision and making solid decisions.

In this chapter a wide array of environmental issues are introduced, and their impacts are related to chemical production and use. The pertinent chemicals and the environmental reactions of those chemicals are discussed. For many environmental problems, the chemicals causing the adverse environmental or health impacts are not the same chemical originally emitted from the production process or from the use of a chemical. Thus, the environment is a complex system with a large number of transport and transformation processes occurring simultaneously. With a basic understanding of environmental issues, the chemical engineer will be able to spot environmental problems earlier and will contribute to the solution of those problems by improving the environmental performance of chemical processes and products. The challenge for future generations of chemical engineers is to develop and master the technical tools and approaches that will integrate environmental objectives into design decisions. With each environmental problem introduced, the chemicals or classes of chemicals implicated in that problem are identified. Whenever possible, the chemical reactions or other mechanisms responsible for the chemical's impact are explained.

These issues are not only of concern to the general public, but are challenging problems for the chemical industry and for chemical engineers. When considering the potential impact of any human activity on the environment, it is useful to regard the environment as a system containing interrelated sub-processes. The environment functions as a sink for the wastes released as a result of human activities. The various subsystems of the environment act upon these wastes, generally rendering them less harmful by converting them into chemical forms that can be assimilated into natural systems. It is essential to understand these natural waste conversion processes so that the capacity of these natural systems is not exceeded by the rate of waste generation and release.

The impact of waste releases on the environment can be global, regional, or local in scope. On a global scale, man-made (anthropogenic) greenhouse gases, such as methane and carbon dioxide, are implicated in global warming and climate change. Hydrocarbons released into the air, in combination with nitrogen oxides originating from combustion processes, can lead to air quality degradation over urban areas and extend for hundreds of kilometres. Chemicals disposed of in the soil can leach into underground water and reach groundwater sources, having their primary impact locally, near to the point of release.

GLOBAL ENVIRONMENTAL ISSUES

Global Energy Issues

The availability of adequate energy resources is necessary for most economic activity and makes possible the high standard of living that developed societies enjoy. Although energy resources are widely available, some such as oil and coal are non-renewable, and others, such as solar, although inexhaustible, are not currently cost effective for most applications. An understanding of global energy usage patterns, energy conservation, and the environmental impacts associated with the production and use of energy are therefore very important. Often, primary energy sources such as fossil fuels must be converted into another form such as heat or electricity.

The global use of energy has steadily risen since the dawn of the industrial revolution. Currently, fossil fuels make up roughly 85 per cent of the world's energy consumption, while renewable sources such as hydroelectric, solar, and wind power account for only about 8 per cent of the power usage. Nuclear power provides roughly 6 per cent of the world energy demand, and its contribution varies from country to country.

Another interesting aspect of energy consumption by industrialised countries and the developing world is the trend in energy efficiency, the energy consumed per unit of economic output. Future chemical engineers will need to recognise the importance of energy efficiency in process design.

Many environmental effects are associated with energy consumption. Fossil fuel combustion releases large quantities of carbon dioxide into the atmosphere. During its long residence time in the atmosphere, CO_2 readily absorbs infrared radiation contributing to global warming. Further, combustion processes release oxides of nitrogen and sulphur oxide into the air where photochemical and/or chemical reactions can convert them into ground level ozone and acid rain. Hydropower energy generation requires widespread land inundation, habitat destruction, alteration in surface and groundwater flows, and decreases the acreage of land available for agricultural use. Nuclear power has environmental problems linked to uranium mining and spent nuclear rod disposal. 'Renewable fuels' are not benign either. Traditional energy usage (wood) has caused widespread deforestation in localised regions of developing countries. Solar power panels require energy-intensive use of heavy metals and creation of metal wastes. Satisfying future energy demands must occur with a full understanding of competing environmental and energy needs.

Global Warming

The atmosphere allows solar radiation from the sun to pass through without significant absorption of energy. Some of the solar radiation reaching the surface of the earth is absorbed, heating the land and water. Infrared radiation is emitted from the earth's surface, but certain gases in the atmosphere absorb this infrared radiation, and redirect a portion back to the surface, thus warming the planet and making life, as we know it, possible. This process is often referred to as the greenhouse effect. The surface temperature of the earth will rise until a radiative equilibrium is achieved between the rate of solar radiation absorption and the rate of infrared radiation emission. Human activities, such as fossil fuel combustion, deforestation, agriculture and large-scale chemical production, have measurably altered the composition of gases in the atmosphere. Some believe that these alterations will lead to a warming of the earth-atmosphere system by enhancement of the greenhouse effect. Figure 13.1 summarises the major links in the chain of environmental cause and effect for the emission of greenhouse gases.

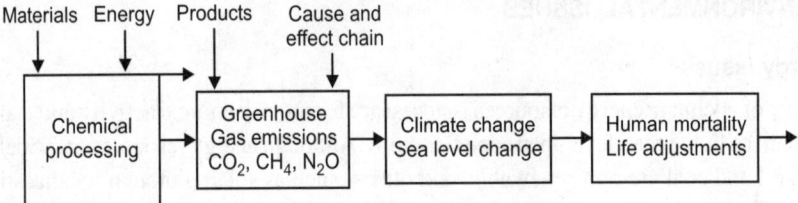

Fig. 13.1. Greenhouse emission from chemical processes and the major cause and environmental effect chain.

The primary greenhouse gases are water vapour, carbon dioxide, methane, nitrous oxide, chlorofluorocarbons, and tropospheric ozone. Water vapour is the most abundant greenhouse gas, but is omitted because it is generally not from anthropogenic sources. Carbon dioxide contributes significantly to global warming due to its high emission rate and concentration.

The major factors contributing to global warming potential of a chemical are infrared absorptive capacity and residence time in the atmosphere. Gases with very high absorptive capacities and long residence times can cause significant global warming even though their concentrations are extremely low. A good example of this phenomenon is the chlorofluorocarbons, which are, on a pound-for-pound basis, more than 1000 times more effective as greenhouse gases than carbon dioxide.

Environmental issues must be considered not only within the context of chemical production but also during other stages of a chemical's life cycle, such as transportation, use by customers, recycling activities and ultimate disposal.

Also the types of procedures that will be used in designing processes that minimise environmental impacts, and the responsibilities of chemical engineers to reduce pollution generation within chemical processes, and briefly notes some of the other professional responsibilities of chemical engineers, i.e. issues dealing with engineering ethics.

As part of their professional responsibilities, engineers should, through their designs, continuously improve the environmental performance of chemical processes.

A major objective for chemical process design is the inclusion of safeguards that minimise the number and severity of accidental releases of toxic chemicals and the incidence of fires and explosions. A number of chemical plant accidents have occurred in the relatively recent past illustrating the importance of integrating safety into process designs. These accidents resulted in the loss of life, permanent disability, and the destruction of chemical plant, process equipment and neighbouring residences. The most famous accidents occurred in Flixborough, England (1974) and Bhopal, India (1984).

Inherently safer design is a fundamentally different approach to chemical process safety. Instead of working with existing hazards in a chemical process and adding layers of protection, the engineer is challenged to reconsider the design and eliminate or reduce the source of the hazard within the process.

GREEN CHEMISTRY

Green chemistry is an approach to the design, manufacture and use of chemical products to intentionally reduce or eliminate chemical hazards. The goal of green chemistry is to create better, safer chemicals while choosing the safest, most efficient ways to synthesise them and to reduce wastes.

Chemicals are typically created with the expectation that any chemical hazards can somehow be controlled or managed by establishing 'safe' concentrations and exposure limits. Green chemistry aims to eliminate hazards right at the design stage. The practice of eliminating hazards from the beginning of the chemical design process has benefits for our health and the environment, throughout the design, production, use/reuse and disposal processes.

One example of the difference between traditional chemistry and green chemistry is the use of petroleum. Today's chemical industry relies almost entirely on non-renewable petroleum as the primary building block to create chemicals. This type of chemical production typically is very energy intensive, inefficient, and toxic — resulting in significant energy use, and generation of hazardous waste. One of the principles of green chemistry is to prioritise the use of alternative and renewable materials including the use of agricultural waste or biomass and non-food-related bioproducts. In general, chemical reactions with these materials are significantly less hazardous than when conducted with petroleum products. Other principles focus on prevention of waste, less hazardous chemical syntheses, and designing safer chemicals including safer solvents. Others focus on the design of chemical products to safely degrade in the environment and efficiency and simplicity in chemical processes.

A transformation to green chemistry techniques would result in safer work-places for industry workers, greatly reduced risks to fenceline communities and safer products for consumers. And because green chemistry processes are more efficient, companies would consume less raw materials and energy as well as save money on waste disposal.

How to design safer chemicals: The more we know about how a chemical's structure causes a toxic effect, the more options are available to design a safer chemical. Chemists now have access to many sources of information to determine the potential toxicity of the molecules they design and the ingredients they choose. Green chemists are trained to integrate this information into the design of molecules to avoid or reduce toxic properties.

For example, they might design a molecule large enough that it is unable to penetrate deep into the lungs, where toxic effects can occur. Or they might change the properties of a molecule to prevent its absorption by the skin or ensure it safely breaks down in the environment.

Green chemists also take a life cycle approach to reduce the potential risks throughout the production process (Fig. 13.2). They work to ensure that a product will pose minimal threats to human health or the environment during production, use, and at the end of its useful life when it will be recycled, or disposed of. A green chemistry approach is one of 'continual improvement, discovery, and innovation' that will bring us ever closer to processes and products that are safe within natural ecosystems. Ultimately a product should safely degrade as a biological nutrient or it should be safely recycled. Converting biomass from wood, prairie grass, plants, or other agricultural raw materials into non-conventional uses is consistent with the principles of green chemistry, which call on the chemist to use renewable materials, reduce energy use, and synthesise chemicals in an environmentally benign manner. Such uses include cleaner-burning fuels, new chemicals for industrial uses, and new animal feeds (Fig. 13.3).

Green computing or green IT, refers to environmentally sustainable computing or IT. It is the study and practice of designing, manufacturing, using, and disposing of computers, servers, and associated subsystems — such as monitors, printers, storage devices, and networking and communications systems — efficiently and effectively with minimal or no impact on the environment. Green IT also strives to achieve economic viability and improved system performance and use, while abiding by our social and ethical responsibilities.

Thus, green IT includes the dimensions of environmental sustainability, the economics of energy efficiency, and the total cost of ownership, which includes the cost of disposal and recycling. It is the study and practice of using computing resources efficiently.

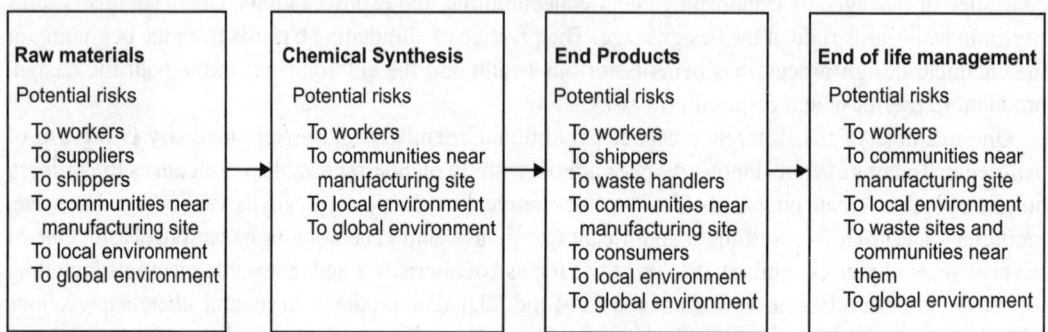

Raw materials
 	
Potential risks
 	
To workers
To suppliers
To shippers
To communities near
  manufacturing site
To local environment
To global environment
	Chemical Synthesis
	
	Potential risks
	
	To workers
	To communities near
	  manufacturing site
	To local environment
	To global environment
		End Products
		
		Potential risks
		
		To workers
		To shippers
		To waste handlers
		To communities near
		  manufacturing site
		To consumers
		To local environment
		To global environment
			End of life management
			
			Potential risks
			
			To workers
			To communities near
			  manufacturing site
			To local environment
			To waste sites and
			  communities near
			  them
			To global environment

Fig. 13.2. The benefits of green chemistry reduce risks all along the life cycle of chemical production and use.

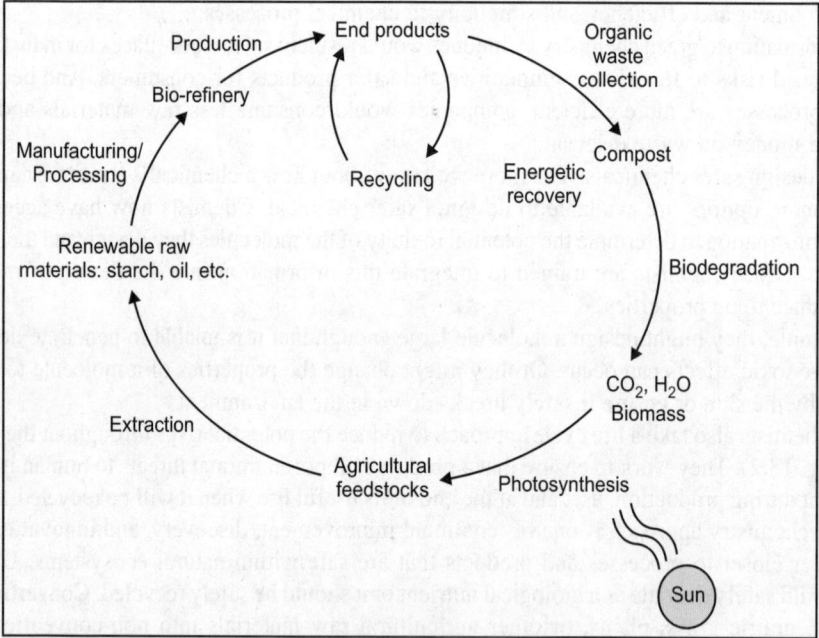

Fig. 13.3. Use of renewable feedstocks.

The goals of green computing are similar to green chemistry; reduce the use of hazardous materials, maximise energy efficiency during the product's lifetime, and promote the recyclability or biodegradability of defunct products and factory waste. Research continues into key areas such as making the use of computers as energy-efficient as possible, and designing algorithms and systems for efficiency-related computer technologies.

BIODEGRADABLE AND ECOFRIENDLY PRODUCTS

Physical and chemical methods of pollution control were always in the forefront because they were easy to understand, easy to control and were reproducible. Biodegradation, the real mechanism of nature of balancing the material, was always found to be incompletely understood, unpredictable and uncontrollable if we have to adapt it in the form of biological treatment methods. A better option then is to modify our materials, processes and products in such a way that we can rely upon the biodegradation in nature and recalcitrance, bioaccumulation problems are overcome.

Manufacturing processes are rapidly changing and biodegradable products are fast replacing man-made, difficult to degrade products. After understanding the nature, its role, its superiority and by knowing the fact that biodegradation is the ultimate fate of any material that enters into environment, it was inevitable for us to change our psychology — industrial ecology — industrial ecosystem — industrial metabolism — going close to nature. Objectives then are to improve upon the method of production, searching for alternative raw materials, recycling, conversion to suitable forms of certain wastes, so that we do not add any material, waste in nature which nature cannot take care of.

Biodegradation mechanisms which occur in the soil, aquatic environment though slow are important for us as they do not involve any cost of treatment when it occurs naturally, and are safer and bring about complete degradation and not mere conversion. Hence incorporating biodegradability or aiming for biodegradability is an obvious approach while carrying out production of different items. The agricultural products, food products, commodities hereafter will carry a label of ecofriendly in most of the countries. The trend is to spread to every material and product. The consumables, coal, oil, petrol, synthetic fibre, plastic which are polluting/non-biodegradable will be replaced slowly by the one which is non-polluting/biodegradable.

Biodegradable waste is a type of waste, typically originating from plant or animal sources, which may be broken down by other living organisms. Waste that cannot be broken down by other living organisms may be called non-biodegradable.

Biodegradable waste can be commonly found in municipal solid waste (sometimes called biodegradable municipal waste, or BMW) as green waste, food waste, paper waste, and biodegradable plastics. Other biodegradable wastes include human waste, manure, sewage, slaughterhouse waste.

Through proper waste management, it can be converted into valuable products by composting, or energy by waste-to-energy processes such as anaerobic digestion and incineration. As part of an integrated waste management system, waste-to-energy processes reduces the emission of landfill gas into the atmosphere. Composting converts biodegradable waste into compost. Anaerobic digestion converts biodegradable waste biogas and soil amendment (digestate). Incineration as well as biogas can be used to generate electricity and/or heat for district heating.

Biodegradable waste is an important substance due to its links with global warming. When it is disposed of in landfills, it breaks down under uncontrolled anaerobic conditions. This produces landfill gas which, if not harnessed, escapes into the atmosphere. Landfill gas contains methane, a more potent greenhouse gas than carbon dioxide. This can cause harmful effects on the environment.

Biodegradable plastics are plastics that will decompose in natural aerobic (composting) and anaerobic (landfill) environments. Biodegradation of plastics can be achieved by enabling micro-organisms in the environment to metabolise the molecular structure of plastic films to produce an inert humus-like material that is less harmful to the environment. They may be composed of either bioplastics, which are plastics whose components are derived from renewable raw materials, or petroleum-based plastics which utilise an additive. The use of bioactive compounds compounded with swelling agents ensures that, when combined with heat and moisture, they expand the plastic's molecular structure and allow the bioactive compounds to metabolise and neutralise the plastic.

Biodegradable plastics typically are produced in two forms: injection moulded (solid, 3D shapes), typically in the form of disposable food service items, and films, typically organic fruit packaging and collection bags for leaves and grass trimmings, and agricultural mulch.

Biodegradable plastics are not a panacea, however, some critics claim that a potential environmental disadvantage of certified biodegradable plastics is that the carbon that is locked up in them is released into the atmosphere as a greenhouse gas. However, biodegradable plastics from natural materials, such as vegetable crop derivatives or animal products, sequester CO_2 during the phase when they are growing, only to release CO_2 when they are decomposing, so there is no net gain in carbon dioxide emissions.

However, certified biodegradable plastics require a specific environment of moisture and oxygen to biodegrade, conditions found in professionally managed composting facilities. There is much debate about the total carbon, fossil fuel and water usage in processing biodegradable plastics from natural materials and whether they have a negative impact to human food supply. Traditional plastics made from non-renewable fossil fuels lock up much of the carbon in the plastic as opposed to being utilised in the processing of the plastic.

The carbon is permanently trapped inside the plastic lattice, and is rarely recycled. There is concern that another greenhouse gas, methane, might be released when any biodegradable material, including truly biodegradable plastics, degrades in an anaerobic (landfill) environment. Methane production from these specially managed landfill environments are typically captured and burned to negate the release of methane in the environment. Some landfills today capture the methane biogas for use in clean inexpensive energy. Of course, incinerating non-biodegradable plastics will release carbon dioxide as well. Disposing of biodegradable plastics made from natural materials in anaerobic (landfill) environments will result in the plastic lasting for hundred of years.

RENEWABLE RESOURCE

A natural resource is a renewable resource if it is replaced by natural processes at a rate comparable or faster than its rate of consumption by humans. Solar radiation, tides, winds and hydroelectricity are in no danger of a lack of long-term availability. Renewable resources may also mean commodities such as wood, paper, and leather, if harvesting is performed in a sustainable manner.

Some natural renewable resources such as geothermal power, fresh-water, timber, and biomass must be carefully managed to avoid exceeding the world's capacity to replenish them. A life cycle assessment provides a systematic means of evaluating renewability.

The term has a connotation of sustainability of the natural environment. Gasoline, coal, natural gas, diesel, and other commodities derived from fossil fuels are non-renewable. Unlike fossil fuels, a renewable resource can have a sustainable yield.

Renewable Energy

Solar energy is the energy derived directly from the Sun. Along with nuclear energy, it is the most abundant source of energy on Earth. The fastest growing type of alternative energy, increasing at 50 per cent a year, is the photovoltaic cell, which converts sunlight directly into electricity. The sun yearly delivers more than 10,000 times the energy that humans currently use.

Wind power is derived from uneven heating of the earth's surface from the sun and the warm core. Most modern wind power is generated in the form of electricity by converting the rotation of turbine blades into electrical current by means of an electrical generator. In windmills (a much older technology) wind energy is used to turn mechanical machinery to do physical work, like crushing grain or pumping water. Hydropower is energy derived from the movement of water in rivers and oceans (or other energy differentials), can likewise be used to generate electricity using turbines, or can be used mechanically to do useful work. It is a very common resource.

Geothermal power directly harnesses the natural flow of heat from the ground. The available energy from natural decay of radioactive elements in the earth's crust and mantle is approximately equal to that of incoming solar energy. Alcohol derived from corn, sugarcane, switchgrass, etc. is also a renewable source of energy. Similarly, oils from plants and seeds can be used as a substitute for non-renewable diesel. Methane is also considered as a renewable source of energy.

Renewable Materials

Agricultural products

Techniques in agriculture which allow for minimal or controlled environmental damage qualify as sustainable agriculture. Products (foods, chemicals, biofuels, etc.) from this type of agriculture may be considered 'sustainable' when processing, logistics, etc. also have sustainable characteristics.

Similarly, forest products such as lumber, plywood, paper and chemicals, can be renewable resources when produced by sustainable forestry techniques.

Water

Water can be considered a renewable material (also non-renewable) when carefully controlled usage, treatment, and release are followed. If not, it would become a non-renewable resource at that location. For example, groundwater could be removed from an aquifer at a rate greater than the sustainable recharge. Removal of water from the pore spaces may cause permanent compaction (subsidence) that cannot be renewed.

RENEWABLE CHEMICALS

The global renewable chemicals market is estimated to reach US $59 billion in 2014 from about US $45 billion in 2009. Though several companies in the chemicals industry were affected by the recent economic crisis, the companies producing renewable chemicals are expected to weather the crisis. The driving force for the renewable chemicals market is the low requirement of capital for both production as well as feedstock. Moreover, consumer demand for green products and governmental support to the industry for reducing dependence on finite non renewable petroleum feedstock as well as reducing greenhouse gas emission has been driving the market for renewable chemicals. The growth of the industrial biotechnology has also contributed to the growth of the overall renewable chemicals market due to their innovations in biocatalysis that finds extensive usage in manufacturing renewable chemicals. Renewable chemicals find industrial, pharmaceutical application as well as in consumer products.

Focus on reducing global greenhouse gas emissions levels has led to an increase in the activities in the field of renewable chemicals. Foreseeing the rising importance of renewable chemicals the major players in the chemicals industry such as Dow, BASF have already rendered an increased attention on this market. Though the market for alcohols in the overall renewable chemicals market accounts for the largest share of the market, polymers is expected to gain the maximum growth rate for the next five years. Renewable chemicals such as polymers are expected to command significant share in the overall polymers market mostly due to their usage in making biodegradable and compostable plastics and consumer goods such as cell phones, laptops, etc. Platform chemicals also play an important role in the renewable chemicals market since they contain multiple functional groups and hence present practical potential for their conversion to families of useful products.

The indepth market estimates and forecast for global renewable chemicals market are as follows:

1. Renewable chemicals — products: Alcohols, organic chemicals, ketones, polymers and other markets.
2. Renewable chemicals — application: Industrial, transportation, textiles, safe food supply, environment, communication, housing, recreation, health and hygiene and other applications.
3. Renewable chemicals — catalysis: Biocatalysis and chemical catalysis.
4. Renewable chemicals — technology: Thermochemical conversion, fermentation and bioconversion, product separation and bioconversion, enzymatic hydrolysis, gasification-fermentation, acid hydrolysis, biochemical-thermochemical, biochem-organisolve, fischer-tropsch diesel, reductive transformation, dehydrative transformation and other technologies.
5. Renewable chemicals — platform chemicals: 1,4-diacids, 2,5-furan dicarboxylic acid, 3-hydroxypropionic acid, aspartic acid, glucaric acid, glutamic acid, itaconic acid, levulinic acid, glycerol and other chemicals.
6. Renewable chemicals — biofeedstock: Starch, cellulose, lignin and oil/fats/protein.
7. Renewable chemicals — source: Plant biomass, animal biomass and marine biomass.

Chemicals from biomass covers resources, chemical composition of biomass, key factors affecting composition, utilisation of wastes, extraction technologies, controlled pyrolysis, fermentation, platform molecules, and green chemical technologies for their conversion to valuable chemicals. In the process smaller volume chemicals could become bulk chemicals as a result of a greater exploitation of biomass products, making it an important resource for academic and industrial scientists and researchers.

As a source of chemicals, biomass has several intrinsic advantages over fossil mass: it is renewable, flexible through crop switching, and adaptable through genetic manipulation. Inflexibility of the fossil mass resource is compensated for by highly effective technology for production of olefins and aromatics, economies of scale, and a highly developed system of conversion products with large markets. Direct and indirect strategies to substitute for petrochemicals are based on ecological succession concepts. A proliferation of lignocellulosic fractionation processes is arising from the need for inexpensive, homogeneous, chemically useful biomass feedstocks.

Fuels and chemicals from biomass: In the era of peak oil and rising greenhouse gas emissions, the world needs renewable fuels and chemicals. In particular, we are investigating the use of hot compressed and supercritical water as a means of upgrading oxygen-containing organics, such as biomass, by preferential removal of oxygen by decarboxylation to yield useful oils and chemicals.

CHEMICALS FROM BIOMASS

Refined processing methods and advanced technologies for biomass conversion enable low-value by-products or waste materials to be transformed into products or waste materials to be transformed into value-added products. Such products include the same chemicals produced from petroleum that are used, for example, as plastics for car components, food additives, clothing fibres, polymers, paints, and other industrial and consumer products. Other replacement products that can be produced from biomass include different chemicals with similar or better properties. The chemicals and products discussed here are capable of reducing the 17 per cent petroleum utilisation number and making a significant contribution toward meeting those goals.

The discussion on chemicals from biomass is limited to large-scale commodity products for which a noticeable impact on the energy market would be made by displacing petroleum. Speciality chemical products whose limited market would have minimal impact on energy markets are not addressed, nor is the use of biomass for material products such as cellulosic fibre for paper or lignocellulosic material for construction materials. Using biomass to produce cellulosic fibre for paper or lignocellulosic material for construction materials. Using biomass to produce a vast array of small-market speciality chemicals based on unique biochemical structures is another part of the overall concept of a sustainable economy but is outside the scope of this chapter. Instead focuses on the chemical structure of various commodity products and their fabrication from biomass. A history of the use of biomass as the basis for chemical products is included for background.

Historical Developments

Biomass has been used to meet the needs of civilisation since prehistory. Several biomass chemical products can be traced to a time when neither the chemical mechanistic transformation process nor the chemical identity of the product were understood. These early products include chemicals still widely used, such as ethanol and acetic acid. A more scientific understanding of chemistry in the 19th currently led to the use of biomass as a feedstock for chemical processes. However, in the 20th century, the more easily processed and readily available petroleum eventually changed the emphasis of chemical process development from carbohydrate feedstock to hydrocarbon feedstock. Through major investments by industry and government, utilisation of petroleum as a chemical production feedstock has been developed to a high level and provides a wide array of materials and products. Now, the expansion of synthetic chemical products from petroleum provides the framework in which to reconsider the use of renewable biomass resources for meeting society's needs in the 21st century. However, only since the 1980s has interest returned to the use of biomass and its chemical structures for the production of chemical products; therefore, much research and development of chemical reactions and engineered processing systems remains to be done (Table 13.1).

Table 13.1. Petroleum-derived products.

Finished motor gasoline
Finished aviation gasline
Jet fuel
Kerosene
Distillate fuel oil
Residual fuel oil
Asphalt and road oil
Petroleum coke
Chemical products
Liquefied petroleum gases (net)
Ethane/ethylene
Propane/Propylene

(Contd ...)

Butane/Butylene
Isobutane/Isobutylene
Naphtha for petrochemicals
Other oils for petrochemicals
Special naphthas
Lubricants
Waxes
Still gases
Miscellaneous products

Early developments

The earliest biomass uses often involved food products. One of the first was ethanol, which was limited to beverage use for centuries. Ethanol's origin is lost in antiquity, but it is generally thought to be an accidental development. Only with an improved understanding of chemistry in the mid-1800s was ethanol identified and its use expanded for solvent applications and as a fuel. Acetic acid, the important component in vinegar, is another accidental product based on fermentation of ethanol by oxidative organisms. A third fermentation product from antiquity, lactic acid, also an accidental product of oxidative fermentation, is commonly used in sauerkraut, yogurt, buttermilk, and sourdough.

Another old-world fermentation method is the production of methane from biomass, a multistep process accomplished by a consortium of bacteria, which is an important part of the cyclical nature of carbon on earth. These processes, now generally referred to as anaerobic digestion, are well-known in ruminant animals as well as in stagnant wet biomass. The methane from these processes has only been captured at the industrial scale for use as a fuel since the 1970s. Additional early uses of biomass involved its structural components. The fibrous matrix (wood) was first used in constructing shelter. Later, the fibrous component was separated for paper products. As a result, two major industries, the wood products industry and the pulp and paper industry, were developed at sites where biomass was available. Biomass also comprised materials used for clothing. Cotton fibres, consisting mainly of cellulose, and flax fibre are examples.

At the beginning of the 20th century, cellulose recovered from biomass was processed into chemical products. Cellulose nitrate was an early synthetic polymer used for 'celluloid' moulded plastic items and photographic film. The nitrate film was later displaced by cellulose acetate film, a less flammable and more flexible polymer. Cellulose nitrate lacquers retained their market, although acetate lacquers were also developed. Cellulose acetate fibre, called Celanese, grew to one-third of the synthetic market in the United States. Injection-moulded cellulose acetate articles became a significant component of the young automobile industry. Today, cellulose-based films and fibres have largely been supplanted by petroleum-based polymers, which are more easily manipulated to produce properties of interest, such as strength, stretch recovery, or permeability.

Wood processing to chemicals by thermal methods (pyrolysis), another practice from antiquity, was originally the method used for methanol and acetic acid recovery but has been replaced by petrochemical processes. One of the earliest synthetic resins, the phenol-formaldehyde polymer Bakelite, was originally produced from phenolic wood extractives. However, petrochemical sources for the starting materials displaced the renewable sources after World War I. More chemically complex products, such as terpenes and rosin chemicals, are still recovered from wood processing but only in a few cases as a by-product from pulp and paper production.

Chemurgy

In the early 20th century, the chemurgy movement blossomed in response to the rise in the petrochemical industry and in support of agricultural markets. The thrust of the movement was that agricultural products and food by-products could be used to produce the chemicals important in modern society. A renewable resource, such as biomass, could be harvested and transformed into products without depleting the resource base. Important industrialists, including Henry Ford, were participants in this movement, which espoused that anything made from hydrocarbons can be made from carbohydrates. Supply inconsistencies and processing costs eventually turned the issue in favour of the petrochemical industry. The major investments for chemical process development based on hydrocarbons were more systematically coordinated within the more rightly held petrochemical industry, whereas the agricultural interests were too diffuse (both geographically and in terms of ownership) to marshal much of an effort. Furthermore, the chemical properties of petroleum, essentially volatile hydrocarbon liquids, allowed it to be more readily processed to chemical products. The large scale of petroleum refining operations, developed to support fuel markets, made for more efficient processing systems for chemical production as well.

Biorefinery Concept

The vision of a new chemurgy-based economy began evolving in the 1980s. The biorefinery vision foresees a processing factory that separates the biomass into component streams and then transforms the component streams into an array of products using appropriate catalytic or biochemical processes, all performed at a scale sufficient to take advantage of processing efficiencies and resulting improved process economics. In essence, this concept is the modern petroleum refinery modified to accept and process a different feedstock and make different products using some of the same unit operations along with some new unit operations. The goal is to manufacture a product slate, based on the best use of the feedstock components, that represents higher value products with no waste. Overall process economics are improved by production of higher value products that can compete with petrochemicals. Some biomass processing plants already incorporate this concept to various degrees.

Wood pulp mill

The modern wood pulp mill represents the optimised, fully integrated processing of wood chips to products. The main process of pulping the wood allows the separation of the cellulosic fibre from the balance of the wood structure. The cellulose has numerous uses, ranging from various paper and cardboard types to chemical pulp that is further processed to cellulose-based chemicals. The lignin and hemicellulose portions of the wood are mainly converted to electrical power and steam in the recovery boiler. In some cases, the lignin is recovered as precipitated sulphonate salts, which are sold based on surfactant properties. In other cases, some lignin product is processed to vanillin favouring. The terpene chemicals can be recovered, a practice limited primarily to softwood processing plants in which the terpene fraction is larger.

Corn wet mill

In the corn wet mill, several treatment and separation processes are used to produce a range of food and chemical products. Initial treatment of the dry corn kernel in warm water with sulphur dioxide results in a softened and protein-stripped kernel and a protein-rich liquor called corn steep liquor. The corn is then milled and the germ separated. The germ is pressed and washed with solvent to recover the corn oil. Further milling and filtering separate the cornstarch from the corn fibre (essentially the seed coat). The starch is the major product and can be refined to a number of grades of starch product. It can also be hydrolysed to glucose. The glucose, in turn, has various uses, including isomerisation to fructose for high-fructose corn syrup; hydrogenation to sorbitol; or fermentation (using some of the steep liquor for nutrients) to a number of products, such as ethanol, lactic acid, or citric acid. The balance of the corn steep liquor is mixed with the corn fibre and sold as animal feed.

Cheese Manufacturing Plant

Cheese manufacturing has evolved into a more efficient and self-contained processing operation in which recovering the whey by-product (as dried solids) is the industry standard instead of disposing of it, which was the practice for many centuries. The whey is a mostly water solution of proteins, minerals, and lactose from the milk following coagulation and separation of the cheese. To make the whey more valuable, ultrafiltration technology is used in some plants to separate the protein from the whey for recovery as a nutritional supplement. Further processing of the ultrafiltration permeate is also used in a few plants to recover the lactose for use as a food ingredient. Developing methods for recovering the mineral components is the next stage in the development process.

Biomass-Derived Chemical Products

For production of chemicals from biomass, there are several processing method categories, which are used based on the product desired: fermentation of sugars to alcohol or acid products; chemical processing of carbohydrates by hydrolysis, hydrogenation, or oxidation; pyrolysis of biomass to structural fragment chemicals; or gasification by partial oxidation or steam reforming to a synthesis gas for final product formation. Various combinations of these processing steps can be integrated in the biorefinery concept. Table 13.2 illustrates the wide range of biomass-derived chemicals. These products include the older, well-known fermentation products as well as new products whose production methods are being developed. Development of new processing technology is another key factor to making these products competitive in the marketplace.

Table 13.2. Chemicals derived from biomass.

Chemical	Derivative chemicals	Uses
Fermentation products		
Ethanol		Fuel
Lactic acid	Poly lactic acid	Plastics
Acetone/butanol/ethanol		Solvents
Citric acid		Food ingredient
Carbohydrate chemical derivatives		
Hydrolysate sugars	Fermentation feedstocks	Fermentation products
Furfural, hydroxymethylfurfural	Furans, adiponitrile	Solvents, binders
Levulinic acid	Methyltetrahydrofuran, δ-amino lactic acid, succinic acid	Solvents, plastics, herbicide/pesticide

(Contd ...)

Chemical	Derivative chemicals	Uses
Polyols	Glycols, glycerol	Plastics, formulations
Gluconic/glucaric acids		Plastics
Pyrolysis products		
Chemicals fractions	Phenolics, cyclic ketones	Resins, solvents
Levoglucosan, levoglucosenone		Polymers
Aromatic hydrocarbons	Benzene, toluene, xylenes	Fuel, solvents
Gasification products		
Synthesis gas	Methanol, ammonia	Liquid fuels, fertiliser
Tar chemicals		Fuels
Development fermentations		
2,3-Butanediol/ethanol		Solvents
Propionic acid		Food preservative
Glycerol, 1,3-propanediol		C3 plastics, formulations
3-Dehydroshikimic acid	Vanillin, catechol, adipic acid	Flavours, plastics
Catalytic/bioprocessing		
Succinic acid	Butanediol, tetrahydrofuran	Resins, solvents
Itaconic acid	Methyl-1,4-butanediol and tetrahydrofuran (or methyltetrahydrofuran)	Resins, solvents
Glutamate, lysine	Pentanediol, pentadiamine	C5 plastics
Plant-derived		
Oleochemicals	Methyl esters, epoxides	Fuel, solvents, binders
Polyhydroxyalkanoates		Medical devices

Modern fermentation products

Fermentation products include ethanol, lactic acid, acetone/butanol/ethanol, and citric acid. Ethanol and lactic acid are produced on a scale sufficient to have an impact on energy markets. Ethanol is primarily used as a fuel, although it has numerous potential derivative products as described later.

Ethanol

The feedstock for the fermentation ranges from corn starch-derived glucose, wheat gluten production by-product starches, and cheese whey lactose to food processing wastes and, until recently, wood pulping by-products. Although some of this ethanol is potable and goes into the food market, the vast majority is destined for the fuel market for blending with gasoline. However, other possible chemical products derived from ethanol include acetaldehyde, acetic acid, butadiene, ethylene, and various esters, some of which were used in the past and may again become economically viable.

Lactic acid

Lactic acid, as currently utilised, is an intermediate-volume speciality chemical used in the food industry for producing emulsifying agents and as a nonvolatile, odour-free acidulant. Its production by carbohydrate fermentation involves any of several *Lactobacillus* strains. The fermentation can produce a desired stereoisomer, unlike the chemical process practiced in some areas of the world. Typical fermentation produces a 10 per cent concentration lactic acid broth, which is kept neutral by addition of calcium carbonate throughout the derived lactic acid for chemical products requires complicated separation and purification, including the reacidification of the calcium lactate salt and resulting disposal of a calcium sulphate sludge by-product. An electrodialysis membrane separation method, in the development stage, may lead to an economical process. The lactic acid can be recovered as lactide by removing the water, which results in the dimerisation of the lactic acid to lactide that can then be recovered by distillation.

Poly lactic acid (PLA) is the homopolymer of biomass-derived lactic acid, usually produced by a ring-opening polymerisation reaction from its lactide form.

Other chemical products that can be derived from lactic acid include lactate esters, propylene glycol, and acrylates. The esters are described as nonvolatile, nontoxic, and biodegradable, with important solvent properties. Propylene glycol can be produced by a simple catalytic hydrogenation of the lactic acid molecule. Propylene glycol is a commodity chemical with a 1 billion pound per year market and can be used in polymer synthesis and as a low-toxicity antifreeze and deicer. Dehydration of lactate to acrylate appears to be a straightforward chemical process, but a high-yield process has yet to be commercialised.

Acetone/butanol/ethanol

Acetone/butanol/ethanol fermentation was widely practiced in the early 20th century before being displaced by petrochemical operations. The two-step bacterial fermentation is carried out by various species of the *Clostridium* genus. The solvent product ratio is typically 3:6:1, but due to its toxicity, the butanol product is the limiting component. The yield is only 37 per cent, and final concentrations are approximately 2 per cent. The high energy cost for recovery by distillation of the chemical products from such a dilute broth is a major economic drawback. Alternative recovery methods, such as selective membranes, have not been demonstrated.

Citric acid

More than 1 billion pounds per year of citric acid is produced worldwide. The process is a fungal fermentation with *Aspergillus niger* species using a submerged fermentation. Approximately one-third of the production is in the United States, and most of it goes into beverage products. Chemical uses will depend on the development of appropriate processes to produce derivative products.

Chemical processing of carbohydrates

Hydrolysate sugars

Monosaccharide sugars can be derived from biomass and used as final products or further processed by catalytic or biological processes to final value-added products. Disaccharides, such as sucrose, can be recovered from plants and used as common table sugar or 'inverted' (hydrolysed) to a mixture of glucose and fructose. Starch components in biomass can readily be hydrolysed to glucose by chemical processes (typically acid-catalysed) or more conventionally by enzymatic processes. Inulin is a similar polysaccharide recovered from sugar beets that could serve as a source of fructose. More complex carbohydrates, such as cellulose or even hemicellulose, can also be hydrolysed to monosaccharides. Acid-catalysed hydrolysis has been studied extensively and is used in some countries to produce glucose from wood. The processing parameters of acidity, temperature, and residence time can be controlled to fractionate biomass polysaccharides into component streams of primarily five- and six-carbon sugars based on the relative stabilities of the polysaccharides. However, fractionation to levels of selectivity sufficient for commercialisation has not been accomplished.

The development of economical enzymatic processes for glucose production from cellulose could provide a tremendous boost for chemicals production from biomass. The even more complex hydrolysis of hemicellulose for monosaccharide production provides a potentially larger opportunity because there are fewer alternative uses for hemicellulose than for cellulose.

Furfural, hydroxymethylfurfural, and derived products

More severe hydrolysis of carbohydrates can lead to dehydrated sugar products, such as furfural from five-carbon sugar and hydroxymethylfurfural from six-carbon sugar. In fact, the careful control of the hydrolysis to selectively produce sugars without further conversion to furfurals is an important consideration in the hydrolysis of biomass when sugars are the desired products. However, furfural products are economically valuable in their own right. The practice of processing the hemicellulose five-carbon sugars in oat hulls to furfural has continued despite economic pressure from petrochemical growth, and it is still an important process, with an annual domestic production of approximately 100 million pounds. Furfural can be further transformed into a number of products, including furfural alcohol and furan.

Levulinic acid and derived products

Levulinic acid (4-oxo-pentanoic acid) results from subsequent hydrolysis of hydroxymethylfurfural. It is relatively stable toward further chemical reaction under hydrolysis conditions. Processes have been developed to produce it from wood, cellulose, starch, or glucose.

Levulinic acid is potentially a useful chemical compound based on its multifunctionality and its many potential derivatives. In the latter half of the 20th century, its potential for derivative chemical products was reviewed numerous times in the chemical literature. Some examples of the useful products described include levulinate esters, with useful solvent properties; δ-aminolevulinic acid, identified as having useful properties as a herbicide and for potentially controlling cancer; hydrogenation of levulinic acid, which can produce γ-valerolactone, a potentially useful polyester monomer (as hydroxyvaleric acid); 1,4-pentanediol, also of value in polyester production; methyltetrahydrofuran, a valuable solvent or a gasoline blending component; and diphenolic acid, which has potential for use in polycarbonate production.

Hydrogenation to polyols

Sugars can readily be hydrogenated to sugar alcohols at relatively mild conditions using a metal catalyst. Some examples of this type of processing include sorbitol produced from glucose, xylitol produced from xylose, lactitol produced from lactose, and maltitol produced from maltose. All these sugar alcohols have current markers as food chemicals. New chemical markets are developing; for example, sorbitol has been demonstrated in environmentally friendly deicing solutions.

In addition, the sugar alcohols can be further processed to other useful chemical products. Isosorbide is a doubly dehydrated product from sorbitol produced over an acid catalyst. Isosorbide has uses in polymer blending, for example, to improve the rigidity of polyethylene terephthalate in food containers such as soda pop bottles.

Hydrogenation of the sugar alcohols under more severe conditions can be performed to produce lower molecular-weight polyols. By this catalytic hydrogenolysis, five- or six-carbon sugar alcohols have been used to produce primarily a three-product slate consisting of propylene glycol, ethylene glycol, and glycerol. Although other polyols and alcohols are also produced in lower quantities, these three are the main products from hydrogenolysis of any of the epimers of sorbitol or xylitol. Product separations and purifications result in major energy requirements and costs. Consequently, controlling the selectivity within this product slate is a key issue in commercialising the technology. Selectivity can be affected by processing conditions as well as catalyst formulations.

Oxidation to gluconic/glucaric acid

Glucose oxidation can be used to produce six-carbon hydroxyacids—either the monoacid, gluconic, or the diacid, glucaric. The structures of these chemicals suggest opportunities for polyester and polyamide formation, particularly polyhydroxypolyamides (i.e. hydroxylated nylon). Oxidation is reportedly performed economically with nitric acid as the catalyst. Specificity to the desired product remains to be overcome before these chemicals can be produced commercially.

Potential of pyrolysis products

Pyrolysis of wood for chemicals and fuels production has been practiced for centuries; in fact, it was likely the first chemical production process known to humans. Early products of charcoal and tar were used not only as fuels but also for embalming, filling wood joints, and other uses. Today, pyrolysis is still important in some societies, but it has largely been displaced by petrochemical production in developed countries. At the beginning of the 20th century, important chemical products were methanol, acetic acid, and acetone. Millions of pounds of these products were produced by wood pyrolysis until approximately 1970, when economic pressure caused by competition with petroleum-derived products became too great. Due to new developments in 'flash' pyrolysis beginning in the 1980s, there are new movements into the market with wood pyrolysis chemicals. However, these products are speciality chemicals and not yet produced on a commodity scale.

Chemical fractions

Most new development work in pyrolysis involves separating the bulk fractions of chemicals in the pyrolysis oil. Heavy phenolic tar can be separated simply by adding water. A conventional organic chemical analytical separation uses base extraction for acid functional types, including phenolics. Combining the water-addition step and the base extraction with a further solvent separation can result in a stream of phenolics and neutrals in excess of 30 per cent of the pyrolysis oil. This stream has been tested as a substitute for phenol in phenol-formaldehyde resins commonly used in restructured wood-based construction materials, such as plywood and particle board. This application allows for the use of the diverse mixture of phenolic compounds produced in the pyrolysis process, which typically includes a range of phenols, alkyl-substituted with one- to three-carbon side chains, and a mix of methoxyphenols (monomethoxy from softwoods and a mix of mono- and dimethoxy phenols from hardwood) with similar alkyl substitution. In addition, there are more complex phenolics with various oxygenated side chains, suggesting the presumed source—lignin in the wood. The lower molecular-weight nonphenolics comprise a significant but smaller fraction of pyrolysis oil as produced in flash processes. Methanol and acetic acid recovery has potential. Hydroxyacetaldehyde, formic acid, hydroxyacetone, and numerous small ketones and cyclic ketones, which may have value in fragrances and flavourings, also have potential.

Levoglucosan and levoglucosenone

Cellulose pyrolysis can result in relatively high yields of levoglucosan or its subsequent derivative levoglucosenone. Both are dehydration products from glucose. Levoglucosan is the 1,6-anhydro product, and levoglucosenone results from removal of two of the three remaining hydroxyl groups of levoglucosan with the formation of a conjugated olefin double bond and a carbonyl group. Levoglucosenone's chiral nature and its unsaturated character suggest uses for this compound in pharmaceuticals and in polymer formation.

Aromatic hydrocarbon compounds

The large fraction of phenolic compounds in pyrolysis oil suggests the formation of aromatics for chemical or fuel use. Catalytic hydrogenation (in this case, hydrodeoxygenation) can be used for removing the hydroxyl group from the phenolic compounds to produce the respective phenyl compound. For example, hydrogenation of phenol or methoxyphenol (guaiacol) would give benzene, and methyl guaiacol would give toluene. This chemistry was studied in the 1980s, when aromatics were being added to gasoline to improve octane number. Due to increased restriction on aromatics in gasoline to meet emission guidelines, this process concept was shelved. Production of such hydrocarbons (and other saturated cyclic hydrocarbons) remains an option for chemical production from wood pyrolysis; however, the economics are a drawback due to the extensive amount of hydrogenation required to convert the oxygenate to a hydrocarbon.

Potential gasification products

Whereas the pyrolysis process results in a product that retains much of the chemical functional character of the biomass feedstock, gasification operates under more severe process conditions and results in products of a relatively simple chemical structure but with wide applications. Gasification usually entails high-temperature partial oxidation or steam reforming of the biomass feedstock with typical products of hydrogen, carbon monoxide and dioxide, and light hydrocarbons. Other gasification processes using catalysis or bioconversion methods can also result in a primarily methane and carbon dioxide product gas.

Synthesis gas to produce chemicals

Synthesis gas produced from biomass can be used for chemical production in much the same way as synthesis gas from steam reforming of natural gas or naphtha. Certain process configurations can be used with proper optimisation or catalysis to control the important ratio of hydrogen to carbon monoxide in the synthesis gas. The synthesis gas can then be used in catalysed reactions to produce methanol or hydrocarbons or even ammonia. Methanol can subsequently be converted by further catalytic steps to formaldehyde, acetic acid, or even gasoline.

Tar chemicals

By-products formed in incomplete gasification include an array of organics ranging from the pyrolysis chemicals to further reacted products, including aromatics and particularly polycyclic aromatics, if a sufficient amount of time at high temperature is applied. As a function of temperature and time at temperature, the composition of the tar can vary from a highly oxygenated pyrolysis oil produced at lower temperature and short residence time to a nearly deoxygenated polycyclic aromatic hydrocarbon produced at high temperature. Because of these high-temperature tar components, including four- and five-aromatic ring structures, mutagenic activity can be significant.

Developmental fermentations

New fermentation processes for additional biomass-derived chemical products are in various developmental stages. These processes range from improvements to known fermentations to new chemical products produced in genetically engineered organisms.

Acetic acid

The acetic acid currently available for chemical uses is petrochemically derived. Acetic acid fermentation, known from antiquity to form vinegar, is an *Acetobacter* partial oxidation of ethanol. More recently developed is the homofermentative conversion of glucose to acetic acid by 'acetogenic' *Clostridium* bacteria. In some cases, 3 mol of acetate can be produced from each mole of six-carbon sugar feedstock, and concentrations of acetate of up to 1.5 per cent can be accumulated. This type of fermentation effectively increases the theoretical yield of acetic acid from glucose by 50 per cent and provides a more reasonable pathway for renewable resources.

Propionic acid

All chemical production of propionic acid, best known as a food preservative, is currently derived from petroleum as a coproduct or by-product of several processes. However, propionic acid can readily be fermented with propionibacterium. The production of esters for use as environmentally benign solvents represents a small but growing market.

Extremophilic lactic acid

As an economical production process, lactic acid fermentation from glucose has led the way into the marketplace for chemicals production. Although the lactic acid process used for PLA production is based on conventional fermentation, the recovery of the lactide product for polymerisation eliminates the neutralisation and reacidification steps typically required. Efforts to improve the process are focused on the development of new organisms that can function under acidic conditions and produce higher concentrations of lactic acid.

These organisms, called extremophiles, are expected to operate at a pH less than 2 in a manner similar to citric acid fermentations. Other work to develop higher temperature-tolerant organisms may also result in reduced production costs.

2,3-Butanediol/ethanol

For this fermentation, a variety of bacterial strains can be used. An equimolar product mix is produced under aerobic conditions, whereas limited aeration will increase the production of the 2,3-butanediol to the level of exclusivity. Either optically active 2,3-butanediol or a racemic mixture can be produced, depending on the bacterial strain. Similarly, the final product concentration, ranging from 2 or 3 to 6–8 per cent, depends on the bacterial strain. Recovering the high-boiling diol by distillation is problematic.

Succinic acid

Succinate production from glucose and other carbohydrates has been well characterised for numerous bacterial organisms, including several succinogenes and the *Anaerobiospirillum succiniproducens*. The biological pathways for these organisms are similar and usually result in significant acetate production. The pathways also include a carbon dioxide insertion step such that potential theoretical yield is greater than 100 per cent based on glucose feedstock. Mutated *Escherichia coli* has also been developed for succinic acid production, with yields as high as 90 per cent shown on the pilot scale and a final succinic acid concentration of 4 or 5 per cent.

As with lactic acid production, maintaining the pH of the fermentor at approximately neutral is vital. The resulting disposal of mineral by-products from neutralisation and reacidification is problematic. Recent advances in membrane separation methods or acid-tolerant fermentation organisms are important for the commercial development of succinic acid production.

Itaconic acid

Itaconic acid is succinic acid with a methylene group substituted onto the carbon chain. It is typically produced in a fungal fermentation at relatively small scale. It is feasible that the economics of this fermentation could be improved in a manner similar to that for citric acid fermentation. As a result, itaconic acid could become available for use in polymer applications as well as for further processing to value-added chemical products.

Glycerol and 1,3-PDO

Bacterial fermentation of glycerol to 1,3-propanediol (PDO) has been studied as a potential renewable chemical production process. The glycerol feedstock could be generated as by-product from vegetable oil conversion to diesel fuel (biodiesel). The anaerobic conversion of glycerol to PDO has been identified in several bacterial strains, with *Clostridium butyricum* most often cited. The theoretical molar yield is 72 per cent, and final product concentration is kinetically limited to approximately 6 or 7 per cent. Genetic engineering of the bacteria to include a glucose to glycerol fermentation is underdevelopment and would result in a single-step bioprocess for PDO directly from glucose. The PDO product is a relatively new polyester source, only recently economically available from petroleum sources. Its improved properties as a fibre include both stain resistance and easier dye application.

3-Dehydroshikimic acid

This interesting building block has been identified as a potential renewable resource. It is a hydroaromatic intermediate in the aromatic amino acid biosynthetic pathway. Its formation in mutated *E. coli* and its recovery in significant yields have been reported. It can be produced from six-carbon sugars with a theoretical yield of 43 per cent, whereas a 71 per cent yield is theoretically possible from five-carbon sugars based on a different metabolic pathway. Further processing of 3-dehydroshikimic acid can lead to protocatechuic acid, vanillin, catechol, gallic acid, and adipic acid.

Catalytic/bioprocessing combinations

Combining bioprocessing systems with catalytic processing systems provides some important opportunities for biomass conversion to chemicals. Although fermentation of biomass feedstocks to useful chemicals can be a direct processing step, in many cases the fermentation product can, in turn, be transformed into a number of value-added chemical products. Thus, the fermentation product acts as a platform chemical from which numerous final products can be derived. Some of these combined bioprocessing platform chemicals and catalytically derived families of value-added chemicals.

Succinic acid derivatives

Two branches of the succinic acid family tree have been described. The first, involving hydrogenation, is similar to hydrogenation of maleic anhydride performed in the petrochemical industry based on butane oxidation. The hydrogenation products from succinic acid include γ-butyrolactone, 1,4-butanediol (BDO), and tetrahydrofuran (THF). Currently, 750 million pounds of both BDO and THF are produced per year. Lactone and THF have solvent markets, whereas BDO is an important polyester resin monomer (polybutylene terephthalate). THF is also used for Spandex fibre production. The second pathway for succinic acid chemical production is through the succinamide to form pyrrolidones. N-methylpyrrolidone is an important low-toxicity, environmentally benign solvent with a growing annual market of more than 100 million pounds, displacing chlorinated hydrocarbon solvents. 2-Pyrrolidone can form the basis for several polymers, including polyvinyl pyrrolidone.

Itaconic acid derivatives

The structural similarity of itaconic acid to that of succinic acid suggests that it could be used to produce similar families of chemical products. Typically, the products would be methylated versions of the succinic-derived products, such as 2-methyl-1,4-butanediol, 3-methyltetrahydrofuran, or 3-methyl-*N*-methylpyrrolidone. Therefore, the final products would include methylated polyesters or methylated Spandex.

Glutamic acid and lysine

Glutamic acid and lysine are major chemical products, with glutamate sold as the sodium salt, monosodium glutamate, and lysine used as an important animal feed supplement. Both compounds are produced industrially by fermentation.

Glutamate is produced as L-glutamic acid at 680 million pounds per year, with 85 per cent of the production in Asia. In the first half of the 20th century, glutamate was produced by hydrolysis of protein, primarily wheat gluten. The current fermentation process, using molasses or another cheap glucose source, had replaced protein hydrolysis processing by 1965. Because of the current cost of glutamate, it is too expensive for use as a platform chemical, and its use for chemical production is dependent on improved fermentation, recovery method modification, and integration of catalytic processing with the existing fermentation process. Worldwide lysine production is only approximately one-third that of glutamate. Thus, the cost of lysine is approximately twice that of glutamate. Consequently, chemical production based on glutamate rather than lysine appears to be more likely.

Glutamate and lysine are interesting as renewable chemical feedstocks because each provides both multifunctionality and a five-carbon backbone. The five-carbon-based polymers are not widely used and are less well-known or understood since there is no easy way to produce five-carbon petroleum products. Processing steps can be envisioned for direct conversion of glutamate either to a five-carbon diacid by deamination (or thereafter to 1,5-pentanediol) for polyester production or to a five-carbon terminal amine/acid for nylon production. Lysine could be decarboxylated to produce a five-carbon diamine or could be deaminated to the amine/acid (the same as that for glutamate) for nylon production.

Plant-derived chemicals

One strategy for the use of biomass for chemicals production is the recovery of chemical products produced directly by plants. This strategy can involve either recovery from existing plants, such as the oils from certain seed crops, or recovery from genetically modified plants in which the chemical is produced specifically for harvesting.

Oleochemicals

The oil recovered from certain seed crops has been an important food product for many centuries. The growth of scale of the processing industry in the case of corn and soyabeans has resulted in a reduction of the cost of the oil to the point that it can be considered for use in commodity chemical products. The use of vegetable oil for fuel as the transesterified methyl esters of the fatty acids derived from the triglycerides, known as biodiesel, has made some market inroads but primarily to meet alternative fuel use requirements and it is not based on demand. An important market breakthrough for renewable resources was the use of soyabean oils in printing inks, instituted in the 1990s, which has been growing steadily. They are also used in toners for photocopy machines and in adhesive formulations. Other uses, such as plasticisers or binders based on epoxide formation from the unsaturated fatty acids, have also been investigated.

Polyhydroxyalkanoates

Polyhydroxyalkanoate (PHA) production has been demonstrated by both fermentative and plant growth methods. PHA is a biodegradable polymer with applications in the biomedical field. Originally, PHA was limited to polyhydroxybutyrate (PHB), but it was found to be too brittle. A copolymer (PHBV) of 75 per cent hydroxybutyrate and 25 per cent hydroxyvalerate is now a commercial product with applications in medical devices. However, the fermentative process has some potentially significant drawbacks that, when considered on a life cycle basis, could result in poorer performance of PHA production from a renewable perspective than that of petroleum-derived polystyrene. To address these drawbacks, which include high energy requirements for cell wall rupture and product recovery, plant-based production of PHA has been investigated using genetic modification to engineer the PHB production path way into *Arabidopsis* as the host plant. Alternatively, PHBV might be produced by other organisms that can use multiple sugar forms from a less-refined biomass feedstock than corn-derived glucose.

Future Biorefinery

The future biorefinery (Fig. 13.4) is conceptualised as an optimised collection of the various process options described previously. In addition to the chemical processing steps for producing value-added chemicals, it will likely incorporate components from existing biorefinery-type operations, such as wood pulp mill, wet corn mill, and cheese manufacturing processes.

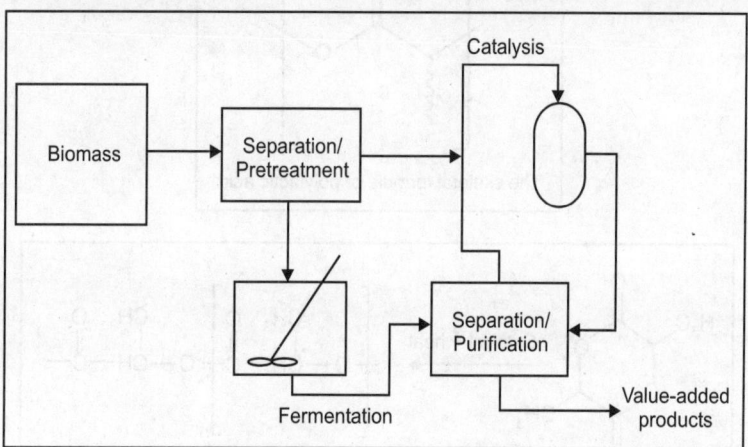

Fig. 13.4. Biorefinery concept for production of value-added chemicals from biomass.

Carbohydrate separation and recovery will likely involve both the five- and six-carbon compounds. These compounds will respond to separations by selective hydrolysis because they have different activities in either chemical or enzymatic hydrolyses. Any oil or protein components liberated through this processing will be considered as potential high-value chemicals. The lignin component, the least developed of the biomass fractions, is often viewed as a fuel to drive the processing systems, but it is also a potential source of aromatic chemical products. Mineral recovery is most likely to be important for maintaining agricultural productivity, with the return of the minerals to the croplands as fertiliser. In order to achieve a true biorefinery status, each biomass component will be used in the appropriate process to yield the highest value product. Therefore, a combination of processing steps, tailored to fit the particular operation, will be used to produce a slate of chemical products, which will be developed in response to the market drivers to optimise feedstock utilisation and overall plant income. In this manner, the return on investment for converting low-cost feedstock to high-value chemical products can be maximised. The future biorefinery will likely have a major impact on society as fossil-derived resources become scarce and more expensive. Indeed, use of biomass is the only option for maintaining the supply of new carbon-based chemical products to meet society's demands. The biorefinery can have a positive environmental impact by deriving its carbon source (via plants) through reducing atmospheric levels of carbon dioxide, an important greenhouse gas. With appropriate attention to the biomass-derived nitrogen and mineral components in the biorefinery processes, a life cycle balance may be achieved for these nutrient materials if their recovery efficiencies can be maintained at high levels.

POLYLACTIC ACID

Polylactic acid or polylactide (PLA) is a biodegradable, thermoplastic, aliphatic polyester derived from renewable resources, such as corn starch (in the US) or sugarcanes (rest of world). Although PLA has been known for more than a century, it has only been of commercial interest in recent years, in light of its biodegradability. The skeletal formula and synthesis of polylactic acid is given below.

The skeletal formula of polylactic acid

Lactide   Polylactide

Ring-opening polymerisation of lactide to polylactide

Bacterial fermentation is used to produce lactic acid from corn starch or cane sugar. However, lactic acid cannot be directly polymerised to a useful product, because each polymerisation reaction generates one molecule of water, the presence of which degrades the forming polymer chain to the point that only very low molecular weights are observed. Instead, lactic acid is oligomerised and then catalytically dimerised to make the cyclic lactide monomer. Although dimerisation also generates water, it can be separated prior to polymerisation. PLA of high molecular weight is produced from the lactide monomer by ring-opening polymerisation using most commonly a stannous octoate catalyst, but for laboratory demonstrations tin(II) chloride is often employed. This mechanism does not generate additional water, and hence, a wide range of molecular weights is accessible.

Polymerisation of a racemic mixture of L- and D-lactides usually leads to the synthesis of poly-DL-lactide (PDLLA) which is amorphous. Use of stereospecific catalysts can lead to heterotactic PLA which has been found to show crystallinity. The degree of crystallinity, and hence many important properties, is controlled by the ratio of D to L enantiomers used.

Due to the chiral nature of lactic acid, several distinct forms of polylactide exist: poly-L-lactide (PLLA) is the product resulting from polymerisation of L,L-lactide (also known as L-lactide). PLLA has a crystallinity of around 37 per cent, a glass transition temperature between 50°–80°C and a melting temperature between 173°–178°C. PLA has similar mechanical properties to PETE polymer, but has a significantly lower maximum continuous use temperature.

Polylactic acid can be processed like most thermoplastics into fibre (for example using conventional melt spinning processes) and film. The melting temperature of PLLA can be increased 40°–50°C and its heat deflection temperature can be increased from approximately 60°C to up to 190°C by physically blending the polymer with PDLA (poly-D-lactide). PDLA and PLLA form a highly regular stereocomplex with increased crystallinity. The temperature stability is maximised when a 50:50 blend is used, but even at lower concentrations of 3–10 per cent of PDLA, there is still a substantial improvement. In the latter case, PDLA acts as a nucleating agent, thereby increasing the crystallisation rate. Biodegradation of PDLA is slower than for PLA due to the higher crystallinity of PDLA. PDLA has the useful property of being optically transparent.

VEHICLE EMISSIONS CONTROL

Vehicle emissions control is the study and practice of reducing the polluting emissions produced by vehicles powered by internal combustion engines.

Motor vehicles produce many different pollutants. The principal pollutants of concern — those that have been demonstrated to have significant effects on human, animal, plant, and environmental health and welfare — include:

1. Hydrocarbons: This class is made up of unburned or partially burned fuel, and is a major contributor to urban smog, as well as being toxic. They can cause liver damage and even cancer.

2. Carbon monoxide (CO): A product of incomplete combustion, carbon monoxide reduces the blood's ability to carry oxygen; overexposure may be fatal.

3. Nitrogen oxides (NO_x): These are generated when nitrogen in the air reacts with oxygen at the high temperature and pressure inside the engine. NO_x is a precursor to smog and acid rain. NO_x is a mixture of NO and NO_2. NO_2 destroys resistance to respiratory infection.

4. Particulates — soot or smoke made up of particles in the micrometer size range: Particulate matter causes respiratory health effects in humans and animals.

5. Sulphur oxides (SO_x): A general term for oxides of sulphur, which are emitted from motor vehicles burning fuel containing a high concentration of sulphur.

Emission standards are requirements that set specific limits to the amount of pollutants that can be released into the environment. Many emissions standards focus on regulating pollutants released by automobiles (motor cars) and other powered vehicles but they can also regulate emissions from industry, power plants, small equipment such as lawn mowers and diesel generators. Frequent policy alternatives to emissions standards are technology standards (which mandate the use of a specific technology and emission trading. Standards generally regulate the emissions of nitrogen oxides (NO_x), sulphur oxides, particulate matter (PM) or soot, carbon monoxide (CO), or volatile hydrocarbons.

ACCIDENTS IN CHEMICAL PROCESS INDUSTRY

Accidents in chemical process industries constitute major threat to property and population because of the magnitude. With rapid development in science and technology, several new innovations have come up and process industries deal with thousands of new materials and several processes. Nevertheless, there are innumerable causes that lead to accidents of major or minor in nature.

The terms 'chemical accident' or 'chemical incident' refer to an event resulting in the release of a substance or substances hazardous to human health and/or the environment in the short- or long-term. Such events include fires, explosions, leakages or releases of toxic or hazardous materials that can cause people illness, injury, disability or death.

While chemical accidents may occur whenever toxic materials are stored, transported or used, the most severe accidents are industrial accidents, involving major chemical manufacturing and storage facilities. The most significant chemical accidents in recorded history was the 1984 Bhopal disaster in India, in which more than 3000 people were killed after a highly toxic vapour, (methyl isocyanate), was released at a Union Carbide pesticides factory.

Efforts to prevent accidents range from improved safety systems to fundamental changes in chemical use and manufacture, referred to as primary prevention or inherent safety.

In the UK, the UK Chemical Reaction Hazards Forum publishes reports of accidents on its web site. These accidents were, at the time, minor in nature, but they could have escalated into major accidents. It is hoped that publishing these incidents will prevent 're-inventing the wheel'.

Dynamic Simulation of Industrial Accidents

In the past, process simulators and accident simulators belonged to two distinct worlds. Dynamic process simulators were mainly applied to process control, process dynamic investigation and safety analysis (in terms of definition of emergency shutdown procedures). Accident simulators were related usually to the steady-state accident investigation, as well as risk analysis and emergency preparedness and response. The biunique interaction between these simulators allows investigating feedbacks and interactions among plants and industrial accidents. Actually, process dynamics affects the source term, i.e. the amount of released components and their chemical and physical properties, while the accident dynamics closes the loop by affecting the plant dynamics. The accident may have a direct effect on the plant dynamics (as in the presence of a heat source, e.g. a pool fire) or may have an indirect effect by influencing the integrity of field operators when the release of a toxic substance occurs.

Dynamic process simulation has become an indispensable and central tool for process design, analysis, and operation in the chemical industry. The dynamic simulation of a chemical process is a step ahead from the steady-state analysis and has some effective and significant advantages. A dynamic simulation of the process allows: checking the control system configurations before applying it to the real plant so as to uncover possible control system errors; training the operators to increase their awareness and skills; planning and testing the start-up and shutdown procedures; increasing the process safety by testing and validating the procedures in a non-destructive environment. The support of a dynamic process simulator to training allows operators gaining experience, facing malfunctions and deviations from the nominal conditions, and becoming aware of the importance of following the correct procedures. The operators can modify virtually the process variables of the dynamic simulation while quantifying the consequences on the plant conditions without incurring into real risks.

Moreover, the operators are trained to cope with plant deviations from nominal conditions due to accidental events that affect either the plant (e.g. emission, release, fire, explosion) or the people working in the plant (e.g. a fire, explosion, toxic gas cloud). Industrial accidents are dynamic phenomena that evolve depending on the environmental conditions and the characteristics of the emitted substance or mixture. Consequently, to simulate a realistic unusual situation in case of accident, both the dynamics of the plant and of the accident should be accounted for and run simultaneously, since they are interrelated and mutually influenced. For instance, process conditions determine the leakage characteristics whilst accident consequences affect the plant variables.

Integration of process and accident simulators

To evaluate simultaneously both the process condition and the accident evolution it is necessary to link the chemical process simulator to the accident simulator. Both the accident and the process dynamics should be simulated in a CPU time that allows achieving a real-time performance. Furthermore, the simulation should be so responsive to perform simulations faster than clock-wall so to shorten the dynamics of transients and catch the attention and participation of operators.

With reference to accident simulation, there is no commercial software based on a single and unified model capable of modelling and simulating interconnected and consequent events, e.g. emission → spreading → evaporation → ignition → fire → dispersion. Actually, there are a number of programs that simulate only some of the aforementioned events and ask the user to specify some input data that can be computed by other software. These programs force the user to pass manually the output of a program to the following one as input data. In addition, it is still missing an automated procedure that links the single pieces of software to get the whole picture of the accidental event. Finally, the commercial software was coded for direct and interactive used by means of a GUI (graphical user interface).

The security and economic stability of many nations and multinational oil companies are highly dependent on the safe and uninterrupted operation of their oil, gas and chemical facilities. One of the most critical impacts that can occur to these operations are fires and explosions from accidental incidents.

Flash fires and explosions are common hazards at a variety of workplaces. These hazards are present in work areas where flammable materials are handled, processed, stored, or in any way present. In the petrochemical industry for example, flash fire can occur at well head sites, collection points, compressor stations, refineries, and petrochemical and plastic plants. In such areas, the potential exists for developing an explosive atmosphere capable of injuring or killing workers and causing extensive property damage.

Industrial flash fires and explosions result from the accidental release and ignition of flammable fuels. The size and duration of the flame that results from this ignition is determined by the amount of fuel available, the efficiency of combustion, and the environmental and physical characteristics of the site of the flash fire or explosion. The temperatures attained by flash fires have been estimated to range from 550° to 1050°C, although higher temperatures are believed to occur. Even the lowest estimated temperature exceeds the temperature at which most regular clothing fabrics burst into flames.

BIOLOGICAL HAZARD

A biological hazard, also known as a biohazard, is an organism or a by-product from an organism that is harmful or potentially harmful to other living things, primarily human beings. Common types of biological hazards include viruses, medical waste, or toxins that were created by a particular organism or micro-organism. The 'biohazard symbol' is a familiar sight in hospitals, and any object that carries it should be treated with extreme caution.

There are four levels of biohazards, classified by the Center for Disease Control and Prevention (CDC) in the United States. A level 1 biological hazard poses the least risk while a level 4 poses the greatest. The CDC has a great number of procedures in place for the prevention of a biological hazard disaster, and many different ways to handle and clean up small and large-scale contaminant disasters.

Level 1 biological hazards consist mainly of bacteria and other micro-organisms, which pose little risk in the case of exposure and can be warded off through the simple use of gloves or a mask. They can generally be disposed of in their own separate trash container without worry and are easily decontaminated. Level 2 biohazards consist of viruses and bacteria that can have a limited detrimental effect on humans, for example they may cause a disease such as salmonella poisoning, hepatitis, measles, Lyme disease and more. People working in the presence of these biological hazards will usually exercise a substantial amount of care in their handling and disposal with proper hand, eye and body protection.

When biological hazards are considered to be at level 3, they become much more serious, since contact with them can now be fatal if left untreated. Examples if biohazards at this dangerous level are anthrax, West Nile Virus, malaria, typhus and more. People that aid victims of biological hazards at this level must be well trained and use very specific safety equipment and clothing so they do not contract a deadly disease. The same is obviously true of level 4 biohazards, for which there are no known treatments, are generally fatal, and are easily spread through contact and through the air. Someone working in the vicinity of level 4 biohazards like the Ebola virus or dengue fever, for example, must use an airtight Hazmat suit with his or her own oxygen supply.

WASTE TREATMENT

Waste treatment refers to the activities required to ensure that waste has the least practicable impact on the environment. In many countries various forms of waste treatment are required by law.

Solid Waste Treatment

The treatment of solid wastes is a key component of waste management. Different forms of solid waste treatment are graded in the waste hierarchy.

Waste-water Treatment

Agricultural waste-water treatment

Agricultural waste-water treatment is treatment and disposal of liquid animal waste, pesticide residues, etc. from agriculture.

Industrial waste-water treatment

Industrial waste-water treatment is the treatment of wet wastes from manufacturing industry and commerce including mining, quarrying and heavy industries.

Sewage treatment

Sewage treatment is the treatment and disposal of human waste. Sewage is produced by all human communities and is often left to compost naturally or is treated using processes that separate solid materials by settlement and then convert soluble contaminants into biological sludge and into gases such as carbon dioxide or methane.

Radioactive waste treatment

Radioactive waste treatment is the treatment and containment of radioactive waste.

Water Pollution

Water pollution is the contamination of water bodies (e.g. lakes, rivers, oceans, groundwater). Water pollution affects plants and organisms living in these bodies of water; and, in almost all cases the effect is damaging either to individual species and populations, but also to the natural biological communities. Water pollution occurs when pollutants are discharged directly or indirectly into water bodies without adequate treatment to remove harmful compounds. Water pollution is a major problem in the global context. It has been suggested that it is the leading worldwide cause of deaths and diseases, and that it accounts for the deaths of more than 14,000 people daily.

Water is typically referred to as polluted when it is impaired by anthropogenic contaminants and either does not support a human use, like serving as drinking water, and/or undergoes a marked shift in its ability to support its constituent biotic communities, such as fish. Natural phenomena such as volcanoes, algae blooms, storms, and earthquakes also cause major changes in water quality and the ecological status of water. Sources of surface water pollution are generally grouped into two categories based on their origin.

Point source pollution refers to contaminants that enter a waterway through a discrete conveyance, such as a pipe or ditch. Examples of sources in this category include discharges from a sewage treatment plant, a factory, or a city storm drain.

Non-point source (NPS) pollution refers to diffuse contamination that does not originate from a single discrete source. NPS pollution is often the cumulative effect of small amounts of contaminants gathered from a large area. The leaching out of nitrogen compounds from agricultural land which has been fertilised is a typical example. Nutrient runoff in stormwater from 'sheet flow' over an agricultural field or a forest are also cited as examples of NPS pollution. The specific contaminants leading to pollution in water include a wide spectrum of chemicals, pathogens, and physical or sensory changes such as elevated temperature and discolouration. While many of the chemicals and substances that are regulated may be naturally occurring (calcium, sodium, iron, manganese, etc.) the concentration is often the key in determining what is a natural component of water, and what is a contaminant. Oxygen-depleting substances may be natural materials, such as plant matter (e.g. leaves and grass) as well as man-made chemicals. Other natural and anthropogenic substances may cause turbidity (cloudiness) which blocks light and disrupts plant growth, and clogs the gills of some fish species.

Many of the chemical substances are toxic. Pathogens can produce waterborne diseases in either human or animal hosts. Alteration of water's physical chemistry includes acidity (change in pH), electrical conductivity, temperature, and eutrophication. Eutrophication is an increase in the concentration of chemical nutrients in an ecosystem to an extent that increases in the primary productivity of the ecosystem. Depending on the degree of eutrophication, subsequent negative environmental effects such as anoxia (oxygen depletion) and severe reductions in water quality may occur, affecting fish and other animal populations.

Water pollution may be analysed through several broad categories of methods: physical, chemical and biological. Most involve collection of samples, followed by specialised analytical tests. Some methods may be conducted *in situ,* without sampling, such as temperature. Government agencies and research organisations have published standardised, validated analytical test methods to facilitate the comparability of results from disparate testing events.

Air Pollution

Air pollution is the introduction of chemicals, particulate matter, or biological materials that cause harm or discomfort to humans or other living organisms, or damages the natural environment into the atmosphere. The atmosphere is a complex dynamic natural gaseous system that is essential to support life on planet earth. Stratospheric ozone depletion due to air pollution has long been recognised as a threat to human health as well as to the earth's ecosystems.

An air pollutant is known as a substance in the air that can cause harm to humans and the environment. Pollutants can be in the form of solid particles, liquid droplets, or gases. In addition, they may be natural or man-made. Pollutants can be classified as either primary or secondary. Usually, primary pollutants are substances directly emitted from a process, such as ash from a volcanic eruption, the carbon monoxide gas from a motor vehicle exhaust or sulphur dioxide released from factories.

Secondary pollutants are not emitted directly. Rather, they form in the air when primary pollutants react or interact. An important example of a secondary pollutant is ground level ozone—one of the many secondary pollutants that make up photochemical smog. Note that some pollutants may be both primary and secondary: that is, they are both emitted directly and formed from other primary pollutants.

Major primary pollutants produced by human activity include:

1. Sulphur oxides (SO_x) — especially sulphur dioxide, a chemical compound with the formula SO_2. SO_2 is produced by volcanoes and in various industrial processes.
2. Nitrogen oxides (NO_x) — especially nitrogen dioxide are emitted from high temperature combustion. Can be seen as the brown haze dome above or plume downwind of cities.
3. Carbon monoxide — is a colourless, odourless, non-irritating but very poisonous gas. It is a product by incomplete combustion of fuel such as natural gas, coal or wood. Vehicular exhaust is a major source of carbon monoxide.
4. Carbon dioxide (CO_2) — a greenhouse gas emitted from combustion but is also a gas vital to living organisms. It is a natural gas in the atmosphere.
5. Volatile organic compounds — VOCs are an important outdoor air pollutant. In this field they are often divided into the separate categories of methane (CH_4) and non-methane (NMVOCs). Methane is an extremely efficient greenhouse gas which contributes to enhanced global warming. Other hydrocarbon VOCs are also significant greenhouse gases via their role in creating ozone and in prolonging the life of methane in the atmosphere, although the effect varies depending on local air quality.
6. Particulate matter — Particulates, alternatively referred to as particulate matter (PM) or fine particles, are tiny particles of solid or liquid suspended in a gas. In contrast, aerosol refers to particles and the gas together. Sources of particulate matter can be man-made or natural. Some particulates occur naturally, originating from volcanoes, dust storms, forest and grassland fires, living vegetation, and sea spray.

The most important environmental measurement is generally the density or concentration of a pollutant. The impact of a particular pollutant is typically assumed or observed to have a direct relation to its prevalence in the system or concentration.

Density or concentration is measured by the ratio of the quantity of a substance per unit volume or mass of the system in which the substance resides. For this purpose, the system is generally defined as a single phase, either air, water, soil, or oil or a nonaqueous phase. To minimise confusion, let us assign letters to each of these phases as shown in Table 13.3. A subscripted letter will be used to indicate that phase.

Table 13.3. Nomenclature for density and concentration.

Concentration symbols		Phase indicator subscripts	
φ	Volume fraction	a	Air
ω	Weight fraction	w	Water
x,y,z	Mole fraction	s	Soil
C	Moles/mass contaminant per volume of fluid	o,n	Oily/nonaqueous phase
W	Moles/mass contaminant per mass of solid/sorbent	b	Animal or human phase (biological phase)
ρ	Moles/mass solvent/sorbent per volume	v	Vapour phase (e.g. in soil)
		pw	Porewater phase (e.g. in soil)
*	Superscript indicates equilibrium	oc	Organic carbon

Some common examples:

C_w^*	Concentration of a contaminant in water, mg/l
C_w	Concentration of a contaminant in water in equilibrium with an adjacent phase (e.g. air), mg/l
ρ_w	Density of water, mg/l
W_s	Mass fraction of contaminant on soil, mg/kg
C_v	Concentration of a contaminant in soil vapour, mg/l
ω_{oc}	Fraction organic carbon in soil, g/g
ρ_{oc}	Density of organic carbon (e.g. in water), mg/l

Thus, the volume of an arbitrarily defined air phase system might be denoted as V_a and that of a similarly defined soil system, V_s. It becomes important to differentiate between the concentration of a component in a phase vs. the density of the phase itself. Let C_a represent the concentration of a component of interest in the air, while ρ_a represents the density of the bulk air phase.

Density or concentration is a continuum quantity. A fluid is a collection of individual molecules. Density or concentration only has a meaning when a measuring volume is sufficiently large that it contains a large number of molecules as shown in Fig. 13.5. The spacing between molecules is of the order of the mean free path between collisions and measuring volumes of this size would be influenced dramatically by the random motion of molecules in and out of the region. Figure 13.5 also depicts the density that would be measured as a function of measuring volume if successively larger measuring volumes were collected in a stationary or time-independent system.

The measuring volume must be very much larger than the mean spacing between molecules before the random motion of the molecules is statistically averaged such that consistent measures of the number of molecules per unit volume are found. Luckily, this does not generally pose a problem as the number of molecules in 1 cm^3 of a gas at 25°C approaches 2.5×10^{19}. Densities in liquids are typically of the order of 1000 times greater and densities in solids are typically slightly greater still. A mole of a substance is the quantity of that substance that contains Avogadro's number of molecules, that is $6.023 \times (10^{23})$ molecules.

This measure of the quantity of a substance is very useful when considering processes or behaviour that depend on the number of molecules of a substance rather than the mass of that substance. Chemical reactions, for example, generally occur at a reaction rate that is a function of the number of molecules per unit volume.

The ratio of reactants to products in a chemical reaction is also a function of molecular ratios and more easily measured in a quantity unit that measures number of molecules rather than mass. In addition, a molecule of a gas at low pressure and a given temperature will fill approximately the same volume, regardless of the mass of the molecule. In particular, at 0°C and 1 atmosphere pressure, the volume of 1 mole of any gas is about 22.4 litres. At 25°C and 1 atmosphere pressure, the volume of 1 mole of any gas is about 24.5 litres. Under each of these conditions, it is often more convenient to work in molecular units; that is moles.

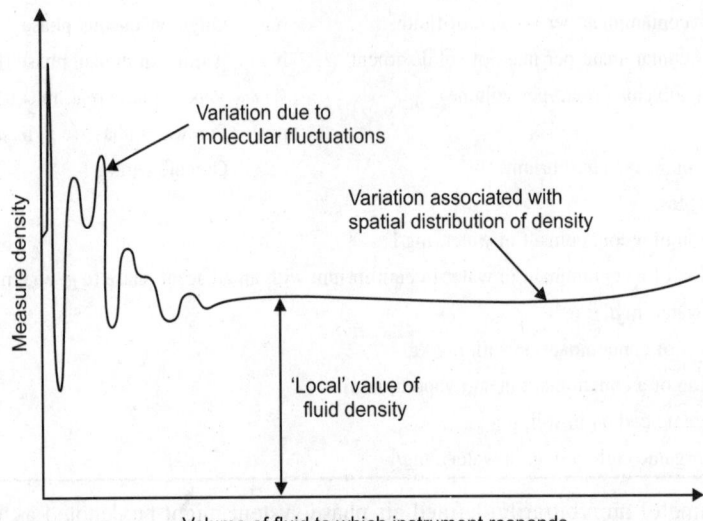

Fig. 13.5. Fluid density as a function of averaging volume indicating that a minimum volume is required to define a continuum.

Another way of defining a mole of a substance is that quantity of the substance whose weight in grams is equal to the molecular weight of the substance. Just 1 mole of atomic oxygen has a mass of about 16 grams while 1 mole of atmospheric oxygen (O_2) has a mass of about 32 grams. Therefore, the number of moles of any substance, n, is its mass m, in grams divided by its molecular weight.

$$n_A = \frac{m_A}{MW_A} \qquad \qquad ... (13.1)$$

Troposphere and Stratosphere

Troposphere

The troposphere is the lowest portion of earth's atmosphere. It contains approximately 75 per cent of the atmosphere's mass and 99 per cent of its water vapour and aerosols.

The chemical composition of the troposphere is essentially uniform, with the notable exception of water vapour. The source of water vapour is at the surface through the processes of evaporation and transpiration. Furthermore, the temperature of the troposphere decreases with height, and saturation vapour pressure decreases strongly as temperature drops, so the amount of water vapour that can exist in the atmosphere decreases strongly with height. Thus, the proportion of water vapour is normally greatest near the surface and decreases with height.

The pressure of the atmosphere is maximum at sea level and decreases with higher altitude. This is because the atmosphere is very nearly in hydrostatic equilibrium, so that the pressure is equal to the weight of air above a given point. The change in pressure with height therefore can be equated to the density with this hydrostatic equation:

$$\frac{dp}{dz} = -\rho g_n = -\frac{mpg}{RT}$$

where,

g_n stands for the standard gravity
ρ stands for density
z stands for height
p stands for pressure
R stands for the gas constant
T stands for temperature in kelvins
m stands for the molar mass.

Since temperature in principle also depends on altitude, one needs a second equation to determine the pressure as a function of height.

The temperature of the troposphere generally decreases as altitude increases. The rate at which the temperature decreases, $-dT/dz$, is called the environmental lapse rate (ELR). The ELR is nothing more the difference in temperature between the surface and the tropopause divided by the height. The reason for this temperature difference is the absorption of the suns energy occurs at the ground which heats the lower levels of the atmosphere, and the radiation of heat occurs at the top of the atmosphere cooling the earth, this process maintaining the overall heat balance of the earth. As parcels of air in the atmosphere rise and fall, they also undergo changes in temperature for reasons described below. The rate of change of the temperature in the parcel may be less than or more than the ELR. When a parcel of air rises, it expands, because the pressure is lower at higher altitudes. As the air parcel expands, it pushes on the air around it, doing work; but generally it does not gain heat in exchange from its environment, because its thermal conductivity is low (such a process is called adiabatic). Since the parcel does work and gains no heat, it loses energy, and so its temperature decreases. (The reverse, of course, will be true for a sinking parcel of air.) The tropopause is the boundary region between the troposphere and the stratosphere. Measuring the temperature change with height through the troposphere and the stratosphere identifies the location of the tropopause. In the troposphere, temperature decreases with altitude.

Stratosphere

The stratosphere is the second major layer of earth's atmosphere, just above the troposphere, and below the mesosphere. It is stratified in temperature, with warmer layers higher up and cooler layers farther down. This is in contrast to the troposphere near the earth's surface, which is cooler higher up and warmer farther down.

The border of the troposphere and stratosphere, the tropopause, is marked by where this inversion begins, which in terms of atmospheric thermodynamics is the equilibrium level. The stratosphere is situated between about 10 km (6 miles) and 50 km (31 miles) altitude above the surface at moderate latitudes, while at the poles it starts at about 8 km (5 miles) altitude.

The stratosphere is layered in temperature because it is heated from above by absorption of ultraviolet radiation from the sun. Within this layer, temperature increases as altitude increases; the top of the stratosphere has a temperature of about 270 K (–3°C or 29.6°F), just slightly below the freezing point of water. This top is called the stratopause, above which temperature again decreases with height. The vertical stratification, with warmer layers above and cooler layers below, makes the stratosphere dynamically stable: there is no regular convection and associated turbulence in this part of the atmosphere.

Commercial airliners typically cruise at altitudes of 9–12 km in temperate latitudes, in the lower reaches of the stratosphere. They do this to optimise jet engine fuel burn, mostly thanks to the low temperatures encountered near the tropopause.

ECOSYSTEM MODEL

Ecosystem models, or ecological models, are mathematical representations of ecosystems. Typically they simplify complex foodwebs down to their major components or trophic levels, and quantify these as either numbers of organisms, biomass or the inventory/concentration of some pertinent chemical element (for instance, carbon or a nutrient species such as nitrogen or phosphorus). Ecosystem models are a development of theoretical ecology that aim to characterise the major dynamics of ecosystems, both to synthesise the understanding of such systems and to allow predictions of their behaviour (in general terms, or in response to particular changes).

Because of the complexity of ecosystems (in terms of numbers of species/ecological interactions), ecosystem models typically simplify the systems they are studying to a limited number of pragmatic components. These may be particular species of interest, or may be broad functional types such as autotrophs, heterotrophs or saprotrophs. In biogeochemistry, ecosystem models usually include representations of nonliving 'resources' such as nutrients, which are consumed by (and may be depleted by) living components of the model (Fig. 13.6).

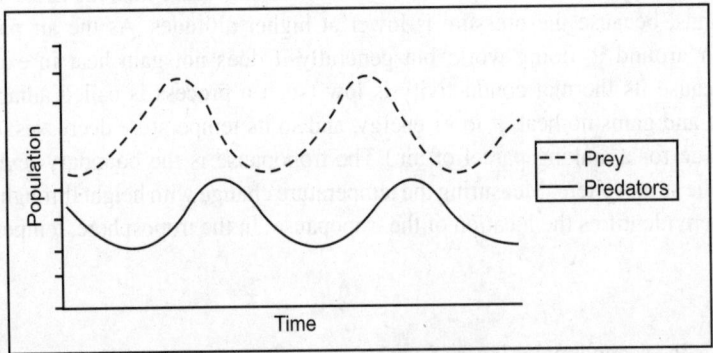

Fig. 13.6. A sample time-series of the Lotka-Volterra model. Note that the two populations exhibit cyclic behaviour, and that the predator cycle lags behind that of the prey.

One of the earliest, and most well-known, ecological models is the predator-prey model of Alfred J. Lotka and Vito Volterra. This model takes the form of a pair of ordinary differential equations, one representing a prey species, the other its predator.

$$\frac{dX}{dt} = \alpha.X - \beta.X.Y$$

$$\frac{dY}{dt} = \gamma.\beta.X.Y - \delta.Y$$

where,

X is the number/concentration of the prey species.

Y is the number/concentration of the predator species.

α is the prey species' growth rate.

β is the predation rate of Y upon X.

γ is the assimilation efficiency of Y.

δ is the mortality rate of the predator species.

Volterra originally devised the model to explain fluctuations in fish and shark populations observed in the Adriatic Sea after the First World War (when fishing was curtailed). However, the equations have subsequently been applied more generally. Although simple, they illustrate some of the salient features of ecological models: modelled biological populations experience growth, interact with other populations (as either predators, prey or competitors) and suffer mortality.

Glossary

Absorption	:	Uptake of a gas into the bulk of a liquid. Gas absorption takes place for example in the liquid of a scrubber tower where an upstreaming gas is washed by a down-going flow of a scrubber solution.
Acrolein	:	Acrolein is a highly toxic, flammable material with extreme lacrimatory properties. At room temperature acrolein is a liquid with volatility and flammability somewhat similar to acetone, but unlike acetone, its solubility in water is limited.
Activated complex	:	An intermediate structure formed in the conversion of reactants to products. The activated complex is the structure at the maximum energy point along the reaction path; the activation energy is the difference between the energies of the activated complex and the reactants.
Activation energy	:	Activation energy can be thought of as the height of the potential barrier (sometimes called the energy barrier) separating two minima of potential energy (of the reactants and products of a reaction).
Additives	:	In the manufacturing process of plastics, polymer is just one constituent. There are other chemicals like impact modifiers, colourants, reinforcements, plasticisers and stabilisers, etc. that give specific properties to the plastics. These are called additives.
Adiabatic	:	Without loss or gain of heat.
Adiabatic operations	:	In adiabatic operation, escapement occurs simultaneously with shifting balance.
Adsorption	:	Attachment of a molecule or atom to a solid surface. Adsorption involves a chemical bond between the adsorbed species and the surface.
Aliphatic	:	Any organic compound in which the main structure is a chain of carbon atoms joined to each other.
Alkali	:	A compound that has the ability to neutralise an acid to form a salt. A substance which is somewhat irritating or corrosive to the skin, eyes and mucous membranes. Turns red litmus paper to blue. Common strong alkalis are sodium and potassium hydroxide.
Alkenes	:	Unsaturated hydrocarbons that contain one or more carbon-carbon double bonds.
Amines	:	Amines are organic compounds containing nitrogen. Here one or more hydrogen atoms are replaced by alkyl groups or other groups where the nitrogen is bonded to a carbon atom in the group. Amine are used in rubber, dyes, pharmaceuticals, and synthetic resins and fibres and in a host of other applications.
Amorphous solid	:	A noncrystalline solid with no well-defined ordered structure.
Analytical integration	:	Analytical integration is an attractive approach to improve the accuracy of the results.

473

Anisotropy	:	Variation of a transport property in different directions in a material. It is often obtained from homogenisation of regular structures, for example, monolithic structures in tubular reactors.
Arbitrary reaction kinetics	:	The rate equation for an arbitrary mth order growth or decay reaction can be expressed in terms of the q-exponential function $\exp_q f = [1 + (1\ q)f]^{1/(1-q)}$, with q equal to m. The analysis suggests that a wide variety of reaction rate (kinetic) processes and models, in chemistry, biology/ecology, unit operations and contaminant transport, are amenable to analysis by Tsallis' non-extensive statistical mechanics.
Aromatics	:	Aromatics, so-called because of their distinctive perfumed smell, are a group of hydrocarbons including, mainly, benzene, toluene and the xylenes. These are basic chemicals used as starting materials for a wide range of consumer products. Almost all aromatics come from crude oil, although small quantities are made from coal.
Arrhenius equation	:	The Arrhenius equation is a simple, but remarkably accurate, formula for the temperature dependence of the rate constant, and therefore, rate of a chemical reaction.
Arrhenius rate equation	:	Expression that relates the rate constant of a chemical reaction to the exponential of the temperature.
Atom	:	Atom is the basic building block of chemistry. Atoms, also called chemical elements, can combine with one another to form compound. It is the smallest unit of matter that cannot be decomposed into simpler substances by ordinary chemical processes.
Automobile antifreeze	:	Any substance that lowers the freezing point of water, protecting a system from the ill effects of ice formation. Antifreezes such as ethylene glycol or propylene glycol commonly added to water in automobile cooling systems prevent damage to radiators.
Azeotrope	:	A specific mixture of components, which at a given pressure cannot be separated by distillation, i.e. the liquid and vapour phases have the same compositions.
Batch reactor	:	Reactor characterised by its operation, which means that the reactor does reaches steady-state.
Biocide	:	Biocides are formulations of one or more active substances which can kill or control viruses, bacteria, algae, moulds or yeasts.
Bipolar plate	:	Electrically conducting plate connected to the anode on one side and to the cathode on the other side in an electrochemical cell.
Boundary layer	:	Region in a fluid close to a solid surface. This region is characterised by large gradients in velocity and is often treated with approximative methods, because it is difficult to geometrically resolve the large gradients found there.
Brinkman equations	:	Extension of Darcy's law in order to include the transport of momentum through shear in porous media flow.
Bubble point	:	Upon heating a liquid mixture, this is the point at which bubbles first appear.
Bubbling fluidised bed (BFB)	:	Bubbling fluid bed (BFB) boiler is especially well-suited for the waste fuels found at paper mills. BFB technology is ideal for repowering an existing facility, recovery boiler conversions, or when changing fuel or firing techniques of an existing power boiler.

Butler-Volmer equation	:	Expression that relates the reaction rate of an electron transfer reaction on an electrode surface to the exponential of the overpotential. The equation can be derived from the Arrhenius rate equation by accounting for the contribution of the electric potential to the activation energy.
Catalyst/Catalysis	:	A substance that increases the rate of a chemical reaction, without being consumed or produced by the reaction. Catalysts speed both the forward and reverse reactions, without changing the position of equilibrium. Enzymes are catalysts for many biochemical reactions.
Catalytic cracking	:	The process of breaking up heavier hydrocarbon molecules into lighter hydrocarbon fractions by use of heat and catalysts.
Category	:	A group of closely related chemicals whose physico-chemical, ecotoxicological or toxicological properties follow a regular pattern because of structural similarity.
Centrifugal compressor	:	A centrifugal compressor compresses air or gas by means of mechanical rotating vanes or impellers.
Chemical reaction	:	A chemical process in which substances are changed into different substances with different properties.
Chemical recycling	:	Chemical recycling is the process of recycling waste products by partially altering their chemical structure.
Chromatography	:	Chromatography is a process for separating mixtures such as gases into their component parts for analytical purposes.
Circulating fluidised bed (CFB)	:	A circulating fluidised bed is a relatively new and evolving technology that has become a very efficient method of generating low-cost electricity while generating electricity with very low emissions and environmental impacts.
Closed vessel	:	A closed vessel is an apparatus with very thick walls, in which propellants under test are burnt.
Collision frequency	:	The average number of collisions that a molecule undergoes each second.
Collision theory/model	:	A theory that explains reaction rates in terms of collisions between reactant molecules.
Complex reactions	:	A chemical reaction that takes place between a metal ion and a molecular or ionic entity known as a ligand that contains at least one atom with an unshared pair of electrons.
Composite	:	A solid material made of two or more different substances, combined to produce a new substance whose properties are superior to the original components in a specific application.
Composite/Parallel reactions	:	A chemical reaction for which the expression for the rate of disappearance of a reactant (or rate of appearance of a product) involves rate constants of more than a single elementary reaction.
Contaminant	:	An impurity not intended to be present in the product that may be introduced through such things as poor cleaning, processing, lack of appropriate environmental and personnel controls during the manufacturing process, handling and distribution.
Continuous reactor	:	Reactor that operates without interruption. This type of reactor is characterised by its steady-state operation.
Continuous stirred tank reactor (CSTR)	:	This may be thought of as a tank to which reactants flow in, and products flow out. In an ideal CSTR the contents of the reactor are uniformly distributed.

Continuum decay reactions	:	The kinematics of continuum decay reactions is derived from that for continuum two-body reactions. The difference is that the initial unstable nucleus plays the role of the incident particle, and the target is null.
Darcy's law	:	Equation that gives the velocity vector as proportional to the pressure gradient. Often used to describe flow in porous media.
Deactivating catalysts	:	The loss of catalytic activity and/or selectivity over time, is of crucial importance in three-way catalysis where catalytic materials are exposed to high temperatures under fluctuating conditions.
Dew point	:	Upon cooling a vapour mixture, this is the point at which droplets of liquid first appear.
Differential reactor	:	Differential flow reactors are typically used for the determination of the kinetics of catalytic reactions. The reactor is operated as a differential reactor.
Diffusion layer	:	Fictitious layer in a fluid close to a solid surface where a chemical reaction takes place. The flux of species perpendicular to the surface in this layer is dominated by diffusion.
Diffusivity	:	The diffusivity of a solute defines the rate of transfer of the solute in a given fluid under the driving force of a concentration gradient.
Dioxins (PCDDs)	:	General name given to 210 organic compounds containing carbon, oxygen and hydrogen with one to eight chlorine atoms.
Dispersion model	:	The dispersion model is used to describe nonideal tubular reactors.
Distillation	:	The process of boiling a liquid and collecting its condensed vapour. This process is used to purify liquids and to separate liquid mixtures.
E. coli bacteria	:	*Escherichia coli* (*E. coli*) are members of a large group of bacterial germs that inhabit the intestinal tract of humans and other warm-blooded animals (mammals, birds).
Electroneutrality condition	:	Condition that states that the sum of charges in a control volume in an electrolyte should be zero.
Electroosmosis	:	Onset of a flow due to the application of an external electric field or due to the formation of an electric field created by ion transport in membranes, for example.
Electrophoresis	:	Migration of charged electrolyte ions in an electric field.
Elementary reaction	:	Compare with net chemical reaction. A reaction that occurs in a single step. Equations for elementary reactions show the actual molecules, atoms, and ions that react on a molecular level.
Emulsifier	:	An emulsifier (or emulsifying agent) is a substance which can be used to produce an emulsion out of two liquids that normally cannot be mixed together (such as oil and water). Emulsifiers are common in foods to maintain consistency within puddings, powders, etc.
Enothermic reactions	:	An endothermic reaction is any chemical reaction that absorbs heat from its environment.
Enzyme	:	Protein or protein-based molecules that speed up chemical reactions occurring in living things. Enzymes act as catalysts for a single reaction, converting a specific set of reactants (called substrates) into specific products. Without enzymes life as we know it would be impossible.
Equilibrium constants from thermodynamics	:	A thermodynamic system is said to be in thermodynamic equilibrium when it is in thermal equilibrium, mechanical equilibrium, radiative equilibrium, and chemical equilibrium. Classical thermodynamics deals with dynamic equilibrium states.

Euler flow	:	Flow at high velocities, where incompressibility of the fluid is of importance whereas the influence of viscous momentum transport is negligible.
Exothermic reactions	:	An exothermic reaction is a chemical reaction that releases energy in the form of heat. It is the opposite of an endothermic reaction.
Extinguishing agents	:	Media suitable for controlling or putting out a fire, when properly applied.
Fast fluidised bed (FFB)	:	Fast fluidised bed is a technique for bringing the high velocity intimate contact with a fine solid in what is essentially an entrsined, dense suspension.
Fick's law	:	The first law relates the concentration gradients to the diffusive flux of a solute infinitely diluted in a solvent. The second law introduces the first law into a differential material balance for the solute.
First order reaction	:	Compare with zero order reaction and second order reaction. The sum of concentration exponents in the rate law for a first order reaction is one. Many radioactive decays are first order reactions.
Flame retardants	:	Any chemical compound used to raise the ignition point of such materials as cloth or plastic, and thereby increase their resistance to combustion.
Flocculant	:	Flocculants are products used in waste treatment to separate unwanted components from water and sludge.
Flow system	:	An alternative to conventional programming languages and architectures which is able to achieve a high degree of parallel computation, in which values rather than value containers are dealt with, and in which all processing is achieved by applying functions to values to produce new values.
Fluid catalytic cracking	:	Fluid catalytic cracking (FCC) is the most important conversion process used in petroleum refineries. It is widely used to convert the high-boiling, high-molecular weight hydrocarbon fractions of petroleum crude oils to more valuable gasoline, olefinic gases and other products.
Free radicals	:	Free radicals are atoms, molecules, or ions with unpaired electrons on an open shell configuration. Free radicals may have positive, negative or zero charge.
Fully developed laminar flow	:	Laminar flow along a channel or pipe that only has velocity components in the main direction of the flow. The velocity profile perpendicular to the flow does not change downstream in the flow.
Half life	:	The half life of a reaction is the time required for the amount of reactant to drop to one half its initial value.
Heat duty	:	Heat absorbed by the process, usually expressed as British thermal unit (BTU)/hr.
Heat of reaction	:	The amount of heat that must be added or removed during a chemical reaction in order to keep all of the substances present at the same temperature.
Heat of temperature	:	Temperature is a number that is related to the average kinetic energy of the molecules of a substance. If temperature is measured in Kelvin degrees, then this number is directly proportional to the average kinetic energy of the molecules. Heat is a measurement of the total energy in a substance. That total energy is made up of not only of the kinetic energies of the molecules of the substance, but total energy is also made up of the potential energies of the molecules.
Heterogeneous	:	Containing more than one phase.
Heterogeneous reaction	:	Reaction that takes place at the interface between two phases.
Homogeneous	:	Containing a single phase.
Homogeneous reaction	:	Reaction that takes place in the bulk of a solution.

Hydrodynamic flow models	:	Hydrodynamics is the study of motion of liquids, and in particular water. A hydrodynamic model is a tool able to describe or represent in some way the motion of water.
Ideal reactors	:	A reactor is a system (volume) with boundaries. Mass may enter and leave across boundary.
Immiscible	:	Two liquids are said to be immiscible, if when added together they do not mix but form two separate liquid phases.
Integrated rate law	:	Rate laws like $d[A]/dt = -k[A]$ give instantaneous concentration changes. To find the change in concentration over time, the instantaneous changes must by added (integrated) over the desired time interval. The rate law $d[A]/dt = -k[A]$ can be integrated from time zero to time t to obtain the integrated rate law $\ln([A]/[A]_0) = -kt$, where, $[A]_0$ is the initial concentration of A.
Isomerisation	:	The chemical process by which a compound is transformed into any of its isomers, i.e. forms with the same chemical composition but with different structure or configuration and, generally different physical and chemical properties.
Kinetic expression/ Rate equation	:	The rate law or rate equation for a chemical reaction is an equation which links the reaction rate with concentrations or pressures of reactants and constant parameters (normally rate coefficients and partial reaction orders).
Laminar flow	:	Laminar flow, sometimes known as streamline flow, occurs when a fluid flows in parallel layers, with no disruption between the layers.
Mass transfer	:	Mass transfer is the net movement of mass from one location to another. Mass transfer is used by different scientific disciplines for different processes and mechanisms.
Maxwell-Stefan equations	:	Set of equations that describe the diffusion of solutes and solvent in a concentrated solution. In such a solution, the solutes interact with each other and with the solvent.
Microbial fermentation	:	Fermentation is the enzymatic decomposition and utilisation of foodstuffs, particularly carbohydrates, by microbes. Fermentation takes place throughout the gastrointestinal tract of all animals, but the intensity of fermentation depends on microbe numbers, which are generally highest in the large bowel.
Microfluids	:	Microfluidics deals with the behaviour, precise control and manipulation of fluids that are geometrically constrained to a small, typically sub-millimetre, scale.
Minimum fluidising velocity	:	A general equation is proposed for predicting the minimum fluidisation velocity of a mixture of particles of various sizes but all of the same shape and density.
Miscible fluids	:	A cutting fluid which is a mixture of water and soluble oil and/or chemical agents.
Molar volume	:	The molar volume, symbol V_m, is the volume occupied by one mole of a substance (chemical element or chemical compound) at a given temperature and pressure.
Molecular/Reaction intermediates	:	A reaction intermediate or an intermediate is a molecular entity that is formed from the reactants (or preceding intermediates) and reacts further to give the directly observed products of a chemical reaction.
Molecules	:	A molecule is defined as an electrically neutral group of at least two atoms in a definite arrangement held together by very strong (covalent) chemical bonds.
Monolithic reactor	:	Catalytic reactor made of one single piece of solid material. Incorporates a catalytic structure in its often porous structure.

Navier-Stokes equations	:	Equations for the momentum balances coupled to the equation of continuity for a Newtonian incompressible fluid.
Nernst-Planck equation	:	Equation that describes the flux of an ion through diffusion, convection, and migration in an electric field. The equation is valid for diluted electrolytes.
Newtonian flow	:	Flow characterised by a constant viscosity or a viscosity that is independent of the shear rate in the fluid.
Nonadiabatic reactions	:	Theoretical explanations of the chemical processes leading to the generation of a mass spectrum are usually given in terms of the characteristics of the potential energy surface of the ground electronic state of the molecular ion.
Nonelementary reactions	:	In these reactions, intermediates may not be observed or quantitated, either because they are presents in very small amounts or because they are unstable.
Nonpolar substances	:	Substances, like fats, are non-polar. They will not dissolve in water. These substances are neutral and there is no excess charge at one end of the molecule.
Numerical integration	:	In numerical analysis, numerical integration constitutes a broad family of algorithms for calculating the numerical value of a definite integral, and by extension, the term is also sometimes used to describe the numerical solution of differential equations.
Order of reaction/ reaction order	:	The order of a reaction is the sum of concentration exponents in the rate law for the reaction. For example, a reaction with rate law $d[C]/dt = k[A]^2[B]$ would be a third order reaction. Noninteger orders are possible.
Packed bed catalyst	:	A packed bed is a hollow tube, pipe, or other vessel that is filled with a packing material. The packing can be randomly filled with small objects like Raschig rings or else it can be a specifically designed structured packing.
Particle size	:	Particle size is a notion introduced for comparing dimensions of solid particles (flecks), liquid particles (droplets), or gaseous particles (bubbles).
Pasteurisation process	:	Pasteurisation is a process of heating a food, usually liquid, to a specific temperature for a definite length of time, and then cooling it immediately.
Plug flow reactor (PFR)	:	This may be thought of as a long pipe to which the reactants flow in, and the products flow out. In an ideal PFR, the components will be distributed axially, but will have uniform radial distribution.
Pneumatic conveying (PC)	:	Pneumatic conveying is widely used for the transport of dry bulk particulate materials.
Poiseuille's law	:	Equation that relates the mass rate of flow in a tube as proportional to the pressure difference per unit length and to the fourth power of the tube radius. The law is valid for fully developed laminar flow.
Polar substances	:	It turns out that molecular substances, like methanol (CH_3OH), and glucose ($C_6H_{12}O_6$), are very soluble in water. These two substances, along with water, are polar substances. Their charge is neutral, but one end of the molecule is positive and the other end is negative.
Pore distribution resistance	:	Shale durability is measured by resistance to slaking in a standard laboratory test. All slaking mechanisms (namely, air-pressure breakage, differential swelling, and dissolution of cementing agents) require that water penetrate the pore space of the shale pieces.
Pressure effects	:	An air pressure difference causes movement of particles. The amount of pressure is determined by the difference of pressure in the air and the amount of area affected.

RANS	:	Reynolds-averaged Navier-Stokes, which implies time averaging of the velocity fluctuations in turbulent flow. The Reynolds' stresses obtained by this averaging have to be expressed with an additional set of equations. Turbulence models like the k-ε and k-ω models belong to this class.
Rate constant (k)	:	A rate constant is a proportionality constant that appears in a rate law. For example, k is the rate constant in the rate law $d[A]/dt = k[A]$. Rate constants are independent of concentration but depend on other factors, most notably temperature.
Rate law/equation	:	A rate law or rate equation relates reaction rate with the concentrations of reactants, catalysts, and inhibitors. For example, the rate law for the one-step reaction $A + B \rightarrow C$ is $d[C]/dt = k[A][B]$.
Reaction intermediate	:	A highly reactive substance that forms and then reacts further during the conversion of reactants to products in a chemical reaction. Intermediates never appear as products in the chemical equation for a net chemical reaction.
Reaction mechanism	:	A list of all elementary reactions that occur in the course of an overall chemical reaction.
Reaction rate	:	A reaction rate is the speed at which reactants are converted into products in a chemical reaction. The reaction rate is given as the instantaneous rate of change for any reactant or product, and is usually written as a derivative (e.g. $d[A]/dt$) with units of concentration per unit time (e.g. mol L^{-1} s^{-1}).
Reactor design	:	Chemical reactors are said to be the heart of a chemical plant. The chemical reactor is used where high value products are produced through chemical transformation.
Reynolds' number	:	This is a number which characterises flow. If the Reynold's number is low (under approximately 1800) then the flow is said to be laminar. This may be thought of as the fluid flowing in layers. If the Reynold's number is high (over approximately 2300) then the flow is said to be turbulent. Turbulent flow is considered to be well mixed.
RTD models	:	Temperature is determined by measuring resistance and then using the RTD's 'R vs T' characteristics to extrapolate temperature.
Second order reaction	:	Compare with zero order reaction and first order reaction. A reaction with a rate law that is proportional to either the concentration of a reactant squared, or the product of concentrations of two reactants.
Single replacement reactions	:	A chemical reaction in which an element replaces one element in a compound.
Solubility equilibrium	:	Solubility equilibrium is a type of dynamic equilibrium. It exists when a chemical compound in the solid state is in chemical equilibrium with a solution of that compound.
Spacetime	:	Spacetime is any mathematical model that combines space and time into a single continuum.
Space velocity	:	The relation between volumetric flow and reactor volume in a chemical reactor.
Specific surface area	:	Internal surface area of a porous structure given in area per unit volume, which yields the unit one over unit length. Often used to characterise the structure of porous catalysts.
Stabiliser	:	A stabiliser is a substance added to another substance to prevent an alteration of its physical state.

Stoichiometry	:	Stoichiometry is a branch of chemistry that deals with calculating the relation between the quantities of substances that take part in a balanced chemical reaction or that combine to form a chemical compound.
Streamline-diffusion stabilisation	:	A numerical technique for stabilisation of the numeric solution to a convection-dominated PDE by artificially adding diffusion in the direction of the streamlines.
Substrate	:	In biochemistry, a substrate is a molecule upon which an enzyme acts.
Surface kinetics	:	Understanding of the complex behaviour of particles at surfaces requires detailed knowledge of both macroscopic and microscopic processes that take place.
Surfactant	:	Surfactants are products used as detergents, dispersing agents, emulsifiers, wetting agents, foaming or anti-foam agents, and solubilisers. They also constitute the raw material for the formulation of household products such as fabric detergents, shampoos, house-cleaning products, as well as industrial auxiliary products for facilitating work in the manufacture of textile, flotation agents for ore, metal working, etc. They are used in other sectors of industry such as food processing, metallurgy, pharmaceuticals and public works.
Switch function	:	Conditional function that gives a smooth onset of a variable, for example from 0 to 1 or from 1 to 0. Often used for phase changes or saturation.
Terminal velocity	:	In fluid dynamics an object is moving at its terminal velocity if its speed is constant due to the restraining force exerted by the air, water or other fluid through which it is moving.
Thermal conductivity	:	Thermal conductivity, k, is the property of a material that indicates its ability to conduct heat. It appears primarily in Fourier's Law for heat conduction. Thermal conductivity is measured in watts per kelvin per metre.
Trickle bed oxidation reactors	:	Trickle-bed reactors are a type of the second classification in which both gas and liquid flow downward through the catalyst bed.
Unimolecular reaction	:	A reaction that involves isomerisation or decomposition of a single molecule.
Vapour pressure	:	Vapour evaporates at the surface of a liquid. The pressure it exerts is known as vapour pressure. The higher the temperature the greater the vapour pressure.
Volatility	:	The tendency or ability of a liquid to pass into the vapour phase; liquids such as alcohol or gasoline, because of their tendency to evaporate rapidly, are called volatile liquids.
Wall function	:	Semi-empirical expression for the anisotropic flow close to a solid surface used in turbulence models. Often based on negligible variations in pressure gradient in the direction tangential to the surface.
Zero order reaction	:	Compare with first order reaction and second order reaction. A reaction with a reaction rate that does not change when reactant concentrations change.

Useful Information

GREEK SYMBOLS

α	:	m^3 wake/m^3 bubble.
δ	:	Volume fraction of bubbles in a BFB.
δ	:	Dirac delta function, an ideal pulse occurring at time $t = 0$ (s^{-1}).
$\delta(t - t_0)$	:	Dirac delta function occurring at time t_0 (s^{-1}).
ε_A	:	Expansion factor, fractional volume change on complete conversion of A.
$\in$	:	Void fraction in a gas-solid system.
$\theta = t/t$	:	Dimensionless time units (–).
K'''	:	Overall reaction rate constant in BFB (m^3 solid/m^3 gas·s).
μ	:	Viscosity of fluid (kg/m·s).
μ	:	Mean of a tracer output curve (s).
π	:	Total pressure (Pa).
ρ	:	Density or molar density (kg/m^3 or mol/m^3).
σ^2	:	Variance of a tracer curve or distribution function (s^2).
τ	:	$V/v = C_{A0}V/F_{A0}$, space-time (s).
τ	:	Time for complete conversion of a reactant particle to product (s).
τ'	:	$\tau = C_{A0}V/F_{A0}$, weight-time (kg·s/m^3).
$\tau', \tau'', \tau''', \tau''''$	:	Various measures of reactor performance.
Φ	:	Overall fractional yield.
ϕ	:	Sphericity.
φ	:	Instantaneous fractional yield.

NOTATION

A	:	Heat transfer area.
A_c	:	Reactor tube cross-sectional area.
A_i	:	Pre-exponential factor for rate constant i.
c_j	:	Concentration of species j.
c_{jf}	:	Feed concentration of species j.
c_{js}	:	Steady-state concentration of species j.
c_{j0}	:	Initial concentration of species j.
C_P	:	Constant-pressure heat capacity.
$\overline{C}_{Pj}$	:	Partial molar heat capacity.
C_{Ps}	:	Heat capacity per volume.
$\hat{C}_P$	:	Constant-pressure heat capacity per mass.

483

$\hat{C}_V$	:	Constant-volume heat capacity per mass.
ΔC_P	:	Heat capacity change on reaction, $\Delta C_P = \sum_j v_j \bar{C}_{pj}$.
E_i	:	Activation energy for rate constant i.
E_k	:	Total energy of stream k.
$\hat{E}_k$	:	Total energy per mass of stream k.
ΔG°	:	Gibbs energy change on reaction at standard conditions.
$\bar{H}_j$	:	Partial molar enthalpy.
$\hat{H}$	:	Enthalpy per unit mass.
ΔH_{Ri}	:	Enthalpy change on reaction, $\Delta H_{Ri} = \sum_j v_{ij} \bar{H}_j$.
ΔH°	:	Enthalpy change on reaction at standard conditions.
I	:	Reaction rate constant for reaction i.
K_i	:	Equilibrium constant for reaction i.
k_m	:	Reaction rate constant evaluated at mean temperature T_m.
$\hat{K}$	:	Kinetic energy per unit mass.
l	:	Tubular reactor length.
m_k	:	Total mass flow of stream k.
n	:	Reaction order.
n_j	:	Moles of species j, $V_R c_j$.
n_r	:	Number of reactions in the reaction network.
n_s	:	Number of species in the reaction network.
N_j	:	Molar flow of species j, $Q c_j$.
N_{jf}	:	Feed molar flow of species j, $Q c_j$.
P	:	Pressure.
P_j	:	Partial pressure of component j.
P_{nj}	:	$P_{nj} = (\partial P/\partial n_j)_{T,V,n_k}$.
$\dot{q}$	:	Heat transfer rate per volume for tubular reactor, $\dot{q} = \frac{2}{R} U^o (T_a - T)$.
Q	:	Volumetric flowrate.
Q_f	:	Feed volumetric flowrate.
$\dot{Q}$	:	Heat transfer rate to reactor, usually modelled as $\dot{Q} = U^o A (T_a - T)$.
r_i	:	Reaction rate for ith reaction.
r_{tot}	:	Total reaction rate $\sum_i r_i$.
R	:	Gas constant.
R_j	:	Production rate of jth species.
t	:	Time.
T	:	Temperature
T_a	:	Temperature of heat medium.
T_m	:	Mean temperature at which k is evaluated.
U^o	:	Overall heat transfer coefficient.
$\hat{U}$	:	Internal energy per mass.
v_k	:	velocity of stream k.
V	:	Reactor volume variable.

$\bar{V}j$	:	Partial molar volume of species j.
V_j°	:	Specific molar volume of species j.
V_R	:	Reactor volume.
ΔV_i	:	Change in volume upon reaction i, $\Sigma_j v_{ij}\bar{V}_j$.
$\dot{W}$	:	Rate work is done on the system.
x_j	:	Number of molecules of species j in a stochastic simulation.
x_j	:	Molecular conversion of species j.
y_j	:	Mole fraction of gas-phase species j.
z	:	Reactor length variable.
α	:	Coefficient of expansion of the mixture, $\alpha = (1/V)(\partial V/\partial T)_{P,\,n_j}$.
ε_i	:	Extent of reaction i.
τ	:	Reactor residence time, $\tau = V_R/Q_f$.
v_{ij}	:	Stoichiometric coefficient for species j in reaction i.
$\bar{v}_i$	:	$\Sigma_j v_{ij}$.
ρ	:	Mass density.
ρ_k	:	Mass density of stream k.
$\hat{\Phi}$	:	Potential energy per mass.

SYMBOLS AND ABBREVIATIONS

BFB	:	Bubbling fluidised bed.
BR	:	Batch reactor.
CFB	:	Circulating fluidised bed.
FF	:	Fast fluidised bed.
LFR	:	Laminar flow reactor.
MFR	:	Mixed flow reactor.
M-M	:	Michaelis-Menten.
mw	:	Molecular weight (kg/mol)
PC	:	Pneumatic conveying.
PCM	:	Progressive conversion model.
PFR	:	Plug flow reactor.
RTD	:	Residence time distribution.
SCM	:	Shrinking-core model.
TB	:	Turbulent fluidised bed.

SUBSCRIPTS

b	:	Batch.
b	:	Bubble phase of a fluidised bed.
c	:	Of combustion.
c	:	Cloud phase of a fluidised bed.
c	:	At unreacted core.
d	:	Deactivation.
d	:	Deadwater, or stagnant fluid

e	:	Emulsion phase of a fluidised bed.
e	:	Equilibrium conditions.
f	:	Leaving or final.
f	:	Of formation.
g	:	Of gas.
i	:	Entering.
l	:	Of liquid.
m	:	Mixed flow.
mf	:	At minimum fluidising conditions.
p	:	Plug flow.
r	:	Reaction or of reaction.
s	:	Solid or catalyst or surface conditions.
0	:	Entering or reference.
θ	:	Using dimensionless time units.

SUPERSCRIPTS

$a, b, ...$	:	Order of reaction.
n	:	Order of reaction.
o	:	Refers to the standard state.

DIMENSIONLESS GROUPS

$\dfrac{D}{uL}$	:	Vessel dispersion number.
$\dfrac{D}{ud}$	:	Intensity of dispersion number.
M_H	:	Hatta modulus.
M_T	:	Thiele modulus.
M_W	:	Wagner-Weisz-Wheeler modulus.
$Re = \dfrac{du\rho}{\mu}$	:	Reynolds number.
$Sc = \dfrac{\mu}{\rho D}$	:	Schmidt number.

References

Allen, G., *Comprehensive Polymer Science*, Pergamon Press, New York.

Aris, Rutherford, *Introduction to the Analysis of Chemical Reactors*, Prentice-Hall, London.

Biesenherger, J.A., *Principles of Polymerisation Engineering*, John Wiley & Sons, New York.

Boudart, Michel, *Kinetics of Chemical Processes*, Applied Science Publishers, London.

Carberry, C., *Chemical and Catalytic Reaction Engineering*, John Wiley & Sons, New York.

Clark, A., *The Theory of Adsorption and Catalysis*, D. Van Nostrand, New York.

Copper, C.D., *Environmental Engineering*, Waveland Press, US.

Denbigh, K.L., *Chemical Reactor Theory*, Cambridge University Press.

Farrauto, R.J., *Fundamentals of Catalytic Processes*, Chapman & Hall.

Felder, R.M., *Elementary Principles of Chemical Processes*, John Wiley & Sons, New York.

Fogler, H.S., *Elements of Chemical Reaction Engineering*, Pergamon Press, New York.

Gardiner, W.C., *Catalytic Chemistry*, John Wiley & Sons, New York.

Gates, B.C., *Chemistry of Catalytic Processes*, McGraw-Hill, New York.

Hill, C.G., *An Introduction to Chemical Engineering Kinetics and Reactor Design*, Heinemann, London.

Himmelblau, D.M., *Basic Principles and Calculations in Chemical Engineering*, Elsevier Scientific Publishing Co., Amsterdam.

Kunii, D., *Fluidization Engineering*, Butterworth, Boston.

Lanny, D.S., *Chemical Reactions Engineering*, Oxford University Press, New York.

Levenspiel, O., *Chemical Reaction Engineering*, Wiley, Singapore.

Lewis, B., *Combustion, Flames and Explosion of Gases*, Academic Press, London.

Mcketta, J., *Encyclopedia of Chemical Processing and Design*, McGraw-Hill, New York.

Meyer, R.A., *Handbook of Chemical Production Processes*, Applied Science Publishers, London.

Perry, R.H., *Chemical Reactor Design and Operations*, Reston Publishing Co., Reston, Virginia.

Peterson, E.E., *Chemical Reaction Analysis*, Academy Press, New York.

Rudd, L. and Herman, S., *Handbook of Chemical Engineering*, Elsevier Scientific Publishing Co., Amsterdam.

Smith, J.M., *Chemical Engineering Kinetics*, McGraw-Hill, London.

Twigg, M.V., *Catalyst Handbook*, Harvard University Press, New York.

Walas, S.M., *Reaction Kinetics for Chemical Engineers*, Pergamon Press, New York.

Westerterp, K.R., *Chemical Reactor Design and Operations*, John Wiley & Sons, New York.

Index